Theory and Analysis of Phased Array Antennas

Theory and Analysis of Phased Array Antennas

NOACH AMITAY

Bell Telephone Laboratories, Incorporated

VICTOR GALINDO

Virginia Polytechnic Institute and State University

CHEN PANG WU

Bell Telephone Laboratories, Incorporated

WILEY-INTERSCIENCE, a Division of John Wiley & Sons, Inc.
New York · London · Sydney · Toronto

Library of Congress Catalogue Card Number: 70-174768

ISBN 0-471-02553-4

Printed in the United States of America.

10 9 8 7 6 5 4 3 2 1

To my wife Joan Amitay, להורי חיים ובלהה אמתי

To my darling wife and daughters, Adele, Stephanie Dee, and Maria Louise Galindo–my devoted Wandering Jews of America

In memory of my parents, Tsu Chai and Li Hsieh Wu

Preface

Modern radar, communications, and radio astronomy place increasing demands for large-aperture antennas which are capable of high-speed scanning and simultaneous operation of multiple functions. An array of radiating elements with independent phase control, that is, a phased array, can provide these capabilities. Most of the methods that have been used in the design of phased arrays have relied on classical array theory, which neglects mutual coupling, approximate analytical methods, experimental techniques, and combinations thereof. In modern array radar antennas which require wide angle scanning, it has been found that the mutual coupling between antenna elements plays a most significant role in determining the array performance. The effects of mutual coupling are not properly accounted for in the approximate analytical methods. The design of large phased arrays, often using many thousands of antenna elements, by experimental methods is a time consuming and very costly procedure. A need exists, therefore, for a general understanding of the performance of phased aray antennas as well as for efficient design methods in which the effects of mutual coupling are incorporated.

The purpose of this book is to present a mathematical approach for the analysis and design of a broad class of phased array antennas and to examine their electromagnetic properties. It is hoped that this book will provide not only a basic understanding of the properties of phased array antennas, but also the analysis and synthesis techniques needed to design these expensive antennas efficiently before large funds are committed for construction.

Emphasis is placed on the analysis of uniformly and closely spaced planar arrays of open ended waveguides. The boundary value problem associated with these arrays is rigorously formulated and expressed by one of several types of integral equations. The integral equations can be solved numerically (and analytically in special cases) with a very high degree of accuracy. The solution provides a complete characterization of the electromagnetic field, from which pertinent array properties, such as the reflection and radiation (including polarization) characteristics and mutual coupling coefficients, are derived. The formulation and numerical methods can be extended to dipoles and other related types of array elements. Since dipole arrays have been treated elsewhere, we shall not dwell upon them here. The pattern synthesis and system aspects of phased array antennas have also been extensively discussed in the literature and therefore will not be considered herein.

In Chapter 1 we introduce the reader to the terminology and concepts of phased arrays, and review various approaches to array analysis. The dependence of the array properties on mutual coupling is demonstrated. In Chapter 2 we discuss Floquet's theorem and utilize it in the formulation of the boundary value problems associated with infinite and periodic arrays. Various integral equations and variational expressions are derived. Methods of solving these equations are presented in Chapter 3. Emphasis is placed on numerical methods which can be handled efficiently by high speed and large storage digital computers. A general discussion of various techniques for ascertaining the validity and accuracy of a solution is presented.

In Chapter 4 we derive an analytical solution to the infinite parallel-plate phased array with thin walls. Certain basic array relationships are presented, and examples of performance data are also given. The characteristics of planar arrays of rectangular waveguides are discussed in Chapter 5. The existence of forced surface wave resonances which cause blindness in phased arrays is demonstrated. For the purpose of illustration, some of the techniques developed in Chapters 2 and 3 are rederived for parallel-plate arrays. In Chapter 6 we examine the effects of dielectric loading on the basic properties of parallel-plate arrays. The improvement of impedance match and the appearance of forced surface wave resonances are considered in detail.

The properties of infinite planar arrays of circular waveguides with and without dielectric loading are investigated in Chapter 7. Again, forced surface wave resonances are observed and their salient features are discussed. The extension of the integral equation formulation to skewed periodic and planar grid arrays is derived. In Chapter 8 we extend the integral equation formulation to uniformly and nonuniformly spaced finite arrays and a certain class of infinite aperiodic arrays. Dispersion relations for periodically modulated surfaces are derived. The relationship between the forced surface wave resonances in infinite arrays and these dispersion relations is demonstrated. The relation between the array element characteristics and the proximity of an element to the edge of a finite array is examined. In the final chapter, we present systematic methods to optimize the array match over a prescribed scan region and frequency band. These methods are quite general and are applicable to any type of array element. Experimental methods to characterize the array terminal relationships are also discussed.

Basically, this book is a research monograph. Although the material presented is drawn from many technical publications, the book is based mainly on work performed at Bell Telephone Laboratories, where a considerable effort has been devoted to phased array research and development.

A bibliography is provided to assist the reader in finding additional information that could not be included in a book of this scope. However, there has been no attempt to make the bibliography complete.

This book is addressed to engineers, physicists, and students who are active or interested in phased arrays and antennas. It is written on an intermediate level. Familiarity with Maxwell's equations, guided wave theory, and elementary antenna theory is assumed. The book may serve as a supplementary text for graduate courses in antennas and applied electromagnetic theory and applications.

We are indebted to our colleagues at Bell Telephone Laboratories for their cooperation and advice. Special thanks are due to Messrs. J. S. Cook and R. Lowell Dr. E. R. Nagelberg for their constructive comments and encouragement. The support of Bell Telephone Laboratories and TRW systems in the preparation of the manuscript is gratefully acknowledged.

We wish also to thank the following organizations for permission to use parts of certain articles which appeared in their various publications: The Institute of Electrical and Electronic Engineers, Inc., the American Telephone and Telegraph Co., and the American Geophysical Union.

NOACH AMITAY
VICTOR GALINDO
CHEN PANG WU

Whippany, New Jersey
Blacksburg, Virginia
August 1971

Contents

Theory and Analysis of Phased Array Antennas

1. Introduction to Array Theory

1. INTRODUCTION

The ever-increasing demands of the space and missile age upon modern radar and communication systems have propelled phased array antennas into the limelight. The need for specialized multifunction operation (i.e., simultaneous surveillance, discrimination, tracking, etc.), coupled with high power, high data rates, and the ability to withstand adverse environmental conditions, has stimulated considerable activity in the research, design, and deployment of phased array radars and antennas. Figures 1.1–1.4 show some of the phased array antennas that have already been designed and constructed. Phased array antenna systems have been deployed on the ground and aboard ships. It is anticipated that phased arrays with miniaturized components will eventually be deployed in space and aboard aircraft as well.

An array antenna refers to an antenna system consisting of more than one antenna element radiating in phase coherence. Horns (open ended waveguides), dipoles, helices, spiral antennas, polyrods, parabolic dishes, and many other types of antennas may constitute the radiating elements. The array concept was originally introduced mainly as a means for achieving a better control of the antenna aperture illumination. In order to obtain high angular resolution, large planar arrays are being utilized.

A periodic planar phased array consists of identical radiating elements which are arranged in a planar and doubly periodic grid. Constant incremental phase shifts (steering phases) are applied between adjacent elements in order to electronically scan the radiation pattern of the array in free space. In this chapter, the significance of analyzing phased arrays by means of a solution to a rigorously formulated boundary value problem is discussed. Also, the weaknesses of various approximate approaches are brought out.

In what may be termed "classical" array theory, the mutual coupling between the array elements is neglected. Thus, the radiation pattern of a finite sized phased array is expressed as a product of the array factor and the radiation pattern of the array element. The array factor determines the beam

Fig. 1.1. MAR/WSMR: Multifunction array radar installed at White Sands Missile Range (courtesy Bell Telephone Laboratories).

Fig. 1.2. Mobile surface to air missile radar (courtesy Hughes Aircraft Company).

Fig. 1.3. AN/FPS-85: Air Force space track radar (courtesy Bendix Communication Division).

shape and sidelobe level of the array radiation pattern and also relates the array geometry and steering phases to the direction cosines of the beam pointing directions. The special functional relationship between the array factor, the steering phases, and the beam directional cosines lends itself to a convenient graphical diagram, known as the grating lobe diagram, which is a useful concept often employed in the discussion of array properties and the design of phased array antennas. This diagram will be discussed in more detail later.

Experimental and theoretical investigations have clearly indicated that mutual coupling cannot be ignored, especially in the case of closely spaced array elements (i.e., spacing of the order of half a wavelength). The mutual coupling strongly affects the radiation and reflection characteristics (as a function of the beam pointing direction) of phased array antennas. Furthermore, in a finite sized array, the radiation patterns and the reflection characteristics of the various elements may depend on their relative positions in the array. They are strongly influenced by the proximity of an element to the

Fig. 1.4. MSR: Missile site radar installed at Meck Island (courtesy Bell Telephone Laboratories).

edge of the array. In the case of "large" planar arrays, however, the bulk of the inner core elements behave almost uniformly, and the important aspects of this behavior may be approximated well by the uniform behavior of elements in an infinite array. Thus the infinite array can serve as a useful model for the analysis of large planar arrays. In this case the array radiation pattern can still be expressed as a product of the array factor and the element radiation pattern, the latter being determined in an infinite array environment. The effects of mutual coupling are taken care of by the element radiation pattern (excited in the array environment with the rest of the elements being properly terminated). Certain general expressions which relate the mutual coupling coefficients, the reflection coefficients, and the radiation power pattern of an element can be derived. In addition the concept of ideally matched arrays is expressed in terms of an ideal element power radiation pattern.

Infinite arrays of narrow slots, dipoles, and current elements have been analyzed in previous years. However, these models, as well as the approximations involved in their analysis, tend to be limited in their applicability in the design of practical elements of closely packed arrays. In this book emphasis is placed upon the analysis of open ended waveguide type array elements. The array boundary value problem associated with these elements can be rigorously formulated and expressed by one of several types of integral equations which will be discussed in the next chapter. The integral equations can be solved numerically (and analytically in special cases) with a very high degree of accuracy and thus provide a powerful tool for the analysis and design of phased array antennas. This approach can be extended to dipoles and other related types of array elements. Furthermore, this analysis has led to several general theories and conclusions concerning all types of array elements which will be discussed in later chapters. The system aspects and the pattern synthesis of phased array antennas have been extensively discussed and referenced in recent literature[1–3] and will not be treated herein.

2. ELEMENTARY ARRAY THEORY

In this section we shall introduce some frequently used definitions and concepts of phased array antennas. A finite array of apertures in an infinite conducting ground plane (Fig. 1.5) will serve as a model. These apertures

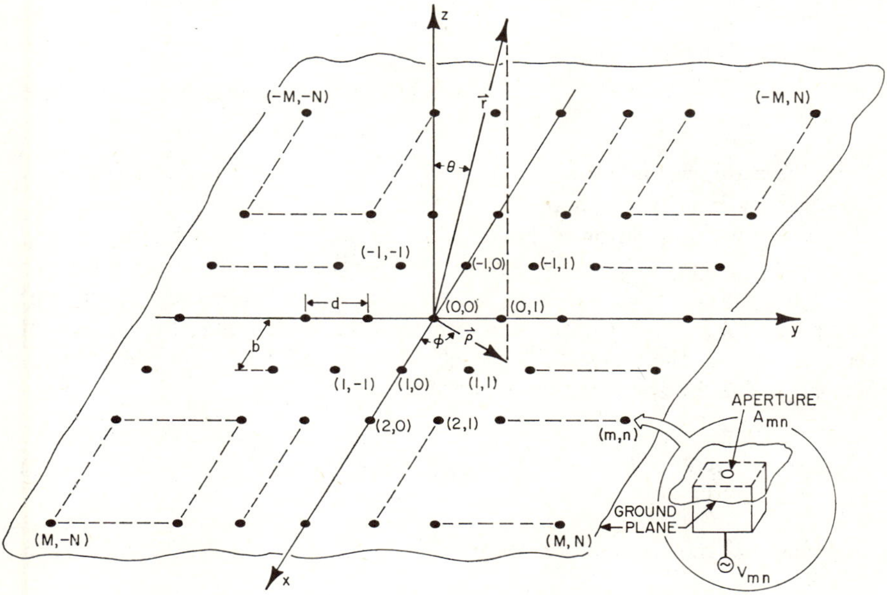

Fig. 1.5. Planar array of apertures in a ground plane.

are arranged in a rectangular grid* and may be excited by waveguides beneath the ground plane. By using Babinet's principle, the results obtained for an aperture array in a ground plane can be carried over to the dual case of current elements in free space. Our particular model will prove useful in pointing out the limitations of certain assumptions which are common to aperture arrays as well as their duals.

2.1 Ordinary Array Theory

Consider a planar array of $(2M + 1) \times (2N + 1)$ apertures embedded in an infinite conducting ground plane at $z = 0$ (Fig. 1.5). These apertures are arranged in a regular rectangular grid as indicated. The location of an element is defined by the indices (m, n), which correspond to a physical location:

$$\boldsymbol{\rho}_{mn} = m\, b\hat{x} + n\, d\hat{y} \tag{1}$$

where $\hat{x}$ and $\hat{y}$ are unit vectors along the x and y axes, while b and d are the separations between adjacent elements (apertures) in the x and y directions. We designate by $\mathbf{E}_{mn}(x, y)$ the tangential electric field at the (m, n)th aperture, A_{mn}, per unit voltage of the generator, V_{mn}. In the source free region above the array ($z > 0$) the electric field, $\mathscr{E}(x, y, z)$, may be regarded as a superposition of plane waves[3] of the form $\mathbf{F}(k_x, k_y) \exp(-j\mathbf{k} \cdot \mathbf{r})$, given by

$$\mathscr{E}(x, y, z) = \frac{1}{4\pi^2} \iint_{-\infty}^{+\infty} \mathbf{F}(k_x, k_y) \exp(-j\mathbf{k} \cdot \mathbf{r})\, dk_x\, dk_y \tag{2}$$

where the propagation vector $\mathbf{k} = k_x\hat{x} + k_y\hat{y} + k_z\hat{z} = \mathbf{k}_t + k_z\hat{z}$ and $|\mathbf{k}| = k$. The quantity $k = 2\pi/\lambda$, where λ is the free space wavelength. The field, (2), satisfies the wave equation when $k_x^2 + k_y^2 + k_z^2 = k^2$. Hence, propagating waves (which carry energy away from the array) correspond to the range of real values of k_z (i.e., $k_x^2 + k_y^2 < k^2$), while for evanescent (decaying) waves $k_x^2 + k_y^2 > k^2$. Collin[3] and others have shown that the tangential electric field at the aperture uniquely determines $\mathbf{F}(k_x, k_y)$ and thereby $\mathscr{E}(x, y, z)$. The far-zone electric field,† $\mathscr{E}(\mathbf{r})$, is given by

$$\mathscr{E}(\mathbf{r})_{|r| \to \infty} = j\, \frac{e^{-jkr}}{2\pi r}\, k_z' \left[\mathbf{F}_t(k_x', k_y') - \frac{\mathbf{F}_t(k_x', k_y') \cdot \mathbf{k}_t'}{k_z'}\, \hat{z} \right] \tag{3}$$

* A rectangular grid is chosen here for simplicity. Any other doubly periodic grid can also be used.

† That is, the electric field at a point $P(r, \theta, \phi)$ far away from the aperture ($r \to \infty$).

where the subscript t denotes the transverse (to z) component of the vectors, and

$$\begin{aligned} k_x' &= k \sin\theta \cos\phi = kT_x \\ k_y' &= k \sin\theta \sin\phi = kT_y \\ k_z' &= k \cos\theta = kT_z \end{aligned} \tag{4}$$

The quantities T_x, T_y, and T_z, which are used extensively in antenna nomenclature, are the directional cosines of $\mathbf{r}$ with respect to the x, y, and z axes of Fig. 1.5. The tangential electric field across the apertures is related to $\mathbf{F}_t$ through the following inverse transform of (2) at the aperture plane ($z = 0$):

$$\mathbf{F}_t(k_x, k_y) = \sum_{m=-M}^{M} \sum_{n=-N}^{N} V_{mn} \iint_{A_{mn}} \mathbf{E}_{mn}(x, y) \exp(j\mathbf{k}_t \cdot \boldsymbol{\rho})\, dx\, dy \tag{5}$$

Thus, (3) and (5) relate the far field to the tangential aperture field.

2.1.1 The Array Factor and Element Factor. The underlying assumption in the ordinary array theory is that the mutual coupling (or interaction) between the various aperture fields can be neglected.* Thus, for the rectangular grid array of Fig. 1.5 with identical elements, the field distribution, $\mathbf{E}_{mn}(x, y)$, is identical to $\mathbf{E}_{00}(x, y)$:

$$\mathbf{E}_{mn}(x, y) = \mathbf{E}_{mn}(x^0 + mb;\, y^0 + nd) = \mathbf{E}_{00}(x^0, y^0)$$

or, equivalently:

$$\mathbf{E}_{mn}(\boldsymbol{\rho}) = \mathbf{E}_{mn}(\boldsymbol{\rho}_{mn} + \boldsymbol{\rho}^0) = \mathbf{E}_{00}(\boldsymbol{\rho}^0) \tag{6}$$

where $\boldsymbol{\rho}^0 = x^0\hat{x} + y^0\hat{y}$ are the coordinates of the aperture A_{00}. Substituting (6) into (5) yields

$$\mathbf{F}_t(k_x', k_y') = \iint_{A_{00}} \mathbf{E}_{00}(x^0, y^0) \times \exp(j\mathbf{k}_t' \cdot \boldsymbol{\rho}^0)\, dx^0\, dy^0 \left[\sum_{m=-M}^{M} \sum_{n=-N}^{N} V_{mn} \exp(j\mathbf{k}_t' \cdot \boldsymbol{\rho}_{mn}) \right] \tag{7}$$

The exponential factor in the brackets arises from the difference in location of the various elements. Combining (3), (4), and (7), we obtain the far-zone electric field:

$$\mathscr{E}(\mathbf{r})_{|r| \to \infty} = \mathbf{f}(T_x, T_y) S_a(T_x, T_y) \tag{8}$$

* The implication of this assumption is that, when only a single aperture is excited in the array, the radiation field is determined solely by the field across this excited aperture. The contributions of the nonzero fields at the other apertures (due to mutual coupling) are neglected. For widely spaced array of very small apertures this assumption may be reasonable, but, as we shall see later, it breaks down for closely spaced arrays.

where

$$S_a(T_x, T_y) = \sum_{m=-M}^{M} \sum_{n=-N}^{N} V_{mn} \exp(j\mathbf{k}'_t \cdot \boldsymbol{\rho}_{mn})$$
$$= \sum_{m=-M}^{M} \sum_{n=-N}^{N} V_{mn} \exp\left[j2\pi\left(m\frac{b}{\lambda}T_x + n\frac{d}{\lambda}T_y\right)\right] \tag{9}$$

and

$$\mathbf{f}(T_x, T_y) = \frac{je^{-jkr}}{2\pi r}[k\sqrt{1 - T_x^{\,2} - T_y^{\,2}} - \hat{z}k(T_x\hat{x} + T_y\hat{y})\cdot]$$
$$\times \iint_{A_{00}} \mathbf{E}_{00}(x^0, y^0)e^{jk(T_x x^0 + T_y y^0)}\, dx^0\, dy^0 \tag{10}$$

The scalar quantity $S_a(T_x, T_y)$ is called the *array factor*, while the vector quantity $\mathbf{f}(T_x, T_y)$ is called the *element factor*.* Equation 8 states the well-known principle of pattern multiplication: the far-zone field is the product of the array and element factors. The array factor may be viewed as the radiation pattern of an array of isotropic elements. The element factor specifies the vector nature of the far-zone field, such as its polarization and orientation. In the usual applications of phased arrays the element factor is broad [i.e., a slowly varying function of (T_x, T_y)] in comparison to the array factor. Consequently, most of the antenna properties—directivity, beamwidth, sidelobe level, etc.—are determined mainly by the array factor. Radiation pattern synthesis consists of the judicious choice of the $\{V_{mn}\}$ (sometimes called the illumination function) in order to achieve a prescribed far-zone field distribution.

So far, we have limited our discussion to transmitting arrays. The results are equally valid, however, for receiving arrays. Utilizing the reciprocity theorem, it can be shown that the antenna receiving cross section is proportional to its radiation pattern.[3,4]

2.1.2 Properties of the Array Factor. The array factor, S_a, is a doubly periodic function of T_x and T_y with periods λ/b and λ/d, respectively. Hence a knowledge of $S_a(T_x, T_y)$ in a rectangular region

$$\left(-\frac{\lambda}{2b} \leq T_x \leq \frac{\lambda}{2b};\ -\frac{\lambda}{2d} \leq T_y \leq \frac{\lambda}{2d}\right)$$

is sufficient to characterize the array factor completely. Usually, a planar array is designed to radiate into a hemispherical region $(0 \leq \theta \leq \pi/2;\ 0 \leq \phi \leq 2\pi)$. In terms of T_x and T_y, this region corresponds to $T_x^2 + T_y^2 \leq 1$,

* The array factor and the element factor are sometimes expressed as functions of the spatial coordinates (θ, ϕ), which are related through (4) to the directional cosines (T_x, T_y).

which defines a circular region of unit radius in the T_x–T_y plane. This region is commonly referred to as the real space (visible region), and its complementary region, $T_x^2 + T_y^2 > 1$, is called the imaginary space (invisible region).* The special case $T_x^2 + T_y^2 = 1$ corresponds to a beam grazing in the plane of the array.

Let us assume that the voltages $\{V_{mn}\}$ have been properly synthesized to produce the prescribed radiation pattern. An array is phased when we modify its illumination by adding a linear phase taper of the form

$$V'_{mn} = V_{mn} e^{j(m\psi_x + n\psi_y)} \tag{11}$$

where the steering phases, ψ_x and ψ_y, are the incremental phase shifts between adjacent elements in the x and y directions, respectively. The corresponding array factor is

$$S'_a(T_x, T_y) = \sum_{m=-M}^{M} \sum_{n=-N}^{N} V_{mn} \times \exp\left\{ j\left[\frac{2\pi mb}{\lambda}\left(T_x + \frac{\psi_x}{2\pi b/\lambda}\right) + \frac{2\pi nd}{\lambda}\left(T_y + \frac{\psi_y}{2\pi d/\lambda}\right) \right] \right\} \tag{12}$$

As can be seen, the new array factor is merely the original one translated[9] (scanned) by T_{x0} and T_{y0} in the T_x and T_y directions, respectively, with

$$T_{x0} = -\frac{\psi_x}{2\pi b/\lambda}, \qquad T_{y0} = -\frac{\psi_y}{2\pi d/\lambda} \tag{13}$$

In many applications the array is required to produce a directive radiation pattern, which is characterized by a lobe of high amplitude (main beam), accompanied by small lobes (sidelobes) which are present because of the finite size of the antenna aperture. We usually normalize the peak of the main beam to unity (0 dB). Let the main beam point toward broadside (i.e., $T_x = 0$ and $T_y = 0$). Upon the application of the linear phase taper, (11), the main beam will be scanned (translated) to a new position (direction), given by (T_{x0}, T_{y0}). If properly controlled steering phases are applied, the main beam can be made to scan the whole space. This is the basis of operation of phased arrays.

2.1.3 The Grating Lobe Diagram. Because of the periodicity of the array factor in the T_x–T_y plane, the shape of the main beam (and associated sidelobes) will be repeated every λ/b and λ/d interval in the T_x and T_y directions, respectively. Any such repetition of the main beam is called a grating lobe.

* In the visible region (8) represents the actual far-zone fields, while in the invisible region (8) can be tied to the energy stored in the electric and magnetic fields of the evanescent waves and the antenna aperture Q^{5-8}.

Thus we can construct in the T_x–T_y plane an infinite grid of grating lobes spaced at λ/b and λ/d apart (Fig. 1.6), which is known as the grating lobe diagram.[9,10] The small solid circle around the origin designates the broadside position of the maximum of the main beam, while the rest of the small circles designate the corresponding position of the maxima of the grating lobes. The circle with unit radius corresponds to the edge of real space (i.e., $\theta = \pi/2$). As can be observed in Fig. 1.6, the smaller the spacing between elements the larger is the separation between the grating lobes. When the main beam is scanned to a new position, T_0 ($T_0 = \sin\theta_0$), as indicated by the arrow in

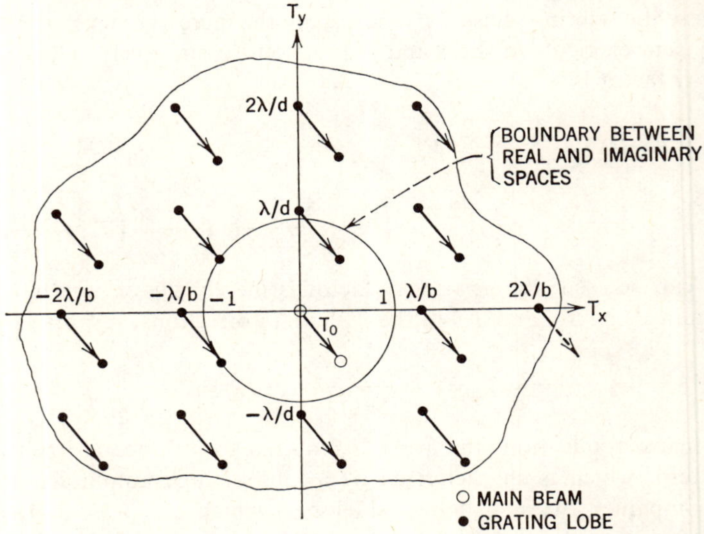

Fig. 1.6. Grating lobe diagram for a rectangular grid array.

Fig. 1.6, the whole grating lobe structure is translated by T_0 as well. As a result, some grating lobes which existed in imaginary space and did not affect the radiation pattern* are now swept into real space, resulting in a multibeam radiation pattern. In Fig. 1.6 there are two grating lobes which lie on the unit circle as a result of the scanning; these are known as grazing or incipient grating lobes.

A modified version of the grating lobe diagram is shown in Fig. 1.7. A

* The presence of grating lobes in the imaginary space actually represents stored energy in the vicinity of the aperture and thus affects the matching characteristic of an antenna. As the main beam and grating lobes are scanned, the match of the array may change, thus manifesting the effects of mutual coupling, which were neglected.

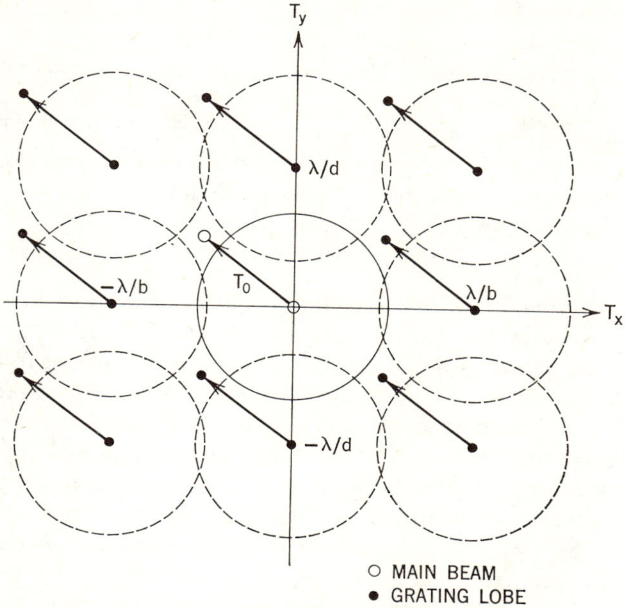

Fig. 1.7. Modified grating lobe diagram.

unit circle (dashed) is drawn around each unscanned grating lobe. These circles define the range of scan positions of the appropriate grating lobes, which correspond to the scanning of the main beam over the entire real space. In Fig. 1.7 one can observe regions of the space which are not intercepted by any circle. When the values of the steering phases are such that the main beam is scanned into these regions (T_0 in Fig. 1.7), neither the main beam nor a grating lobe exists in real space. For reasonably sized arrays having narrow beams with low sidelobes, this situation would correspond to nearly a total reflection of the incident signal (in either the transmission or the reception mode of the array). In an infinite array a total reflection of the incident signal would occur. Regions of the real space which are common to one or more circles would correspond to scanning directions in which a single beam or multiple beams can exist in real space. In most applications this multibeam situation is avoided either by limiting the range of scanning angles (scan region) or by decreasing the interelement spacing. The simple relation between the maximum scan angle in the T_x or T_y direction, θ_{max}, for which grating lobes do not appear in real space, and the element spacing is[10]

$$\frac{b}{\lambda}, \frac{d}{\lambda} < \frac{1}{1 + |\sin \theta_{max}|} \tag{14}$$

For a *prescribed* scan region it is desirable, in practice, to minimize the number of elements necessary to occupy an array aperture of a given size. The reason is that the directivity of an array is approximately proportional to its physical size, while the cost is related to the number of elements used. Thus, depending on the shape of this scan region in the T_x–T_y plane, we can arrange the array elements in an optimum doubly periodic grid* that will minimize the number of elements needed to maintain a single beam within the scan region. For a circular scan region in the T_x–T_y plane, for example, an equilateral triangular grid constitutes the optimum.[11]

2.1.4 Beam Broadening as a Function of Scan. The resolution of a radar system is limited by the width of its antenna beam. If this beamwidth varies with the beam pointing direction, the resolution becomes a function of the scan angle. Thus far we have studied the array factor, S_a, in the directional cosine space and have observed the periodicity and invariance of S_a (except for a translation) under a linear phase taper excitation. Therefore, the (T_x, T_y) coordinate system seems to be a natural one to describe certain aspects of phased array properties. To examine the characteristics of the array factor $S(\theta, \phi)$ with scan in the spherical coordinate systems (θ, ϕ), we construct $S(\theta, \phi)$ from $S_a(T_x, T_y)$ by projecting $S_a(T_x, T_y)$ onto the unit hemisphere (Fig. 1.8). This permits immediate interpretation of the radiation pattern for any desired scan angle in terms of the spherical coordinates (θ, ϕ).

The directional cosine (T_x, T_y) coordinates are related through a nonlinear transformation, (4), to the spatial coordinates (θ, ϕ). Therefore, a distortion of the beam shape occurs in the (θ, ϕ) coordinates with the application of linear phase taper to the array excitation. To visualize the beam distortion that occurs with scanning, consider the pencil (shaped) beam of a uniformly illuminated square array. The 3 dB contour of this pencil beam (which can be associated with the beamwidth) in the T_x–T_y plane is practically a circle[9,13] (Fig. 1.8). As this beam is scanned away from the array normal, although the 3 dB contour remains invariant in the T_x–T_y plane (circles along the T_y axis), the projection of the 3 dB contour onto the unit hemisphere becomes very distorted, as seen in Fig. 1.8. When the beam is scanned close to grazing, the beam becomes eccentric,[9] that is, the beam maximum, at θ_0, is not centered between the two half-power points, θ_1 and θ_2 (Fig. 1.8). For narrow pencil beams which are not scanned closer than several beamwidths to grazing ($\theta = \pi/2$) and for most practical aperture illuminations, it was shown[14] that the beamwidths in the θ and ϕ directions, B_θ and B_ϕ, are independent of the azimuthal scan angle, ϕ_0. The beam distortion is given by

$$B_\theta \approx \frac{B_0}{\cos \theta_0} \tag{15}$$

* The optimum grid does not have to be rectangular. The array factor of a nonrectangular grid can be obtained from (7) with an appropriate modification[11,12] of ρ_{mn}, Eq. 1.

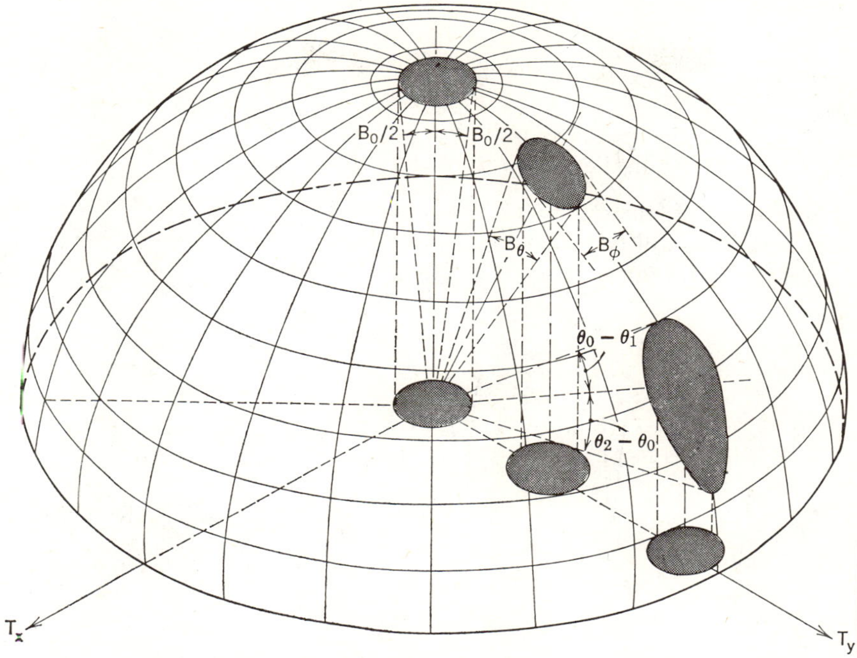

Fig. 1.8. Beamwidth and eccentricity of the scanned beam.

where B_0 is the beamwidth at broadside. The beam broadening factor, $1/\cos\theta_0$, can be linked to the reduction in the effective (projected) aperture area of the array with scan.

3. MUTUAL COUPLING EFFECTS AND THE INFINITE ARRAY CONCEPT

So far only the array factor has been discussed. When antennas with nonisotropic radiation patterns are used as the radiating elements in an array, the array pattern can be expressed as the product of the element factor and the array factor. This statement, of course, is based on the assumption that the mutual coupling between the various array elements can be neglected. We shall now examine the validity of this assumption, especially for closely spaced elements ($b, d \sim \lambda/2$) in the array.

3.1 Inadequacy of the Ordinary Array Theory Model

When many radiating elements are placed in an array, some mutual coupling generally exists. The effects of such mutual coupling among elements in an array, by and large, are threefold:

1. The radiation impedance of an element in an array differs from its free space value and depends on the beam scanning, that is, the phasing of the array.
2. There is a change in the radiation pattern.
3. Deterioration in the polarization characteristics occurs.

The reason why the radiation impedance of an antenna in a phased array varies with scan will become clear later. The radiation impedance variation produces an impedance mismatch between the feed and the antenna. This

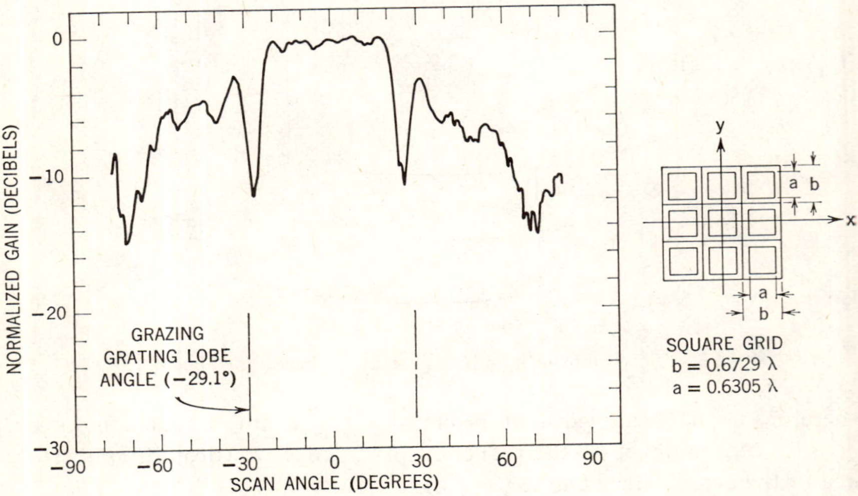

Fig. 1.9. *E*-plane radiation pattern of the center element in a 19 × 19 square waveguide array.

mismatch is manifested by a reduction of the array efficiency and the appearance of spurious lobes.[10,15,16] In a finite array, because each element is not affected by mutual coupling in the same way, the radiation impedances and patterns depend also, to a certain extent, on the element location (this subject is discussed further in Chapter 8). Consequently the principle of pattern multiplication may not be applicable.

One of the severe effects of mutual coupling, which has been observed experimentally, is shown in Figs. 1.9 and 1.10. Deep and sharp dips appear in the radiation pattern of the center element of each array when the rest of the elements are ideally terminated. When such a waveguide element is

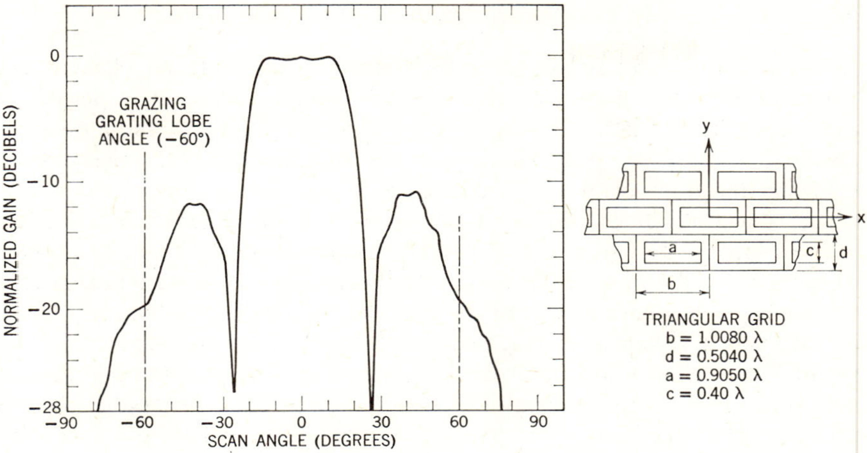

Fig. 1.10. *H*-plane radiation pattern of the center element in a 7 × 7 rectangular waveguide array—triangular grid.

isolated in a ground plane with no other elements present, we observe no sharp dips, but rather a smooth radiation pattern. This is due to the relatively small (in terms of wavelength) waveguide aperture. Consequently, the acute difference between the radiation patterns of the elements inside and outside the array environment is a direct result of the mutual coupling and constitutes evidence of its importance.

The significance of this phenomenon may be further appreciated by considering the following situation. When the whole array is excited and the main beam is scanned to the angle corresponding to the occurrence of these sharp dips, the center elements (and most of the other array elements as well) will be unable to radiate any significant portion of the available power, resulting in a severe mismatch between the feed and the antenna aperture. In fact almost all the incident power will be reflected. Large arrays are referred to as being "blind" at these scan directions. It is important to note that the "blind" scan directions occur in the visible region where only the main beam radiates (no radiating grating lobes). This effect of course reduces the usable scan region of the array. A considerable amount of discussion is devoted to these "blind" spots in later chapters.

The ordinary array theory does not explain nor does it predict the various effects of mutual coupling. Therefore a more precise model for the quantitative analysis and understanding of phased arrays is needed.

3.2 The Uniformly Illuminated Infinite Array Model

The radiation and reflection characteristics of an antenna element are determined by the particular array environment in which it is placed. If more elements are added to the array, the characteristics of the individual elements will be changed. The degree of change depends on the type of elements and their location and arrangement. An element near the center of an array will be affected differently from one near the edge. If the size of an array is increased gradually in a regular manner, an element near the center of the array will eventually be very little affected by the increase in size, and the effect should become negligible if the array becomes sufficiently large. When such a situation is reached, the elements near the center may be regarded as being in an infinite array environment because the addition of still more elements to the array will essentially not alter their radiation and reflection characteristics. Conversely, the radiation properties of an element in an infinite array environment may be said to closely approximate the characteristics of an element near the center of a sufficiently large array. A large array in which a high percentage of the elements "see" an infinite array environment may be treated as a (quasi) infinite array. How large an array is needed to approximate an infinite array and how far removed from the edge an element must be in order to be treated as an infinite array element will be discussed in Chapters 4, 8, and 9.

A convenient and useful illumination function for the infinite array is the uniform illumination (with a linear phase taper), given by

$$V'_{mn} = V e^{j(m\psi_x + n\psi_y)} \tag{16}$$

For a plane wave incident on a receiving infinite array, (13) and (16) describe the relationship between the impressed fields or voltages (at a given reference plane) in the various elements (or their feed lines). In a uniformly illuminated infinite array, each element is affected by mutual coupling in exactly the same way. Therefore the radiation and reflection characteristics of all the elements are identical, and the principle of pattern multiplication is applicable. The element factor to be used in this case is the radiation pattern of a single element embedded in an infinite array of identical elements passively terminated in the impedance of the feed line.

The infinite array model will be used extensively in this book. It provides a basis for various experimental design methods. This model also allows rigorous formulation and analysis of the boundary value problems associated with a broad class of phased arrays with meaningful and practical applications. In addition, the case of nonuniform illumination and certain finite array edge aspects can be derived from the solution of the uniformly illuminated infinite array.

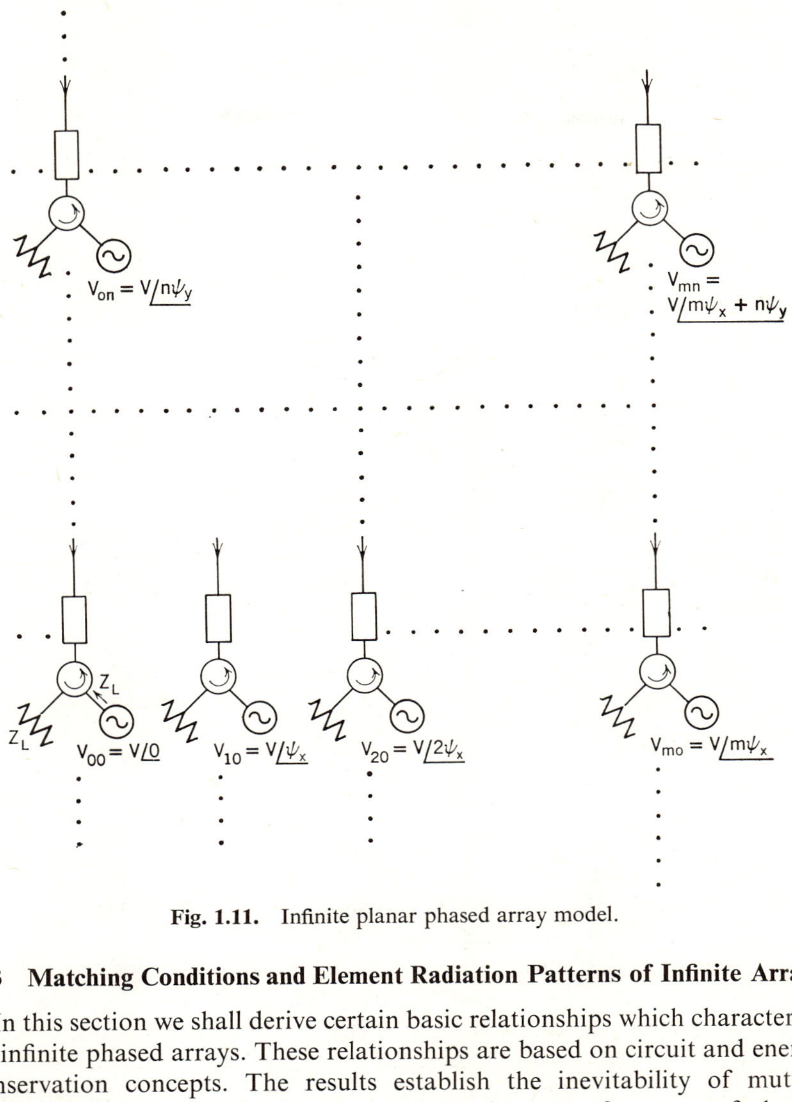

Fig. 1.11. Infinite planar phased array model.

3.3 Matching Conditions and Element Radiation Patterns of Infinite Arrays

In this section we shall derive certain basic relationships which characterize all infinite phased arrays. These relationships are based on circuit and energy conservation concepts. The results establish the inevitability of mutual coupling and also provide criteria for the optimum performance of phased array antennas.

3.3.1 Matching Conditions and Terminal Relationships. Consider the uniformly illuminated infinite planar array depicted schematically in Fig. 1.11. Since the array is infinite, all the elements "see" identical environments. The elements may be arranged in a square or hexagonal (equilateral triangular)

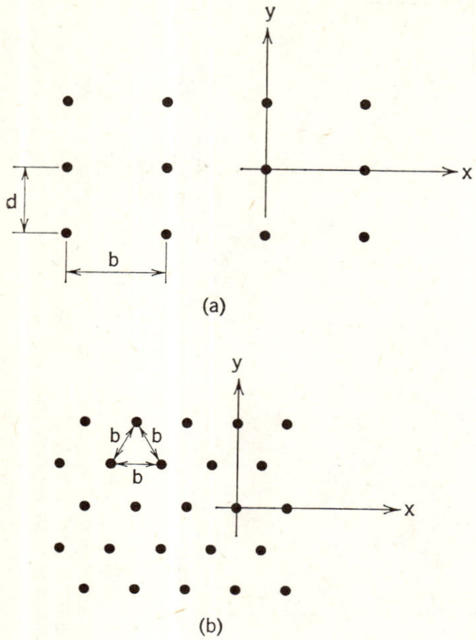

Fig. 1.12. (*a*) Rectangular grid arrangement of elements. (*b*) Hexagonal (equilateral triangular) grid arrangement of elements.

grid* (Fig. 1.12). They are fed by means of ideal and symmetrical circulators† from independent voltage sources, given by (16). The power injected from the source into each circulator is constant and is given by

$$P_e = \frac{V^2}{Z_L} \tag{17}$$

where Z_L is the input and load resistance of the circulator. All reflections caused by mismatches within an element channel, as well as energy coupled from the rest of the elements into this channel, will be dissipated in the Z_L of the terminated port of the circulator.

Let us now define unconditional and conditional matches[17] of phased arrays. A phased array is considered unconditionally matched to free space when the total power radiated from it equals the power fed into it independently of ψ_x and ψ_y. An array is considered conditionally matched when the total power radiated equals the total power fed in only for a certain range of values of ψ_x and ψ_y.

The motivation for choosing these definitions becomes clear when we consider the grating lobe diagrams of Figs. 1.13 and 1.14. In very large arrays

* The following discussion is equally valid for any periodic grid.

† The ideal circulators are introduced for convenience and clarity. The discussion can proceed just as well, however, without the circulators.

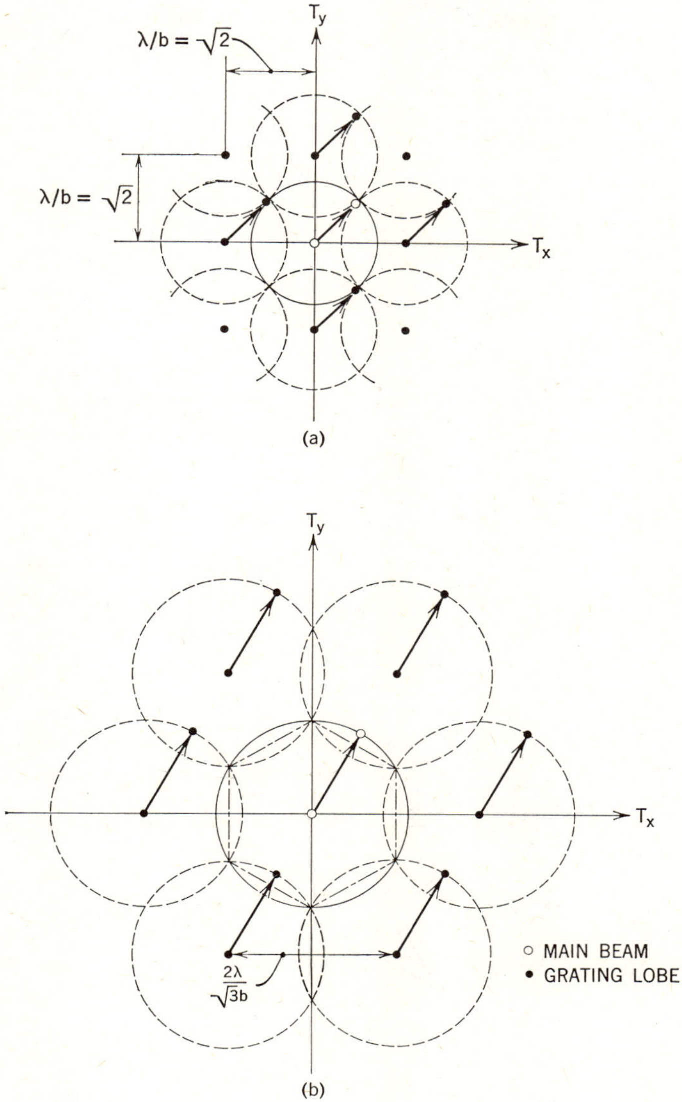

Fig. 1.13. Grating lobe diagram. (*a*) Square grid, $b/\lambda = 1/\sqrt{2}$. (*b*) Hexagonal grid, $b/\lambda = \frac{2}{3}$.

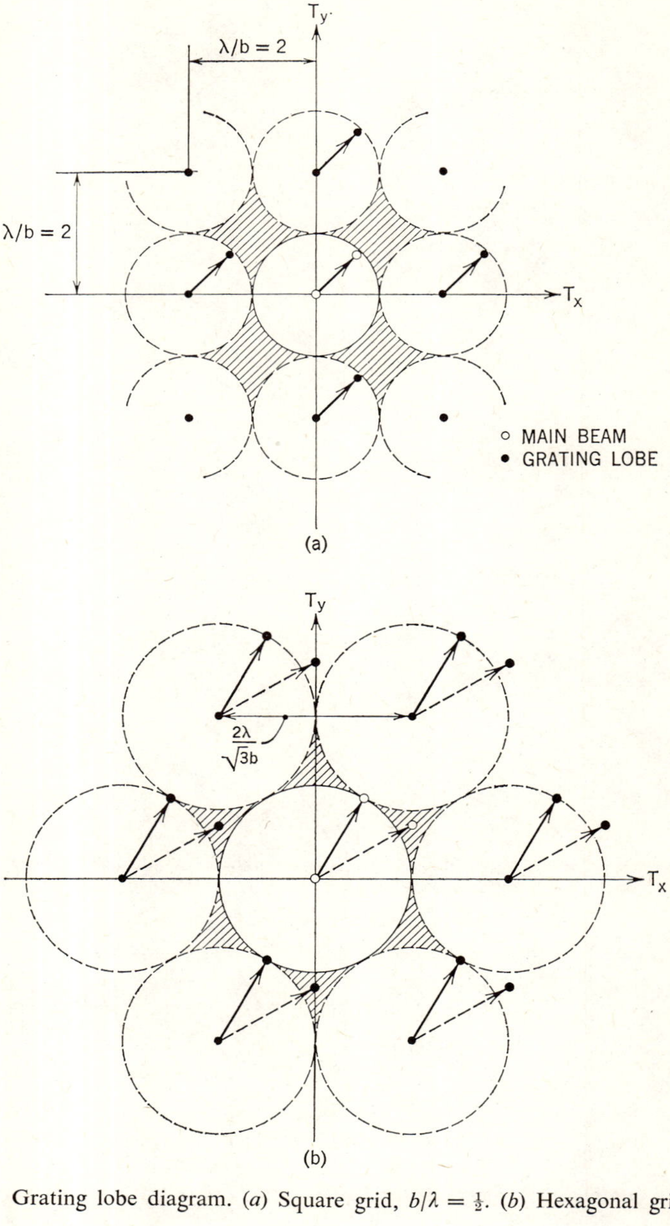

Fig. 1.14. Grating lobe diagram. (*a*) Square grid, $b/\lambda = \frac{1}{2}$. (*b*) Hexagonal grid, $b/\lambda = 1/\sqrt{3}$.

the power radiated is concentrated within the immediate vicinity of the main beam.* Thus, if the spacing between elements is such that, independently of ψ_x and ψ_y, at least the main beam or one grating lobe is always present within the unit circle (i.e., in real space), the power radiated from the array might, at least conceptually, stay constant and equal to the power fed in. This is shown in Fig. 1.13 for $b/\lambda = 1/\sqrt{2}$ and $b/\lambda = \frac{2}{3}$ in the cases of square and hexagonal grids, respectively. If the elements in the array are not coupled to one another, the power radiated is independent of ψ_x and ψ_y and can be made equal to the input power by independently matching the array elements. On the other hand, for smaller spacing between elements, such as is shown in Fig. 1.14, there are certain ranges of values of ψ_x and ψ_y for which neither the main beam nor any grating lobes will be present in the real space. Consequently the total power radiated must vanish, while the total available power will be dissipated in the loads Z_L. For this range of values of element spacing only a conditional match is feasible. Moreover mutual coupling *must* exist among the array elements. Thus a necessary condition for an unconditional match is $b \geq \lambda/\sqrt{2}$ or $b \geq 2\lambda/3$ for square[18] and hexagonal[17] grids, respectively.

To understand the implication of unconditional and conditional matches more precisely, let us consider the following experiment. Let the (m, n)th element be excited by a voltage V in a passive array (i.e., the rest of the generators are turned off). If $\mathscr{C}_{pq}{}^{mn}$ denotes the coupling to the (p, q)th element when a unit voltage excites the (m, n)th element, then $V\mathscr{C}_{pq}{}^{mn}$ is the voltage developed across Z_L in the terminated port of the (p, q)th element. The same voltage (except for a phase factor) will exist in the (p, q)th element feed line. The mutual coupling coefficient, $\mathscr{C}_{pq}{}^{mn}$ (i.e., the scattering matrix coefficient), is a complex quantity to account for phase differences and is a function of the wavelength, the type of array element, and the geometry. Because of reciprocity it is clear that $\mathscr{C}_{pq}{}^{mn} = \mathscr{C}_{mn}{}^{pq}$. Moreover, as a result of the periodicity of the array grid,

$$\mathscr{C}_{pq}{}^{mn} = \mathscr{C}_{p'q'}^{m'n'} \qquad \text{when} \quad m - p = m' - p' \quad \text{and} \quad n - q = n' - q' \tag{18}$$

Therefore, we shall focus our attention on the $(p = 0, q = 0)$ element and redesignate the coupling coefficients with respect to it as

$$\mathscr{C}_{00}{}^{mn} = \mathscr{C}_{mn} \tag{19}$$

Once the set of all the coupling coefficients, $\{\mathscr{C}_{mn}\}$, is known, the coupling coefficients of the rest of the elements are automatically specified. When the

* For infinite arrays the array radiation pattern is a periodic Dirac delta function of the scan angle.

(0, 0)th element is excited in a passive array, the fractional [relative to P_e of (17)] dissipated power in the array is

$$\eta_p = \frac{(V^2/Z_L)\sum_{(m)}\sum_{(n)}|\mathscr{C}_{mn}|^2}{P_e} = \sum_{(m)}\sum_{(n)}|\mathscr{C}_{mn}|^2 \tag{20}$$

The factor η_p is a measure of the radiation efficiency of the element. The gain of the element is the product[3,4] of the directivity and $(1 - \eta_p)$.

When all the generators are turned on with the proper steering phases, (16), the (active) reflection coefficient, R_a, referred to Z_L in the terminated port, is given by

$$R_a(\psi_x, \psi_y) = \sum_{(m)}\sum_{(n)}\mathscr{C}_{mn}e^{j(m\psi_x+n\psi_y)} \tag{21}$$

In (21) R_a is a double Fourier series which relates the active reflection coefficient to the coupling coefficients:*

$$\mathscr{C}_{mn} = \frac{1}{4\pi^2}\int_{-\pi}^{\pi}\int_{-\pi}^{\pi}R_a(\psi_x, \psi_y)e^{-j(m\psi_x+n\psi_y)}\,d\psi_x\,d\psi_y \tag{22}$$

By using Parseval's theorem in (21) we can relate[18] η_p to R_a by

$$\eta_p = \frac{1}{4\pi^2}\int_{-\pi}^{\pi}\int_{-\pi}^{\pi}|R_a(\psi_x, \psi_y)|^2\,d\psi_x\,d\psi_y \tag{23}$$

The fractional dissipated power per element of the active array is

$$\eta_a = |R_a|^2 = \left|\sum_{(m)}\sum_{(n)}\mathscr{C}_{mn}e^{j(m\psi_x+n\psi_y)}\right|^2 \tag{24}$$

From the above equations it is clear that for an unconditional match $\mathscr{C}_{mn} = 0$ and $\eta_p = 0 = \eta_a$. For a conditional match $\mathscr{C}_{mn} \neq 0$, while $\eta_a = 0$ for a certain range of values of ψ_x and ψ_y (destructive interference) and is equal to unity for the complementary values of ψ_x and ψ_y (constructive interference). An ideal conditional match is shown in Fig. 1.15 for a square grid array with $b = \lambda/2$.

* The mutual coupling (scattering) coefficients, $\{\mathscr{C}_{mn}\}$, determine the terminal relationships of the array. An alternative representation for the terminal relationships is through the mutual admittance (or impedance) matrix, $\|Y_{mn}\|$ (or $\|Z_{mn}\|$), where

$$Y_a(\psi_x, \psi_y) = \sum_{(m)}\sum_{(n)}Y_{mn}e^{j(m\psi_x+n\psi_y)}$$

or

$$Z_a(\psi_x, \psi_y) = \sum_{(m)}\sum_{(n)}Z_{mn}e^{j(m\psi_x+n\psi_y)}$$

and

$$\frac{1}{Z_a(\psi_x, \psi_y)} = Y_a(\psi_x, \psi_y) = \frac{1 - R_a(\psi_x, \psi_y)}{1 + R_a(\psi_x, \psi_y)}$$

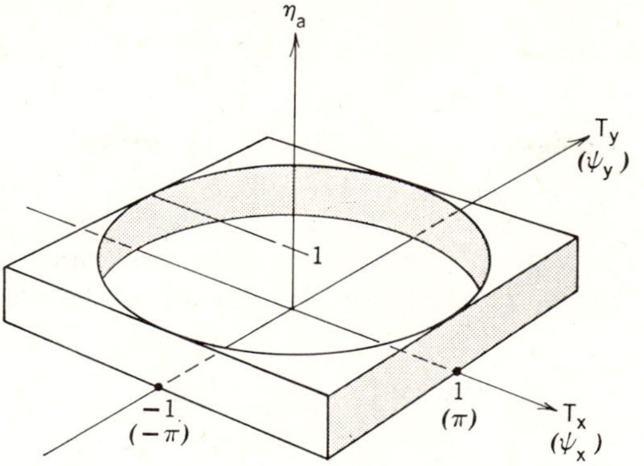

Fig. 1.15. Fractional dissipated power, η_a, in a square grid array, $b/\lambda = \frac{1}{2}$. Ideal conditional match.

3.3.2 Ideal Element Radiation Patterns. Here we shall examine the restrictions placed on the array element radiation pattern if a match is to be achieved. Consider the rectangular grid array of Figs. 1.11 and 1.12*a*. Let the far-zone electric field radiation pattern per unit voltage of the generator of a singly excited element in a passive array be $\mathscr{E}_1(\theta, \phi)$ or $\mathscr{E}_1(T_x, T_y)$. The total far-zone field of the phased array, $\mathscr{E}_t$, is the linear superposition of all individual element patterns with the proper phasing, (16), between them and is given by

$$\mathscr{E}_t = \mathscr{E}_1(T_x, T_y) S_a(T_x - T_{x0}, T_y - T_{y0}) \tag{25}$$

Here S_a is the array factor, (9), while (T_{x0}, T_{y0}) define the scanned beam directional cosines, which are related through (13) to the steering phases. The total power, P_t, radiated by the array into the visible space consisting of a hemisphere is

$$\begin{aligned} P_t &= \frac{1}{Z_0} \int_0^{\pi} \int_0^{2\pi} |\mathscr{E}_t|^2 \sin\theta \, d\theta \, d\phi \\ &= \frac{1}{Z_0} \iint_{T_x^2 + T_y^2 \leq 1} |\mathscr{E}_t|^2 \frac{dT_x \, dT_y}{\cos\theta} \end{aligned} \tag{26}$$

where

$$|\mathscr{E}_t|^2 = \mathscr{E}_t \cdot \mathscr{E}_t^*, \quad Z_0 = \sqrt{\mu/\varepsilon}, \quad \text{and} \quad \cos\theta = \sqrt{1 - T_x^2 - T_y^2}.$$

The array factor of an array of M columns and N rows is given by[4]

$$S_a(T_x - T_{x0}, T_y - T_{y0}) = V \frac{\sin(M\pi b/\lambda)(T_x - T_{x0})}{\sin(\pi b/\lambda)(T_x - T_{x0})} \frac{\sin(N\pi d/\lambda)(T_y - T_{y0})}{\sin(\pi d/\lambda)(T_y - T_{y0})} \tag{27}$$

For a very large array ($M, N \to \infty$), the array factor approaches a series of Dirac delta functions located at $T_x - T_{x0} = p\lambda/b$, $T_y - T_{y0} = q\lambda/d$:

$$\sum_{p=-\infty}^{\infty} \sum_{q=-\infty}^{\infty} \delta\left(T_x - T_{x0} - \frac{p\lambda}{b}\right) \delta\left(T_y - T_{y0} - \frac{q\lambda}{d}\right)$$

For a rectangular grid with both b/λ and d/λ less than half (i.e., only one beam can exist for any real scan direction), (26) can be rewritten as

$$P_t = \frac{|\mathscr{E}_1(T_{x0}, T_{y0})|^2}{Z_0 \cos\theta_0} \lim_{M,N\to\infty} \iint_{T_x^2+T_y^2\leq 1} |S_a(T_x - T_{x0}, T_y - T_{y0})|^2 \, dT_x \, dT_y$$

$$= \frac{V^2 |\mathscr{E}_1(T_{x0}, T_{y0})|^2}{Z_0 \cos\theta_0} \lim_{M,N\to\infty} \int_{-\infty}^{+\infty}\int_{-\infty}^{+\infty} \frac{\sin^2 (M\pi b/\lambda)(T_x - T_{x0})}{[(\pi b/\lambda)(T_x - T_{x0})]^2}$$

$$\times \frac{\sin^2 (N\pi d/\lambda)(T_y - T_{y0})}{[(\pi d/\lambda)(T_y - T_{y0})]^2} d(T_x - T_{x0})\, d(T_y - T_{y0})$$

$$= \lim_{M,N\to\infty} \frac{MNV^2}{Z_0} \frac{\lambda^2}{bd} \frac{|\mathscr{E}_1(T_{x0}, T_{y0})|^2}{\cos\theta_0} \tag{28}$$

For larger element spacing where l' grating lobes corresponding to l' delta functions can exist in real space, (28) becomes[19]

$$P_t = \lim_{M,N\to\infty} \frac{MNV^2}{Z_0} \frac{\lambda^2}{bd} \sum_{l=0}^{l'} \frac{|\mathscr{E}_1(T_{xl}, T_{yl})|^2}{\cos\theta_l} \tag{29}$$

The power radiated by each individual element in the array, P_a, can be related to the power input to the element by utilizing (17), (24), and (28):

$$P_a = \frac{P_t}{MN} = \frac{V^2}{Z_L}(1 - \eta_a) = \frac{V^2}{Z_0} \frac{\lambda^2}{bd} \frac{|\mathscr{E}_1(T_{x0}, T_{y0})|^2}{\cos\theta_0}, \quad b, d \leq \frac{\lambda}{2} \tag{30}$$

For the hexagonal grid array of Fig. 1.12*b*, a superscript h is added to the appropriate quantities; P_a^h is given by[17]

$$P_a^h = \frac{V^2}{Z_L}(1 - \eta_a^h) = \frac{V^2}{Z_0} \frac{2\lambda^2}{\sqrt{3}\, b^2} \frac{|\mathscr{E}_1(T_{x0}, T_{y0})|^2}{\cos\theta_0}, \quad b \leq \frac{\lambda}{\sqrt{3}} \tag{31}$$

where $b \leq \lambda/\sqrt{3}$ ensures only a single beam in the real space. Whenever the element radiation power pattern (in a passive array environment) assumes the following form:

$$|\mathscr{E}_1(T_x, T_y)|^2 = B \cos \theta = B\sqrt{1 - T_x^2 - T_y^2} \tag{32}$$

where B is a proportionality constant, P_a stays constant with scanning, and (conceptually at least) a conditional match can be achieved. The radiation pattern in (32) may be defined as that of an ideal element. The power gain pattern[3,4] of an element is defined as

$$g_e(\theta, \phi) = 4\pi \frac{(1 - \eta_p)\,|\mathscr{E}_1(\theta, \phi)|^2}{\int_0^{\pi/2}\int_0^{2\pi} |\mathscr{E}_1(\theta, \phi)|^2 \sin\theta \, d\theta \, d\phi} \tag{33}$$

with

$$\frac{V^2}{Z_L}(1 - \eta_p) = \frac{V^2}{Z_0}\int_0^{\pi/2}\int_0^{2\pi} |\mathscr{E}_1(\theta, \phi)|^2 \sin\theta \, d\theta \, d\phi \tag{34}$$

Recall that η_p is related to the coupling coefficients by (20) and (23). By utilizing (24), (30), (31), (33), and (34), we can express the single-element gain pattern in terms of the array grid and active reflection coefficient, R_a, as

$$g_e(\theta, \phi) = 4\pi \frac{bd}{\lambda^2}[1 - |R_a(\theta, \phi)|^2]\cos\theta; \qquad \text{rectangular grid,} \quad b, d \leq \frac{\lambda}{2}$$

$$g_e^h(\theta, \phi) = 2\pi \frac{\sqrt{3}\, b^2}{\lambda^2}[1 - |R_a^h(\theta, \phi)|^2]\cos\theta; \qquad \text{hexagonal grid,} \quad b \leq \frac{\lambda}{\sqrt{3}} \tag{35}$$

For a conditionally matched element (i.e., $R_a = 0 = R_a^h$ for $T_x^2 + T_y^2 \leq 1$), the ideal gain, g_{em}, is

$$\begin{aligned} g_{em} &= 4\pi \frac{bd}{\lambda^2}\cos\theta; && \text{rectangular grid,} \quad b, d \leq \frac{\lambda}{2} \\ g_{em}^h &= 2\pi\sqrt{3}\left(\frac{b}{\lambda}\right)^2 \cos\theta; && \text{hexagonal grid,} \quad b \leq \frac{\lambda}{\sqrt{3}} \end{aligned} \tag{36}$$

For greater element spacings, multiple beams can exist in real space. The element power pattern has to be modified, therefore, in the sectors of the hemisphere where two beams appear simultaneously in order to maintain a

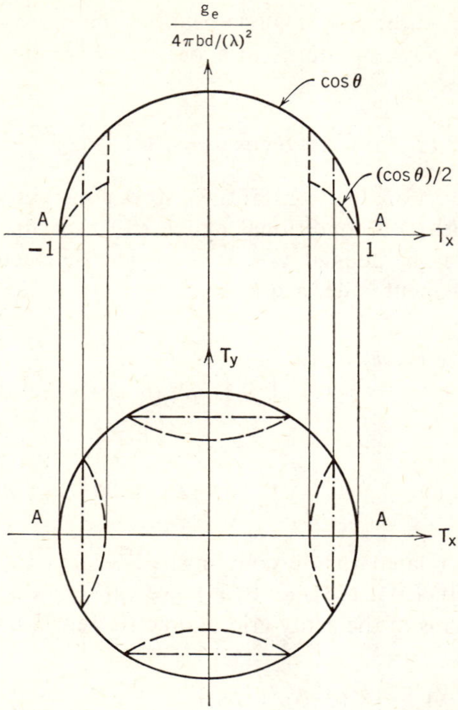

Fig. 1.16. Ideal element normalized gain in a passive array; square grid, $\frac{1}{2} < b/\lambda < 1/\sqrt{2}$.

constant radiated power. The modified pattern is shown in Figs. 1.16 and 1.17. The basic $\cos\theta$ curve (solid) is a semicircle. The areas defined by the dashed arcs and the unit circle correspond to a $(\cos\theta)/2$ pattern.[20] An alternative pattern[19] is obtained by chopping the $\cos\theta$ pattern along the dash-dotted lines. The gain of a conditionally matched element as a function of the spacing is obtained[18] in a similar fashion by utilizing the patterns of Figs. 1.16 and 1.17:

$$g_{em} = 4\cos\theta \begin{cases} \pi\left(\dfrac{b}{\lambda}\right)^2, & \dfrac{b}{\lambda} \le \dfrac{1}{2} \\[2ex] \dfrac{(\pi/4) - \cos^{-1}(\lambda/2b) + (\lambda/2b)\sqrt{1-(\lambda/2b)^2}}{(\lambda/2b)^2}, & \dfrac{1}{2} \le \dfrac{b}{\lambda} \le \dfrac{1}{\sqrt{2}} \\[2ex] 1 \text{ (unconditional match)}, & \dfrac{b}{\lambda} \ge \dfrac{1}{\sqrt{2}} \end{cases} \tag{37}$$

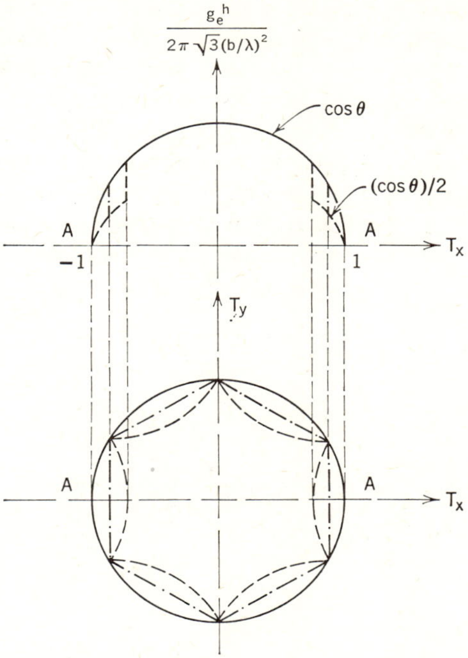

Fig. 1.17. Ideal element normalized gain in a passive array; hexagonal grid, $b/\lambda = \frac{2}{3}$.

for a square grid. For a hexagonal grid the ideal gain is

$$g_{em}{}^h = 4\cos\theta \begin{cases} \dfrac{\sqrt{3}\,\pi}{2}\left(\dfrac{b}{\lambda}\right)^2, & \dfrac{b}{\lambda} \leq \dfrac{1}{\sqrt{3}} \\[2ex] \dfrac{\sqrt{3}}{2}\left(\dfrac{b}{\lambda}\right)^2\left[\pi - 6\cos^{-1}\dfrac{\lambda}{\sqrt{3}\,b}\right] + \sqrt{9(b/\lambda)^2 - 3}, & \dfrac{1}{\sqrt{3}} \leq \dfrac{b}{\lambda} \leq \dfrac{2}{3} \\[2ex] 1 \text{ (unconditional match)}, & \dfrac{b}{\lambda} \geq \dfrac{2}{3} \end{cases} \tag{38}$$

Later in this book it is shown that a broad class of physically realizable array elements exhibits a match performance which is inconsistent with the ideal conditional or unconditional matches. Specifically, for this class of arrays the reflection coefficient does not exhibit the discontinuity shown in Fig. 1.15 in the case of conditional match. Moreover, the reflection coefficient cannot be independent of the scan angle in the vicinity of grazing beams in the case of unconditional match.

It has been shown that for a certain class of infinite phased array of periodic current sheets[43] the reflection coefficient cannot vanish over a continuous range of values of the scan angle. Nonetheless, the reflection coefficient can vanish at a denumerably infinite number of values of the scan angle if one interconnects all the array elements in a particular fashion.[44]

These facts do not diminish the importance of the conditional and unconditional match concepts. As we shall see later, the ideal element gain is an important yardstick in judging array performance. It is unlikely that a small aperture antenna element (in a passive array environment) can be made to

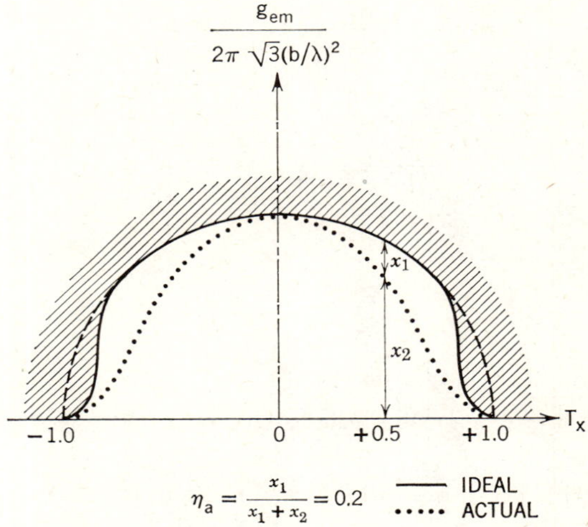

Fig. 1.18. Ideal and actual elements normalized gain (cross section); hexagonal grid, $b/\lambda = \frac{2}{3}$.

produce the abrupt patterns of Figs. 1.16 and 1.17 (unless the coupling between the various apertures is strong enough to increase considerably the effective aperture of the element). The only requirement of the pattern is to accommodate various lobes in such a way that the total radiated power is constant and is shared by these lobes. A pattern such as that shown in Fig. 1.18 is perhaps more plausible. The boundary of the hatched area represents a physical restriction on the gain of a single element in the array environment, no matter what its gain is outside the array environment.

The element radiation pattern measured in a passive array environment can provide some indication of the array match as a whole. Insofar as the element radiation pattern falls short of that demanded by the element spacing (boundary of hatched area), the fraction by which it falls short in any given direction is a direct measure of the power reflected when the beam is steered to that direction. Even if only the relative power pattern of the element is known (the directivity), some indication of the mismatch can be obtained by inscribing the measured element pattern in the ideal pattern, as shown by the dotted curve of Fig. 1.18. If the array is indeed matched at broadside, only about 80% of this power will be radiated when the beam is steered to 30° off broadside ($\sin 30° = 0.5$), while 20% will be reflected and dissipated.

Phased arrays are usually required to be steered only in a certain region of the hemisphere; thus only within this region need the actual and ideal element patterns coincide. Meeting this requirement will produce a match within the scanned region, while in the rest of the hemisphere a mismatch can occur should the beam be steered there at all.

The equations and conclusions derived in this section express the limitations that simple physics seems to demand of phased arrays, based on energy conservation. However, the possibility of achieving the ideal element radiation pattern, which is not quite obvious, should be established through a rigorous field theory analysis. Moreover the precise reflection and radiation characteristics (with scanning) of an actual array element can be obtained only through the solution of the boundary value problem of the array, or through elaborate and costly experimentation.

4. PHASED ARRAYS AS BOUNDARY VALUE PROBLEMS

In this section we shall review briefly various approximate methods which have been used for the evaluation of mutual coupling effects in phased array antennas. We shall indicate the weaknesses of some of these methods and establish the need for a precise analysis of phased arrays. We shall also point out how such an approach can be utilized in the design of practical array elements.

4.1 The Mutual Impedance Approach

The mutual impedance between two antennas is defined as the open circuit voltage induced in the terminals of one antenna when the other antenna is excited by a unit current. Several analytical methods[3,20–25] have been developed for the determination of the mutual impedance.* These approaches are based on the assumption that the terminal current of an

* In some cases it is more convenient to express the pertinent antenna parameters in terms of the mutual admittances. The mutual admittance between two antennas is defined as the current induced in the short-circuited terminals of one antenna when the other antenna is excited by a unit voltage.

antenna determines completely the current distribution (or field) on the antenna and thereby its radiation pattern.

In applying the mutual impedance approach to phased array problems, the terminal relation of each element is related to those of other elements by the mutual impedance (or admittance). This leads to a formulation of the problem in terms of a multiport circuit problem. Thus, the terminal currents are determined for specified terminal voltages or vice versa. The approach is suitable for arrays of finite size. Although the mutual impedances should be determined when the elements are placed in the particular array environment under consideration, they are rather difficult to calculate. As a result, the mutual impedances (or admittances) between pairs of array elements have been taken as the mutual impedances between these pairs of elements *outside* the array environment. This first-order interaction approach does not account for the contribution of the rest of the open or short-circuited array elements, which may be substantial in some cases.

Calculations of the driving point impedances, based on these assumptions, have been carried out for a number of dipole arrays both with and without a ground plane.[27–30] In these calculations, the mutual impedances are determined by assuming either a constant or a sinusoidal current distribution on the dipoles. The effects of mutual coupling on the gain of finite dipole arrays and the driving point impedances of the various elements have been studied[29,30] as a function of the interelement spacing and the height above a ground plane. Measured element gain and beamwidth showed reasonable agreement with the calculated results.

The applicability of this approximate mutual impedance method is limited, however, to elements which are small in terms of wavelength. Approximate analytical expressions for the mutual impedances are available for thin dipoles and (their duals) narrow slots in a ground plane. It should be pointed out that for closely spaced arrays of dipoles the feed network (which is usually neglected in the analysis) constitutes a significant fraction of the dimensions of the dipole. Therefore, the results so obtained should be used only qualitatively.[26]

4.2 The Infinite Array-Periodic Structure Approach

One important reason for considering the infinite array model is that it can be treated as a periodic structure. Various analytical methods are available for the analysis of periodic structures.

Because of the periodicity of the grid, the active element aperture field (or current) has to be evaluated only in a single periodic cell when the array is excited uniformly, (16). The active reflection (or impedance) and radiation characteristics (including gain, axial ratio, and tilt angle) are directly related to the aperture field (or current). The effects of mutual coupling are automatically taken into account in this approach. Moreover, from a knowledge of the radiation and reflection characteristics of the infinite array we can obtain meaningful, and often exact, results and insight into certain aspects of finite and nonperiodically excited arrays with similar elements. As we shall see later, this approach is most suitable for a rigorous formulation and analysis of the boundary value problems associated with a broad class of practical phased array elements.

4.2.1 Method of Assumed Aperture Distribution. One approximate method of field analysis of phased arrays is based on combining the periodic properties of the array with an assumed aperture field (or current) distribution in the array elements. Accurate results can often be obtained for certain physical quantities, such as impedance, if they are cast in a variational form, provided that there is no appreciable difference between the actual and the assumed distributions.

On the basis of the observation that the current distribution (in a uniformly illuminated array) repeats periodically along two major coordinate axes, Wheeler[31] applied image principles to the array, using electric and magnetic (fictitious) walls at certain planes of symmetry. In this way he reduced the problem to an equivalent one of radiation from a current distribution in a waveguide and determined the active radiation resistance of a dipole or a thin slot array element for broadside radiation. The effects of the feed structure were not accounted for in this analysis.

Edelberg[32] and Edelberg and Oliner[33] have extended the approach taken by Wheeler to include the effects of scanning in an infinite array of small and narrow slots located on a ground plane. Each slot is excited separately by a rectangular waveguide. While Oliner and Edelberg properly apply the periodic boundary conditions appropriate to the doubly periodic structure, they assume a scan-invariant sinusoidal electric field distribution in the narrow slot. The results are readily extended to dipole arrays[26] in free space by using Babinet's principle.

Stark[34] made an independent calculation for an electric dipole array located in free space or above a ground plane. The current distribution on the dipole was also assumed to be sinusoidal and scan independent. Again, the radiation impedance was obtained from energy consideration without accounting for the feed network.

It is interesting to note that in this method, as well as in the mutual impedance approach of Section 4.1, identical aperture distributions were assumed. Calculations of the radiation impedance by these two methods show good agreement for infinitesimal and thin half-wave dipole arrays[26,35] located $\lambda/4$ above a ground plane.

The applicability of Stark's method is also limited to either thin dipoles or narrow slots in the ground plane. Moreover, more rigorous recent analyses of dipole arrays[36,37] have shown that for certain practical ranges of dipole size and spacing the assumed sinusoidal current distribution is inadequate. In order to obtain meaningful results, more accurate knowledge of the currents is necessary. In these alternative methods,[36,37] the current distribution is obtained accurately from the solution of the pertinent integral equation.

4.2.2 The Waveguide Simulator. An interesting and novel idea which has developed in part from the previous analysis and earlier work[45] is the waveguide simulator. The periodicity of the array grid, combined with certain required symmetries of the grid, array element, and mode of excitation, leads to the derivation of an equivalent waveguide transition by utilizing image principles. This waveguide transition simulates the performance of the array under certain scan conditions (hence the name waveguide simulator[38–40]) and is a useful experimental tool. In principle, every scan angle of an infinite array[41] can be simulated by a junction of a finite number of array elements and a cylindrical rectangular (or sometimes triangular) waveguide with metallic walls (provided that certain basic symmetries of the grid, element, and mode of excitation exist).

Unfortunately, each combination of scan angle and mode of excitation requires a different size of simulator. Moreover, only a limited number of angles can be simulated in practice by a reasonably small waveguide simulator. Therefore, the array characteristics can be obtained at only a limited number of isolated scan angles. Certain anomalous forced resonances, such as those manifested by the "blind" spots (see Section 3.1), may go undetected, therefore, if one relies totally on a simulator in the design of phased array elements.

4.2.3 Integral Equation Formulation. The methods discussed in Sections 4.1 and 4.2.1 provide approximate analytical solutions for the reflection and radiation characteristics of arrays of (electrically) very small elements. Through these models, qualitative understanding of array performance was gained in the early days of phased array research and development. Needless to say, much more information is required to attain a good understanding of the reflection and radiation characteristics of arrays of other types of elements.

The present day requirements of modern radar and communication, with regard to high power handling capabilities, have ruled out the use of (electrically) small elements. In addition, more accurate quantitative data about array performance are necessary before the costly construction of array antennas is undertaken. Hence, more accurate methods for the analysis and synthesis of practical phased array elements are needed.

The formulation of electromagnetic boundary value problems in terms of integral equations has long been established. It provides a useful starting point for an accurate analysis of phased arrays. One of the attractive features of this formulation is the fact that the boundary conditions of the problem are intrinsically accounted for in the functional form of the integral equation. As in most practical problems, exact analytical solutions of the pertinent integral equations are scarce. However, concurrently with the rapid advances in numerical analysis, many powerful tools have been developed for obtaining accurate numerical solutions of complex integral equations. In particular, because of the rapid growth and availability of large storage and rapid access digital computers, accurate numerical solutions of complex problems can now be obtained.

One of the most important classes of practical array elements is the open ended waveguide element. Parallel-plate, rectangular, circular, and coaxial waveguides can be easily manufactured and are suitable, from both a mechanical and an environmental point of view, for many array applications. The boundary value problem associated with open ended waveguide arrays can be rigorously formulated and expressed in terms of several types of integral equations. The waveguide feed and mode of excitation, as well as the use of protective radomes (consisting of dielectric materials), can be accounted for precisely in this formulation. The integral equation can be solved numerically (and analytically in special cases) with a high degree of accuracy. This approach has been found to constitute a powerful tool for the analysis and design of phased array antennas.

It should be pointed out that certain other methods for the formulation of open ended waveguide arrays (including thin slots), whereby a specific numerical or approximate analytical method is applied to obtain the solution, are equivalent to the integral equation approach. However, formulation of the integral equation allows the flexibility of choosing the optimum method of solution for the particular type of problem at hand.

Although emphasis is placed in this book on open ended waveguide arrays, the formulation and numerical methods can be extended to dipoles and other related types of array elements. In fact, the integral equation approach has already been applied to the analysis of infinite[36,37] and finite[42] arrays of dipoles. Since this subject has been treated elsewhere, we shall not dwell upon it here.

REFERENCES

1. R. C. Hansen (ed.), *Microwave Scanning Antennas*, Vol. II: *Array Theory and Practice*, Academic Press, New York, 1966.

2. R. C. Hansen (ed.), *Microwave Scanning Antennas*, Vol. III: *Array Systems*, Academic Press, New York, 1966.

3. R. E. Collin and F. J. Zucker, *Antenna Theory*, Part I, McGraw-Hill Book Company, New York, 1969.

4. S. Silver (ed.), *Microwave Antenna Theory and Design*, M.I.T. Radiation Laboratory Series, Vol. 12, McGraw-Hill Book Company, New York, 1949.

5. H. G. Booker and P. C. Clemmow, "The Concept of an Angular Spectrum of Plane Waves and Its Relation to That of Polar Diagram and Aperture Distribution," *Proc. IEE*, Vol. 97, Part III, pp. 11–17, January 1950.

6. R. E. Collin and S. Rothschild, "Reactive Energy in Aperture Fields and Aperture Q," *Can. J. Phys.*, Vol. 41, pp. 1967–1979, December 1963; See also *IEEE Trans. Antennas and Propagation*, Vol. AP-15, pp. 565–569, July 1967.

7. D. R. Rhodes, "On a Fundamental Principle in the Theory of Planar Antennas," *Proc. IEEE*, Vol. 52, pp. 1013–1021, September 1964.

8. G. Borgiotti, "Fourier Transforms Method of Aperture Antennas Problem," *Alta Freq.*, Vol. 32, pp. 196–204, November 1963.

9. W. H. Von Aulock, "Properties of Phased Arrays," *Proc. IEEE*, Vol. 48, pp. 1715–1727, October 1960.

10. J. Blass and S. J. Rabinowitz, "Mutual Coupling in Two-Dimensional Arrays," *IRE WESCON Convention Record*, Vol. I, Part I, pp. 134–150, 1957.

11. E. D. Sharp, "A Triangular Arrangement of Planar Array Elements That Reduces the Number Needed," *IEEE Trans. Antennas and Propagation*, Vol. AP-9, pp. 126–129, March 1961.

12. Y. T. Lo and S. W. Lee, "Affine Transformation and Its Application to Antenna Arrays," *IEEE Trans. Antennas and Propagation*, Vol. AP-13, pp. 890–896, November 1965.

13. R. W. Bickmore, "A Note on the Effective Aperture of Electronically Scanned Arrays," *IEEE Trans. Antennas and Propagation*, Vol. AP-6, pp. 194–196, April 1958.

14. R. S. Elliott, "Beamwidth and Directivity of Large Scanning Arrays," Part I, *Microwave J.*, Vol. 6, pp. 53–60, December 1963; Part II, *Microwave J.*, Vol. 7, pp. 74–82, January 1964.

15. L. A. Kurtz and R. S. Elliott, "Systematic Errors Caused by the Scanning of Antenna Arrays: Phase Shifters in the Branch Lines," *IEEE Trans. Antennas and Propagation*, Vol. AP-4, pp. 619–627, October 1956.

16. L. A. Kurtz, R. S. Elliott, S. Wehn, and W. L. Flock, "Mutual Coupling Effects in Scanning Dipole Arrays," *IEEE Trans. Antennas and Propagation*, Vol. AP-9, pp. 433–443, September 1961.

17. N. Amitay, J. S. Cook, R. G. Pecina, and C. P. Wu, "On Mutual Coupling and Matching Conditions in Large Planar Phased Arrays," *IEEE G-AP International Symposium Digest*, pp. 150–156, 1964; Also *Bell Tel. System Monograph* 5047, February 1965.

18. P. W. Hannan, "The Element-Gain Paradox for a Phase-Array Antenna," *IEEE Trans. Antennas and Propagation*, Vol. AP-12, pp. 423–433, July 1964.

19. J. L. Allen et al., "Phased Array Radar Studies, July 1961–January 1963," Lincoln Laboratories, M.I.T. *Tech. Rep.* 299, February 1963.

20. P. S. Carter, "Circuit Relations in Radiating Systems and Applications to Antenna Problems," *Proc. IRE*, Vol. 20, pp. 1004–1041, June 1932.

21. H. G. Booker, "Slot Aerials and Their Relation to Complementary Wire Aerials (Babinet's Principle)," *J. IEE* (*London*), Part IIIA, Vol. 93, pp. 620–626, 1946.

22. J. D. Kraus, *Antennas*, McGraw-Hill Book Company, New York, 1950.

23. R. W. P. King, *Theory of Linear Antennas*, Harvard University Press, Cambridge, Mass., 1956.

24. H. E. King, "Mutual Impedance of Unequal Length Antennas in Echelon," *IEEE Trans. Antennas and Propagation*, Vol. AP-5, pp. 306–313, July 1957.

25. H. C. Baker and A. H. LaGrone, "Digital Computation of the Mutual Impedance between Thin Dipoles," *IEEE Trans. Antennas and Propagation*, Vol. AP-10, pp. 172–178, March 1962.

26. A. A. Oliner and R. G. Malech, in *Microwave Scanning Antennas* (R. C. Hansen, ed.), Vol. II, Academic Press, New York, 1966, Chapters 2–4.

27. E. A. Blasi and R. S. Elliott, "Scanning Antenna Arrays of Discrete Elements," *IEEE Trans. Antennas and Propagation*, Vol. AP-7, pp. 435–436, October 1959.

28. P. S. Carter, Jr., "Mutual Impedance Effects in Large Beam Scanning Arrays," *IEEE Trans. Antennas and Propagation*, Vol. AP-8, pp. 276–285, May 1960.

29. J. L. Allen et al., "Phased Array Radar Studies, July 1960–July 1961," Lincoln Laboratories, M.I.T., *Tech. Rep.* 236, November 1961.

30. J. L. Allen, "Gain and Impedance Variations in Scanned Dipole Arrays," *IEEE Trans. Antennas and Propagation*, Vol. AP-10, pp. 566–572, September 1962.

31. H. A. Wheeler, "The Radiation Resistance of an Antenna in an Infinite Array or Waveguide," *Proc. IRE*, Vol. 36, pp. 478–487, April 1948.

32. S. Edelberg, "The Effects of Mutual Coupling on the Scanning of Two-Dimensional Arrays," *Proc. Electronic Scanning Symposium*, AFCRC, AD152409, 1958.

33. S. Edelberg and A. A. Oliner, "Mutual Coupling Effects in Large Antenna Arrays," Part I: "Slot Arrays," *IEEE Trans. Antennas and Propagation*, Vol. AP-8, pp. 286–297, May 1960.

34. L. Stark, "Radiation Impedance of a Dipole in Infinite Arrays," Hughes Aircraft Co., *Tech. Rept.* FL60-230, May 1960; also *Radio Science*, Vol. 1, pp. 361–379, March 1966.

35. B. L. Diamond, "Phased Array Radar Studies, January 1963–July 1964," Lincoln Laboratories, M.I.T., *Tech. Rept.* 38, pp. 239–298, March 1965.

36. A. L. Van Koughnett, and J. L. Yen, "Properties of a Cylindrical Antenna in an Infinite Planar or Colinear Array," *IEEE Trans. Antennas and Propagation*, Vol. AP-15, pp. 750–757, November 1967.

37. V. W. H. Chang, "Infinite Phased Dipole Arrays," *Proc. IEEE*, Vol. 56, pp. 1892–1900, November 1968.

38. P. W. Hannan, P. J. Meier, and M. A. Balfour, "Simulation of Phased Array Antenna Impedance in Waveguide," *IEEE Trans. Antennas and Propagation*, Vol. AP-11, pp. 715–716, November 1963.

39. P. W. Hannan and M. A. Balfour, "Simulation of a Phased Array Antenna in Waveguide," *IEEE Trans. Antennas and Propagation*, Vol. AP-13, pp. 342–353, May 1965.

40. M. A. Balfour, "Phased Array Simulators in Waveguide for a Triangular Arrangement of Elements," *IEEE Trans. Antennas and Propagation*, Vol. AP-13 pp. 475–476, May 1965.

41. V. Galindo and C. P. Wu, "Integral Equations and Variational Expressions for Arbitrary Scanning of Regular Infinite Arrays," *IEEE Trans. Antennas and Propagation*, Vol. AP-14, pp. 392–394, May 1966.

42. R. W. P. King, R. B. Mack, and S. S. Sandler, *Arrays of Cylindrical Dipoles*, Cambridge University Press, New York, 1968.

43. D. Varon and G. J. Zysman, "Some Properties and Limitations of Electronically Steerable Phased Array Antennas," *Bell System Tech. J.*, Vol. 46, pp. 1561–1568, September 1967.

44. P. W. Hannan, "Proof that a Phased Array Antenna Can Be Impedance Matched for All Scan Angles," *Radio Science*, No. 3, pp. 361–369, March 1967.

45. R. E. Collin, *Field Theory of Guided Waves*, McGraw-Hill Book Company, New York, 1960.

2. General Formulation of the Boundary Value Problem

The approach to the analysis of phased arrays in which mutual coupling effects between array elements are ignored has been shown to be generally inadequate. However, a reasonably accurate mathematical accounting of coupling is possible for large arrays assumed to be periodic and infinite in extent. In fact, we will observe in later chapters that even finite sized arrays are amenable to certain mathematical approaches. The infinite array, however, is particularly amenable to analysis because Floquet's theorem is applicable to the problem.

Floquet's theorem will allow us to describe the fields in the open region outside *any single* element of the waveguide array ($z \geq 0$ in Fig. 2.1) in terms of a complete orthogonal set of modes. These "Floquet" modes and the use of the natural waveguide modes permit us to formulate the phased array boundary value problem in terms of several different types of integral equations. These vector two-dimensional integral equations are generally applicable to a very broad class of phased array antennas, including, with some extension, wire (dipole) type arrays of elements in addition to waveguide elements.

1. INFINITE ARRAYS AND FLOQUET'S THEOREM

Since the application of Floquet's theorem to the analyses presented here and later in the book is of central importance, we will derive this theorem in some detail.

1.1 Infinite Geometry of Waveguide Arrays

Floquet's theorem is essentially an extension of the Fourier series theorem for periodic functions. The extension permits a modal description of any field or function which repeats itself periodically except for a multiplicative

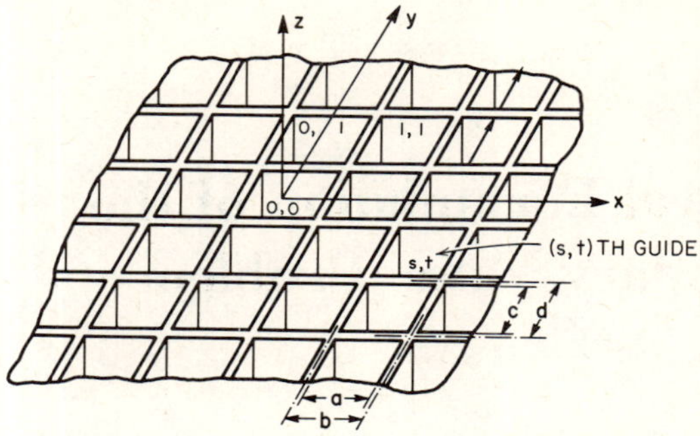

Fig. 2.1. An infinite array of waveguides.

exponential factor. Such a "periodic" function is an appropriate description for the field in the vicinity of an infinite periodic phased array that is excited uniformly in amplitude but with a linearly varying phase.

Consider the rectangular grid of waveguides in Fig. 2.1 (the elements may be rectangular, as shown, or of any cross section). If the (s, t)th waveguide element is excited by one of its modes with coefficient

$$V_{00}e^{-j(s\psi_x + t\psi_y)} \tag{1}$$

where ψ_x and ψ_y are fixed interelement phase shifts in the x and y directions, respectively, then this linear taper will excite a radiated beam in the direction (θ, ϕ) given by

$$\begin{aligned} \psi_x &= kb \sin \theta \cos \phi \\ \psi_y &= kd \sin \theta \sin \phi \end{aligned} \tag{2}$$

where $k = 2\pi/\lambda$, λ being the free space wavelength.

As will be illustrated later, we need restrict neither the analysis nor Floquet's theorem to waveguide elements. Wire antennas, dipoles, or any type of element can be treated similarly if excited by a linearly tapered phase. In fact, we shall see in Chapter 8 how nonlinear phase tapers and amplitudes may be treated.

1.2 Field Representation ($z \geq 0$)—Floquet's Theorem

We assume now that the waveguide (or other type) array elements are excited by modes with coefficients given by (1). We wish to determine a suitable field representation for the fields in the region ($z \geq 0$). In effect, we shall derive Floquet's theorem.

Any rectangular field component is well known to be a solution of the homogeneous scalar Helmholtz equation in this region:

$$(\nabla^2 + k^2)\xi(x, y, z) = 0 \tag{3}$$

where

$$\nabla^2 = \frac{\partial^2}{\partial x^2} + \frac{\partial^2}{\partial y^2} + \frac{\partial^2}{\partial z^2}$$

For waveguides, ξ generally represents either E_z or H_z. It is frequently assumed that the field varies with z, the direction of propagation, as

$$\xi(x, y, z) = e^{-j\Gamma z}\xi(x, y) \tag{4}$$

and that on the waveguide perimeter either $E_z = \xi = 0$ or $\partial H_z/\partial n = \partial\xi/\partial n = 0$, where n = perimeter normal.

Now with the excitation given by (1) and the periodic symmetry of the array structure, we observe that *all* field components, or ξ, must behave as

$$\xi(x + b, y + d, z) = \xi(x, y, z)e^{-j(\psi_x + \psi_y)} \tag{5}$$

This is the required behavior of the fields in any transverse plane (z = constant). The z-directed behavior is assumed (similar to waveguides) to be such that

$$\xi(x, y, z) = \xi(x, y)e^{-j\Gamma z} \tag{6}$$

Equation 3 thus becomes

$$[\nabla_t^2 + (k^2 + \Gamma^2)]\xi(x, y) = 0 \tag{6a}$$

where

$$\nabla_t^2 = \frac{\partial^2}{\partial x^2} + \frac{\partial^2}{\partial y^2}$$

with the phase tapered *periodic* boundary condition, (5).

If we apply the technique of separation of variables to (6), we find

$$\left(\frac{\partial^2}{\partial x^2} + k_x^2\right) f(x) = 0 \tag{7}$$

for the x dependence and

$$\left(\frac{\partial^2}{\partial y^2} + k_y^2\right) g(y) = 0 \tag{8}$$

for the y dependence, where k_x and k_y are separation constants such that

$$\Gamma^2 = k^2 - k_x^2 - k_y^2 \tag{9}$$

We note first, considering the x dependence only, that if

$$f(x + b) = f(x)e^{-j\psi_x}$$

then

$$F(x) = f(x)e^{j(\psi_x/b)x}$$

is a periodic function since

$$F(x + b) = f(x + b)e^{j(\psi_x/b)x}e^{j\psi_x} = F(x)$$

Hence $F(x)$ can be represented by the following Fourier series:

$$F(x) = \sum_{m=-\infty}^{\infty} A_m e^{j(2\pi m/b)x}$$

and

$$f(x) = \sum_{m=-\infty}^{\infty} A_m e^{j[(2\pi m - \psi_x)/b]x} \tag{10}$$

Each term of this series also satisfies (7), so that in general

$$k_x = k_{x_m} = \frac{2\pi m - \psi_x}{b} \tag{11}$$

Similarly, for the separable y dependence, we obtain

$$k_y = k_{y_n} = \frac{2\pi n - \psi_y}{d} \tag{12}$$

so that

$$\Gamma_{mn}^{\ 2} = k - \left(\frac{2\pi m - \psi_x}{b}\right)^2 - \left(\frac{2\pi n - \psi_y}{d}\right)^2 \tag{13}$$

and

$$\xi_{mn}(x, y, z) = e^{jk_{x_m}x}e^{jk_{y_n}y}e^{-j\Gamma_{mn}z} \tag{14}$$

The above periodic solution of (3) subject to boundary condition (5) is, essentially, a statement of Floquet's theorem.

1.3 Vector Floquet Modes and Admittances

The transverse dependent part of ψ can be orthonormalized so that

$$\xi_{mn}(x, y) = \sqrt{\frac{1}{bd}}\, e^{jk_{x_m}x}e^{jk_{y_n}y} \tag{15}$$

and

$$\iint_{A_p} \xi_{mn}\xi^*_{m'n'}\, dx\, dy = \delta_{mm'}\delta_{nn'} \tag{16}$$

with the Kronecker delta, $\delta_{mm'}$, equal to 1 for $m = m'$ and to 0 for $m \neq m'$, and A_p denoting the area of any single periodic cell on the face of the array ($b \times d$ for the rectangular array in Fig. 2.1). Equation 15 represents an orthonormal set of *scalar* Floquet modes.

An orthonormal complete set of *vector* Floquet modes can be derived from the scalar modes in much the same manner as vector TE and TM modes are derived from a scalar Hertzian potential in waveguide mode theory.[1] A complete set of vector TE and TM Floquet modes is derived next from the scalar Floquet modes, (15).

TE Modes. If $\xi_{mn}(x, y) = H_{z_{mn}}(x, y)$ and $E_z = 0$, then the corresponding transverse ($\hat{x}$ and $\hat{y}$ components) fields are derived from $H_{z_{mn}}$ by

$$\mathbf{E}_{t_{mn}} = \frac{j\omega\mu}{k^2_{r_{mn}}} \hat{z} \times \nabla_t H_{z_{mn}}$$

and

$$\mathbf{H}_{t_{mn}} = \frac{j\Gamma_{mn}}{k^2_{r_{mn}}} \nabla_t H_{z_{mn}}$$

where

$$k^2_{r_{mn}} = k_{x_m}{}^2 + k_{y_n}{}^2 = k^2 - \Gamma_{mn}{}^2 \tag{17}$$

and

$$\nabla_t = \hat{x}\frac{\partial}{\partial x} + \hat{y}\frac{\partial}{\partial y} = j(\hat{x}k_{x_m} + \hat{y}k_{y_n})$$

The vector modal functions derived above are orthogonal but not normalized. To obtain an *orthonormalized* vector mode, we define

$$\mathbf{E}_{t_{mn}} = \frac{\omega\mu}{k_{r_{mn}}} \mathbf{\Psi}_{1_{mn}}(x, y)$$

and obtain

$$\mathbf{\Psi}_{1_{mn}} = \sqrt{\frac{1}{bd}} \left\{ \frac{k_{y_n}\hat{x} - k_{x_m}\hat{y}}{k_{r_{mn}}} \right\} e^{j(k_{x_m}x + k_{y_n}y)} \tag{18}$$

as an orthonormal mode, where the subscript 1 preceding the *mn* subscript denotes a TE mode. These modes have the property

$$\iint_{A_p} \mathbf{\Psi}_{1_{mn}} \cdot \mathbf{\Psi}^*_{1_{m'n'}}\, dx\, dy = \delta_{mm'}\delta_{nn'} \tag{18a}$$

We note that $\mathbf{E}_{t_{mn}}$ and $\mathbf{H}_{t_{mn}}$ are related by

$$\hat{z} \times \mathbf{H}_{t_{mn}} = \left(\frac{\Gamma_{mn}}{\omega\mu}\right) \mathbf{E}_{t_{mn}}$$

and denote this ratio as the TE modal admittance:

$$Y_{1_{mn}} = \frac{\Gamma_{mn}}{\omega\mu} \tag{19}$$

TM Modes. For the TM mode case, we let $\xi_{mn}(x, y) = E_{z_{mn}}(x, y)$ and $H_z = 0$; the transverse fields are then derivable by[1]

$$\mathbf{E}_{t_{mn}} = \frac{-j\Gamma_{mn}}{k_{r_{mn}}^2} \nabla_t E_{z_{mn}}$$

and

$$\mathbf{H}_{t_{mn}} = \frac{j\omega\varepsilon}{k_{r_{mn}}^2} \hat{z} \times \nabla_t E_{z_{mn}}$$

As before, we define a TM orthonormal mode by letting

$$\mathbf{E}_t = \frac{\Gamma_{mn}}{k_{r_{mn}}} \mathbf{\Psi}_{2mn}$$

where

$$\mathbf{\Psi}_{2mn} = \sqrt{\frac{1}{bd}} \left\{ \frac{k_{x_m}\hat{x} + k_{y_n}\hat{y}}{k_{r_{mn}}} \right\} e^{j(k_{x_m}x + k_{y_n}y)} \tag{20}$$

the subscript 2 denoting a TM mode. The set of TE and TM free space modes have the property

$$\iint_{A_p} \mathbf{\Psi}_{pmn} \cdot \mathbf{\Psi}^*_{p'm'n'}\, dx\, dy = \delta_{pp'}\delta_{mm'}\delta_{nn'} \tag{20a}$$

where the Kronecker delta, $\delta_{pp'}$, indicates that the TE and TM vector modes are mutually orthogonal.

The TM modal admittance is found from

$$\hat{z} \times \mathbf{H}_{t_{mn}} = \frac{\omega\varepsilon}{\Gamma_{mn}} \mathbf{E}_{t_{mn}}$$

as

$$Y_{2mn} = \frac{\omega\varepsilon}{\Gamma_{mn}} \tag{21}$$

Physically, the vector mode functions, $\mathbf{\Psi}_{pmn}$, are TE or TM plane waves which propagate (or decay) away from the aperture plane ($z = 0$). The z-directed propagation constant, Γ_{mn}, is real for propagating waves, while a purely imaginary value of Γ_{mn} corresponds to a decaying (evanescent) wave.

1.4 Grating Lobe Diagram

As was pointed out in Chapter 1, the grating lobe diagram is a useful device for relating the interelement phasing and spacing of an array to the number and direction of radiated lobes (beams). Our study of Floquet modes now enables us to study and apply the grating lobe diagram in a very precise way. We shall see that this diagram is a useful graphical tool for studying the behavior of Γ_{mn} as a function of the phasing (ψ_x, ψ_y) or direction cosines (T_x, T_y), and that a "radiated beam" or "grating lobe" corresponds to a Floquet mode, which may propagate in the z direction by virtue of Γ_{mn} being real.

The z-directed propagation constant, Γ_{mn}, of the (m, n)th mode was found as

$$\left.\begin{aligned} \Gamma_{mn} &= \pm\sqrt{k^2 - [(2\pi m - \psi_x)/b]^2 - [(2\pi n - \psi_y)/d]^2} \\ \text{or} \qquad \Gamma_{mn} &= \pm k\sqrt{1 - [m(\lambda/b) - T_x]^2 - [n(\lambda/d) - T_y]^2} \end{aligned}\right\} \tag{22}$$

where $T_x = \sin\theta\cos\phi$ and $T_y = \sin\theta\sin\phi$ are the beam pointing direction cosines for interelement phase shifts ψ_x and ψ_y. The sign of Γ_{mn} is $+$ for a radiated wave and $-$ for an incoming wave. If the (m, n)th Floquet mode does not propagate, $\Gamma_{mn} = -j\,|\Gamma_{mn}|$, and this mode is evanescent and decays away from the waveguide aperture ($z > 0$).

The critical phase (ψ_x, ψ_y) or scan directions (T_x, T_y) for which a radiated beam becomes "invisible" or points at grazing incidence are determined by setting $\Gamma_{mn} = 0$. If we write

$$\Gamma_{mn} = k\sqrt{1 - T_{xm}^2 - T_{yn}^2} = 0 \quad \text{or} \quad 1 = T_{xm}^2 + T_{yn}^2 \tag{23}$$

where

$$T_{xm} = m\frac{\lambda}{b} - T_x \quad \text{and} \quad T_{yn} = n\frac{\lambda}{d} - T_y$$

then the grazing angle of the (m, n)th Floquet mode is clearly found by setting

$$T_{xm} = \sin\theta_{mn}\cos\phi_{mn}, \quad T_{yn} = \sin\theta_{mn}\sin\phi_{mn}, \quad \text{and}$$
$$\sin^2\theta_{mn} = T_{xm}^2 + T_{yn}^2 = 1$$

(i.e., $\theta_{mn} = 90°$). Note that θ_{mn} and ϕ_{mn} are the pointing directions of the (m, n)th Floquet mode for a phasing (ψ_x, ψ_y) of the array.

As a function of the pointing direction cosines of the "principle" radiated beam, $T_x = \sin\theta\cos\phi$ and $T_y = \sin\theta\sin\phi$ ($\theta = \theta_{00}$, $\phi = \phi_{00}$), condition 23 is

$$\left(m\frac{\lambda}{b} - T_x\right)^2 + \left(n\frac{\lambda}{d} - T_y\right)^2 = 1 \tag{24}$$

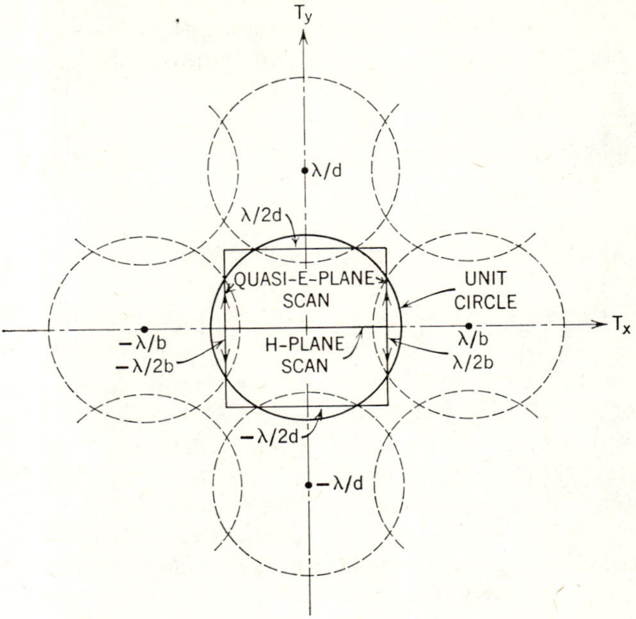

Fig. 2.2. Grating lobe diagram for $T_x = \sin\theta\cos\phi$, $T_y = \sin\theta\sin\phi$, $d/\lambda = b/\lambda = 0.5714$. Solid circle: boundary between "real" and "imaginary" space. Dashed circle: repetitive symmetry of grating lobe structure.

Equation 24 describes a displaced circle of unit radius in the T_x–T_y plane. If the values of T_x and T_y describe a point within the (m, n)th displaced circle, then Γ_{mn} is real and the (m, n)th pair of Floquet modes is radiated. Otherwise it is evanescent for these values of $(T_x$–$T_y)$.

The displacement between *centers* of adjacent circles, λ/b in the T_x direction and λ/d in the T_y direction, may be less than 2. In this case, one or more circles may overlap so that more than one lobe or beam may propagate for a given direction or phase ($\psi_x = k\,bT_x$, $\psi_y = k\,dT_y$). Such a situation is shown in Fig. 2.2, where both $\lambda/2b$ and $\lambda/2d$ are less than 1. Illustrated in this figure, for this spacing of elements, are phasings or scan directions where one, two, and no beams propagate.

Certain properties of the array can be gleaned directly from this spacing information. For angles in which no beams or modes can propagate, an incident waveguide mode ($z < 0$) upon the array aperture ($z = 0$) will be totally reflected. Later it will be shown that the impedances and reflection properties of all arrays have a certain discontinuous behavior at pointing directions on the unit or grating lobe circles.

2. INTEGRAL EQUATIONS FOR THE INFINITE PHASED ARRAY

We now proceed to formulate the problem of the phased array of Fig. 2.1 (with arbitrary waveguide elements) as a mathematically precise electromagnetic boundary value problem. Our starting points are the structure of Fig. 2.1, Maxwell's equations, and the appropriate field boundary conditions.[1]

In place of Maxwell's equations and the usual boundary conditions, we may describe the fields inside the waveguides in terms of a complete set of the modal solutions of Maxwell's equations appropriate to a given waveguide. In the free space region ($z \geq 0$) we may use the Floquet modal description of the fields as given in Section 1. The unknown coefficients of these modes may then be determined by application of the appropriate *continuity* conditions for the aperture tangential fields, $\mathbf{E}_t$ and $\mathbf{H}_t$, at $z = -\infty$, $z = 0$ and $z = +\infty$.

The conditions for the fields at $z = +\infty$ and $z = -\infty$ are determined by the radiation condition or terminal conditions of the waveguides. If, as we shall assume, the waveguides are terminated at $z = -\infty$ in their characteristic impedance* (as is true also for free space at $z = +\infty$), then, except for any incident field, the modal fields must be *outgoing* from the waveguide aperture-free space junction ($z = 0$). We shall always assume that an incident field is present so that we will, in general, be treating the problem as a scattering problem, wherein the fields are scattered at the aperture junction. As we shall see, the incident field may originate from inside the waveguides (the active transmitting array) or from the free space region (the passive receiving array).

After applying the radiation conditions, application of the continuity conditions at $z = 0$ will lead to a solution for the unknown field modal coefficients. Except for a small class of problems, however, this *formal* solution cannot be obtained without approximation. Although a discussion of exact and approximate solutions is reserved for Chapter 3, it is important to point out here that the applicability and flexibility of many approximate techniques rely on established mathematical methods for solving certain types of integral equations. It is not insignificant, therefore, that we choose to formulate the scattering problem as an integral equation. We will, therefore, express the continuity condition at $z = 0$, the aperture, in several different forms of integral equation. The integral equations derived will, implicitly, contain all the boundary and radiation conditions.

* With no loss of generality, we will assume that the waveguides of Fig. 2.1 are terminated in their characteristic impedance. In Chapter 8, we shall see that solutions for other terminations are readily derived from this solution.

2.1 Integral Equation for the Unknown *E* Field

We will proceed to derive a Fredholm integral equation of the first kind with $\mathbf{E}_t$, the tangential aperture electric field, as the unknown variable. We will treat a waveguide or aperture type of array like that illustrated in Fig. 2.1, wherein a typical element (of any cross section) may appear as shown in Fig. 2.3.

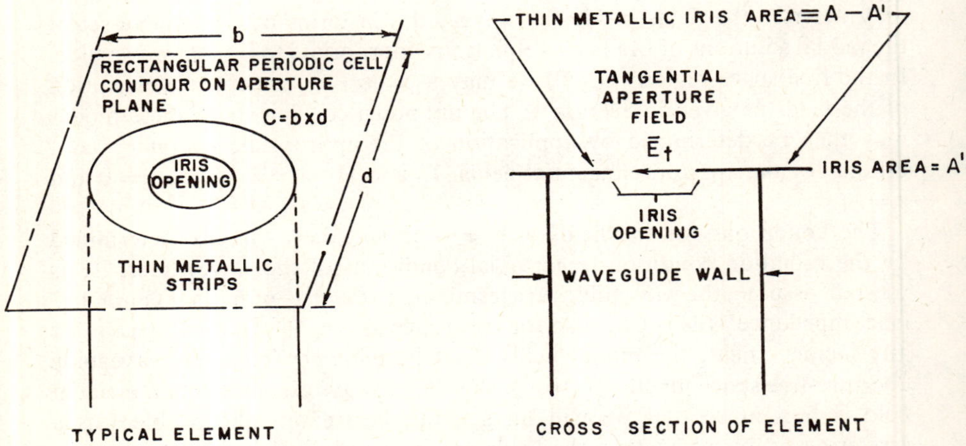

Fig. 2.3.

Note that we have allowed for the possibility of including a thin metallic iris across the waveguide opening. The resulting aperture opening, with the iris present, is designated as A', and the waveguide cross-sectional area as A. We should note that a thick metallic iris could also be allowed, as could many other more complex geometries. However, a thick iris would result in a pair of coupled integral equations, and we shall exclude such equations unless the resultant geometry has special significance.

In this regard, it is possible to include dielectric materials in the geometry, in the forms of sheaths outside the waveguides and plugs inside the waveguides, and still obtain a single uncoupled integral equation. We shall defer a detailed discussion of such geometries to Chapter 6.

Finally, it should be noted that the waveguide cross-sectional area, A, and the iris opening, A', could, in fact, represent a composite of several waveguides in a periodic cell, as shown in Fig. 2.4.

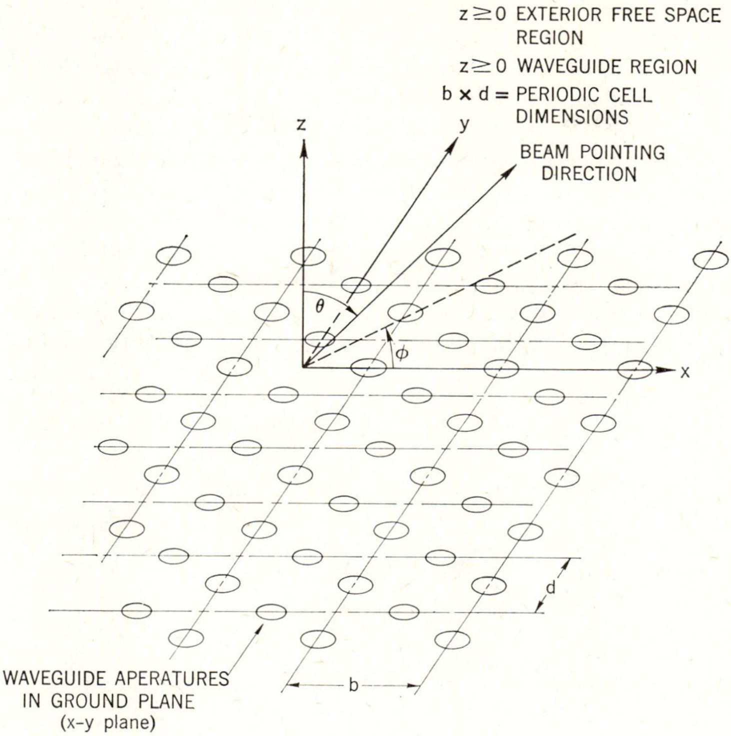

Fig. 2.4. Composite waveguide planar array—a basic periodic cell.

The transverse electric field, $\mathscr{E}_t = \hat{x}\mathscr{E}_{t_x} + \hat{y}\mathscr{E}_{t_y}$, inside of the (q, s)th waveguide may be expressed as a sum of the vector orthonormal waveguide modes, $\mathbf{\Phi}_{pqs}$ ($p = 1 \rightarrow$ TE modes, $p = 2 \rightarrow$ TM modes):

$$\mathscr{E}_t(x, y, z) = \left[\sum_{i=1}^{J} A_i(e^{-j\gamma_i z} + R_i e^{+j\gamma_i z})\mathbf{\Phi}_i(x, y) + \sum_{i=J+1}^{\infty} v_i \mathbf{\Phi}_i(x, y) e^{-j\gamma_i z}\right] e^{-j(q\psi_x + s\psi_y)} \tag{25}$$

In (25) we have, for simplicity, allowed for a *reordering* of the waveguide modes:

$$\mathbf{\Phi}_{pqs} \rightarrow \mathbf{\Phi}_i \tag{25a}$$

according to a single index (i). This notation will also prove useful in Chapter 7, when certain approximate solutions are sought. The $A_i : i = 1, \ldots, J$ are the *known* or *given* coefficients of the first J modes, which are incident upon the aperture at $z = 0$. Usually these modes will be propagating, that is, $\gamma_i = |\gamma_i|$ for $i = 1, \ldots, J$. However, it is possible to allow some of these modes to be evanescent, $\gamma_i = -j\,|\gamma_i|$, although in such cases the impedance and scattering matrices must be carefully redefined. The R_i are the unknown reflection coefficients referred to the aperture, $z = 0$, of each of these first J modes. The v_i are the amplitudes of the remaining modes, which must be determined, together with R_i, for a complete solution to the problem. Since the fields repeat periodically in the Floquet sense, from cell to cell (i.e., with a phase shift ψ_x in the x direction and ψ_y in the y direction), it will be sufficient to discuss only the fields in the $(q = 0, s = 0)$th cell.

We note that the field described by (25) satisfies the radiation condition at $z = -\infty$. Since we must ultimately satisfy the continuity condition at $z = 0$, we obtain

$$\mathbf{E}_t(x, y, z = 0^-) = \sum_{i=1}^{J} (A_i + R_i)\mathbf{\Phi}_i + \sum_{i=J+1}^{\infty} v_i \mathbf{\Phi}_i \tag{26}$$

Application of the orthonormality properties of the $\mathbf{\Phi}_i$ over the waveguide cross section, A,

$$\iint_A \mathbf{\Phi}_i \cdot \mathbf{\Phi}_k \, dx\, dy = \delta_{ik} \tag{27}$$

(δ_{ik} = Kronecker delta), allows us to express the coefficients in terms of the (unknown) tangential aperture electric field:*

$$\iint_A \mathbf{\Phi}_i \cdot \mathbf{E}_t \, dx\, dy = \begin{cases} A_i + R_i, & i \leq J \\ v_i, & i > J \end{cases} \tag{28}$$

The transverse magnetic field, $\mathscr{H}_t$, in the waveguides can be found from either Maxwell's equations or, as shown in Section 1 of this chapter for the Floquet modes, by the use of the modal admittances. Modes traveling in the $+z$ direction will, of course, have a transverse magnetic field pointing (for that mode) oppositely to that of the same mode traveling in the $-z$ direction. We therefore obtain for $\mathscr{H}_t$

$$-\hat{z} \times \mathscr{H}_t(x, y, z) = \sum_{i=1}^{J} (A_i e^{-j\gamma_i z} - R_i e^{+j\gamma_i z}) y_i \mathbf{\Phi}_i - \sum_{i=J+1}^{\infty} v_i y_i \mathbf{\Phi}_i e^{j\gamma_i z} \tag{29}$$

where the y_i are the waveguide modal admittances. They are defined as $y_i = \gamma_i/\omega\mu$ for TE modes and $y_i = \omega\varepsilon/\gamma_i$ for TM modes. We observe that for propagating modes the y_i are always positive real. For evanescent modes

* When the iris is present, the area of integration reduces to A' on account of the boundary condition which requires the vanishing of $\mathbf{E}_t$ on the iris, A–A'.

we find y_i given as[3]

$$y_i = \frac{\gamma_i}{\omega\mu} = +j\frac{|\gamma_i|}{\omega\mu} \qquad \text{for} \quad \text{TE modes}$$

and

$$y_i = \frac{\omega\varepsilon}{\gamma_i} = -j\frac{\omega\varepsilon}{|\gamma_i|} \qquad \text{for} \quad \text{TM modes} \tag{30}$$

Note, however, that we have *not* precluded the possibility of using *complex* μ or ε in (30). For example, it is frequently useful to represent lossy dielectrics by a complex ε. In this circumstance the equations for y_i remain valid except that γ_i is complex.

At the aperture we obtain for the magnetic field

$$\begin{aligned} -\hat{z} \times \mathbf{H}_t(x, y) &= -\hat{z} \times \mathscr{H}_t(x, y, z = 0^-) \\ &= \sum_{i=1}^{J} (A_i - R_i) y_i \mathbf{\Phi}_i - \sum_{i=J+1}^{\infty} v_i y_i \mathbf{\Phi}_i \end{aligned} \tag{31}$$

It is instructive to consider at this point the representation of $\mathbf{E}_t$ and $\mathbf{H}_t$ as *integrals* by use of the coefficient relationship, (28). We note that (26) becomes

$$\mathbf{E}_t(x, y) = \sum_{i=1}^{\infty} \mathbf{\Phi}_i \iint_A \mathbf{\Phi}_i \cdot \mathbf{E}_t \, dx' \, dy' \tag{32}$$

If we now consider the possibility of *reversing the order* of integration and summation in (32), we find that the classical* mathematical theorems[4] do not permit this reversal of the order of operations because of the possibility of $\mathbf{E}_t$ being an unbounded function. That is, it is well known[5] that at the corner or edge of the waveguide wall (at $z = 0$) the normal component of $\mathbf{E}_t$ has a singularity or goes to infinity in magnitude (differently according to whether the corner is a 90° edge or a knife edge). Nevertheless, by introducing the concept of the delta function or identity operator,[6] it is possible to reverse the operations in (32) so that

$$\mathbf{E}_t(x, y) = \iint_A \left[\sum_{i=1}^{\infty} \mathbf{\Phi}_i(x, y) \mathbf{\Phi}_i(x', y') \right] \cdot \mathbf{E}_t(x', y') \, dx' \, dy' \tag{33}$$

We note that the sum of dyads[6] represents a delta function by virtue of the completeness of the $\mathbf{\Phi}_i$ functions over A:

$$\bar{\bar{I}} \, \delta(x - x') \, \delta(y - y') = \sum_{i=1}^{\infty} \mathbf{\Phi}_i(x, y) \mathbf{\Phi}_i(x', y') \tag{34}$$

* Classical mathematical theorems, as opposed to those which employ the theory of distributions.

where $\bar{\bar{I}}$ is a unit three-dimensional dyadic (or matrix) operator. The integral operator,

$$\iint_A \left[\sum_{i=1}^{\infty} \mathbf{\Phi}_i(x, y)\mathbf{\Phi}(x', y') \right] \cdot \text{——} \, dx' \, dy' \tag{35}$$

is an identity operator[6] on vector functions.

With these concepts in mind, we can write the electric field at the aperture, (26), as

$$\mathbf{E}_t(x, y) = \sum_{i=1}^{J} (A_i + R_i)\mathbf{\Phi}_i + \iint_A \left[\sum_{J+1}^{\infty} \mathbf{\Phi}_i \mathbf{\Phi}_j \right] \cdot \mathbf{E}_t \, dx' \, dy' \tag{36}$$

and, with a simple extension of these concepts, the magnetic field, (31), can be written as

$$-\hat{z} \times \mathbf{H}_t(x, y) = \sum_{i=1}^{J} (A_i - R_i) y_i \mathbf{\Phi}_i - \iint_A \left[\sum_{J+1}^{\infty} y_i \mathbf{\Phi}_i \mathbf{\Phi}_i \right] \cdot \mathbf{E}_t \, dx' \, dy' \tag{37}$$

The operator in (37) may be viewed as an *admittance* operator, as opposed to the unit operator of (36).

The integral forms, (36) and (37), are actually not necessary for the derivation of the integral equation or for obtaining either exact (where possible) or approximate solutions. They are useful conceptually, however, since they lead to a *standard* form of integral equation with *singular* kernels.

The transverse electric field in the free space region ($z \geq 0$) may be represented by the complete set of vector (Floquet type) modes discussed earlier. Since the only excitation or incident field is from the waveguides, these modes must be outward going or traveling in the $+z$ direction at $z = \infty$. We obtain

$$\mathscr{E}_t(x, y, z \geq 0) = \sum_{p=1}^{2} \sum_{m=-\infty}^{\infty} \sum_{n=-\infty}^{\infty} V_{pmn} \mathbf{\Psi}_{pmn}(x, y, z) e^{-i\Gamma_{mn} z} \tag{38}$$

where the V_{pmn} are unknown coefficients in this representation of the electric field. At $z = 0^+$, we can use the orthonormality of the free space modes, (20a), to express*

$$V_{pmn} = \iint_A \mathbf{\Psi}^*_{pmn}(x, y) \cdot \mathbf{E}_t(x, y) \, dx \, dy \tag{39}$$

In (39) we have used the *same field designation*, $\mathbf{E}_t$, for the region $z = 0^+$ as we did in the expression for the location $z = 0^-$ (Eq. 28). Equations 28 and 39, taken together, constitute an expression for the *continuity of E_t across the aperture* at $z = 0$. We need only equate expressions for $\mathbf{H}_t$ across the

* See footnote on page 48.

aperture in order to satisfy all the boundary conditions and arrive at an integral equation representation.

An expression for $\mathscr{H}_t\,(x, y, z \geq 0)$ is obtained, as described earlier, by the use of the free space modal admittances:

$$-\hat{z} \times \mathscr{H}_t(x, y, z \geq 0) = \sum_{p=1}^{2} \sum_{-\infty}^{\infty} \sum_{-\infty}^{\infty} V_{pmn} Y_{pmn} \mathbf{\Psi}_{pmn} e^{-i\Gamma_{mn}z} \tag{40}$$

At the aperture, using (39), we obtain

$$-\hat{z} \times \mathbf{H}_t(x, y, z = 0^+) = \iint_A \left[\sum_{p=1}^{2} \sum_{-\infty}^{\infty} \sum_{-\infty}^{\infty} Y_{pmn} \mathbf{\Psi}_{pmn} \mathbf{\Psi}^*_{pmn}\right] \cdot \mathbf{E}_t \, dx' \, dy' \tag{41}$$

In order to account for the possible presence of an iris at the waveguide aperture (Fig. 2.3), we need to set $\mathbf{E}_t = 0$ in the area A–A'. This boundary condition is readily satisfied by replacing A by A', that is,

$$A \rightarrow A' \tag{42}$$

in (37) and (41).

Finally, we equate these two expressions for $\hat{z} \times \mathbf{H}_t$ *over the area* A' only, and the ranges of (x, y) and (x', y') are defined over the same region, A'. By making use of (28)† and the identity

$$(A_i - R_i) y_i \mathbf{\Phi}_i = 2 A_i y_i \mathbf{\Phi}_i - y_i \iint_{A'} \mathbf{\Phi}_i \cdot \mathbf{E}_t \, dx' \, dy' \tag{43}$$

we obtain a useful form of integral equation for $\mathbf{E}_t$:

$$2 \sum_{i=1}^{J} A_i y_i \mathbf{\Phi}_i = \iint_{A'} \left[\sum_{i=1}^{\infty} y_i \mathbf{\Phi}_i(x, y) \mathbf{\Phi}_i(x', y') + \sum_{p=1}^{2} \sum_{-\infty}^{\infty} \sum_{-\infty}^{\infty} Y_{pmn} \mathbf{\Psi}_{pmn}(x, y) \mathbf{\Psi}^*_{pmn}(x', y')\right] \cdot \mathbf{E}_t(x', y') \, dx' \, dy' \tag{44}$$

Among other important properties of (44) that will be discussed is the fact that its admittance operator (or kernel):

$$\iint_{A'} \left[\sum_{i=1}^{\infty} y_i \mathbf{\Phi}_i \mathbf{\Phi}_i + \sum_{p=1}^{2} \sum_{-\infty}^{\infty} \sum_{-\infty}^{\infty} Y_{pmn} \mathbf{\Psi}_{pmn} \mathbf{\Psi}^*_{pmn}\right] \cdot \text{——} \, dx' \, dy' \tag{45}$$

† It may seem that in the case of a metallic iris (28) cannot be utilized because

$$\iint_{A'} \mathbf{\Phi}_i \mathbf{\Phi}_k \, dx \, dy \neq \delta_{ik}.$$

Nevertheless, since $\mathbf{E}_t = 0$ over A–A', (28) remains valid.

does not change if the *incident magnetic field*, one-half the *free* term on the left-hand side of (44), changes. The free term on the left can be changed to include incident free space modes or *any* incident field.

2.2 Integral Equation for the Unknown H Field

An integral equation with the tangential magnetic field at the aperture, $\mathbf{H}_t$, as the unknown can be derived in a manner very similar to that used for the electric field, $\mathbf{E}_t$. As opposed to using modal admittances and obtaining an admittance integral operator as in (44), we use modal *impedances* and obtain an integral impedance operator. The principal difference in regard to the expression for $\mathbf{H}_t$ is, however, that we cannot obtain a single integral equation embodying all boundary conditions, as in (44), when an iris (Fig. 2.3) is present in the aperture. The difficulty arises because of the discontinuity of $\mathbf{H}_t$ across the iris region, A–A'.

For the case with no iris we can write expressions for $\mathbf{H}_t(x, y, z = 0^-)$ and $\mathbf{H}_t(x, y, z = 0^+)$ which, if projected to $z = -\infty$ and $z = +\infty$, respectively, will satisfy the far-zone ($z = \pm\infty$) boundary conditions. We use the set $\{\mathbf{\Phi}_i\}$, which is complete over A, the set $\{\mathbf{\Psi}_{pmn}\}$, which is complete over A, *and* also the unit cell ($b \times d$ in Fig. 2.3) to obtain the expressions

$$\mathbf{H}_t(x, y, z = 0^-) = \sum_{q=1}^{J}(A_q - R_q)\mathbf{\Phi}_q(x, y) + \sum_{q=J+1}^{\infty} i_q\mathbf{\Phi}_q(x, y) \tag{46}$$

and

$$\mathbf{H}_t(x, y, z = 0^+) = \sum_{p=1}^{2}\sum_{m=-\infty}^{\infty}\sum_{n=-\infty}^{\infty} I_{pmn}\mathbf{\Psi}_{pmn}(x, y) \tag{47}$$

In (46), as before, the A_q denote *given* excitation coefficients for the incident field. The R_q, i_q, and I_{pmn} are modal coefficients to be determined. By the use of the orthonormality relations, (20a) and (27), we can write (46) and (47) as integrals:

$$\mathbf{H}_t(x, y, z = 0^-) = \sum_{q=1}^{J}(A_q - R_q)\mathbf{\Phi}_q(x, y) + \sum_{q=J+1}^{\infty}\mathbf{\Phi}_q(x, y)\iint_A \mathbf{\Phi}_q \cdot \mathbf{H}_t\, dx'\, dy' \tag{48}$$

and

$$\mathbf{H}_t(x, y, z = 0^+) = \sum_{p=1}^{2}\sum_{-\infty}^{\infty}\sum_{-\infty}^{\infty}\mathbf{\Psi}_{pmn}\iint_C \mathbf{\Psi}^*_{pmn} \cdot \mathbf{H}_t\, dx'\, dy' \tag{49}$$

where C denotes the area of the unit cell.

We note, now, that $\mathbf{H}_t(x, y, z = 0^-)$ is *zero* in the region C–A, while $\mathbf{H}_t(x, y, z = 0^+)$ is *not zero* in this region. The tangential induced currents on the ground plane of the array are not zero in general. Hence, in (48) and (49), by using the same expression $\mathbf{H}_t$ under the integral sign, we have implicitly defined

$$\begin{aligned}\mathbf{H}_t &\equiv \mathbf{H}_t(x, y, z = 0^+)\\ &\equiv \mathbf{H}_t(x, y, z = 0^-)\end{aligned} \tag{50}$$

We have also, in (48) and (49), implicitly satisfied the continuity of tangential $\mathbf{H}_t$ across the open aperture, A.

If an iris existed in A, such that $A \neq A'$ (see Fig. 2.3), it would not be possible to use the same expression for $\mathbf{H}_t$ in (48) and (49) except over the region A'. Actually, $\mathbf{H}_t(z = 0^+)$ and $\mathbf{H}_t(z = 0^-)$ are not continuous across the iris, A–A'. As a result we cannot obtain a simple integral equation for $\mathbf{H}_t$. In any practical case, if $A \neq A'$, we can, of course, solve for $\mathbf{E}_t$ first from (44), and then obtain $\mathbf{H}_t$ from $\mathbf{E}_t$. For the present case, we assume $A = A'$.

If we extend the definition of the waveguide modes over the entire unit cell so that

$$\mathbf{\Phi}_q(x, y) = 0 \qquad \text{for} \quad (x, y) \text{ in } C - A \tag{51}$$

then we can write (48) as

$$\mathbf{H}_t(x, y, z = 0^-) = \sum_{q=1}^{J}(A_q - R_q)\mathbf{\Phi}_q(x, y) + \sum_{q=J+1}^{\infty} \mathbf{\Phi}_q(x, y)\iint_C \mathbf{\Phi}_q \cdot \mathbf{H}_t \, dx' \, dy' \tag{52}$$

Expressions for the tangential electric field can be obtained from (49) and (52) by utilizing the appropriate modal impedances with the result

$$\begin{aligned}\hat{z} \times \mathbf{E}_t(x, y, z = 0^-) = &\sum_{q=1}^{J}(A_q + R_q)z_q\mathbf{\Phi}_q(x, y)\\ &- \iint_C \left[\sum_{q=J+1}^{\infty} z_q\mathbf{\Phi}_q\mathbf{\Phi}_q\right] \cdot \mathbf{H}_t \, dx' \, dy'\end{aligned} \tag{53}$$

and

$$\hat{z} \times \mathbf{E}_t(x, y, z = 0^+) = \iint_C \left[\sum_{p=1}^{2}\sum_{-\infty}^{\infty}\sum_{-\infty}^{\infty} Z_{pmn}\mathbf{\Psi}_{pmn}\mathbf{\Psi}^*_{pmn}\right] \cdot \mathbf{H}_t \, dx' \, dy' \tag{54}$$

where $z_q = 1/y_q$ and $Z_{pmn} = 1/Y_{pmn}$.

In (53) it is clear from the definitions of $\mathbf{\Phi}_i$ that $\mathbf{E}_t = 0$ on the ground plane C–A. If we set $\mathbf{E}_t$ in (53) equal to $\mathbf{E}_t$ in (54), we will satisfy the continuity requirements (across $z = 0$) for $\mathbf{E}_t$ throughout the unit cell area, C. This leads, after minor manipulations, to

$$2\sum_{q=1}^{J} A_q z_q \mathbf{\Phi}_q = \iint_C \left[\sum_{q=1}^{\infty} z_q\mathbf{\Phi}_q\mathbf{\Phi}_q + \sum_{p=1}^{2}\sum_{-\infty}^{\infty}\sum_{-\infty}^{\infty} Z_{pmn}\mathbf{\Psi}_{pmn}\mathbf{\Psi}^*_{pmn}\right] \cdot \mathbf{H}_t \, dx' \, dy' \tag{55}$$

an integral equation for $\mathbf{H}_t$ over the entire unit cell region, C. As for the equation for $\mathbf{E}_t$, (44), the left-hand side of (55) contains twice the *incident electric* field. A different incident field will alter only this free term. The *impedance* operator in (55) is characteristic of the phased array *at a particular scan angle* defined by the $\mathbf{\Psi}_{pmn}$. Equations 44 and 55 are Fredholm integral equations of the first kind.

2.3 Array Scattering and Coupling Coefficient Parameters

We will assume, for the moment, that we can obtain a solution, either exact or approximate, to the integral equation, (44), for $\mathbf{E}_t$ or to (55) for $\mathbf{H}_t$. Once the aperture field solution is obtained, it is possible to derive from $\mathbf{E}_t$ or $\mathbf{H}_t$ various scattering and impedance parameters which are useful for characterizing the array.

For example, it is possible to derive a scattering matrix[7] of finite order to define the phased array. If J modes can propagate within a waveguide of the array and K free space modes can propagate in the free space region ($z > 0$), the characteristic scattering matrix will be of the $(J + K)$th order. Practical arrays are generally designed so that there are only one propagating mode inside the waveguide and only one radiated beam. Under such a situation, we have a 2×2 scattering matrix to characterize the array aperture:

$$[S] = \begin{bmatrix} S_{11} & S_{12} \\ S_{21} & S_{22} \end{bmatrix}$$

Notice that these scattering matrices are functions of the steering phases (ψ_x, ψ_y). A complete characterization of the scattering matrix terms would require solving (44) [or (55)] separately for each different propagating waveguide and propagating free space mode incident upon the aperture at $z = 0$. It is important to recall, however, that the admittance operator of (44) or the impedance operator of (55) is not altered as the incident field (one-half the free term) is changed. Of course, once the scattering matrix is obtained, it is a straightforward procedure to obtain the active impedance or admittance matrix for the array.[8]

In addition to the scattering parameters, we will discuss some other parameters of importance, including element coupling coefficients and polarization parameters of the radiated field.

Once $\mathbf{E}_t$ is determined, approximately or exactly from (44), it is always possible to obtain the coefficients, v_i, from (28). With v_i thus determined, the aperture magnetic field, $\mathbf{H}_t$, is obtainable from (31). The near and far-zone fields, $k\,|z| \gtrsim 0$ and $k\,|z| \gg 1$, are then obtainable from (25) and (29). (If desired, E_z and H_z can always be obtained from Maxwell's divergence equation.) Hence, once $\mathbf{E}_t$ is found at the aperture, the *total* field everywhere is obtainable. [The same is generally true if we start with $\mathbf{H}_t$, a solution of (55), in this and what follows.]

The reflection coefficient, referred to $z = 0$ in phase, is also found from (28) for $i = 1, \ldots, J$. If only one propagating mode is exciting the array ($A_1 \neq 0$, $A_i = 0$, $i = 2, \ldots, J$), even though more than one mode may propagate in the waveguides, the *input admittance* for that mode is found as

$$y_{\text{in}_1} = \frac{1 - R_1}{1 + R_1} \tag{56}$$

where y_{in_1} is *normalized* to y_1. Of course, the R_i are the diagonal components of the scattering matrix.

A very important property of the reflection coefficients can be proved if we utilize the periodicity property of the reflection coefficients, R_i, that is, if we make use of the fact that the R_i are *periodic* functions of ψ_x and ψ_y, the element-to-element phase shifts in the x and y directions. That the R_i are periodic functions of ψ_x and ψ_y follows from the fact that the array excitation is not altered by advancing or retarding the phases of the excitation signals by integral multiples of 2π. It is clear, first of all, that the complete behavior of the array can be determined if we restrict

$$-\pi \leq \psi_x \leq \pi, \qquad -\pi \leq \psi_y \leq \pi \tag{57}$$

We will show, further, that the reflection coefficients have the property

$$R_i(\psi_x, \psi_y) = R_i(-\psi_x, -\psi_y) \tag{58}$$

independently of array symmetry properties. This *reflection symmetry* of the array reflection coefficient is a consequence of the Lorentz reciprocity theorem[1] and the periodicity in the array geometry.

Since $R(\psi_x, \psi_y)$ is a doubly periodic function, we may express R as an exponential Fourier series:

$$R(\psi_x, \psi_y) = \sum_{s=-\infty}^{\infty} \sum_{t=-\infty}^{\infty} \mathscr{C}_{00}{}^{st} e^{i(s\psi_x + t\psi_y)} \tag{59}$$

The coefficients, $\mathscr{C}_{00}{}^{st}$, of this series may be interpreted in terms of the diagram of Fig. 2.1. If we allow a *single* waveguide to be excited in this figure, the (0, 0)th waveguide, by a single mode, we find that the amplitude of this mode that is excited, by mutual coupling, in the (s, t)th waveguide is $\mathscr{C}_{00}{}^{st}$. By superposition, then, it is possible to excite all the infinite number of waveguides as a phased array with differential phase shifts ψ_x and ψ_y to arrive at (59). This mutual coupling coefficient is illustrated diagrammatically in Fig. 2.5.

If we excite the (s, t)th waveguide by a waveguide mode of unit amplitude, the wave coupled into the (0, 0)th guide is of amplitude $\mathscr{C}_{00}^{-s,-t}$. By the Lorentz

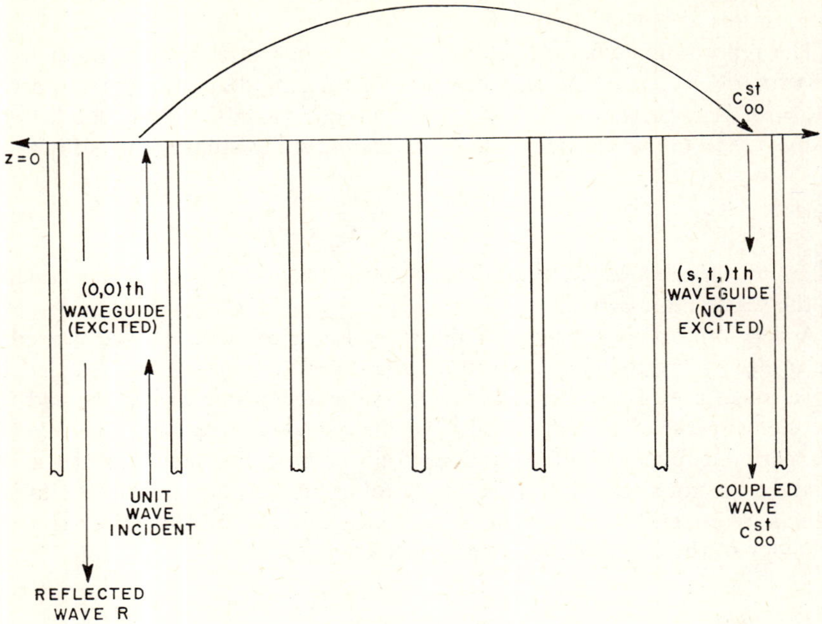

Fig. 2.5. Mutual coupling coefficients.

reciprocity theorem, it is clear that

$$\mathscr{C}_{00}^{s,t} = \mathscr{C}_{00}^{-s,-t} = \mathscr{C}_{st}^{\ \ 00} \tag{60}$$

independently of waveguide or array symmetries. It is now straightforward, by the use of (59) and (60), to show the reflection symmetry property stated in (58).

It is worth mentioning, at this point, that an equation for y_{in} can be written which is similar to (59):

$$y_{\text{in}}(\psi_x, \psi_y) = \sum_{s=-\infty}^{\infty} \sum_{t=-\infty}^{\infty} y_{00}{}^{st} e^{i(s\psi_x + t\psi_y)} \tag{61}$$

since y_{in} is also a doubly periodic function of ψ_x and ψ_y. The Fourier coefficients, $y_{00}{}^{st}$, are the *mutual admittances*. That is, a *single* waveguide may be excited with a unit *magnetic* field mode, $\mathbf{\Phi}_1$, and open circuits placed at $z = 0$ for all other waveguides [in the appropriate mode by short circuiting at $z = -(n\lambda_g + \lambda_g/4)$, $n =$ any integer]. The inverse of the coefficient of the coupled *electric* field mode, at $z = 0$, is then the mutual admittance, $y_{00}^{s,t}$.

Another quantity of significance is the far-zone or radiated field of the phased array. It is frequently desirable to obtain the radiated electric field in terms of the E_θ and E_ϕ components. If we are interested in the radiated field of the principal lobe (in the presence of multiple or grating lobes), or only one lobe is radiated, the radiated field is given by

$$\mathscr{E}_t(z \to \infty) = \left(\sum_{p=1}^{2} V_{p_{00}} \boldsymbol{\Psi}_{p_{00}} \right) e^{-i\Gamma_{00} z} \tag{62}$$

for the $\hat{x}$ and $\hat{y}$ components and by

$$E_z = \frac{1}{\Gamma_{00}} \nabla_t \cdot \mathbf{E}_t \tag{62a}$$

for the $\hat{z}$ component (obtained from the divergence relation). If we decompose this field into θ and ϕ components (and drop the z-dependence term), we get

$$E_\theta = \frac{1}{\cos\theta} \iint_{A'} \boldsymbol{\Psi}^*_{200} \cdot \mathbf{E}_t \, dx\, dy = \frac{V_{200}}{\cos\theta} \tag{63}$$

and

$$E_\phi = -\iint_{A'} \boldsymbol{\Psi}^*_{1_{00}} \cdot \mathbf{E}_t \, dx\, dy = V_{1_{00}} \tag{64}$$

It is interesting to note that the ϕ component is proportional only to the TE Floquet mode and that the θ component is proportional only to the TM Floquet mode.

The transmission coefficients of the array scattering matrix, T_θ and T_ϕ, are obtained by appropriate normalization of the radiated field components. The normalization is determined by the conservation of energy condition or, equivalently, the unitary requirement[8] for the scattering matrix. If the array is excited by one waveguide mode, with admittance y_1 (or by two modes with the same admittance, as, for example, two degenerate circularly polarized TE_{11} circular waveguide modes[9]), the incident power is then given by y_1. Similarly, powers carried by $\boldsymbol{\Psi}_{1_{00}}$ and $\boldsymbol{\Psi}_{2_{00}}$ modes are $Y_{1_{00}} |E_\phi|^2$ and $Y_{2_{00}} |E_\theta|^2$, respectively:

$$Y_{1_{00}} |V_{1_{00}}|^2 = Y_{1_{00}} |E_\phi|^2 \quad \text{and} \quad Y_{2_{00}} |V_{2_{00}}|^2 = Y_{2_{00}} \cos^2\theta \, |E_\theta|^2$$

The normalized transmission coefficients are given by

$$T_\phi = \sqrt{Y_{2_{00}}/y_1}\, E_\theta \tag{65}$$

and

$$T_\phi = \sqrt{Y_{1_{00}}/y_1}\, E_\phi \tag{66}$$

The transmission coefficients, T_θ and T_ϕ, are functions of θ and ϕ, the beam pointing directions, as derived from the differential phase shifts ψ_x and ψ_y [see (2), this chapter]. It will be shown in Chapter 4 that $T_\theta(\theta, \phi)$ and $T_\phi(\theta, \phi)$ are proportional to the radiation pattern of the infinite array when only *one element is excited*. This is a result of the facts that the propagating Floquet modes are plane waves and that the *array factor* of the infinite array is a delta function of angular direction.

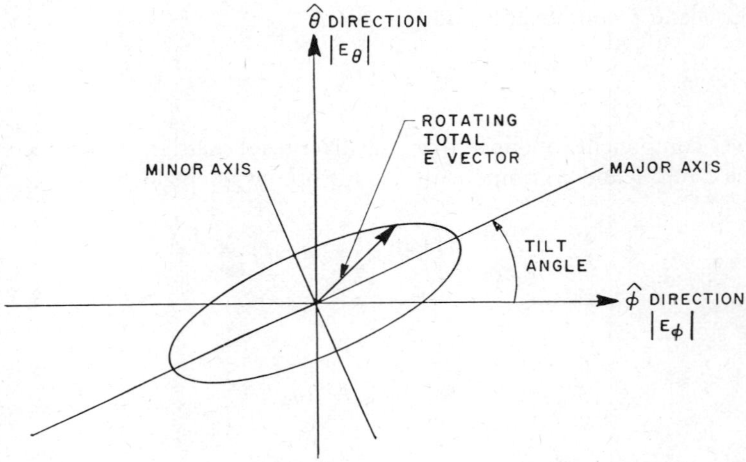

Fig. 2.6. Polarization ellipse.

The radiated field may also be expressed in circularly polarized components. With an $e^{+j\omega t}$ time dependence the right and left circularly polarized components are

$$E_R = \frac{1}{\sqrt{2}}(-jE_\theta + E_\phi) \quad \text{and} \quad E_L = \frac{1}{\sqrt{2}}(jE_\theta + E_\phi) \tag{67}$$

Two important polarization parameters are then readily derivable from (67). When E_θ and E_ϕ are not in phase, the resultant electric vector will rotate at an angular velocity, ω, and will vary in amplitude. In general, the rotating vector *tip* will transcribe an ellipse (see Fig. 2.6) that can be defined by the ratio of the *minor* axis to the *major* axis of the ellipse, called the *axial ratio*, and the *tilt angle* of the major axis with respect to the $\hat{\phi}$ axis (which is parallel to the array plane). These parameters may be expressed simply in terms of the circularly polarized components, (67), as

$$\text{Axial ratio} = \frac{\big|\, |E_R| - |E_L| \,\big|}{\big|\, |E_R| + |E_L|} \tag{68}$$

and

$$\text{Tilt angle} = \frac{\text{phase of } (E_R) - \text{phase of } (E_L)}{2} \tag{68a}$$

The multimode excitation of the array, as defined by $\sum_{i=1}^{J} A_i \boldsymbol{\Phi}_i$ in (36), can lead to algebraic complications in determining some array parameters, although, formally speaking, multimode excitation introduces no conceptual difficulties. A particularly important multimode excitation deserves some special attention, however. Consider excitation of the waveguides with *two* modes. Frequently, these two modes will be degenerate such as the circular waveguide TE_{11} modes. These may be characterized as horizontal ($\hat{x}$ polarized) $\text{TE}_{11} \equiv \boldsymbol{\Phi}_1$ and vertical ($\hat{y}$ polarized) $\text{TE}_{11} \equiv \boldsymbol{\Phi}_2$ modes. These two "linearly" polarized modes constitute an orthogonal two-port feed for the waveguide. By a simple second-order unitary transformation, it is always possible to find two new orthogonal modes or ports to describe the waveguide feed. If

$$A_1\boldsymbol{\Phi}_1 + A_2\boldsymbol{\Phi}_2 \tag{69}$$

is the *desired* relative excitation of the original two ports, then a new *normalized* mode can be defined as

$$\boldsymbol{\Phi}_{1N} = \frac{A_1}{(|A_1|^2 + |A_2|^2)^{1/2}}\boldsymbol{\Phi}_1 + \frac{A_2}{(|A_1|^2 + |A_2|^2)^{1/2}}\boldsymbol{\Phi}_2 \tag{70}$$

The normalized *orthogonal* mode or port is then given by

$$\boldsymbol{\Phi}_{2N} = \frac{|A_2|}{(|A_1|^2 + |A_2|^2)^{1/2}}\boldsymbol{\Phi}_1 + \frac{A_1^* |A_2|}{A_2^*(|A_1|^2 + |A_2|^2)^{1/2}}\boldsymbol{\Phi}_2 \tag{71}$$

In the case of circular (or square) waveguides, wherein $\boldsymbol{\Phi}_1$ and $\boldsymbol{\Phi}_2$ are essentially linearly polarized modes, circularly polarized modes are easily defined by letting $A_1 = \pm jA_2$, the sign determining the preferred sense of polarization.

2.4 Integral Eigenvalue Equation, Variational Principles, and Waveguide Simulators

A parameter of particular significance is the normalized input admittance to an array element under a given modal excitation. The input admittance is related to the reflection coefficient by (56). It is of interest, therefore, that an integral eigenvalue equation, with the input admittance y_{in} as the eigenvalue, can be formulated.

This eigenvalue equation, which will be derived shortly, is significant in several respects. As is well known,[6] an eigenvalue equation, wherein the pertinent operator has appropriate properties, can be used to form a *variational* expression for the unknown eigenvalue. In a variational expression we may use a first-order approximation for the unknown field, $\mathbf{E}_t$, and thus obtain a second-order or more accurate approximation for the eigenvalue: y_{in} in this case. We will find, furthermore, that the conditions which we must impose on the array geometry so as to obtain a variational expression are identical to those required to construct a *waveguide* simulator.*[10–13] This derivation will also show the relationship between a problem of the interior type (wherein boundary walls are conducting) and the periodic wall problem of the phased array.

We will assume, for convenience, that only a single waveguide mode is incident upon the aperture; $J = 1$ and $A_1 = 1$ in (44). Since the effects of more than one incident mode can be added by superposition, this will not reduce the generality of what follows. By the use of (28), we can readily manipulate the integral equation (44) into the eigenvalue equation form:

$$y_{\text{in}_1} y_1 \mathbf{\Phi}_1(\mathbf{r}) \iint_{A'} \mathbf{\Phi}_1 \cdot \mathbf{E}_t \, da'$$

$$= \iint_{A'} \left[\sum_{i=2}^{\infty} y_i \mathbf{\Phi}_i(\mathbf{r}) \mathbf{\Phi}(\mathbf{r}') + \sum_{p=1}^{2} \sum_{-\infty}^{\infty} \sum_{-\infty}^{\infty} Y_{pmn} \mathbf{\Psi}_{pmn}(\mathbf{r}) \mathbf{\Psi}^*_{pmn}(\mathbf{r}') \right] \cdot \mathbf{E}_t(\mathbf{r}') \, da' \quad (72)$$

where the eigenvalue y_{in_1} is the input admittance given by (56). The vector $\mathbf{r}$ denotes a position coordinate (x, y) in A', and da' denotes a differential area at $\mathbf{r}'$. This eigenvalue equation is of the *generalized* eigenvalue equation form:

$$y_{\text{in}_1} \mathcal{O}_1 \mathbf{E}_t = \mathcal{O} \mathbf{E}_t \quad (73)$$

where the operators $\mathcal{O}_1$ and $\mathcal{O}$ are defined as

$$\mathcal{O}_1 = y_1 \iint_{A'} \mathbf{\Phi}_1(\mathbf{r}) \mathbf{\Phi}(\mathbf{r}') \cdot \text{——} \, da' \quad (74)$$

and

$$\mathcal{O} = \iint_{A'} \left[\sum_{i=2}^{\infty} y_i \Phi_i(\mathbf{r}) \mathbf{\Phi}_i(\mathbf{r}') + \sum_{p=1}^{2} \sum_{-\infty}^{\infty} \sum_{-\infty}^{\infty} Y_{pmn} \mathbf{\Psi}_{pmn}(\mathbf{r}) \mathbf{\Psi}^*_{pmn}(\mathbf{r}') \right] \cdot \text{——} \, da' \quad (74a)$$

In the usual form of eigenvalue equation, $\mathcal{O}_1$ is the identity operator. Since, in this particular generalized eigenvalue equation, $\mathcal{O}_1$ is a single dyad operator,

* A waveguide simulator is a rectangular (sometimes triangular) waveguide that is used to cover a prescribed number of waveguides in the $z \geq 0$ region of the array. With the waveguide terminated in its characteristic impedance the free space region is *simulated* for a particular beam pointing direction. The array symmetry conditions which allow such a simulation and the mathematical justification are discussed further.

it can be shown[6,14] that there is but one eigenvalue and eigenfunction, $\mathbf{E}_t$, solution to (69). This is expected since, in general, only one solution to (44) exists also.[6,16] More important, however, is the fact that the operator $\mathcal{O}$ is not in a complex symmetric form, and as a consequence it is not obvious that a variational expression exists for (72) under certain circumstances.

We will now transform (72) into a *complex-symmetric* form so that a variational expression can be derived. We shall examine the conditions which allow us to make this transformation, which can actually be effected in at least two ways. The first way[17] that will be presented will illustrate the connection between accomplishing such a transformation and the possible construction of a waveguide simulator. In order to carry out this transformation, we require that the *waveguide array structure be invariant* under the reflection or symmetry group operators,[8] F_x, F_y, and R_z, where

$$\begin{aligned} F_x &= \text{reflection in } y\text{–}z \text{ plane} \\ F_y &= \text{reflection in } x\text{–}z \text{ plane} \\ R_z &= 180^\circ \text{ rotation around } z \text{ axis} = F_x F_y = F_y F_x \end{aligned} \tag{75}$$

If these symmetry conditions are satisfied, we may phase the array so that *four* simultaneous beams are radiated, as shown in Fig. 2.7, in the directions

$$\theta_1 = \theta_2 = \theta_3 = \theta_4 \equiv \theta$$

$$\phi_1 \equiv \phi, \qquad \phi_2 = \pi - \phi, \qquad \phi_3 = \pi + \phi, \qquad \phi_4 = -\phi$$

or in terms of the corresponding direction cosines

$$\begin{aligned} (T_{x1}, T_{y1}) &= (T_x, T_y) \\ (T_{x2}, T_{y2}) &= (-T_x, T_y) \\ (T_{x3}, T_{y3}) &= (-T_x, -T_y) \\ (T_{x4}, T_{y4}) &= (T_x, -T_y) \end{aligned} \tag{76}$$

with resulting electric field distributions

$$\begin{aligned} \mathscr{E}_{t1} &= \{\mathscr{E}_x(x, y, z), \mathscr{E}_y(x, y, z)\} \equiv \mathbf{E}_t \\ \mathscr{E}_{t2} &= \{-\mathscr{E}_x(-x, y, z), \mathscr{E}_y(-x, y, z)\} \\ \mathscr{E}_{t3} &= \{-\mathscr{E}_x(-x, -y, z), -\mathscr{E}_y(-x, -y, z)\} \\ \mathscr{E}_{t4} &= \{\mathscr{E}_x(x, -y, z), -\mathscr{E}_y(x, -y, z)\} \end{aligned} \tag{77}$$

The total electric field in the exterior or $z \geq 0$ region is found, from (38) and

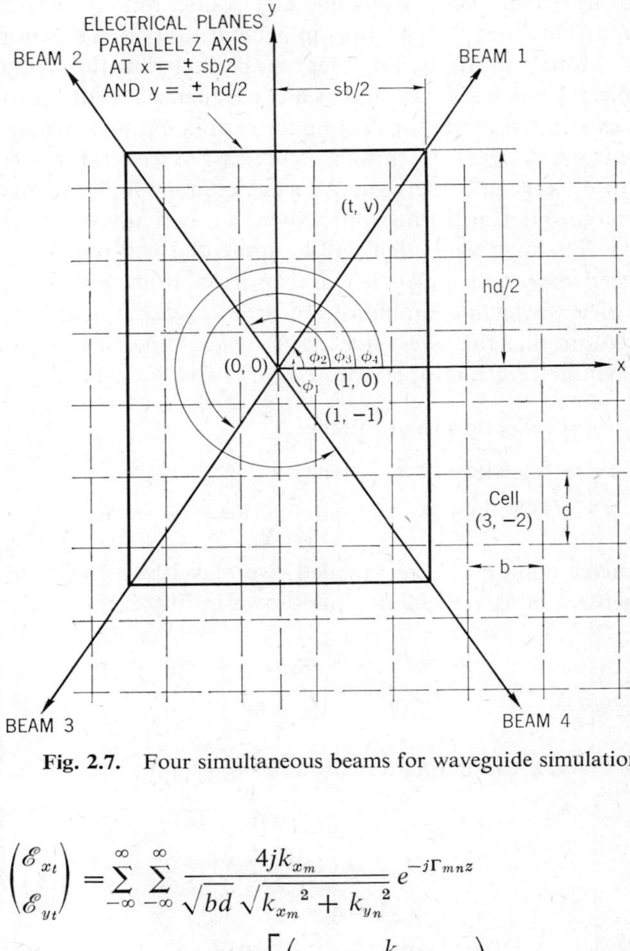

Fig. 2.7. Four simultaneous beams for waveguide simulation.

(77), as

$$\mathscr{E}_t = \begin{pmatrix} \mathscr{E}_{xt} \\ \mathscr{E}_{yt} \end{pmatrix} = \sum_{-\infty}^{\infty} \sum_{-\infty}^{\infty} \frac{4jk_{x_m}}{\sqrt{bd}\sqrt{k_{x_m}^{\;2} + k_{y_n}^{\;2}}} e^{-j\Gamma_{mn}z} \times \begin{bmatrix} \left(V_{2mn} + \dfrac{k_{y_n}}{k_{x_m}} V_{1mn}\right) \sin k_{x_m}x \cos k_{y_n}y \\ \left(V_{2mn} \dfrac{k_{y_n}}{k_{x_m}} - V_{1mn}\right) \cos k_{x_m}x \sin k_{y_n}y \end{bmatrix} \tag{78}$$

with k_{x_m} and k_{y_n} given by (11) and (12), we note that at the incremental phase shifts (or scan angles) given by

$$\psi_x = \psi_{x_{fs}} = \frac{\pi f}{s} \quad \text{and} \quad \psi_y = \psi_{y_{gh}} = \frac{\pi g}{h} \tag{79}$$

$$(f, g) = \text{odd integers}, \qquad (s, h) = \text{integers}$$

the total electric field satisfies (see Fig. 2.7)

$$\begin{aligned} \mathscr{E}_{xt} &= 0 \qquad \text{at} \quad y = \pm \frac{hd}{2} \\ \mathscr{E}_{yt} &= 0 \qquad \text{at} \quad x = \pm \frac{sb}{2} \end{aligned} \tag{80}$$

As a result, we may introduce metallic walls at these locations without disturbing the field distribution. When this is done, we have a situation in which one or more waveguides radiate into a larger waveguide. Thus, a waveguide transition equivalent to an infinite array under some special scan conditions is obtained. Waveguide simulation of the phased array is possible at the angles[18] determined by (79).

From (78), it is possible to define a set of *real orthonormal* modes, $\boldsymbol{\Psi}'_{pmn}$, for the exterior $z \geq 0$ region. These modes are, however, orthogonal over the enlarged rectangular region defined by (80), depicted in Fig. 2.7, and designated A_s (for area of simulator). It is clear at this point that if a simulator construction is possible then the problem is essentially of the interior type and a variational formulation is therefore possible.[19] We will carry this further, however.

It is also possible to appropriately redefine a new set of waveguide modes that are orthonormal over the enlarged region, A_s. We first order the waveguide modes with triple subscripts [see (25a)] so that

$$\boldsymbol{\Phi}_i \rightarrow \boldsymbol{\Phi}_{pmn}$$

and choose the modal functions so that

$$\boldsymbol{\Phi}_{pmn}(-x, -y) = (-1)^{m+n}\boldsymbol{\Phi}_{pmn}(x, y) \tag{81}$$

(this is always possible). A set of modes that are orthonormal over A_s can then be defined as*

$$\boldsymbol{\Phi}'_{pmn} = \boldsymbol{\Phi}_{pmn}\begin{Bmatrix} j\cos\psi_x t \\ \sin\psi_x t \end{Bmatrix} \cdot \begin{Bmatrix} j\cos\psi_y v \\ \sin\psi_y v \end{Bmatrix} \quad \text{for} \quad \begin{Bmatrix} m & \text{odd} \\ m & \text{even} \end{Bmatrix} \cdot \begin{Bmatrix} n & \text{odd} \\ n & \text{even} \end{Bmatrix} \tag{82}$$

where for each $\boldsymbol{\Phi}_{pmn}$ the choice between the upper or lower factor in each of the two brackets is determined by the symmetry of the components of the mode as indicated.

With the total fields in the interior and exterior regions expressed in terms of the new eigenmodes, (78) and (82), we may write an eigenvalue equation

* The integers (t, v) denote the cell or waveguide position within the larger simulator waveguide (Fig. 2.7). Hence (x, y) in the expressions for $\boldsymbol{\Phi}_{pmn}(x, y)$ is taken with respect to the (t, v) cell. Note also that the choice of subscripts in (81) is possible for any *symmetric* waveguide.

for the array as

$$y_{\text{in}_1}(-1)^{m'+n'} y_{p'm'n'} \boldsymbol{\Phi}'_{p'm'n'} \iint_{A_s} \boldsymbol{\Phi}'_{p'm'n'} \cdot \mathbf{E}'_t \, da'$$

$$= {\sum_p}' {\sum_m}' {\sum_n}' (-1)^{m+n} y_{pmn} \boldsymbol{\Phi}'_{pmn} \iint_{A_s} \boldsymbol{\Phi}'_{pmn} \cdot \mathbf{E}'_t \, da'$$

$$+ \sum_p \sum_m \sum_n Y_{pmn} \boldsymbol{\Psi}'_{pmn} \iint_{A_s} \boldsymbol{\Psi}'_{pmn} \cdot \mathbf{E}'_t \, da' \quad (83)$$

where the subscripts $p'm'n'$ denote the indices of the incident mode (formerly $\boldsymbol{\Phi}_i = \boldsymbol{\Phi}_1$), and the field E'_t is the field E_t over the entire area A_s with all four beams radiated. The primed sums, $\sum'\sum'\sum'$, denote omission of the incident mode in the sum. The input admittance, y_{in_1}, is identical to that in (72) and (73).

The kernel of (83) is now complex-symmetric. On this basis, it is possible[3,6] to write a variational principle for y_{in_1}:

$$y_{\text{in}_1}(-1)^{m'+n'} y_{p'm'n'} \left(\iint_{A_s} \boldsymbol{\Phi}'_{p'm'n'} \cdot \mathbf{E}'_t \, da \right)^2$$

$$= {\sum_p}' {\sum_m}' {\sum_n}' (-1)^{m+n} y_{pmn} \left(\iint_{A_s} \boldsymbol{\Phi}'_{pmn} \cdot \mathbf{E}'_t \, da \right)^2$$

$$+ \sum_p \sum_m \sum_n Y_{pmn} \left(\iint_{A_s} \boldsymbol{\Psi}'_{pmn} \cdot \mathbf{E}'_t \, da \right)^2 \quad (84)$$

It is convenient to relate the integrals in (84) to those in (72) for the field, $\mathbf{E}_t$, under the conventional array excitation of a single radiated beam (beam 1 in Fig. 2.7), wherein the integration area is over a single waveguide aperture, A'. As a result, (84) can be written as

$$y_{\text{in}_1}(-1)^{m'+n'} y_{p'm'n'} \left(\iint_{A'} \boldsymbol{\Phi}_{p'm'n'} \cdot \mathbf{E}_t \, da \right)^2$$

$$= {\sum_p}' {\sum_m}' {\sum_n}' (-1)^{m+n} y_{pmn} \left(\iint_{A'} \boldsymbol{\Phi}_{pmn} \cdot \mathbf{E}_t \, da \right)^2$$

$$+ \sum_p \sum_m \sum_n Y_{pmn} \left(\iint_{A'} \boldsymbol{\Psi}^*_{pmn} \cdot \mathbf{E}_t \, da \right)^2 \quad (85)$$

In practical applications it is sometimes desirable to determine the equivalent circuit parameters of a T or π network representing the transition from the waveguide interior region to the exterior waveguide simulated region. Such a variational expression for these parameters is derivable by placing a short-circuit (electric) plane at some distance, z, from the aperture in the $z > 0$ region (see Ref. 1). To obtain all the parameters, several z values must be used. The advantage of this variational expression is that a *real* symmetric kernel for the operator is obtained instead of a *complex* symmetric kernel, and hence a Rayleigh-Ritz[1,6] procedure may be applied for the solution. Complex R can then be determined from the circuit parameters. Furthermore, if the real symmetric operator is either positive or negative definite,[6] it is possible to obtain bounds to the accuracy of an approximately derived solution. This condition, however, occurs for only very few practical problems.

An alternative approach to obtaining a complex-symmetric kernel for the array eigenvalue equation exists.[20,26] However, the connection between the periodic phased array problem and the waveguide simulator is not made clear by this method. Nevertheless, it will be seen that the same symmetry conditions are required to obtain a complex-symmetric kernel.

We note first, from (18) and (20), that the Floquet modes vary functionally as

$$\mathbf{\Psi}_{pmn}(\mathbf{r}) \sim \exp(-j\mathbf{k}_{mn} \cdot \mathbf{r}) = [\exp(+jk_{x_m}x + k_{y_n}y)]^* \tag{86}$$

where the asterisk signifies the complex conjugate. We may then make the substitution

$$(\mathbf{r}) \rightarrow (-\mathbf{r}) \tag{87}$$

in (72) and obtain

$$y_{\text{in}_1} y_1 \mathbf{\Phi}_1(-\mathbf{r}) \int_{A'} \mathbf{\Phi}_1(\mathbf{r}') \cdot \mathbf{E}_t \, da'$$

$$= \iint\limits_{A'} \left[\sum_{i=2}^{\infty} y_i \mathbf{\Phi}_i(-\mathbf{r}) \mathbf{\Phi}_i(\mathbf{r}') + \sum_p \sum_m \sum_n Y_{pmn} \mathbf{\Psi}^*_{pmn}(\mathbf{r}) \mathbf{\Psi}^*_{pmn}(\mathbf{r}') \right] \cdot \mathbf{E}_t(\mathbf{r}') \, da \tag{88}$$

If we assume now that the $\mathbf{\Phi}_i$ have the following property:

$$\mathbf{\Phi}_i(\mathbf{r}) = \pm \mathbf{\Phi}_i(-\mathbf{r}) \tag{89}$$

then the admittance operator of (88) is complex-symmetric. If the waveguides do not satisfy (89), however, (88) is not complex-symmetric and the symmetry conditions, (75), are also not satisfied. We shall illustrate this point further by considering a nonsymmetrical waveguide iris at the aperture, $z = 0$, as shown in Fig. 2.8.

In this case, the variable $\mathbf{r}' \in A'$ and $\mathbf{r} \in A'$, where the symbol $\in$ denotes "exists in or "belongs to." However, in (88) we have $\mathbf{r}' \in A'$ and $-\mathbf{r} \in A \neq A'$. Hence the operator in (88), which includes *both* the integration and the

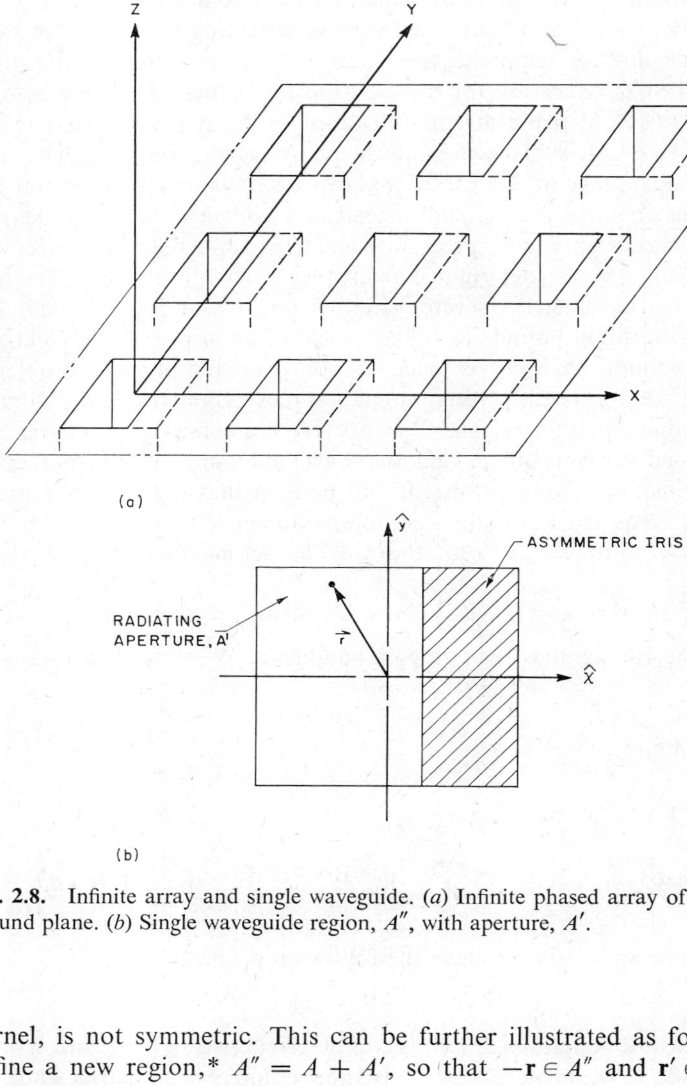

Fig. 2.8. Infinite array and single waveguide. (*a*) Infinite phased array of waveguides in ground plane. (*b*) Single waveguide region, A'', with aperture, A'.

kernel, is not symmetric. This can be further illustrated as follows. If we define a new region,* $A'' = A + A'$, so that $-\mathbf{r} \in A''$ and $\mathbf{r}' \in A''$, by defining functions

$$\begin{aligned} k'(\mathbf{r}') &= 1 \qquad \text{in} \quad A' \text{ and zero outside} \\ k(\mathbf{r}) &= 1 \qquad \text{in} \quad A \text{ and zero outside} \end{aligned} \tag{90}$$

* The region A'' is the entire waveguide region as shown in Fig. 2.1. It is the *smallest* region containing both the ranges of $\mathbf{r}'$ and $-\mathbf{r}$, namely, A' and A.

then the new integral operator and kernel, as defined by

$$2y_1\boldsymbol{\phi}_1(-\mathbf{r}) = \iint_{A''} \left\{ \left[\sum_{=1}^{\infty} y_i \boldsymbol{\Phi}_i(-\mathbf{r})\boldsymbol{\Phi}_i(\mathbf{r}') + \sum_p \sum_m \sum_n^{\infty} Y_{pmn} \boldsymbol{\Psi}^*_{pmn}(\mathbf{r})\boldsymbol{\Psi}^*_{mn}(\mathbf{r}') \right] k(\mathbf{r})k'(\mathbf{r}') \right\} \cdot \mathbf{E}_t(\mathbf{r}')\, da' \tag{91}$$

is clearly not symmetric because of the k functions.

Hence, so far as can be determined to date,[20] the existence of a variational principle for a phased array scanned at any angle is predicated on the assumption that certain reflection symmetries exist, (75), within the array that are consistent with the theoretical possibility of constructing a waveguide simulator. This means that the presence of asymmetric obstacles, irises, or waveguides in the aperture of a phased array element precludes both the construction of a waveguide simulator for the array and the use of a variational expression in the analysis of the array.

It should be noted that a solution of (44) or (88) by the method of moments[21] can be shown to be *equivalent*[22] to a solution of (85), the variational form. Hence it is not necessary to put (44) or (88) into the special form of (85) to obtain a stationary approximate solution. This will be discussed further in Chapter 3.

2.5 Other Forms of Integral Equations and Variational Expressions

The Fredholm integral equations of the first kind, (44) and (45), are only one form of integral equation that can be written to describe the same array. Although the solutions for $\mathbf{E}_t$ and $\mathbf{H}_t$ are unique, the form of equation is not. Analytic manipulation of (44), for example, by use of the orthonormality properties of the waveguide modes, $\boldsymbol{\Phi}_i$, and the free space modes, $\boldsymbol{\Psi}_{pmn}$, can lead to alternative forms of integral equations, each perhaps having a different physical interpretation, and each perhaps more amenable to certain methods of approximate solution.

One particularly useful and common form of integral equation is the Fredholm integral equation of the *second* kind. As we shall illustrate shortly, this form of equation is particularly suited to iterative type approximate solutions. For illustrative purposes we will consider the Fredholm first equation, (44), with only a single incident mode ($A_1 = 1, A_i = 0$ for $i > 0$).* We make use of the identity operator, (34) or (35), to obtain

$$y_1\mathbf{E}_t = y_1 \iint_{A'} \sum_{i=1}^{\infty} \boldsymbol{\Phi}_i\boldsymbol{\Phi}_i \cdot \mathbf{E}_t\, da' \tag{92}$$

* As was mentioned earlier, the multimode solution can always be found by superposition of the solutions for all modes separately incident upon the aperture.

This equation, when added to (44), immediately gives us a Fredholm integral equation of the second kind[17] for $\mathbf{E}_t$:

$$\mathbf{E}_t = 2\mathbf{\Phi}_1 - \iint_{A'} \left[\sum_{i=1}^{\infty} \left(\frac{y_i - y_1}{y_1} \right) \mathbf{\Phi}_i \mathbf{\Phi}_i + \sum_{p=1}^{2} \sum_{m=-\infty}^{\infty} \sum_{n=-\infty}^{\infty} Y_{pmn} \mathbf{\Psi}_{pmn} \mathbf{\Psi}^*_{pmn} \right] \cdot \mathbf{E}_t \, da' \tag{93}$$

Equation 93 is in a suitable form for an iterative solution;[21] in particular, a Neumann[6] iterative solution is possible. We may write (93) in an *operator* notation as

$$(\Pi + \mathcal{O}_N)\mathbf{E}_t = 2\mathbf{\Phi}_1 \tag{94}$$

where Π is a unit operator [see (35), for example] and $\mathcal{O}_N$ is the integral operator of (93), namely,

$$\mathcal{O}_N = \iint_{A'} \sum_{i=1}^{\infty} \left(\frac{y_i - y_1}{y_1} \right) \mathbf{\Phi}_i \mathbf{\Phi}_i + \sum_{p=1}^{2} \sum_m \sum_n Y_{pmn} \mathbf{\Psi}_{pmn} \mathbf{\Psi}^*_{pmn} \cdot \text{——} \tag{95}$$

The solution to (94) may be formally written[21,23] as

$$\mathbf{E}_t = \sum_{q=0}^{\infty} (-1)^q \mathcal{O}_N{}^q (2\mathbf{\Phi}_1) \tag{96}$$

The convergence of the series, (96), is not guaranteed. However, we shall not examine this method of solution in this chapter; we intend only to show a motivation for choosing the integral form of (93).

Another significant class of forms for (44) involves transforming this integral equation into a *discrete* representation in the Hilbert space sense.[6] In (44), for example, we must determine, in an exact solution, the unknown function, $\mathbf{E}_t$, at a continuum of points defined by the variable $\mathbf{r}$ over A'. If we knew the discrete, yet infinite, set of values of v_i and of $A_i + R_i$ in (26) and (28), we would, of course have $\mathbf{E}_t$ equally well defined. The utility of the integral equation is not its representation in the $\mathbf{r}$ coordinate, but rather the fact that it incorporates *all* boundary conditions compactly and serves as a useful starting point for other *equivalent* representations. In the sense that $\mathbf{r}$ is a *continuous coordinate* and $\mathbf{E}_t(\mathbf{r})$ is the desired function, so is $\mathbf{\Phi}_i$ a *discrete coordinate* and $\mathbf{E}_t(\mathbf{\Phi}_i) \equiv v_i$ $(i = 1, \ldots, \infty)$ an equivalent function to $\mathbf{E}_t(\mathbf{r})$.

The transformation of (44), (55), or (93) to an equivalent discrete form is particularly useful when a finite discrete approximation by the Ritz or Galerkin[21] method is desired. A fuller discussion of this type of approximation will be found in Chapter 3.

2.6 Integral Equation for Single-Element Excitation

It will be recalled, from (59) and Fig. 2.5, that, if the *element* coupling coefficients are obtained from either an analysis or an experiment with only one waveguide element of the array excited, a solution to the *phased array* problem of all elements excited can be found by superposition. A full discussion of this aspect of the problem will be presented in Chapter 8. For the present, it is of interest to note that an integral equation for the aperture tangential fields can be derived for the problem with only a single element excited. Furthermore, a variational principle for the element coupling coefficients can also be written.

The procedure for deriving the integral equation and variational expression (V.E.) for the array with a single element excited is similar to that for the case with all elements excited. We will perform, for simplicity, the derivation for the thick-walled parallel-plate linear array[24] illustrated in Fig. 2.9. The

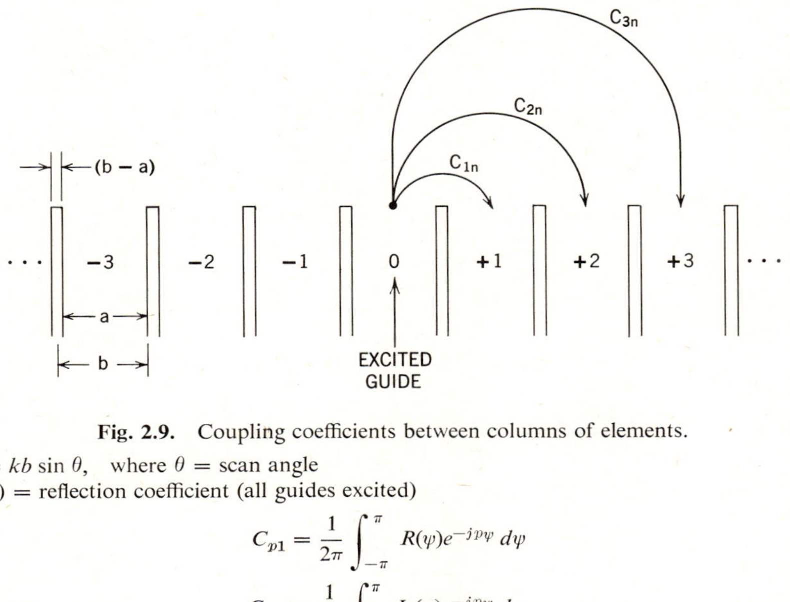

Fig. 2.9. Coupling coefficients between columns of elements.

$\psi = kb \sin \theta$, where θ = scan angle

$R(\psi)$ = reflection coefficient (all guides excited)

$$C_{p1} = \frac{1}{2\pi} \int_{-\pi}^{\pi} R(\psi) e^{-jp\psi} \, d\psi$$

$$C_{pn} = \frac{1}{2\pi} \int_{-\pi}^{\pi} I_n(\psi) e^{-jp\psi} \, d\psi$$

I_n = modal coefficients of interior field (all guides excited)

extension to planar arrays of other element types can be carried out in a straightforward manner (e.g., see Refs. 24 and 25). It should be noted that the requirement for symmetric array elements is necessary here, as it was for the previously developed V.E., (85).

Consider now that only the zeroth element (Fig. 2.9) is excited by an incident wave of unit amplitude. The fields inside the waveguides may be expanded into normal modes, whereas the fields in the free space region may be related to the Fourier transform of the aperture field. Let

$$\phi_{pn}(x) = \begin{cases} \phi_n(x - pb), & |x - pb| < \dfrac{a}{2} \\ 0, & \dfrac{a}{2} \leq |x - pb| \end{cases} \tag{97}$$

be the nth orthonormal modal function in the pth waveguide, where $\phi_n(x)$ is the nth modal function in the zeroth guide. It is assumed here that the waveguides are of width a and have their centers b units apart. Let $\mathscr{C}_{pn}$ be the current coupling coefficient from the excited guide to the nth mode in the pth removed guide. The magnetic field across the entire aperture ($-\infty < x < \infty$) is then given by

$$H_x(x) = \sum_{p=-\infty}^{\infty} \left[(\delta_{p0} - C_{p1})\phi_{p1}(x) + \sum_{n=2}^{\infty} \phi_{pn}(x) \int_{pb-a/2}^{pb+a/2} \phi_{pn}(x')H_x(x')\, dx' \right] \tag{98}$$

for $z = 0^-$, and by

$$H_x(x) = \frac{1}{2\pi} \int_{-\infty}^{\infty} dx' \int_{-\infty}^{\infty} dk_x e^{jk_x(x-x')} H_x(x') \tag{99}$$

for $z = 0^+$. The electric field across the entire aperture ($-\infty < x < \infty$) is then given by

$$E_y(x) = \sum_{p=-\infty}^{\infty} \left[(-\delta_{p0} - C_{p1})Z_1\phi_{p1}(x) + \sum_{n=1}^{\infty} Z_n\phi_{pn}(x) \int_{pb-a/2}^{pb+a/2} \phi_{pn}(x')H_x(x')\, dx' \right] \tag{100}$$

for $z = 0^-$, and by

$$E_y(x) = -\frac{\omega\mu}{2\pi} \int_{-\infty}^{\infty} dx' \int_{-\infty}^{\infty} dk_x \frac{e^{jk_x(x-x')}}{\sqrt{k^2 - k_x^{\,2}}} H_x(x') \tag{101}$$

for $z = 0^+$, where the modal impedances can be written as

$$Z_n = \frac{\omega\mu}{\sqrt{k^2 - (n\pi/a)^2}}$$

Imposing the continuity conditions across the aperture by equating (98) to (99) and (100) to (101) leads to Fredholm integral equations of the first and second kinds for the tangential aperture magnetic field:

$$2Z_1\phi_{01}(x) = \sum_{p=-\infty}^{\infty} \sum_{n=1}^{\infty} Z_n\phi_{pn}(x)I_{pn} + \frac{\omega\mu}{2} \int_{-\infty}^{\infty} H_0^{(2)}(k\,|x - x'|)H_x(x')\, dx' \tag{102}$$

and

$$H_x(x) = 2\phi_{01}(x) - \sum_{p=-\infty}^{\infty} \sum_{n=2}^{\infty} \frac{Z_n - Z_1}{Z_1} \phi_{pn}(x)I_{pn} - \frac{\omega\mu}{2Z_1} \int_{-\infty}^{\infty} H_0^{(2)}(k\,|x - x'|)H_x(x')\, dx' \tag{103}$$

where

$$I_{pn} = \int_{pb-a/2}^{pb+a/2} \phi_{pn}(x')H_x(x')\,dx' \tag{103a}$$

and $H_0^{(2)}(x)$ is the zeroth-order Hankel function of the second kind. Equation 103 may be rearranged in an eigenvalue equation form:

$$\begin{aligned}\frac{1+\mathscr{C}_{01}}{1-\mathscr{C}_{01}} Z_1\phi_{01}(x)\int_{-(a/2)}^{a/2} \phi_{01}(x')H_x(x')\,dx' &= \sum_{p=-\infty}^{\infty}{}' Z_1\phi_{p1}(x)\int_{-(a/2)}^{a/2} \phi_{p1}(x')H_x(x')\,dx' \\ &+ \sum_{p=-\infty}^{\infty}\sum_{n=2}^{\infty} Z_n\phi_{pn}(x)\int_{-(a/2)}^{a/2} \phi_{pn}(x')H_x(x')\,dx' \\ &+ \frac{\omega\mu}{2}\int_{-\infty}^{+\infty} H_0^{(2)}(k\,|x-x'|)H_x(x')\,dx'\end{aligned} \tag{104}$$

where Σ' denotes omission of $p = 0$ in the sum, and $Z_1(1+\mathscr{C}_{01})/(1-\mathscr{C}_{01})$ is the normalized input impedance of the excited element. Since the kernel of (104) is complex-symmetric, a V.E. is readily found for the input impedance and hence for $\mathscr{C}_{01}$. Thus,

$$\begin{aligned}\frac{1+\mathscr{C}_{01}}{1-\mathscr{C}_{01}} Z_1 I_{01}^{\;2} &= \sum_{p=-\infty}^{\infty}{}' Z_1 I_{p1}^{\;2} + \sum_{p=-\infty}^{\infty}\sum_{n=2}^{\infty} Z_n I_{pn}^{\;2} \\ &+ \frac{\omega\mu}{2}\int_{-\infty}^{\infty}\int_{-\infty}^{\infty} H_x(x)H_0^{(2)}(k\,|x-x'|)H_x(x')\,dx'\,dx\end{aligned} \tag{105}$$

with I_{pn} given by (103a).

In order to obtain a V.E. for $\mathscr{C}_{p1}$ when $p \neq 0$, we consider the excitation of two elements, the zeroth and the qth Let $E_y^{\,T}(x')$ and $H_x^{\,T}(x')$ be the aperture electric and magnetic fields under this condition. Then,

$$\begin{aligned}H_x^{\,T}(x) &= \sum_{p=-\infty}^{\infty} (\delta_{p0} + \delta_{pq} - R_p)\phi_{p1}(x) \\ &+ \sum_{p=-\infty}^{\infty}\sum_{n=2}^{\infty} \phi_{pn}(x)\int_{-\infty}^{\infty} \phi_{pn}(x')H_x^{\,T}(x')\,dx'\end{aligned} \tag{106}$$

and

$$\begin{aligned}E_y^{\,T}(x) &= \sum_{p=-\infty}^{\infty} (-\delta_{p0} - \delta_{pq} - R_p)\phi_{p1}(x) \\ &+ \sum_{p=-\infty}^{\infty}\sum_{n=2}^{\infty} Z_n\phi_{pn}(x)\int_{-\infty}^{\infty} \phi_{pn}(x')H_x^{\,T}(x')\,dx' \\ &= \frac{\omega\mu}{2}\int_{-\infty}^{\infty} H_0^{(2)}(k\,|x-x'|)H_x^{\,T}(x')\,dx'\end{aligned} \tag{107}$$

where* $R_p = \mathscr{C}_p + \mathscr{C}_{p-q}$. Combining (106) and (107) yields

$$\frac{1+R_0}{1-R_0} Z_1\phi_{01}(x)\int_{pb-(a/2)}^{pb+(a/2)} \phi_{01}(x')H_x{}^T(x')\,dx' + \frac{1+R_q}{1-R_q} Z_1\phi_{q1}(x)\int_{pb-(a/2)}^{pb+(a/2)} \phi_{q1}(x')H_x{}^T(x')\,dx'$$
$$= \sum_{p=-\infty}^{\infty(0,q)} Z_1\phi_{p1}(x)\int_{pb-(a/2)}^{pb+(a/2)} \phi_{p1}(x')H_x{}^T(x')\,dx' + \sum_{p=-\infty}^{\infty}\sum_{n=2}^{\infty} Z_n\phi_{pn}(x)\int_{pb-(a/2)}^{pb+(a/2)} \phi_{pn}(x')H_x{}^T(x')\,dx' + \frac{\omega\mu}{2}\int_{-\infty}^{\infty} H_0^{(2)}(k\,|x-x'|)H_x{}^T(x')\,dx' \tag{108}$$

where $\sum_{p=-\infty}^{\infty(0,q)}$ denotes the exclusion of 0 and q from the summation.

Since $R_0 = R_q = \mathscr{C}_0 + \mathscr{C}_q$, the left-hand side of (108) may be written† as $(\phi_{01}I_{01}{}^T + \phi_{q1}I_{q1}{}^T)$ times $Z_1(1 + \mathscr{C}_0 + \mathscr{C}_q)/(1 - \mathscr{C}_0 - \mathscr{C}_q)$, the latter being the input impedance of either excited element, the zeroth or the qth. A variational expression is readily obtainable again because the kernel of (108) is complex-symmetric. Thus,

$$Z_1\frac{1+\mathscr{C}_0+\mathscr{C}_q}{1-\mathscr{C}_0-\mathscr{C}_q}([I_{01}{}^T]^2 + [I_{q1}{}^T]^2)$$
$$= \sum_{p=-\infty}^{\infty(0,q)} Z_1[I_{p1}{}^T]^2 + \sum_{p=-\infty}^{\infty}\sum_{n=2}^{\infty} Z_n[I_{pn}{}^T]^2 + \frac{\omega\mu_0}{2}\int_{-\infty}^{\infty}\int_{-\infty}^{\infty} H_x{}^T(x)H_0^{(2)}(k\,|x-x'|)H_x{}^T(x')\,dx'\,dx \tag{109}$$

Additional reduction in the form of (109) is possible if the symmetry about the element centered between $p = 0$ and $p = q$ is further utilized.

REFERENCES

1. R. E. Collin, *Field Theory of Guided Waves*, McGraw-Hill Book Company, New York, 1960.
2. N. Amitay and V. Galindo, "The Analysis of Circular Waveguide Phased Arrays," *Bell System Tech. J.*, Vol. 47, No. 9, November 1968.
3. N. Marcuvitz (ed.), *Waveguide Handbook*, M.I.T. Radiation Laboratory Series, Vol. 10, McGraw-Hill Book Company, New York, 1951.

* Where the second subscript is omitted in $\mathscr{C}_{pn} = \mathscr{C}_p$, it is assumed $\mathscr{C}_{pn} = \mathscr{C}_{p1}$, the dominant mode coefficient.

† The term $I_{pn}{}^T$ is defined as I_{pn} [see (103a)], with $H_x{}^T$ substituted for H_x.

4. E. C. Titchmarsh, *The Theory of Functions*, Oxford University Press, London, 1952, Paragraphs 1.7, 13.5, and 13.53 (pp. 419–421).

5. A. J. Sommerfeld, *Optics*, Academic Press, New York, 1954.

6. B. Friedman, *Principles and Techniques of Applied Mathematics*, John Wiley and Sons, New York, 1960.

7. E. C. DuFort, "Finite Scattering Matrix for an Infinite Antenna Array," *Radio Science*, Vol. 2, No. 1, pp. 19–27, January 1967.

8. C. G. Montgomery, R. H. Dicke, and E. M. Purcell, *Principles of Microwave Circuits*, Radiation Laboratory Series, Vol. 8, McGraw-Hill Book Company, New York, 1948.

9. N. Amitay and V. Galindo, "A Note on the Radiation Characteristics and Forced Surface Wave Phenomena in Triangular Grid Circular Waveguide Phased Arrays," *IEEE Trans. Antennas and Propagation*, Vol. AP-16, No. 6, pp. 760–762, November 1968.

10. P. W. Hannan, P. J. Meier, and M. A. Balfour, "Simulation of Phased Array Antenna Impedance in Waveguide," *IEEE Trans. Antennas and Propagation*, Vol. AP-11, No. 6, November 1963.

11. M. A. Balfour, "Phased Array Simulators in Waveguide for a Triangular Arrangement of Elements," *IEEE Trans. Antennas and Propagation*, Vol. AP-13, No. 3, pp. 475–476, May 1965.

12. P. W. Hannan and M. A. Balfour, "Simulation of a Phased-Array Antenna in Waveguide," *IEEE Trans. Antennas and Propagation*, Vol. AP-13, No. 3, pp. 342–353, May 1965.

13. M. A. Balfour, "Active Impedance of a Phased Array Antenna Element Simulated by a Single Element in Waveguide," *IEEE Trans. Antennas and Propagation*, Vol. AP-15, No. 2, pp. 313–314, March 1967.

14. F. B. Hildebrand, *Methods of Applied Mathematics*, Prentice-Hall, Englewood Cliffs, N.J., 1952.

15. N. Amitay, V. Galindo, and C. P. Wu, "On Certain Eigenvalue Problems in E. M. Wave Scattering," private memoranda, available on request.

16. J. A. Stratton, *Electromagnetic Theory*, McGraw-Hill Book Company, New York, 1941, pp. 486–487.

17. V. Galindo and C. P. Wu, "Integral Equations and Variational Expressions for Arbitrary Scanning of Regular Infinite Arrays," *IEEE Trans. Antennas and Propagation*, Vol. AP-14, No. 3, pp. 392–394, May 1966.

18. R. E. Collin, "A Note on Waveguide Image Techniques," Case Institute of Technology, Cleveland, Ohio, *Sci. Rept.* 19, AD-250931, November 1960.

19. D. S. Saxon, "Notes on Lectures by Julian Schwinger—Discontinuties in Waveguides," to be published shortly; see also Ref. 3.

20. V. Galindo, "On the Stationary Properties of the Integral Equations of Infinite Phased Arrays," *IEEE Trans. Antennas and Propagation*, Vol. AP-17, No. 3, May 1969.

21. L. V. Kantorovich and V. I. Krylov, *Approximate Methods of Higher Analysis*, Interscience Publishers, New York, 1964.

22. D. S. Jones, "A Critique of the Variational Method in Scattering Problems," *IRE Trans. Antennas and Propagation*, Vol. AP-4, pp. 297–301, May 1956.

23. N. Amitay and V. Galindo, "Application of a New Method for Approximate Solutions and Error Estimates to Waveguide Discontinuity and Phased Array Problems," *Radio Science*, Vol. 3, No. 8, pp. 830–843, August 1968.

24. V. Galindo and C. P. Wu, "A Variational Expression for the Dominant Mode Coupling Coefficients Between the Elements in an Infinite Array," *IEEE Trans. Antennas and Propagation*, Vol. AP-14, No. 5, pp. 637–639, September 1966.

25. V. Galindo and C. P. Wu, "Numerical Solution for an Infinite Phased Array of Rectangular Waveguides with Thick Walls," *IEEE Trans. Antennas and Propagation*, Vol. AP-14, pp. 149–158, March 1966.

26. C. P. Wu, "Note on Integral Equations and Variational Expressions for Arbitrary Scanning of Regular Infinite Arrays," *IEEE Trans. Antennas and Propagation*, Vol. AP-16, pp. 136–138, January 1968.

3. Methods of Solution

1. INTRODUCTION

In Chapter 2 we derived the integral equations for the boundary value problems of infinite waveguide arrays. An outstanding feature of such equations is that the kernels are composed of infinite sums representing the contributions from the wave modes of two distinctly different "waveguides." Because of the complexity of the problem, equations of this type generally do not permit analytical solutions to be obtained except for one or two isolated cases. Therefore, in discussing the methods of solutions in this chapter, emphasis will be placed on numerical techniques.

In order to obtain equations suitable for numerical computation, the integral equations are first converted into matrix equations by using the method of moments. Once the matrix equations are obtained, they may then be solved by standard techniques, notably by the method of matrix inversion. Although inversion of matrices can be carried out quite easily with the aid of present day digital computers, the practicability of obtaining a solution depends on many factors, such as the ease with which the matrix equation can be derived and the economy of carrying out the computation of the matrix elements. Several choices of functional representation have proved effective. In particular, the use of waveguide modes as a basis has been found to offer, in many cases, distinct advantages over the choice of other bases.

The matrix equation may be obtained also by applying the Rayleigh–Ritz procedure to the variational expression of the input impedance or admittance of the array. The conditions under which a variational principle exists were discussed in Chapter 2. It will be shown that direct application of the method of moments to the integral equation and the Rayleigh–Ritz procedure lead to the same matrix equation. The implication is that solutions obtained through the moments method automatically satisfy the reciprocity relation regardless of the accuracy of the approximate solutions. This has important consequences for the verification of the validity of solutions.

An alternative approach that can be used in solving the integral equation is to approximate the kernel by a degenerate kernel. That an integral equation with a degenerate kernel can be reduced to a system of algebraic equations is well known from the theory of integral equations. If suitable bases for representation are used in the approximation, the system of equations obtained by this method may be shown to coincide with those derived by the moments method. Thus, the degenerate kernel approximation is equivalent to using the moments method.

An important consideration in using numerical techniques is the determination of both the correctness and the accuracy of the results. A number of useful procedures for ascertaining the validity of solutions have been developed and tried; these will be discussed in detail.

A special array in which the radiators are thin-walled parallel plates can be analyzed exactly by either the Wiener-Hopf or the residue calculus technique. The exact solutions obtainable by such methods prove to be useful in many respects, and Section 6 is devoted to a discussion of these techniques. Since the equations appropriate for the residue calculus method may be developed from the integral equations by the proper choice of basis, it seems natural to consider this method after the method of moments is discussed.

As an alternative approach to the straightforward matrix inversion method, the matrix equation may be rearranged to allow a Neumann series expansion. The series are found to converge rapidly in most situations. Moreover, the method is also useful in obtaining tighter error bounds than conventional approaches have previously yielded.

2. GENERAL DISCRETE BASIS REPRESENTATION

An electromagnetic array boundary value problem may be formulated in terms of integral equations in various forms. It may be expressed as a Fredholm integral equation of the first or of the second kind. It can also be cast into a form which lends itself to a solution by an iterative procedure. For the solution of array problems, equations of the first kind have been found to be more convenient. We shall be concerned mostly with this type of equations here.

Symbolically, the Fredholm integral equation of the first kind may be written in operator form (for an introduction to the theory of operator equations see, e.g., either of the books listed in Ref. 1) as

$$y = Kx \tag{1}$$

where K is a linear operator *determined solely by the array geometry*, y is a known function proportional to the excitation, and x is the unknown function which we wish to determine. Usually, x and y are vector quantities, and K is a dyadic or tensor.*

Associated with an operator is a space called the domain of definition, S_D, that is, the set of functions over which the operator operates. (Since we are dealing with integral operators, boundary conditions are automatically incorporated. In the case of differential operators, the boundary conditions must be specified in order to properly define the domain of definition.) When an operator operates on a function, it generates a new function. The collection of all the functions generated by the operator operating on each member in its domain of definition is called the range of the operator, S_L. By a linear operator, we mean that, given two functions, x_1 and x_2, and two arbitrary constants, α_1 and α_2, the following relation holds:

$$K(\alpha_1 x_1 + \alpha_2 x_2) = \alpha_1 K x_1 + \alpha_2 K x_2$$

Suppose now that we find by some means a set of linearly independent functions which spans† the space S_D, and another set which spans the space S_L. Let these sets be denoted by $\{f_n : f_n \in S_D\}$ and $\{g_n : g_n \in S_L\}$, respectively. The elements in each set may be assumed, for convenience, to be orthonormalized. This is not a restriction, because if they are not, a linearly independent set of functions can always be orthonormalized by using the Gram–Schmidt process.

Since the unknown x is an element in S_D, we may write it as a linear combination of f_n. Thus

$$x = \sum a_n f_n \tag{2}$$

where a_n are unknown coefficients to be determined. The representation mostly requires an infinite sum when exact solutions of (1) are sought. In this case, the sets of functions $\{f_n\}$ and $\{g_n\}$ have to be complete in their respective space also. For approximate solutions, on the other hand, the sum usually is finite for practical reasons. The set of functions $\{f_n\}$ need not necessarily be complete, although it is important that certain elements be included in order to obtain meaningful results. This point will be explained

* Given a mathematical problem, two important questions naturally arise: Is there a solution to the problem (the existence question), and if so is the solution unique? The proof of the existence of a solution to (1) is generally very difficult and is beyond the scope of the present work. Since we are dealing with a physical problem, however, we would expect that, if the problem is well posed, a solution exists. As to the uniqueness of solution, it suffices to point out that the starting point of the problem formulation is Maxwell's equations. It has been proved that solutions to the Maxwell equations are unique.

† A set of functions is said to "span" a space of functions if any functions in the space may be represented, according to an appropriate measure of the space, as a linear combination of the set of functions.

in more detail later when we discuss practical aspects of the problem. We consider the case of exact solution first.

When (2) is substituted into (1), we find by virtue of the linearity of the operator that

$$y = \Sigma a_n K f_n \tag{3}$$

The difference between the left- and right-hand sides of (3) is a zero element in S_L. One way of assuring this is to require the element $(y - \Sigma a_n K f_n)$ to be orthogonal to all the elements in a set spanning S_L. Thus,

$$\langle g_m, y - \Sigma a_n K f_n \rangle = 0 \qquad \text{for} \quad \text{all } m \tag{4a}$$

or equivalently

$$\langle g_m, y \rangle = \Sigma \langle g_m, K f_n \rangle a_n, \quad m = 1, 2, \ldots \tag{4b}$$

This is a system of equations which may be conveniently written in matrix form. Let the square matrix $[K_{mn}]$ and the column vectors **a** and **y** be defined as

$$[K_{mn}] = \begin{bmatrix} \langle g_1, Kf_1 \rangle & \langle g_1, Kf_2 \rangle & \cdots \\ \langle g_2, Kf_1 \rangle & \langle g_2, Kf_2 \rangle & \\ \langle g_3, Kf_1 \rangle & \langle g_3, Kf_2 \rangle & \\ \cdot & \cdot & \\ \cdot & \cdot & \\ \cdot & \cdot & \end{bmatrix}$$

$$\mathbf{a} = \begin{bmatrix} a_1 \\ a_2 \\ a_3 \\ \cdot \\ \cdot \\ \cdot \end{bmatrix}, \quad \text{and} \quad \mathbf{y} = \begin{bmatrix} \langle g_1, y \rangle \\ \langle g_2, y \rangle \\ \langle g_3, y \rangle \\ \cdot \\ \cdot \\ \cdot \end{bmatrix}$$

Equation (4) then becomes

$$\mathbf{y} = [K_{mn}]\mathbf{a} \tag{5}$$

The integral equation is thus converted into a system of algebraic equations. Whether the system can be effectively solved depends on many factors. Since the system is most likely to be infinite in order with infinitely many unknowns, with the exception of one or two isolated cases, no method seems available to effect an exact solution. In order to render the system tractable, it is necessary to introduce approximations. Instead of the infinitely many functions which may be needed for proper representation of the unknown function, we use a finite number, say N, of functions to approximate the unknown. In place of (2), then, we let

$$x \approx \sum_{n=1}^{N} a'_n f_n \tag{6}$$

The unknown coefficients, a'_n, will be different in general from the true solution, a_n. They will vary as functions of N also. Substitution of (6) into (1) yields

$$y \approx \sum_{n=1}^{N} a'_n K f_n \tag{7}$$

The right and left sides of (7) are no longer equal because of the approximation involved. The difference between the two sides of the equation then represents an error. In order to minimize this error in the space spanned by the set of functions g_m, $m = 1, 2, \ldots, N$, both sides are required to have the same projection on the set $\{g_m\}$. Or, equivalently, the difference between the left and right sides is required to be orthogonal to each member of the set $\{g_m\}$. When this is the case, the process yields a system of N equations in N unknowns, which again may be put compactly into a matrix form. Thus,

$$\mathbf{y}' = [K_{N_{mn}}]\mathbf{a}' \tag{8}$$

where

$$[K_{N_{mn}}] = \begin{bmatrix} \langle g_1, Kf_1\rangle & \langle g_1, Kf_2\rangle & \cdots & \langle g_1, Kf_N\rangle \\ \langle g_2, Kf_1\rangle & \langle g_2, Kf_2\rangle & \cdots & \langle g_2, Kf_N\rangle \\ \cdot & \cdot & & \cdot \\ \cdot & \cdot & & \cdot \\ \cdot & \cdot & & \cdot \\ \langle g_N, Kf_1\rangle & \langle g_N, Kf_2\rangle & \cdots & \langle g_N, Kf_N\rangle \end{bmatrix}$$

$$\mathbf{a}' = \begin{bmatrix} a'_1 \\ a'_2 \\ \cdot \\ \cdot \\ \cdot \\ a'_N \end{bmatrix}, \quad \text{and} \quad \mathbf{y}' = \begin{bmatrix} \langle g_1, y\rangle \\ \langle g_2, y\rangle \\ \cdot \\ \cdot \\ \cdot \\ \langle g_N, y\rangle \end{bmatrix}$$

The matrix is now of finite order, so that a solution can be obtained by matrix inversion. Let $[K_{N_{mn}}]^{-1}$ denote the inverse of the matrix $[K_{N_{mn}}]$. We may then formally write the solution as

$$\mathbf{a}' = [K_{N_{mn}}]^{-1}\mathbf{y}' \tag{9}$$

We may arrange the set of functions $\{f_n\}$ as a row vector,

$$\mathbf{f}^T = [f_1, f_2, \ldots, f_N]$$

where the superscript T denotes transposition.

The approximate solution takes the compact form

$$x \approx \mathbf{f}^T\mathbf{a}' = \mathbf{f}^T[K_{N_{mn}}]^{-1}\mathbf{y}' \tag{10}$$

Notice that the matrix equation obtained by using the finite representation (7) is, in effect, the same as the result of appropriately truncating the infinite matrix equation, (5). Intuitively, we know that the approximation will improve as N is increased. In particular, if $\{f_n\}$ and $\{g_n\}$ form complete sets, the approximate solution should approach the exact one as $N \to \infty$. Naturally, the larger N is, the bigger the matrix gets, and its inversion becomes increasingly difficult as well as more costly. How large N must be for a given accuracy depends on the specific problem.

The process of reducing an operator equation to a matrix equation is generally known as the *method of moments*.[2,3] For the special case in which $\{f_n\} = \{g_n\}$, the procedure is referred to as the Galerkin or Ritz-Galerkin method. The set $\{f_n\}$ that was used to represent the unknown function, x, is called the basis or the expansion functions, and the set $\{g_n\}$ is called the testing or weighting functions. Although the basis and weighting functions can be quite arbitrary, their choice is very important from the standpoint of accuracy and efficiency. Economically, the feasibility of tackling a problem may sometimes depend on the ease with which the matrix elements can be evaluated. Another consideration is the conditioning of the matrix,[29] which is related to the accuracy with which the matrix can be inverted. A well-conditioned matrix can be inverted accurately with ease, whereas an ill-conditioned matrix may require special treatment in order to obtain reliable results. Generally speaking, in choosing the bases, it is desirable to incorporate as much prior knowledge about the problem as possible.

The actual inversion of matrices can be carried out with the help of high speed digital computers. Nowadays, computer programs for matrix inversion are widely available in many computation libraries. The mathematics of matrix inversion is discussed in various books on matrix algebra.

3. VARIATIONAL PRINCIPLE AND METHOD OF MOMENTS

In many physical problems, one is interested in the determination of certain quantities which depend on the scalar product between a given function, say y', and the solution, x, of (1), rather than on the solution x itself, namely,*

$$A = \langle y', x \rangle \tag{11}$$

In particular, the given function may be related to the incident wave, y, so that the quantity of interest is

$$B = \langle y, x \rangle \tag{12}$$

When the operator K is symmetric, variational expressions may be derived for the quantity of interest, A or B. The symmetry condition may be expressed by the following relation. For a pair of functions, u and v, in the Hilbert space,

$$\langle Ku, v \rangle = \langle u, Kv \rangle$$

Let x' be the solution to the equation

$$Kx' = y' \tag{13}$$

that is, the solution to (1) with y replaced by y'. The symmetry of the operator then implies that

$$A = \langle y', x \rangle = \langle Kx', x \rangle = \langle x', Kx \rangle = \langle x', y \rangle \tag{14}$$

This relation is a mathematical statement of the reciprocity theorem. Using (14), we can multiply the right-hand side of (11) by the ratio $\langle y, x' \rangle / \langle x', Kx \rangle$ to obtain

$$A = \frac{\langle y', x \rangle \langle y, x' \rangle}{\langle x', Kx \rangle} \tag{15}$$

The advantage of rewriting (11) in this form is that (15) is a variational expression for A. That is, (15) is stationary for independent small variations of x and x' about their correct values, which are solutions to (1) and (13), respectively. In other words, to calculate A, we need to know the solution x from (1) or x' from (13). In general, these are quite difficult to obtain. Suppose now we can obtain by some means approximate solutions to these equations. If the approximate solutions differ from the exact ones by an amount in the order of ε, where ε is a small quantity, then the calculation of A using (15) will give an error of the order of ε^2. Therefore, the better the solutions are approximated, the closer the result will be to the true value of A. The functions which are used in the estimation of A are often called trial functions. It should be cautioned that a poor choice of trial functions may

* The scalar product defined herein does not conjugate one of the two functions, as is customarily done in mathematics and physics. Such a definition is more convenient for electromagnetic scattering problems. When conjugation of functions is needed, it will be shown explicitly by an asterisk (*).

degrade rather than improve the accuracy of the estimate. In practice, the stationary property of (15) is used to determine the approximate solutions to (1) and (13). This is called the Rayleigh–Ritz procedure; it will be discussed in more detail shortly. After the approximate solutions are obtained, they are substituted back into (15) for the evaluation of A.

It turns out that the process of obtaining the approximate solutions based on the stationary property of (15) is equivalent to the method of moments; that is, the application of the Rayleigh–Ritz procedure to the variational expression (15) and the application of the moments method to (1) and (13) lead to the same system of equations. Therefore, when approximate solutions obtained by the moments method are subsequently used in the variational expression, no improvement in the accuracy of the quantity A will result. A further implication is that the approximate solutions obtained by the moments method automatically satisfy the reciprocity relation;[2,6] this will be discussed next.

The variational expression of (15) was originally due to Levine and Schwinger and hence is sometimes referred to as the Levine–Schwinger principle.[7] In the special case wherein the parameter of interest is given by (12), eq. (13) becomes identical to (1). The variational expression then takes the form

$$B = \frac{\langle x, y\rangle^2}{\langle x, Kx\rangle} \tag{16}$$

There are alternative forms of variational expressions. Those due to Ritz are given by

$$A = \langle x, y'\rangle + \langle y, x'\rangle - \langle Kx, x'\rangle \tag{17}$$

and

$$B = 2\langle x, y\rangle - \langle Kx, x\rangle \tag{18}$$

To verify that the application of the variational principle and of the method of moments leads to the same result, let us consider the simpler case using (18). Equation (18) shows that B is a functional of x, namely, the complex quantity B depends on the function x used in the expression. Let x be approximated as

$$x \approx \sum_{i=1}^{N} \tilde{\alpha}_i f_i \tag{19}$$

By substituting (19) into (18), the functional becomes a function of N unknown variables, $\tilde{\alpha}_i$:

$$\begin{aligned} F(\tilde{\alpha}_1, \tilde{\alpha}_2, \ldots, \tilde{\alpha}_N) &= \left\langle \sum_{i=1}^{N} \tilde{\alpha}_i K f_i, \sum_{j=1}^{N} \tilde{\alpha}_j f_j \right\rangle - 2\left\langle \sum_{i=1}^{N} \tilde{\alpha}_i f_i, y \right\rangle \\ &= \sum_{i=1}^{N} \sum_{j=1}^{N} \tilde{\alpha}_i \tilde{\alpha}_j \langle K f_i, f_j\rangle - 2 \sum_{i=1}^{N} \tilde{\alpha}_i \langle f_i, y\rangle \end{aligned} \tag{20}$$

Since (20) is a stationary expression, its variation relative to any arbitrary variation in the $\tilde{\alpha}_i$'s must be zero. Thus,

$$\frac{\partial F}{\partial \tilde{\alpha}_i} = 0 \qquad \text{for} \quad i = 1, 2, \ldots, N$$

This step yields

$$\sum_{j=1}^{N} \langle Kf_i, f_j \rangle \tilde{\alpha}_j = \sum_{j=1}^{N} \langle f_i, Kf_j \rangle \tilde{\alpha}_j = \langle f_i, y \rangle, \quad i = 1, 2, \ldots, N \tag{21}$$

This is the Rayleigh–Ritz procedure for determining the unknowns, $\tilde{\alpha}_i$. Equations (21) form a set of N equations in terms of N unknowns. Upon comparing (8) for $\{g_n\} = \{f_n\}$ and (21), we immediately observe that identical results are obtained by the two alternative procedures. Therefore, the moments method and the variational technique lead to the same result, as stated.

It is also possible to start out from (16) and derive a system of equations based on the stationary property of the expression, as shown by Jones.[6] Again, an identical system of equations results from this approach. The procedure, moreover, may be extended to the more general case of (15). Thus, the approximate solutions obtained by the method of moments actually satisfy the reciprocity relation exactly, independently of how close the approximate solutions are to the true solution.

4. CHOICE OF BASES

4.1 Reordering of the Modes

We now consider the integral equations pertinent to the infinite waveguide arrays derived in Chapter 2. In general, two alternative equations are available; both are Fredholm equations of the first kind. One of them has the tangential aperture electric field as the unknown; the other, the tangential magnetic field. As has been discussed, the kernels of these integral equations are composed of two infinite dyadic sums contributed, respectively, by the interior waveguide modes and the exterior Floquet modes. According to the nomenclature commonly adopted for the classification of wave modes, the modal functions usually carry triple subscripts, one for distinguishing between TE and TM modes and two for the mode order. Although there are infinitely many modes of both the TE and the TM type, since the set of modes is denumerable, it is permissible to renumber them with a single subscript. In fact, it is desirable to do so, because considerable convenience in bookkeeping may be gained when the integral equations are converted into sets of algebraic equations and also when the resulting algebraic equations are implemented for computation.

Since the approximate solution for the system of equations necessarily requires truncation at some point, it is important to rearrange the modes so that all the "important" ones will be included in the computation. Generally, it is not obvious which modes are important and which ones are not. In phased array problems, normally only modes of low modal indices are involved in the transport of power. Since it may be expected from physical considerations that the coupling between modes with nearly the same values of propagation constant will be stronger than the coupling between modes with widely different values, a useful method is to arrange the modes according to the ascending order (in absolute value) of the z-directed propagation constants. When there are mode degeneracies, that is, more than one mode with the same propagation constant, but different polarizations, it is important that the degeneracy be properly accounted for. A practical example will be encountered in Chapter 7 when circular waveguide arrays are considered.

Let the sets of interior waveguide modes and exterior Floquet modes be denoted, respectively, by $\{\mathbf{\Phi}_j(\mathbf{r})\}$ with the corresponding modal admittances y_j, and $\{\mathbf{\Psi}_m(\mathbf{r})\}$ with the corresponding modal admittances Y_m. The integral equation with the aperture electric field as the unknown for an array excited by a mode $\mathbf{\Phi}_{j'}$ is given by

$$2y_{j'}\mathbf{\Phi}_{j'}(\mathbf{r}) = \int_A \left\{ \sum_{j=1}^{\infty} y_j \mathbf{\Phi}_j(\mathbf{r})\mathbf{\Phi}_j(\mathbf{r}') + \sum_{m=1}^{\infty} Y_m \mathbf{\Psi}_m(\mathbf{r})\mathbf{\Psi}_m^*(\mathbf{r}') \right\} \cdot \mathbf{E}_t(\mathbf{r}')\, d\mathbf{r}' \qquad \text{for} \quad \mathbf{r} \in A \qquad (22)$$

Here, $\mathbf{E}_t(\mathbf{r})$ is the unknown aperture electric field that is to be determined. Notice that the limit of integration extends over the aperture A and also that the equation is *valid only in A*. This is to be distinguished from the equation for the magnetic field, which is valid over a unit cell.

The integral equation will now be converted into a set of algebraic equations by the method of moments for computation.[8,9] Although, as discussed earlier, the choice of bases in this process may be quite arbitrary, considerable advantages can be gained if a judicious selection is made. In the case of phased array integral equation (22), the expression for the kernel suggests that the interior modal functions are especially suitable as bases for two important reasons. First, they satisfy the same boundary conditions as those required by the components of $\mathbf{E}_t$ that are tangent to the circumference of the waveguide cross section. Conceivably, it is possible, therefore, to obtain accurate solutions with fewer terms than would be necessary if functions which do not satisfy the boundary conditions are used. Second, since the modal functions appear in the kernel, the orthogonality among them can be utilized to simplify considerably the computation of the matrix elements. Furthermore, the waveguide modal functions integrate in closed form with the Floquet modes in most cases. This is a distinct numerical advantage.

Let

$$\mathbf{E}_t(\mathbf{r}) \approx \sum_{p=1}^{N} v_p' \mathbf{\Phi}_p(\mathbf{r}) \tag{23}$$

The primes are used in v_p to indicate that this is an approximate modal expansion. Substituting (23) into (22) and then taking moments* with respect to $\mathbf{\Phi}_q(\mathbf{r})$, $q = 1, 2, \ldots, N$, we find

$$\sum_{p=1}^{N} A_{qp} v_p' = 2y_{j'} \delta_{j'q}, \quad q = 1, 2, \ldots, N \tag{24}$$

In these equations,

$$A_{qp} = y_q \delta_{qp} + \sum_{m=1}^{\infty} Y_m C_{qm} C_{pm}^* \tag{24a}$$

where

$$\delta_{qp} = \text{the Kronecker delta} = \begin{cases} 1, & q = p \\ 0, & q \neq p \end{cases}$$

and

$$C_{pm} = \int_A \mathbf{\Phi}_p(\mathbf{r}') \cdot \mathbf{\Psi}_m(\mathbf{r}') \, d\mathbf{r}' \tag{24b}$$

are the *intermodal coupling coefficients*. They are to be distinguished from the *mutual coupling coefficients* between the elements, which will be discussed in later chapters. Equation (24) is a set of N equations for the N unknowns, $\{v_p'\}$.

We observe that the interior modal admittances appear only in the diagonal elements, A_{qq}, whereas the exterior ones occur in all the elements.

Equations (24) may be conveniently put into matrix form as

$$[A_{qp}]\mathbf{v}' = \mathbf{f} \tag{25}$$

* It should be noted that it is permissible in principle to take moments with the $\mathbf{\Psi}_{pmn}$ functions. Because the range of integration is over A instead of C, however, we cannot take advantage of the orthogonality property of the $\mathbf{\Psi}$ functions to simplify the expression for the matrix elements. An exception occurs when A is identical to C, a condition which exists for the thin-walled rectangular waveguide array. In the integral equation for the tangential magnetic field, since the domain of definition is C, it is more advantageous to use the $\mathbf{\Psi}_{pmn}$ functions as the basis, for the same reasons.

where

$$\mathbf{v}' = \begin{bmatrix} v_1' \\ v_2' \\ \cdot \\ \cdot \\ \cdot \\ v_N' \end{bmatrix} \quad \text{and} \quad \mathbf{f} = 2y_{j'} \begin{bmatrix} 0 \\ 0 \\ \cdot \\ \cdot \\ \cdot \\ 1 \\ 0 \\ \cdot \\ \cdot \\ \cdot \\ 0 \end{bmatrix}$$

The solution of (25) may be effected by inverting the matrix $[A_{qp}]$ and may be written formally as

$$\mathbf{v}' = [A_{qp}]^{-1}\mathbf{f} \quad \text{or} \quad v_q' = \sum_{p=1}^{N} A_{qp}{}^{-1} f_p = 2y_{j'} A_{qj'}^{-1} \tag{26}$$

where A_{qp}^{-1} are the elements of $[A_{qp}]^{-1}$. The solutions v_p' are the modal voltages of the various modes which exist in the waveguide. The aperture electric field is obtained approximately as a linear combination of these modes, as expressed in (23). The reflection coefficient of the incident mode is given by

$$v_{j'}' = 1 + R_{j'} \quad \text{or} \quad R_{j'} = 2y_{j'} A_{j'j'}^{-1} - 1 \tag{27}$$

The transmission coefficients into other waveguide modes, after proper normalization,* are given by

$$t_n = \sqrt{y_n / y_{j'}}\, v_n' \tag{27a}$$

To determine the transmission coefficient into the free space region, we make use of the continuity of the tangential electric field across the aperture A and write

$$\sum_{q=1}^{N} v_q' \mathbf{\Phi}_q \approx \sum_{m=1}^{M} V_m' \mathbf{\Psi}_m \tag{28}$$

An approximate sign is used in (28) because only a finite number of modes are used to describe the field. Making use of the orthonormality property of the modal functions, we obtain

$$V_m' \approx \int_A \left\{ \sum_{q=1}^{N} v_q' \mathbf{\Phi}_q(\mathbf{r}) \right\} \cdot \mathbf{\Psi}_m^*(\mathbf{r}')\, d\mathbf{r}' = \sum_{q=1}^{N} C_{qm}^* v_q' \tag{29}$$

* The normalization is done in such a way that the resulting scattering matrix is unitary for lossless junctions.

Thus, the normalized transmission coefficient into the mth Floquet mode is

$$T_m = \sqrt{Y_m/y_{j'}}\, V'_m \tag{30}$$

Since only the propagating modes can carry energy away from the array interface to form radiation beams, these are the modes in which we are interested to characterize the properties of the array. For the calculation of the modal coefficients of such modes, (29) is quite adequate when N is chosen sufficiently large.

The integral equation for a receiving array with an incident wave corresponding to the $\mathbf{\Psi}_{m'}(\mathbf{r})$ mode is given by

$$2Y_{m'}\mathbf{\Psi}_{m'}(\mathbf{r}) = \int_A \left\{ \sum_{j=1}^{\infty} y_j \mathbf{\Phi}_j(\mathbf{r})\mathbf{\Phi}_j(\mathbf{r}') + \sum_{m=1}^{\infty} Y_m \mathbf{\Psi}_m(\mathbf{r})\mathbf{\Psi}_m^*(\mathbf{r}') \right\} \cdot \mathbf{E}_t(\mathbf{r}')\, d\mathbf{r}' \tag{31}$$

By applying a procedure similar to that described above, (31) is converted into a linear system of algebraic equations as

$$\sum_{p=1}^{N} A_{qp} v'_p = 2Y_{m'} C_{qm'} \tag{32}$$

or in matrix form as

$$[A_{qp}]\mathbf{v}' = \mathbf{f}' \tag{32a}$$

where

$$\mathbf{f}' = 2Y_{m'} \begin{bmatrix} C_{1m'} \\ C_{2m'} \\ \cdot \\ \cdot \\ \cdot \\ C_{Nm'} \end{bmatrix}$$

Notice that the change of excitation results in a modification of the free term only. Since the form of the kernel is determined entirely by the geometry of the problem, it is not affected by the change of the incident wave. The solution of (32a) again may be written formally as

$$\mathbf{v}' = [A_{qp}]^{-1}\mathbf{f}' \tag{33}$$

or in component form as

$$v'_q = \sum_{p=1}^{N} A_{qp}{}^{-1} f'_p = 2Y_{m'} \sum_{p=1}^{N} A_{qp}{}^{-1} C_{pm'} \tag{33a}$$

The reflection coefficient of the mode $\mathbf{\Psi}_{m'}(\mathbf{r})$ may be obtained by first calculating the modal voltage pertinent to this Floquet mode according to (29). Thus,

$$V'_{m'} = 1 + R_{m'} \tag{34}$$

Transmission coefficients may be calculated by using formulas similar to (27a) and (30).

We next turn our attention to the integral equation which has the tangential magnetic field as the unknown function. The derivation of such equations was discussed in Chapter 2. We first consider the situation in which the array is excited for transmission. The integral equation is given by

$$2z_{j'}\mathbf{\Phi}_{j'}(\mathbf{r}) = \int_C \left\{ \sum_{j=1}^{\infty} z_j \mathbf{\Phi}_j(\mathbf{r})\mathbf{\Phi}_j(\mathbf{r}') + \sum_{m=1}^{\infty} Z_m \mathbf{\Psi}_m(\mathbf{r})\mathbf{\Psi}_m^*(\mathbf{r}') \right\} \cdot \mathbf{H}_t(\mathbf{r}')\, d\mathbf{r}' \qquad \text{for } \mathbf{r} \in C \quad (35)$$

This equation differs from (22) in several respects. The modal impedances, instead of the modal admittances, appear in the kernel. As a result, this kernel converges differently from the one in (22). Also, the limit of integration extends over the entire unit cell in contrast to just over the aperture, as in the case of (22).

In order to reduce (35) to a matrix equation suitable for computation, we set

$$\mathbf{H}_t(\mathbf{r}) \approx \sum_{p=1}^{M} I'_p \mathbf{\Psi}_p(\mathbf{r}) \qquad (36)$$

We use $\{\mathbf{\Psi}\}$ instead of $\{\mathbf{\Phi}\}$ as a basis because the latter are defined only in A while $\mathbf{H}_t(\mathbf{r})$ is nonvanishing in $C \supseteq A$. Thus representation by $\{\mathbf{\Phi}\}$ would be inappropriate unless A and C coincide.

When (36) is substituted into (35) and then moments are taken with $\mathbf{\Psi}_q^*(\mathbf{r})$, $q = 1, \ldots, M$, the following system of equations is obtained:[8]

$$\sum_{p=1}^{M} B_{qp} I'_p = g_q, \quad p = 1, 2, \ldots, M \qquad (37)$$

where

$$\begin{aligned} B_{qp} &= \sum_{j=1}^{\infty} z_j C_{jq}^* C_{jp} + Z_q \delta_{qp} \\ g_q &= 2z_{j'} C_{j'q}^* \end{aligned} \qquad (37a)$$

with C_{jq} as given in (24b). Equation (37) is a set of M equations for the M unknowns, I'_p. The matrix elements are now composed of the modal impedances and the same intermodal coupling coefficients as we found earlier. This is to be contrasted with the use of the aperture electric field as the unknown, in which the modal admittances are involved in the composition of the matrix elements. Another point of interest is that the exterior modal admittances are present only in the diagonal elements, whereas the interior modal admittances appear in all the matrix elements.

Arranging (37) in matrix form and then inverting the matrix, we obtain the modal current of the tangential magnetic field in terms of the exterior

Floquet modes. The modal coefficients of the waveguide modes may be calculated by using the continuity condition similar to (28). The result is

$$i_q \approx \sum_{p=1}^{M} C_{qp} I'_p \tag{38}$$

Since all the necessary modal coefficients have been determined, it is straightforward to calculate the reflection and transmission properties of the array.

The matrix elements of either (24a) or (37a) contain *infinite* sums, which are made up from the product between modal admittances or impedances and the intermodal coupling coefficients. It can be shown, at least for rectangular and circular waveguide arrays, for which explicit expressions of C_{jq} are available, that these are convergent series. However, summing the series to convergence with five or more places of accuracy usually requires the inclusion of many terms. As we shall see presently, this is equivalent to using more modes from one region than the other. On the other hand, in ordinary arrays, the aperture size is generally comparable to the size of the basic unit cell. It thus seems reasonable to expect that to properly describe the tangential field at the array interface about the same number of modes will be required, irrespective of the region from which the modes are chosen. This implies that the sums appearing in the matrix elements may be truncated at a number that is comparable to the order of the matrix. For example, in (24a), the summation may be terminated at $m = M$ with, say, $M \approx N$. Similarly, in (37a), we may sum up to $j = N$ with $N \approx M$. Obviously, in doing so, another source of approximation is introduced into the solution. Under the conditions stated above, however, this approximation should not severely impair the accuracy of the solution.

When the waveguide and the basic unit cell occupy very different areas, the number of modes to use from each region may become strongly dependent on the geometric factors. It is interesting to note that, when a finite number of modes, say N waveguide modes and M Floquet modes, are used, solutions based on the electric field equation and on the magnetic field equation become identical. Specifically, solving system (24) with the matrix elements (24a) summed up to M, and solving (37) with the matrix elements (37a) summed up to N, will yield exactly the same answer.

4.2 Mode Matching Technique

The systems of equations (24) and (37) were obtained in two steps. First, the array boundary value problem was formulated rigorously in terms of integral equations with *either* the tangential electric or the tangential magnetic field as the unknown function. The method of moments was then applied, using appropriate bases to reduce the integral equations to systems of algebraic equations. The same sets of equations can be derived also by a process called the mode matching technique,[10,11] wherein modal expansions for the scattered fields are made for each region, as one would do in deriving the integral equation. Application of the continuity conditions at the interface between the two regions gives two separate relations between the expansion coefficients, one from the continuity of the electric field and the other from the continuity of the magnetic field. These relations may be put into the form of a system of equations in terms of infinitely many unknowns. Usually it is assumed that only a finite number of modes are excited at the aperture A. This is a reasonable approximation if sufficiently many modes are included. The result then is a finite set of equations.

Consider an infinite array of which $\mathbf{\Phi}_j(\mathbf{r})$ are the vector modal functions for the waveguide region, and $\mathbf{\Psi}_m(\mathbf{r})$ are the Floquet modes. When the array is excited by a mode $\mathbf{\Phi}_{j'}(\mathbf{r})$, the tangential electric and magnetic fields at the aperture may be written as

$$\begin{aligned} \mathbf{E}_t(\mathbf{r}) &\approx (1 + R_{j'})\mathbf{\Phi}_{j'}(\mathbf{r}) + \sum_{j=1}^{N}{}^{(j')} v'_j \mathbf{\Phi}_j(\mathbf{r}) \approx \sum_{m=1}^{M} V'_m \mathbf{\Psi}_m(\mathbf{r}) \\ -\hat{z} \times \mathbf{H}_t(\mathbf{r}) &\approx y_{j'}(1 - R_{j'})\mathbf{\Phi}_{j'}(\mathbf{r}) - \sum_{j=1}^{N}{}^{(j')} y_j v'_j \mathbf{\Phi}_j(\mathbf{r}) \approx \sum_{m=1}^{M} Y_m V'_m \mathbf{\Psi}_m(\mathbf{r}) \end{aligned} \tag{39}$$

Notice that approximate signs are used in these expressions, because only finite numbers of modes are included. The magnetic field equation is valid over only the waveguide aperture, whereas the electric field equation is valid over an entire unit cell with $\mathbf{E}_t = 0$ over C–A. Multiplying the first equation by $\mathbf{\Psi}_p^*(\mathbf{r})$ and integrating over C, we find by virtue of the mode orthonormality

$$V'_p \approx (1 + R_{j'})C^*_{j'p} + \sum_{j=1}^{N} C^*_{jp} v'_j, \quad p = 1, 2, \ldots, M \tag{40a}$$

where C_{jp} are the intermodal coupling coefficients given in (24b). Similarly, by multiplying the second equation of (39) by $\mathbf{\Phi}_n(\mathbf{r})$ and integrating over A, we obtain

$$y_{j'}(1 - R_{j'})\delta_{j'n} - (1 - \delta_{j'n}) y_n v'_n \approx \sum_{m=1}^{M} Y_m V'_m C_{nm}, \quad n = 1, 2, \ldots, N \tag{40b}$$

By treating v'_n, $n = 1, \ldots, N$, with $v'_{j'} = R_{j'}$, and V'_m, $m = 1, \ldots, M$ as $N + M$ independent unknown coefficients, we may write (40a) and (40b) in a compact matrix form as

$$[\Lambda]\boldsymbol{\sigma} = \mathbf{x} \tag{41}$$

where

$$
[\Lambda] = \begin{array}{r}
\\
M\left\{\vphantom{\begin{array}{c}C^*_{11}\\C^*_{12}\\\vdots\\C^*_{1M}\end{array}}\right. \\
N\left\{\vphantom{\begin{array}{c}y_1\\0\\\vdots\\0\end{array}}\right.
\end{array}
\left[\begin{array}{cccc|cccc}
\multicolumn{4}{c|}{\overbrace{\hspace{10em}}^{N}} & \multicolumn{4}{c}{\overbrace{\hspace{10em}}^{M}} \\
C^*_{11} & C^*_{21} & \cdots & C^*_{N1} & -1 & 0 & \cdots & 0 \\
C^*_{12} & C^*_{22} & \cdots & C^*_{N2} & 0 & -1 & & 0 \\
\vdots & & & & \vdots & & & \vdots \\
C^*_{1M} & C^*_{2M} & \cdots & C^*_{NM} & 0 & & & -1 \\
\hdashline
y_1 & 0 & \cdots & 0 & Y_1C_{11} & Y_2C_{12} & \cdots & Y_MC_{1M} \\
0 & y_2 & & & Y_1C_{21} & Y_2C_{22} & \cdots & \\
\vdots & & & & \vdots & & & \\
0 & & & y_N & Y_1C_{N1} & & & Y_MC_{NM}
\end{array}\right]
\tag{41a}
$$

$$
\boldsymbol{\sigma} = \begin{bmatrix} v'_1 \\ v'_2 \\ \vdots \\ v'_{j'-1} \\ R_{j'} \\ \vdots \\ v'_N \\ V'_1 \\ V'_2 \\ \vdots \\ V_M \end{bmatrix}, \quad \text{and} \quad \mathbf{x} = \begin{bmatrix} -C^*_{j'1} \\ -C^*_{j'2} \\ \vdots \\ -C^*_{j'M} \\ 0 \\ 0 \\ \vdots \\ y_j \\ 0 \\ \vdots \\ 0 \end{bmatrix}
\tag{41b}
$$

This is a set of $M + N$ equations from which the $M + N$ unknowns may be uniquely determined. An interesting feature of the matrix $[\Lambda]$ is that it is partitioned into four submatrices, of which the two off-diagonal submatrices are diagonal. This partitioning suggests that (40a) and (40b) may be rearranged as

$$\mathbf{V}' = [R]\mathbf{v}' + \mathbf{b} \tag{42a}$$

$$\mathbf{d} - [y]\mathbf{v}' = [S][Y]\mathbf{V}' \tag{42b}$$

This is a coupled set of equations in which the unknown vectors are

$$\mathbf{V}' = \begin{bmatrix} V_1' \\ V_2' \\ \cdot \\ \cdot \\ \cdot \\ V_M' \end{bmatrix} \quad \text{and} \quad \mathbf{v}' = \begin{bmatrix} v_1' \\ v_2' \\ \cdot \\ \cdot \\ \cdot \\ v_{j'-1}' \\ R_{j'} \\ \cdot \\ \cdot \\ \cdot \\ v_N' \end{bmatrix} \tag{43a}$$

Also, $[R]$ and $[S]$ are $M \times N$ and $N \times M$ matrices, respectively, and are given by

$$[R] = \begin{bmatrix} C_{11}^* & C_{21}^* & \cdots & C_{N1}^* \\ C_{12}^* & C_{22}^* & \cdots & \\ \cdot & & & \\ \cdot & & & \\ \cdot & & & \\ C_{1M}^* & C_{2M}^* & \cdots & C_{NM}^* \end{bmatrix} \tag{43b}$$

and

$$[S] = \begin{bmatrix} C_{11} & C_{12} & \cdots & C_{1M} \\ C_{21} & C_{22} & \cdots & \\ \cdot & & & \\ \cdot & & & \\ \cdot & & & \\ C_{N1} & C_{N2} & \cdots & C_{NM} \end{bmatrix} \tag{43c}$$

The admittance matrices have only nonvanishing diagonal elements:

$$[y] = \begin{bmatrix} y_1 & 0 & \dots & 0 \\ 0 & y_2 & & \\ \cdot & & & \\ \cdot & & & \\ \cdot & & & \\ 0 & & & y_N \end{bmatrix}$$

$$[Y] = \begin{bmatrix} Y_1 & 0 & \dots & 0 \\ 0 & Y_2 & & \\ \cdot & & & \\ \cdot & & & \\ \cdot & & & \\ 0 & 0 & \dots & Y_M \end{bmatrix}$$

and

$$\mathbf{b} = \begin{bmatrix} C^*_{j'1} \\ C^*_{j'2} \\ \cdot \\ \cdot \\ \cdot \\ C_{j'M} \end{bmatrix}, \quad \text{and} \quad \mathbf{d} = \begin{bmatrix} 0 \\ \cdot \\ \cdot \\ \cdot \\ 0 \\ y_{j'} \\ 0 \\ \cdot \\ \cdot \\ \cdot \\ 0 \end{bmatrix}$$

By substituting $\mathbf{V}'$ from (42a) into (42b), we eliminate $\mathbf{V}'$ to obtain an equation for $\mathbf{v}'$ alone. After slight rearrangement, the result is

$$([y] + [S][Y][R])\mathbf{v}' = \mathbf{d} - [Y][S]\mathbf{b} \tag{44}$$

The matrix $([y] + [S][Y][R])$ is of rank N, and there are N unknown components in $\mathbf{v}'$. In component form, we find that the qth equation is given by

$$\sum_{p=1}^{N} \left\{ y_q \delta_{qp} + \left(\sum_{m=1}^{M} Y_m C_{qm} C^*_{pm} \right) \right\} v'_p = y_{j'} \delta_{qj'} - \sum_{m=1}^{M} Y_m C_{qm} C^*_{j'm} \tag{44a}$$

By redefining $v'_{j'} = 1 + R_{j'}$, (44a) may be rewritten as

$$\sum_{p=1}^{N} \left\{ y_q \delta_{qp} + \left(\sum_{m=1}^{M} Y_m C_{qm} C^*_{pm} \right) \right\} v'_p = 2 y_{j'} \delta_{qj'} \tag{44b}$$

This equation is identical to (24), which was obtained from the integral equation for the aperture electric field by the method of moments, wherein the waveguide modal functions are chosen as the bases. Once the solutions for $\mathbf{v}'$ are obtained, $\mathbf{V}'$ may be calculated by using (42a), which can be rearranged to show that it is the same as (29).

Alternatively, we may also eliminate $\mathbf{v}'$ from (42a) and (42b) to obtain an equation for $\mathbf{V}'$ alone as

$$([Z] + [R][z][S])\mathbf{I}' = [R][z]\mathbf{d} + \mathbf{b} \tag{45a}$$

where the following relations have been used:

$$[Z] = [Y]^{-1} = \begin{bmatrix} 1/Y_1 & 0 & \cdots & 0 \\ 0 & 1/Y_2 & & \\ \cdot & & & \\ \cdot & & & \\ \cdot & & & \\ 0 & 0 & \cdots & 1/Y_M \end{bmatrix}$$

$$[z] = [y]^{-1}, \quad \text{and} \quad \mathbf{I}' = [Y]\mathbf{V}'$$

Equation (45a) is a set of M equations for the M unknown (I')'s. The qth equation is given by

$$\sum_{p=1}^{M}\left\{Z_q\delta_{qp} + \sum_{n=1}^{N} z_n C_{nq}^* C_{np}\right\} I_q' = 2C_{j'q}^* \tag{45b}$$

Comparing (45b) with (37), we observe that these two equations are identical except for the factor $z_{j'}$ on the right-hand side. This difference is due to the assumption of incident waves. In obtaining (37) the incident wave has unit modal current for the mode $\mathbf{\Phi}_{j'}$, while for (45b) the incident wave has unit modal voltage for the same mode. The ratio is exactly the modal impedance, $z_{j'}$. Since the system is linear, this ratio is merely a proportionality factor between solutions from the two systems.

The above derivations have demonstrated that the moments method, as applied to the electric field or the magnetic field equation, is identical to the mode matching technique, provided that in both procedures the same number of modes are used. The question of how many modes from each region should be included in the computation in order to obtain an accurate solution is still not fully investigated. Further research certainly is needed to clarify this point.

The advantages of the integral equation approach are that it permits a clear perspective of how approximations are introduced, and that it offers the flexibility to obtain alternative matrix equations by choosing different expansion and testing functions. The latter point is particularly important because different functions may be used to meet different specific needs.

4.3 Partitioning Problem

For certain applications, it may be desirable to use waveguides of different sizes as the radiating elements. These waveguides may be arranged so that a basic unit cell encompasses a composite of two or more waveguides. An example is shown in Fig. 3.1. A possible application of such an arrangement is to use small waveguides which may either propagate a wave or be below cut-off to serve as a choke, thus providing reactive loading for the array aperture. (Another important application can be found in Chapter 7).

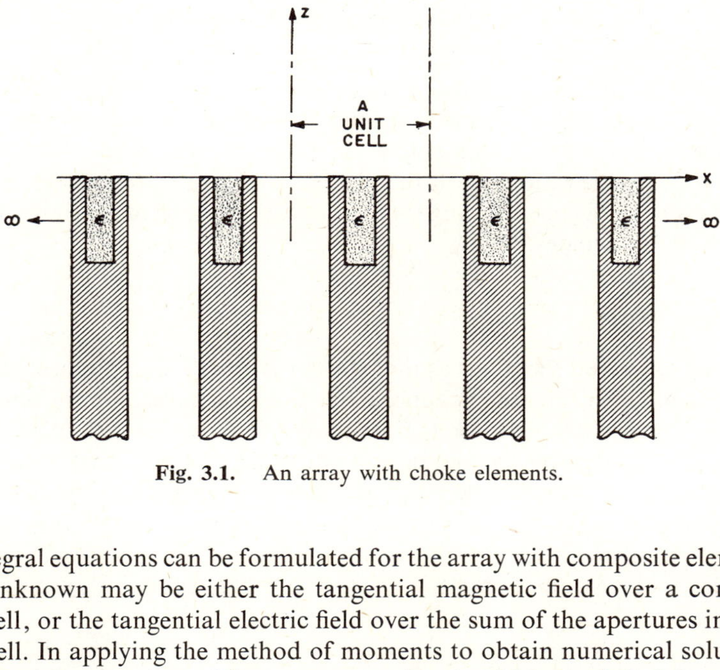

Fig. 3.1. An array with choke elements.

Integral equations can be formulated for the array with composite elements. The unknown may be either the tangential magnetic field over a complete unit cell, or the tangential electric field over the sum of the apertures inside a unit cell. In applying the method of moments to obtain numerical solutions, however, the question of how many modes should be included from each waveguide in a unit cell will naturally arise. This may be termed the partitioning problem. Relatively few data are available in this regard. The information obtained from the scattering by the bifurcation of a rectangular waveguide, however, may serve to illustrate the importance of this problem. The geometry is shown in Fig. 3.2.

When the bifurcation is an infinitesimally thin metallic septum, the problem is amenable to an exact analytical solution in which the effects of all the modes excited by the discontinuity can be accounted for. If, instead, only a finite number of modes from each region are used for an approximate solution

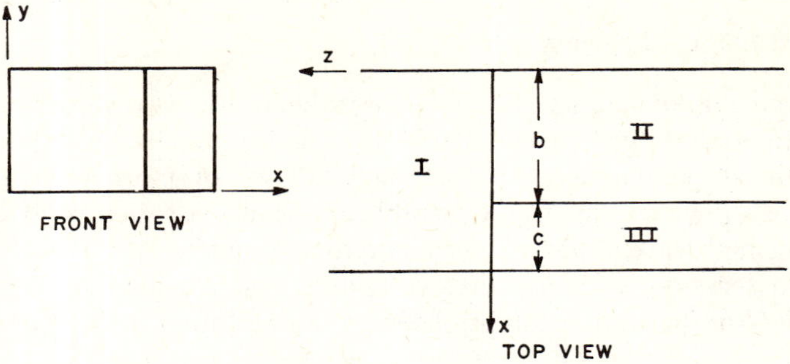

Fig. 3.2. Front and top views of a bifurcated waveguide

and then the number is allowed to grow to infinity, Mittra[12] has shown that the solution depends entirely on the manner in which the limit is taken. Specifically, it was shown that only when the numbers of modes from regions II and III have the ratio b/c will the process converge to the correct answer. There is other evidence showing that the partition problem may also arise in iris-loaded waveguides.[13]

4.4 Pulse Basis

The use of modal functions as the bases in the method of moments implies that the unknown function is approximated by a combination of functions, each of which is defined over the entire aperture or unit cell. We consider next an alternative approximation using different types of functions.

Consider, for example, the integral equation with the electric field as an unknown, as given in (22). The equation is reproduced here for convenience:

$$2y_{j'}\mathbf{\Phi}_{j'}(\mathbf{r}) = \int_A \left\{ \sum_{j=1}^{\infty} y_j \mathbf{\Phi}_j(\mathbf{r})\mathbf{\Phi}_j(\mathbf{r}') + \sum_{m=1}^{\infty} Y_m \mathbf{\Psi}_m(\mathbf{r})\mathbf{\Psi}_m^*(\mathbf{r}') \right\} \cdot \mathbf{E}_t(\mathbf{r}')\, d\mathbf{r}', \quad \mathbf{r} \in A \tag{22}$$

Suppose that the waveguide is divided into many subsections or zones.[14,15] There are many different ways in which this can be done. A straightforward method of using lines parallel to the rectangular coordinate axes is illustrated in Fig. 3.3. If the maximum linear dimension of the subsections is small in terms of the wavelength, it is reasonable to expect that the field within any subsection will remain more or less constant. Thus, only a constant complex vector is needed to describe the field in each subsection, although this complex vector is still an unknown. A possible exception may occur near the periphery of the aperture, where a sharp edge may render the field singular. Rapid variation within a short distance is possible in such cases.

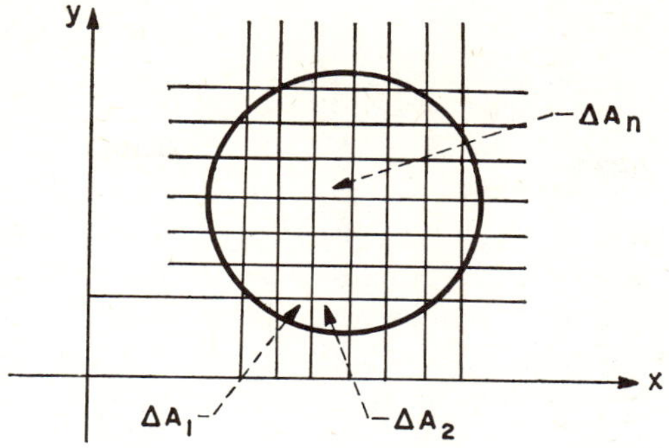

Fig. 3.3. Subdivision of a circular waveguide aperture

Methods of treating such situations will be discussed shortly.

Let us define a set of functions $\{f_n(x, y)\}$:

$$f_n(x, y) = \begin{cases} 1, & (x, y) \in \Delta A_n \\ 0, & \text{elsewhere} \end{cases} \tag{46}$$

Let $\mathbf{e}_n$ be the complex vector denoting the field in ΔA_n. The unknown tangential electric field may then be expressed as

$$\mathbf{E}_t(\mathbf{r}) \approx \sum_{n=1}^{N} \mathbf{e}_n f_n \tag{47}$$

Substituting (47) into (22), and recognizing that $\mathbf{e}_n$ are constant vectors, we find

$$2y_{j'}\boldsymbol{\Phi}_{j'}(\mathbf{r}) \approx \sum_{n=1}^{N} \mathbf{e}_n \int_{\Delta A_n} \left\{ \sum_{j=1}^{\infty} y_j \boldsymbol{\Phi}_j(\mathbf{r})\boldsymbol{\Phi}_j(\mathbf{r}') + \sum_{m=1}^{\infty} Y_m \boldsymbol{\Psi}_m(\mathbf{r})\boldsymbol{\Psi}_m^*(\mathbf{r}') \right\} d\mathbf{r}' \tag{48}$$

where the area of integration has been reduced from A to ΔA_n on account of the definition in (46). The equation is now only approximately satisfied because of (47). The approximation should improve, however, as the area is divided into finer subsections. The next step is to take moments of (48), using a suitable set of functions. We may use, for example, the same set of functions, (46). Thus, multiplying (48) by $f_m(x, y)$ and integrating over A, we obtain the system of equations as

$$\sum_{n=1}^{N} \overline{\overline{M}}_{mn} \cdot \mathbf{e}_n = \mathbf{d}_m, \quad m = 1, 2, \ldots, N \tag{49}$$

where

$$\overline{\overline{M}}_{mn} = \int_{\Delta A_m} d\mathbf{r} \int_{\Delta A_n} d\mathbf{r}' \left\{ \sum_{j=1}^{\infty} y_j \boldsymbol{\Phi}_j(\mathbf{r})\boldsymbol{\Phi}_j(\mathbf{r}') + \sum_{p=1}^{\infty} Y_p \boldsymbol{\Psi}_p(\mathbf{r})\boldsymbol{\Psi}_p^*(\mathbf{r}') \right\} \tag{49a}$$

and

$$\mathbf{d}_m = 2y_{j'} \int_{\Delta A_m} \mathbf{\Phi}_{j'}(\mathbf{r})\, d\mathbf{r} \tag{49b}$$

The double bar over M_{mn} indicates that these are dyadic quantities. Because of the vector nature of the unknowns (there are two components for each $\mathbf{e}_n$), (49) actually is a set of $2N$ equations for $2N$ unknowns. This can be seen by writing the equations in component forms. The system of equations can be solved in the usual fashion, such as by matrix inversion.

Another way of reducing the functional equation, (48), to a system of linear equations is to force it to be an equality at N points with one point lying inside each subsection. Thus, letting $\mathbf{r}_m \in \Delta A_m$, we require that

$$2y_{j'}\mathbf{\Phi}_{j'}(\mathbf{r}_m) = \sum_{n=1}^{N} \mathbf{e}_n \int_{\Delta A_n} \left\{ \sum_{j=1}^{\infty} y_j \mathbf{\Phi}_j(\mathbf{r}_m)\mathbf{\Phi}_j(\mathbf{r}') + \sum_{p=1}^{\infty} Y_p \mathbf{\Psi}_p(\mathbf{r}_m)\mathbf{\Psi}_p^*(\mathbf{r}') \right\} d\mathbf{r}',$$

$$m = 1, 2, \ldots, N \tag{50}$$

This step is equivalent to choosing a set of delta functions, $\delta(\mathbf{r} - \mathbf{r}_m)$, as testing functions. In the latter approach, because we match both sides of (48) at a selected number of points, the process is often referred to as a point matching technique. It is not difficult to show that in so doing we are actually enforcing the boundary condition at a finite number of points across the aperture. The matrix elements obtained by the point matching method have one integration less than those derived by using (46) as testing functions. In exterior type scattering problems, this difference can be very significant, because kernels associated with such problems usually do not permit integration of the kernels to be performed in closed form. As a result, numerical techniques have to be used in calculating the matrix elements.

The evaluation of the matrix elements involves the integration of the modal functions, $\{\mathbf{\Phi}_i\}$ and $\{\mathbf{\Psi}_m\}$, over the subsections. This step is quite easy in rectangular waveguide arrays because only trigonometric functions are involved. It can be rather tedious and time consuming, however, in other cases, as in circular waveguide arrays, where numerical integration of cylinder functions is necessary. Another important consideration in this approach is the convergence of the series which appear in each matrix element. The behavior of modal admittances or impedances is such that they grow with modal indices for one type of mode and decay for the other type. When the problem involves only one type of mode, it is possible to choose between the electric field equation and the magnetic field equation to ensure a rapid convergence of the series. In cases where both types of modes are present, the use of either formulation will entail a similar convergence problem. Specific examples will be discussed in Chapter 5.

The approximation introduced in (48) by choosing functions in (46) as the basis is essentially equivalent to replacing the integral by its Riemann sum. There are many other methods of approximating an integral, some of the most frequently used being the trapezoidal rule, Simpson's rule, and Gauss's quadrature. One of the rules may yield more accurate approximation than others.[16] In general, these rules may be interpreted as applying certain weights to the different subintervals. Thus let $\{D_n : n = 1, 2, \ldots\}$ be the weighting coefficients. (These are known numbers for the various integration rules.) In place of (48) we then have

$$2y_{j'}\mathbf{\Phi}_{j'}(\mathbf{r}) \approx \sum_{n=1}^{N} D_n \mathbf{e}_n \int_{\Delta A_n} \left\{ \sum_{j=1}^{\infty} y_j \mathbf{\Phi}_j(\mathbf{r})\mathbf{\Phi}_j(\mathbf{r}') + \sum_{m=1}^{\infty} Y_m \mathbf{\Psi}_m(\mathbf{r})\mathbf{\Psi}_m^*(\mathbf{r}') \right\} d\mathbf{r}' \quad (51)$$

The reduction of (51) to a system of linear equations follows the procedure discussed above.

4.5 Utilization of a Priori Knowledge

It has already been pointed out that in applying the method of moments it is desirable to choose the bases so that any a priori knowledge about the unknown function is incorporated to as great a degree as possible. The type of information about the tangential aperture fields which may be known before solving the problem is the edge behavior of the fields or their asymptotic behavior with respect to a particular parameter.

The boundary condition which requires $\mathbf{E}_t$ tangent to the aperture to be zero is automatically satisfied when the waveguide modal functions are used as a basis. In applications in which sharp edges are present, the normal (to the edge) electric and magnetic fields have to behave in an appropriate singular manner near the edges. The Fourier series of a singular function is usually slowly convergent. In other words, using only a finite number of terms in a series gives a poor representation of a function which possesses singularities. In order to improve the rate of convergence, we may write the unknown field as the sum of a singular part and a regular part.[17,18] The singular part has the functional dependence describing the known singular behavior near the edges with one or more unknown coefficients. The regular part is a well behaved function, for which relatively few terms are needed in a series expansion for an accurate representation. For a given error tolerance, this method may appreciably reduce the number of terms which would otherwise be necessary.

To illustrate the idea, we consider the array depicted in Fig. 3.4. Suppose that the waveguides are excited in TEM modes and that the array is scanned in the E plane. The problem is formulated in such a way that the aperture electric field, E_x, appears as the unknown; it has a singularity of the form

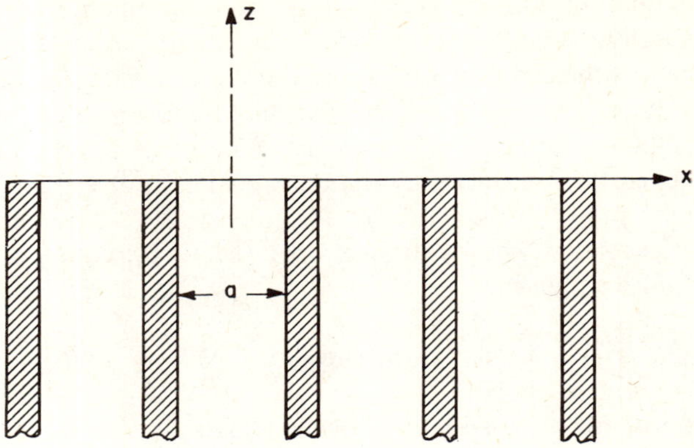

Fig. 3.4. An array of waveguides with sharp edges

$(x \mp a/2)^{-1/3}$ as x approaches $\pm a/2$. To extract the singular part, the aperture field may be written as

$$E_x(x) = A_1\left(x - \frac{a}{2}\right)^{-1/3} + A_2\left(x + \frac{a}{2}\right)^{-1/3} + \mathscr{E}_x(x) \tag{52}$$

where A_1 and A_2 are unknown coefficients and $\mathscr{E}_x(x)$ is a well behaved function. The latter function may in turn be approximated as a finite sum of waveguide modes. Thus, letting $\Phi_n(x)$ be the orthonormal modes of the waveguide, we may write

$$E_x(x) \approx A_1\left(x - \frac{a}{2}\right)^{-1/3} + A_2\left(x + \frac{a}{2}\right)^{-1/3} + \sum_{q=1}^{N} a_q\Phi_q(x) \tag{52a}$$

This expression is substituted into the appropriate integral equation, and then the moments of the resulting approximate equation are taken to generate a system of equations. In the process, integrations of the first two terms of (52a) against the kernel of the integral equation often have to be performed numerically. Since the third term is an expansion of a well behaved function, it is possible that fewer terms are needed for an adequate representation. As a result, we may be able to obtain results with comparable accuracy and yet with fewer terms than are yielded by the usual procedure, where the singular part of the field is not extracted for special consideration. On the other hand, because of the numerical integration necessary in using (52a), the computation of the matrix elements will take more time than before. This factor will have to be weighted against the reduction of the matrix size as a trade-off.

5. VERIFICATION OF THE VALIDITY OF SOLUTIONS

Numerical solutions of physical problems often involve approximations in the mathematical procedure. The translation of mathematical formula into computational algorithm can sometimes be very involved. Moreover, computers are strictly *passive* devices which can only faithfully follow their users' instructions. Therefore, it is necessary to implement certain check procedures in the computation to ascertain the *correctness* and *accuracy* of the solutions. Many procedures have been developed for this purpose:[8,19]

1. Reciprocity relation.[6]
2. Conservation of energy.[21]
3. Convergence test.
4. Use of different bases.[19]
5. Comparison with results obtained by different methods.[14]
6. Boundary conditions.[11,17]
7. Experimental verification.[14,19,20]
8. Other checks.

We shall discuss in the following sections some of the advantages and disadvantages of each method. It should be noted that very often a computer program is designed to solve a problem over a wide range of parameters. In fact, this is one of the major advantages of computer methods. The validity check, on the other hand, is performed within some limited selective set of values. Thus no checks can be considered as providing absolutely foolproof assurance, and it is often desirable to use more than one method of check in order to enhance the degree of confidence in the results.

5.1 Reciprocity Relation

The solution of the matrix equation derived from the integral equation by the method of moments was shown to satisfy the reciprocity relation exactly. This is true regardless of whether or not a good solution is obtained. Hence, reciprocity is *a necessary, but not a sufficient, condition* which the solution must satisfy. The fulfillment of this relation therefore should not be interpreted as a sign that accurate results are obtained. Nevertheless, it is a useful check in the early stages of computation as a gross indicator of the adequacy of the algorithm and of any excessive round-off errors. The implementation of this procedure usually can be carried out relatively easily and does not involve too much additional computation. Hence the incorporation of the procedure is desirable.

5.2 Conservation of Energy

Conservation of energy requires that the power incident upon a passive lossless scatterer must be equal to the total power carried away by the scattered waves. In the case of phased arrays, the incident power of the exciting mode is shared among the various scattered propagating modes, which exist both inside and outside the array elements. At first glance, it may seem reasonable to assume that, if numerically obtained solutions satisfy the conservation of energy law, this may be an indication that accurate solutions have been obtained. Unfortunately, since the computation of power involves only the magnitudes, not the phases, of the modal coefficients, it turns out that this requirement is not stringent enough; there are many ways whereby the incident power may be divided into scattered powers while the total power remains conserved. This means that the power carried by each individual mode may be incorrect and yet the sum of these powers may still add up to the incident power.

It is an inherent property of the Ritz-Galerkin method that the solutions are automatically adjusted in such a way that the conservation of energy law is always satisfied. A proof of this statement is given in Appendix I. As in the case of reciprocity, conservation of energy can serve only as a necessary check for any gross errors in the program. It cannot be used as a measure of the correctness and accuracy of the solution.

5.3 Convergence Test

Since, in converting the integral equation to a matrix equation, only a finite number of functions have been used to approximate the unknown function, the first step in testing the validity of the results is to determine the number of functions needed for a satisfactory approximation. This is usually done by systematically and gradually increasing the number of functions used in the expansion and observing how the solution converges. The solution of the integral equation in general gives the field distribution across the array aperture (or the current around the scatter). Although the field distribution is useful in providing insight into the problem, very often it is not the solution itself, but rather related physical quantities such as the input impedance and the scattering cross section, that are of interest for engineering purposes. These quantities may be expressed as a certain "average" (or scalar product) of the solution. Thus the convergence of these averages is of primary concern.

The manner in which the solutions or some averages of the solutions converge depends very much on the properties of the operator and the basis of representation. In problems in which only one type of mode is excited, that is, either TE or TM modes are involved, the convergence is usually monotonic. This is related to the fact that the modal impedances (or admittances) of the higher-order modes are monotonic functions of the modal indices. In the more general situation in which both TE and TM modes are present, the convergence is usually oscillatory (but with diminishing amplitudes). It should be remembered that, as the number of functions used in the approximation is increased, the size of the matrix gets larger. As a result, more core storage is needed, and a longer computer time is required to invert the matrix. The number of steps necessary for inverting an $n \times n$ matrix[4] is proportional to n^3. Thus, the larger the matrix to be inverted, the more prone it is to round-off errors.

As a general rule, it is desirable to select, for test purposes, a set of parameters as representative as possible. The number of functions used for approximation should be increased to such an extent that a further increase will cause only a very small change in the quantities of interest. The amount of change permissible is dictated by the degree of accuracy desired; it can be fractions of a per cent or a few per cent. In practical computations, a wide range of parameters frequently has to be covered. Computation time then may become an important consideration. It is also true that usually tolerances of a few per cent are allowed in engineering practices. Therefore, it is reasonable to use matrix sizes which will yield results compatible with engineering accuracy requirements. To ensure reasonably uniform accuracy, however, some spot check of the computation should be performed. Whenever possible, special attention should be given to the set of parameters for which a rapid variation of the computed results is observed.

5.4 Use of Different Bases

As mentioned above, the rate of convergence of the solution depends on the basis of representation. If the set of functions chosen to approximate the unknown function can incorporate as much as possible the independent properties of the latter, it may be expected that fewer functions will be needed to obtain a reasonably good solution. From the standpoint of ascertaining the correctness of the solution, if the problem can be solved concurrently by using two different bases, and the results agree, there is good reason to believe that both approaches are satisfactory. The more different the bases are, the higher is the probability that the solution will be acceptable if the results agree. The only drawback to this type of check is that it requires solving the problem twice, including the development of algorithms and computer program debugging. This can be a rather time consuming process. Hence the merits of this approach should be contrasted with those of other methods.

5.5 Comparison with Results Obtained by Different Methods

Sometimes for special sets of parameters the given problem can be related to some equivalent problem for which different methods may be applied to obtain known accurate solutions.[15,24,25] As an example, consider the case of a parallel-plate array scanned in the E plane. When the scan angle is zero, it is clear by the method of images, that the problem can be viewed as a step discontinuity between two parallel-plate waveguides. The latter problem has been solved by a number of different techniques,[28] and the results are available for comparison. Similarly, when the same array is scanned in the H plane, by choosing the steering phase $\psi = \pi$, the problem reduces to an H-plane step discontinuity between parallel-plate waveguides. Again, solutions by different methods are available[28] and can be used for comparison. Another situation which allows such a comparison is that in which a dielectric material is used to load the array. By setting the dielectric constant equal to 1, the results for no dielectric loading should be recovered.

Exterior type scattering problems can be treated in such a way that scatterers of arbitrary shapes may be attacked in one formulation. The shape of the scatterer is fed into the program as input data. It is well known that exact analytical solutions are available for a number of particular geometries such as cylinders, spheres, and spheroids. The exact solutions may serve as standards of comparison for numerically obtained solutions Hence, by choosing special values for some parameters, it is sometime possible to find a situation for which alternative methods may be used to obtain solutions. Under such circumstances, an independent check of the results for these special parameters can be made.

5.6 Boundary Conditions

In interior type problems including phased arrays and waveguide discontinuities, the integral equations can be formulated with either the tangential electric field or the tangential magnetic field as the unknown. The solution from either equation gives one of the unknown functions. The other unknown function can be calculated from the first solution by using Maxwell's equations. The results calculated for the two regions must satisfy the continuity conditions at the interface. The degree to which this boundary condition is satisfied may be used as a measure of the accuracy of the results.[11,17] It is generally difficult to find an absolute relationship between the degree of satisfaction of the boundary condition and the correctness of some engineering parameter such as the "reflection" coefficient in a scattering problem. The value of this test is therefore considerably diminished without this "absolute" measure.

5.7 Experimental Verification

A more direct method is to compare the numerical results with experiment. For scattering problems, scatterers of desired shapes may be constructed for testing. In the case of infinite phased arrays, waveguide simulators can be utilized. The number of scan angles which can be effectively simulated is quite limited, however. This approach thus provides a sparse spot check.

An alternative method is to construct a reasonably sized array and to measure the mutual coupling between the elements as well as the element patterns. Once the mutual couplings are obtained, computer simulation can be performed to evaluate the reflection coefficients versus the scan. This is based on the principle of linear superposition. The range of scan for computer simulation is unlimited. The computer simulation of scanning the array is an alternative to the use of an active array in which the beam steering mechanism is fully implemented. The latter approach can be quite costly and time consuming.

5.8 Other Checks

There are conditions under which a given array may be viewed as being arranged in a rectangular grid with certain element spacings and polarizations or, alternatively, in a triangular grid with different element spacings and polarization. Examples of these situations will be encountered in Chapters 5 and 7. When such conditions exist, solutions for the two alternative array geometries should yield the same results. This possibility may be exploited for checking the numerical algorithm.

The symmetries in the array geometry may also be utilized for such purposes. For example, in certain planes of scan, say the E or H plane, the array geometry may be such that a certain mode should not be excited. Verification of this fact in the numerical solution would give some indication of the adequacy of the program.

6. FUNCTION-THEORETIC CANONICAL SOLUTION

So far we have been discussing approximate techniques for numerically solving integral equations based on the method of moments, because phased array problems are generally too complex to be attacked by purely analytical means. However, there is an array which is quite unique in that, under idealized conditions, the problem is amenable to an exact solution by either the function-theoretic[23,24] or the Wiener–Hopf technique.[25,26] We shall briefly outline these methods below.

In the special array to be discussed, parallel-plate waveguides are the radiating elements. The geometrical arrangement of the array is similar to the one depicted in Fig. 3.4. Integral equations may be formulated for the array boundary value problem by the methods developed in Chapter 2. A derivation for this specific array is discussed in Chapter 4 also. The integral equations can be converted into matrix equations for solution in the usual manner except for the choice of bases. Ordinarily, when the electric field equation is considered, using the interior waveguide modes as both the expansion functions and the testing functions has been found effective. Similarly, certain advantages and conveniences can be gained by choosing the exterior Floquet modes for the conversion of the magnetic field equation. It is generally disadvantageous to use one type of mode for expansion and the other type for testing, because the matrix elements so obtained are more complex than those resulting from the use of only one type of mode. When the waveguide walls are assumed to be infinitesimally thin, however, we have the reverse situation. Because the waveguide aperture now occupies the same area as the basic unit cell, the choice of mixed bases can result in considerable simplification of the matrix elements. Furthermore, the functional form of the matrix elements is such that an exact analytical solution can be found for the entire infinite system of equations.

For illustrative purposes, let the array be excited by the lowest TM mode (also a TEM mode) for an E-plane scan. (The case of TE mode excitation may be analyzed in a similar manner.) The integral equation with the aperture electric field as the unknown then takes the following form:

$$2y_0'\Phi_0'(y) = \int_{-d/2}^{d/2} \left\{ \sum_{q=0}^{\infty} y_q'\Phi_q'(y)\Phi_q'(y') + \sum_{m=-\infty}^{\infty} Y_m'\Psi_m'(y)\Psi_m'^*(y') \right\} E_y(y')\, dy' \quad (53)$$

The unknown electric field may be expanded as

$$E_y(y) = \sum_{n=0}^{\infty} v_n'\Phi_n'(y) \quad (54)$$

Substituting (54) into (53) and then weighting the resulting equation with $\{\Psi_q^*(y),\, q = 0, \pm 1, \pm 2, \ldots\}$, we obtain after some algebraic manipulation (see Chapter 4)

$$\sqrt{2/d}\, \frac{\sin(\psi/2)}{\gamma_0'(\Gamma_q' + \gamma_0')} = \sum_{n=0}^{\infty} \frac{1}{\Gamma_q' - \gamma_n'} B_n, \quad q = 0, \pm 1, \pm 2, \ldots \quad (55)$$

where the unknowns $\{B_n\}$ are related to $\{v_n'\}$ by

$$B_n = (-1)^{n/2}\sqrt{(2 - \delta_{0n})/d}\, \frac{v_n''}{\gamma_n'} \begin{cases} \sin\dfrac{\psi}{2} & \text{for } n \text{ even} \\ \cos\dfrac{\psi}{2} & \text{for } n \text{ odd} \end{cases} \quad (55a)$$

with

$$v'_0 = 1 + v''_0 \qquad \text{and} \qquad v'_n = v''_n, \quad n > 0$$

The set of equations in (55) has a very special form. The coefficient of B_n, $1/(\Gamma'_q - \gamma'_n)$, is particularly simple and depends only on the difference $(\Gamma'_q - \gamma'_n)$. This type of equation may be solved by the function-theoretic technique, also known as residue calculus. In this method, an appropriate integral of a special analytic function is introduced, and then the residue contribution from the integral is related to the unknown coefficients, B_n.

Consider the following contour integral:

$$\int_{C_s} \frac{g(w)}{w - \Gamma'_q}\, dw \tag{56}$$

where $g(w)$ is a special analytic function required to have the following properties:

1. $g(w)$ has simple poles at $w = -\gamma'_0$ and γ'_m, $m = 0, 1, 2, \ldots$.
2. $g(w)$ has simple zeros at $w = \Gamma'_q$, $q = 0, \pm 1, \pm 2, \ldots$.
3. $|g(w)| \to 0$ as $|w| \to \infty$.
4. $\lim_{w \to -\gamma_0'} (w + \gamma'_0) g(w) = -\sqrt{2/d}\, \dfrac{\sin \psi/2}{\gamma'_0}$.

In (56) C_s denotes a closed simple contour surrounding the first s poles. By the Cauchy residue theorem, the contour integral is equal to $2\pi j$ times the sum of residue contributions from the enclosed poles. When C_s grows to infinity, because of conditions (3) and (4), the contour integral yields

$$\sum_{n=0}^{\infty} \frac{r(\gamma'_n)}{\gamma'_n - \Gamma'_q} + \sqrt{2/d}\, \frac{\sin \psi/2}{\gamma'_0} \frac{1}{\gamma'_0 + \Gamma'_q} = 0 \tag{57}$$

where

$$r(\gamma'_n) = \lim_{w \to \gamma_n'} (w - \gamma'_n) g(w)$$

A comparison between (55) and (57) shows that formally

$$B_n = r(\gamma'_n) = \lim_{w \to \gamma_n'} (w - \gamma'_n) g(w) \tag{58}$$

Therefore, the determination of the unknown modal coefficients, B_n, becomes that of finding a suitable analytic function, $g(w)$. The details of constructing such a function are discussed in Appendix II.

The applicability of the residue calculus method obviously is very limited. Its usefulness has recently been extended, however, by a modified residue calculus method,[30] which allows the parallel-plate array with finite wall thickness to be treated.

The thin-walled parallel-plate array may be solved also by the Wiener–Hopf technique. A rather detailed development of the application of this technique to a similar array problem can be found in Chapter 4.

7. ITERATIVE METHOD

Our primary task in obtaining solutions of phased array boundary value problems has been to invert the operator equation, (1). That is, for a given y, we want to determine K^{-1}, the inverse of K. The solution to (1) can then be written as

$$x = K^{-1}y \tag{59}$$

Moreover, once the inverse operator, K^{-1}, is obtained, solutions for different y's corresponding to different excitations may be readily found. The questions concerning the existence of the inverse to an operator are beyond the scope of this book. Finding the inverse operator generally is very difficult so that we must often resort to approximate techniques such as the method of moments to solve the problem. In addition to the approximation introduced by truncation when the moments method is used, inversion of matrices on the computer also may give rise to round-off errors, because of the many steps of algebraic manipulations which are needed. Hopefully, these errors can be kept small in actual computations.

In view of the inherent approximation involved, and the fact that matrix inversion can be time consuming, it sometimes may prove desirable to consider other methods such as an iterative procedure for solution. Let us rewrite (1) as

$$Kx = (I + M)x = y \quad \text{or} \quad x = y - Mx \tag{60}$$

where I is an identity operator. Suppose that we find by some means an approximate solution, x_0. Substituting it into the right side of (60) yields a new approximate solution, x_1. We hope that the new solution is a better approximation to the true solution than the previous one, so that by repeating the process we may eventually converge to the right solution. This process is called iteration.[1] Whether or not the procedure converges depends on the properties of the operator M. A sufficient condition for convergence of the iterative procedure will be given presently. For the time being, let us assume that Mx is in some sense "small." It then seems reasonable to neglect it and to regard y as the zeroth-order approximate solution, that is,

$$x_0 \approx y \tag{61}$$

Notice that a subscript 0 is introduced to indicate the number of iterations which have been performed. Substituting (61) into (60) gives

$$x_1 \approx y - Mx_0 = y - My \tag{62}$$

Continuing the process n times, we find that

$$x_n = \{I - M + M^2 + \cdots + (-1)^n M\}y \tag{63}$$

Thus, the iterative procedure generates a solution in series form. If the process is continued indefinitely, we eventually have an infinite series, called the Neumann series. An immediate question concerning such a solution is whether the series converges; if it does, when does it converge and does it converge to the right solution? It is shown in many books in mathematics[1,2,4] that, if $\|M\| < 1$,* the process is absolutely convergent. It should be emphasized that this condition is a *sufficient but not a necessary* one. When this condition is satisfied, not only does the series solution converge, but also the convergence is independent of the initial approximation used. Of course, the closer the initial guess is to the true solution, the faster the convergence will be. The problem in the iterative approach, in addition to the question of whether the series converges, is to find a good initial approximation that will yield a quick convergence. One hopes that good approximate solutions can be obtained by only a few iterations. Notice that (63) may also be obtained by writing (60) as $x = (I + M)^{-1}y$ and then formally expanding $(I + M)^{-1}$ into a binomial series.

The concept of Neumann series expansion may also be applied to improve the accuracy of approximate solutions. For example, suppose that an approximate solution to (1), x_N, has been obtained by solving

$$y = K_N x_N \tag{64}$$

where K_N denotes an N-dimensional degenerate operator approximation[1] to K. [Here it is assumed that (1) is approximated in a subspace in which y is completely contained, so that $y = y_N$. This is the case in many phased array and waveguide discontinuity problems. Otherwise, slightly different results are obtained. See the footnote on p. 110.]

By rewriting (1) as†

$$y = [K_N + (K - K_N)](x - x_N) + Kx_N$$

and using (64), we find

$$x_N - x = (I + L_N)^{-1} L_N x_N \tag{65}$$

* $\|M\|$ denotes the norm of the operator M. (See Ref. 1.)

† Note that K and K_N must have the same dimension for the theory to be valid. We assume that this is the case here. A method of extending the dimensions of K_N and x_N will be discussed shortly.

where $L_N = K_N^{-1}(K - K_N)$. Expanding $(I + L_N)^{-1}$ formally into a binomial series leads to an alternative Neumann series as*

$$x = \sum_{q=0}^{\infty} (-1)^q L_N{}^q x_N \tag{66}$$

In order for the series to converge, it is required that $\|L_N\| < 1$. Again, this is a sufficient but not a necessary condition. The expression shows that, starting from an approximate solution, one may use the series expression to calculate the solution to the original problem. This approach is particularly useful as a means for upgrading the accuracy of an already determined approximate solution. Suppose that replacing (1) by

$$y = K_P x_P \tag{67}$$

with P sufficiently large, would yield approximate solutions which are within the desired accuracy. By following a procedure similar to the above, we may express x_P in terms of x_N as

$$x_P = \sum_{q=0}^{\infty} (-1)^q (L_{PN})^q x_N \tag{68}$$

where $L_{PN} = K_N^{-1}(K_P - K_N)$. Thus, starting from an initial approximate solution, a more refined solution may be calculated by using (68).

Notice that in obtaining (67) and (68), since the approximate kernel K_N is usually of lower dimension than K or K_P, the dimension must be appropriately increased (or, equivalently, the range of K_N extended). This can be done in many different ways. A convenient method of generating the extended K_N is to make its left upper $N \times N$ submatrix equal to the original K_N, and to augment it by making the diagonal elements in the rest of K_N equal to those of K or K_P. The off-diagonal elements are set equal to zero. That is, let

$$\begin{aligned} K_{NE_{ij}} &= K_{N_{ij}} && \text{for} \quad i, j \leq N \\ &= K_{P_{ii}}\delta_{ij} \quad \text{or} \quad K_{ii}\delta_{ij} && \text{for} \quad i > N \quad \text{or} \quad j > N \end{aligned} \tag{69}$$

where $K_{NE_{ij}}$ are the elements of the extended K_N, etc. Similarly, x_N may be extended to x_{NE} by $x_{NE} = \{\{x_N\}, 0, 0, \ldots, 0\}$. This choice makes the inversion of the extended K_N no more difficult than inverting the original one.

Let us now define the sequence of vectors $(x)^i$, $i = 1, 2, \ldots,$ as the partial

* In the event that y is not completely contained in the subspace used in (64), the series should be modified to

$$x = \sum_{q=0}^{\infty} (-1)^q L_N{}^q (K_N^{-1} y) \qquad \text{with} \quad x_N \to (K_N^{-1} y)$$

sum of the Neumann series in (66), that is,

$$\begin{aligned}(x)^0 &= x_N \\ (x)^1 &= x_N - L_N x_N \\ &\ \vdots \\ (x)^s &= \sum_{q=0}^{s} (-1)^q L_N{}^q x_N\end{aligned} \tag{70}$$

Each vector may be expressed in component form as

$$(x)^r = \{(x)_1{}^r, (x)_2{}^r, (x)_3{}^r, \ldots\}$$

It is interesting to note that the first correction is zero for the first N components of $(x)^1 - (x)^0$, namely,

$$(x)^1 - (x)^0 = \{0, 0, \ldots, 0, \Delta x_{N+1}, \Delta x_{N+2}, \ldots\} \tag{71}$$

where $\Delta x_{N+1} = (x)^1_{N+1} - (x)^0_{N+1}$, etc. Therefore, it is necessary to sum the Neumann series beyond the first term in order to obtain improvement in the first N components of the approximate solution.

The expression in (65) is particularly noteworthy. It gives the deviation of the approximate solution from the correct one solely in terms of the operator, L_N, and the approximate solution, x_N, itself.

There are many other ways in which the matrix equation obtained from (1) may be rearranged to permit an iterative procedure. For example, we can split the matrix K as

$$K = N - P \tag{72}$$

Equation (1) may then be rearranged to read

$$Nx = Px + y \tag{73}$$

This expression enables us to generate a new vector by substituting a known one on the right side. The success of this method requires that the determinant of N be nonzero, that is, N is nonsingular. Thus, starting from an arbitrary vector x_0, we obtain

$$\begin{aligned}x_1 &= N^{-1}Px_0 + N^{-1}y \\ x_2 &= N^{-1}Px_1 + N^{-1}y = N^{-1}PN^{-1}Px_0 + N^{-1}P(N^{-1}y + y) \\ &\ \vdots\end{aligned} \tag{74}$$

Clearly, the matrix N should be chosen so that its inverse can be easily determined. A sufficient condition for the process to converge is that the magnitude of the largest eigenvalue of the matrix $N^{-1}P$ be less than 1. This condition is also necessary if the procedure is to converge independently of the initial guess, x_0.

There are many different ways whereby the original matrix can be split. The specific choice of $N_{ij} = K_{ii}\delta_{ij}$, $P_{ij} = K_{ii}\delta_{ij} - K_{ij}$ is called Jacobi or simultaneous iteration. The Gauss–Seidel or successive iteration refers to the situation in which

$$N_{ij} = \begin{cases} K_{ij}, & i \geq j \\ 0, & i < j \end{cases}$$

The particular method of constructing the extended K_N as given in (69) is sometimes called block iteration.

APPENDIX 1 THE CONSERVATION OF ENERGY AND THE METHOD OF MOMENTS

To prove that the conservation of energy law is inherently satisfied by the approximate Ritz–Galerkin method, let us consider the equation for the aperture electric field as given in (22). The equation is reproduced here for convenience:

$$2y_{j'}\mathbf{\Phi}_{j'}(\mathbf{r}) = \int_A \left\{ \sum_{j=1}^{\infty} y_j \mathbf{\Phi}_j(\mathbf{r})\mathbf{\Phi}_j(\mathbf{r}') + \sum_{m=1}^{\infty} Y_m \mathbf{\Psi}_m(\mathbf{r})\mathbf{\Psi}_m^*(\mathbf{r}') \right\} \cdot \mathbf{E}_t(\mathbf{r}')\, d\mathbf{r}' \tag{22}$$

It is recalled that this equation is derived for the situation in which the array is excited from the waveguide side by a $\mathbf{\Phi}_{j'}(\mathbf{r})$ mode.

Suppose now that the method of moments is used to convert (22) into a matrix equation, using $\boldsymbol{\eta}_i(\mathbf{r})$, $i = 1, 2, \ldots, I$, as the expansion functions and $\boldsymbol{\eta}_i^*(\mathbf{r})$, $i = 1, 2, \ldots, I$, as the testing functions. The result is

$$\begin{pmatrix} \cdot \\ \cdot \\ \cdot \\ d_i \\ \cdot \\ \cdot \\ \cdot \end{pmatrix} = [L_{ji}] \begin{pmatrix} \cdot \\ \cdot \\ \cdot \\ a_i \\ \cdot \\ \cdot \\ \cdot \end{pmatrix} \tag{A.1}$$

where

$$d_j = 2y_{j'}(\mathbf{\Phi}_{j'}, \boldsymbol{\eta}_j^*) \tag{A.2a}$$

$$L_{ji} = \sum_{n=1}^{N} y_n(\boldsymbol{\eta}_j^*, \mathbf{\Phi}_n)(\mathbf{\Phi}_n, \boldsymbol{\eta}_i) + \sum_{m=1}^{M} Y_m(\boldsymbol{\eta}_j^*, \mathbf{\Psi}_m)(\mathbf{\Psi}_m^*, \boldsymbol{\eta}_i) \tag{A.2b}$$

and a_i are the unknown expansion coefficients. Notice that, in obtaining the set of (A.1), a further approximation of the kernel is introduced by truncating the two series at N and M, respectively. This step has been shown to be equivalent to using N waveguide modes and M Floquet modes in the mode matching technique, namely, the aperture electric field is expressed as

$$\mathbf{E}_g \approx (1 + R_{j'})\mathbf{\Phi}_{j'}(\mathbf{r}) + \sum_{n=1}^{N}{}^{(j')} b'_n \mathbf{\Phi}_n(\mathbf{r}) \approx \sum_{i=1}^{I} a_i \boldsymbol{\eta}_i(\mathbf{r}) \tag{A.3a}$$

and

$$\mathbf{E}_a \approx \sum_{m=1}^{M} A'_m \mathbf{\Psi}_m(\mathbf{r}) \approx \sum_{i=1}^{I} a_i \boldsymbol{\eta}_i(\mathbf{r}) \tag{A.3b}$$

Subscripts g and a have been introduced on $\mathbf{E}$ to designate reference to the fields in the waveguide and array sides, respectively. The magnetic fields, $\mathbf{H}_g$ and $\mathbf{H}_a$, are given by

$$-\hat{z} \times \mathbf{H}_g = y_{j'}(1 - R_{j'})\mathbf{\Phi}_{j'}(\mathbf{r}) - \sum_{n=1}^{N} y_n b'_n \mathbf{\Phi}_n(\mathbf{r}) \tag{A.4a}$$

$$-\hat{z} \times \mathbf{H}_a = \sum_{m=1}^{M} Y_m A'_m \mathbf{\Psi}(\mathbf{r}) \tag{A.4b}$$

From (A.3a) and (A.3b) we may express the unknown modal functions in terms of a_i as

$$1 + R_{j'} = \sum_{i=1}^{I} (\boldsymbol{\eta}_i, \mathbf{\Phi}_{j'}) a_i \tag{A.5a}$$

$$b'_n = \sum_{i=1}^{I} (\boldsymbol{\eta}_i, \mathbf{\Phi}_n) a_i \tag{A.5b}$$

and

$$A'_m = \sum_{i=1}^{I} (\boldsymbol{\eta}_i, \mathbf{\Psi}^*_m) a_i \tag{A.5c}$$

The powers carried toward and away from the aperture may be evaluated from the z-directed Poynting vector. Because of the continuity of the tangential electric and magnetic fields at the aperture we have

$$\int_A \hat{z} \cdot \mathbf{E}_g \times \mathbf{H}_g^* \, d\mathbf{r}' = \int_A \hat{z} \cdot \mathbf{E}_a \times \mathbf{H}_a^* \, d\mathbf{r} \tag{A.6}$$

Substituting for the $\mathbf{E}$ and $\mathbf{H}$ from (A.3a), (A.3b), (A.4a), and (A.4b) and then using (A.5a)–(A.5c), we find

$$2 \sum_{i=1}^{I} y_{j'} a_i (\mathbf{\Phi}_{j'}, \boldsymbol{\eta}_i)$$

$$= \sum_{i=1}^{I} \sum_{j=1}^{I} a_i a_j^* \left[\sum_{n=1}^{N} y_n^* (\boldsymbol{\eta}_i, \mathbf{\Phi}_n)(\mathbf{\Phi}_n, \boldsymbol{\eta}_j^*) + \sum_{m=1}^{M} Y'^*_m (\boldsymbol{\eta}_i, \mathbf{\Psi}^*_m)(\mathbf{\Psi}_m, \boldsymbol{\eta}_j^*) \right] \tag{A.7}$$

Comparing (A.1) and (A.7), we recognize that the latter is a quadratic form derivable from the former as

$$\{\ldots a_i \ldots\} \begin{pmatrix} \cdot \\ \cdot \\ \cdot \\ d_i \\ \cdot \\ \cdot \\ \cdot \end{pmatrix}^* = \{\ldots a_i \ldots\} \, [L_{ji}]^* \begin{pmatrix} \cdot \\ \cdot \\ \cdot \\ a_i \\ \cdot \\ \cdot \\ \cdot \end{pmatrix}^* \tag{A.8}$$

We have thus shown that the choice of $L_{ji} = (\boldsymbol{\eta}_j^*, L\boldsymbol{\eta}_i)$, where L denotes the operator, leads to the conservation of energy. It should be noted that the reciprocity is not satisfied by this choice when $\{\boldsymbol{\eta}_i\}$ are complex. In many practical applications, however, since real $\{\boldsymbol{\eta}_i\}$ are used, the approximate solutions obtained by the Ritz–Galerkin method satisfy both reciprocity and conservation of energy.

APPENDIX II THE CONSTRUCTION OF AN ANALYTICAL FUNCTION IN THE RESIDUE CALCULUS METHOD

The function $g(w)$ is chosen to have the following form:

$$g(w) = \frac{p(w)(w - \Gamma_0') \prod_{q=1}^{\infty} (w - \Gamma_q')(w - \Gamma_{-q}')(d/2q\pi)^2 e^{wd/q\pi}}{(w + \gamma_0')(w - \gamma_0') \prod_{n=1}^{\infty} (\gamma_n' - w)(d/n\pi) e^{wd/n\pi}} \tag{A.9}$$

where $p(w)$ is an entire function to be determined later. In choosing $g(w)$, the distribution of zeros and poles has been taken in accordance with conditions 1 and 2 in Section 6. The exponential factors in the infinite products are introduced for the sake of rendering the latter convergent. The presence of $p(w)$ is necessary to make $g(w)$ satisfy condition 3, that is, $|g(w)| \to 0$ as $|w| \to \infty$. To determine the functional form of $p(w)$ the following observations are helpful; namely, for large m,

$$\gamma_m' \approx \frac{m\pi}{d} \quad \text{and} \quad \Gamma_m' \approx \frac{2m\pi - \psi}{d}$$

Hence $g(w)$ differs by a bounded function in the neighborhood of the point at infinity from

$$g_1(w) = \frac{p(w)(w - \Gamma_0') \prod_{q=1}^{\infty} \left(w - \dfrac{2q\pi - \psi}{d}\right)\left(w - \dfrac{2q\pi + \psi}{d}\right)\left(\dfrac{d}{2q\pi}\right)^2 e^{wd/q\pi}}{(w + \gamma_0')(w - \gamma_0') \prod_{n=1}^{\infty} \left(\dfrac{n\pi}{d} - w\right)\left(\dfrac{d}{n\pi}\right) e^{wd/n\pi}}$$

$$= \frac{p(w)(w - \Gamma_0')\Gamma\left(\dfrac{wd - \psi}{2\pi}\right)\Gamma\left(\dfrac{wd + \psi}{2\pi}\right) \sin\left(\dfrac{wd - \psi}{2}\right) \sin\left(\dfrac{wd + \psi}{2}\right)}{\pi(w + \gamma_0')(w - \gamma_0')\Gamma\left(\dfrac{wd}{\pi}\right) \sin wd} \tag{A.10}$$

where $\Gamma(w)$ is the gamma function. Thus, it is necessary only to consider the growth of $g_1(w)$ as $w \to \infty$ in order to determine the behavior of $g(w)$ for large w. As w tends to infinity, the first term of Sterling's formula for the gamma function,

$$\Gamma(z) \sim \sqrt{2\pi/z}\; e^{z \ln z - z}$$

can be used. Substitution of the asymptotic expression of the gamma function into $g_1(w)$ leads to the following asymptotic form:

$$g_1(w) \sim \frac{p(w)}{w^{3/2}} e^{-(wd/\pi) \ln 2}$$

To ensure that $g_1(w)$ has an algebraic growth as w tends to infinity, $p(w)$ is chosen to be $Ke^{(wd/\pi) \ln 2}$, where K is a constant which is to be chosen by imposing condition 4. The choice of $p(w)$ leaves $g_1(w)$ and consequently $g(w)$ to behave as $1/w^{3/2}$ as w tends to infinity. This can be shown to be the proper form for the field to satisfy the edge condition (Ref. 27). Now $g(w)$ becomes

$$g(w) = Ke^{(wd/\pi) \ln 2} \frac{(w - \Gamma_0') \prod_{q=1}^{\infty} (w - \Gamma_q')(w - \Gamma_{-q}')(d/2q\pi)^2 e^{wd/q\pi}}{(w + \gamma_0')(w - \gamma_0') \prod_{n=1}^{\infty} (\gamma_n' - w)(d/n\pi) e^{wd/n\pi}} \tag{A.11}$$

Condition 4 requires that

$$\lim_{w \to \gamma_0'} (w + \gamma_0')g(w) = -\sqrt{2/d}\, \frac{\sin \psi/2}{\gamma_0'}$$

Thus,

$$K = -\sqrt{2/d}\,\frac{\sin \psi/2}{\gamma_0'}\,e^{(\gamma_0' d/\pi)\ln 2}\cdot\frac{2\gamma_0'}{\gamma_0' + \Gamma_0'}$$
$$\times \frac{\prod_{n=1}^{\infty}(\gamma_n' + \gamma_0')(d/n\pi)e^{-\gamma_0' d/n\pi}}{\prod_{q=1}^{\infty}(\gamma_0' + \Gamma_q')(\gamma_0' + \Gamma_{-q}')(d/2q\pi)^2 e^{-\gamma_0' d/q\pi}} \tag{A.12}$$

Using this expression for K, we find that

$$g(w) = -\sqrt{2/d}\,\frac{\sin \psi/2}{\gamma_0'}\,e^{(w+\gamma_0')(d/\pi)\ln 2}$$
$$\times \frac{2\gamma_0'}{\gamma_0' + \Gamma_0'}\cdot\frac{w - \Gamma_0'}{(w + \gamma_0')(w - \gamma_0')}\prod_{n=1}^{\infty}\frac{(\gamma_n' + \gamma_0')}{(\gamma_n' - w)}\,e^{-(\gamma_0' + w)d/n\pi}$$
$$\times \prod_{q=1}^{\infty}\frac{(w - \Gamma_q')(w - \Gamma_{-q}')}{(\gamma_0' + \Gamma_q')(\gamma_0' + \Gamma_{-q}')}\,e^{(w+\gamma_0')d/q\pi} \tag{A.13}$$

The unknown coefficients, B_n, may be obtained from $g(w)$ by (58). The results then give v_n' by use of (55a). In fact, v_n' are the modal coefficients of the unknown aperture field expressed in terms of the normal modes of the waveguide.

We can also calculate the modal coefficients of the aperture field in terms of the exterior Floquet modes. To do so, we make use of the boundary conditions at the aperture, which require the continuity of the tangential field components. Thus

$$(1 + R)\Phi_0'(y) + \sum_{n=1}^{\infty} v_n'\Phi_n'(y) = \sum_{m=-\infty}^{\infty} V_m'\Psi_m'(y) \tag{A.14a}$$

$$-y_0'(1 - R)\Phi_0'(y) + \sum_{n=1}^{\infty} y_n' v_n'\Phi_n'(y) = -\sum_{m=-\infty}^{\infty} Y_m' V_m'\Psi_m'(y) \tag{A.14b}$$

Multiplying (A.14a) and (A.14b) by $\Psi_q'^*(y)$ and integrating over the waveguide aperture, $(0, d)$, we find by virtue of the orthonormality property among (Ψ')'s that

$$V_q' = (1 + R)C_{0q}^* + \sum_{n=1}^{\infty} C_{nq}^* v_n' \tag{A.15a}$$

and

$$-Y_q' V_q' = -y_0'(1 - R)C_{0q}^* + \sum_{n=1}^{\infty} C_{nq}^* y_n' v_n' \tag{A.15b}$$

These equations may be combined to yield

$$-2Y_q' V_q' = -(Y_q' + y_0')C_{0q}^* + (-Y_q' + y_0')RC_{0q}^* + \sum_{n=1}^{\infty}(y_n' - Y_q')C_{nq}^* v_n' \tag{A.16}$$

Substituting for the expressions for y'_n, Y'_q, and C_{nq} and simplyfying yield

$$V'_q = (-1)^{q-1}\left(\frac{2q\pi - \psi}{d}\right)\left[-\sqrt{2/d}\,\frac{\sin \psi/2}{\gamma'_0(\Gamma'_q - \gamma'_0)} + \sum_{n=0}^{\infty} \frac{B_n}{\Gamma'_q + \gamma'_n}\right] \tag{A.17}$$

Comparison of the quantity inside the square bracket with (57) and (58) suggests that the former may be evaluated by considering the following integral:

$$\int_{C_s} \frac{g(w)}{w + \Gamma'_q}\, dw \tag{A.18}$$

Now, $w = -\Gamma'_q$ is a new pole of the integral. Again, by the Cauchy residue theorem, when C_s is allowed to grow to infinity, the integral yields

$$g(-\Gamma'_q) + \sum_{n=0}^{\infty} \frac{r(\gamma'_n)}{\gamma'_n + \Gamma'_q} - \sqrt{2/d}\,\frac{\sin \psi/2}{\gamma'_0} \cdot \frac{1}{\Gamma'_q - \gamma'_0} = 0$$

Thus, V'_q may be determined from the same analytic function, $g(w)$, as

$$V'_q = (-1)^q\left(\frac{2q\pi - \psi}{d}\right) g(-\Gamma'_q) \tag{A.19}$$

REFERENCES

1. B. Friedman, *Principles and Techniques of Applied Mathematics*, John Wiley and Sons, New York, 1960; B. Z. Vulikh, *Introduction to Functional Analysis for Scientists and Technologists*, Addison Wesley Publishing Company, Reading, Mass., 1963.
2. L. V. Kantorovitch and V. I. Krylov, *Approximate Methods of Higher Analysis*, Interscience Publishers, New York, 1958.
3. R. F. Harrington, *Field Computation by Moments Method*, The Macmillan Company, New York, 1968.
4. E. Isaacson and H. B. Keller, *Analysis of Numerical Methods*, John Wiley and Sons, New York, 1966.
5. V. H. Rumsey, "The Reaction Concept in Electromagnetic Theory," *Phys. Review*, Series 2, Vol. 94, pp. 1483–1491, June 1954.
6. D. S. Jones, "A Critique of the Variational Method in Scattering Problems," *IRE Trans. Antennas and Propagation*, Vol. AP-4, pp. 297–301, May 1956.
7. H. Levine and J. Schwinger, "On the Theory of Diffraction by an Aperture in an Infinite Plane Screen I and II," *Phys. Review*, Vol. 74, pp. 958–974, October 1948, and Vol. 75, pp. 1423–1432, May 1948.
8. C. P. Wu and V. Galindo, "Surface-Wave Effects on Dielectric Sheathed Phased Arrays of Rectangular Waveguides," *Bell System Tech. J.*, Vol. 47, pp. 117–142, January 1968.

9. N. Amitay and V. Galindo, "Application of a New Method for Approximate Solutions and Error Estimates to Waveguide Discontinuity and Phased Array Problems," *Radio Science*, Vol. 3 (New Series), pp. 830–843, August 1968.

10. P. J. B. Clarricoats and K. R. Slinn, "Numerical Solution of Waveguide-Discontinuity Problems," *Proc. IEE*, Vol. 114, pp. 878–886, July 1967.

11. W. J. Cole, E. R. Nagelberg, and C. M. Nagel, "Interactive Solution of Waveguide Discontinuity Problems," *Bell System Tech. J.*, Vol. 46, pp. 649–722, March 1967.

12. R. Mittra, "Relative Convergence of the Solution of a Doubly Infinite Set of Equations," *J. Research*, Series D, Vol. 67D, pp. 245–254, March–April 1963.

13. P. H. Masterman, P. J. B. Clarricoats, and C. D. Hannaford, "Computer Methods of Solving Waveguide-Iris Problems," *Electronics Letters*, Vol. 5, pp. 23–25, January 1969.

14. K. K. Mei and J. G. Van Bladel, "Scattering by Perfectly Conducting Rectangular Cylinders," *IEEE Trans. Antennas and Propagation*, Vol. AP-11, pp. 185–192, March 1963.

15. V. Galindo and C. P. Wu, "Numerical Solutions for an Infinite Phased Array of Rectangular Waveguides with Thick Walls," *IEEE Trans. Antennas and Propagation*, Vol. AP-14, pp. 149–158, March 1966.

16. F. B. Hildebrand, *Methods of Applied Mathematics*, Prentice-Hall, Englewood Cliffs, N. J., pp. 444–460, 1952.

17. A. R. Neureuther and K. Zaki, "Numerical Solutions of Electromagnetic Boundary Value Problems by Means of the Asymptotic Anticipation Method," *Radio Science*, Vol. 3 (New Series), pp. 1158–1167, Dedember 1968.

18. S. Abdelmesshih and G. Sinclair, "Treatment of Singularities in Scattering from Perfectly Conducting Polygonal Cylinders—a Numerical Technique," *Can. J. Phys.*, Vol. 45, pp. 1305–1318, March 1967.

19. V. Galindo and C. P. Wu, "Dielectric Loaded and Covered Rectangular Waveguide Phased Arrays," *Bell System Tech. J.*, Vol. 47, pp. 93–116, January 1968.

20. N. Amitay and V. Galindo, "The Analysis of Circular Waveguide Phased Arrays," *Bell System Tech. J.*, Vol. 47, pp. 1903–1932, November 1968.

21. N. Amitay and V. Galindo, "On Energy Conservation and the Method of Moments in Scattering Problems," *IEEE Trans. Antennas and Propagation*, Vol. AP-17, pp. 747–751, November 1969.

22. N. Amitay, J. S. Cook, R. G. Pecina, and C. P. Wu, "Mutual Coupling and Matching Conditions in Large Plannar Phased Arrays," *1964 PTGAP International Symposium Program and Digest*, pp. 150–156.

23. F. Berz, "Reflection and Refraction of Microwaves at a Set of Parallel Metallic Plates," *Proc. IEE* (*London*), Vol. 98, Part III, pp. 47–55, January 1951.

24. C. P. Wu and V. Galindo, "Properties of a Phased Array of Rectangular Waveguides with Thin Walls," *IEEE Trans. Antennas and Propagation*, Vol. AP-14, pp. 163–172, March 1966.

25. J. F. Carlson and A. E. Heins, "The Reflection of an Electromagnetic Wave by an Infinite Set of Plates," *Quarterly Appl. Math.*, Vol. 4, pp. 313–329, January 1947.

26. B. Noble, *Wiener-Hopf Technique*, Pergamon Press, New York, 1958.

27. R. E. Collin, *Field Theory of Guided Waves*, McGraw-Hill Book Company, New York, 1960.

28. N. Marcuvitz (ed.), *Microwave Handbook*, M.I.T. Radiation Laboratory Series, Vol. 10, McGraw-Hill Book Company, New York, p. 59.

29. G. Forsyth and C. Moler, *Computer Solutions of Linear Algebraic Systems*, Prentice-Hall, Englewood Cliffs, N.J., 1967.

30. G. F. Van Blaricum, Jr., and R. Mittra, "A Modified Residue Calculus Technique for Solving a Class of Boundary Value Problems," Part II: "Waveguide Phased Arrays, Modulated Surfaces, and Diffraction Gratings," *IEEE Trans. Microwave Theory and Techniques*, Vol. MTT-17, pp. 310–318, June 1969.

4. The Canonical Rectangular Array of Thin-Walled Rectangular Waveguides

1. INTRODUCTION

The integral equations for phased array problems in general are quite complex and cannot be solved by purely analytical techniques. There is, however, one notable exception, which occurs in an array of rectangular waveguides arranged in a rectangular grid. When it is assumed that the waveguide walls are infinitesimally thin and that the beam is scanned in either the E or the H plane, the complex vector two-dimensional integral equation for such an array simplifies and becomes scalar one dimensional. In applying the method of moments to convert the integral equation into an infinite set of algebraic equations, it is possible to choose the basis and weighting functions so that the resulting algebraic equations are amenable to exact solutions by the function-theoretic technique. One of the advantages of obtaining closed form solutions is that such solutions may be used to calculate, to the desired degree of accuracy, numerical answers which are useful as a standard for canonical comparison with those obtained by numerical techniques, thereby checking the correctness and ascertaining the accuracy of the numerically obtained solutions. More importantly, examination of exact solutions often leads to the extraction of important qualitative information about the problem, which may be more difficult to discern from numerically obtained data. Many important array properties will be so analyzed and discussed.

In addition to the potential usefulness discussed above, rectangular waveguide arrays offer other advantages also. Such an array, because of its structural simplicity, can be conveniently used as a prototype for experimental investigations of array properties or for testing new ideas. Of course, properly designed rectangular waveguide arrays are used also in practical applications.

From the principle of linear superposition, it follows that the coupling coefficients between the elements are related to the array reflection coefficient by the Fourier series transform. This relationship may be used to calculate the coupling coefficients from the reflection coefficients and vice versa. The numerical results show that the coupling decays monotonically as a function of distance, and for large distances the rate of decay is inversely proportional to the 3/2 power of the distance.

This asymptotic behavior of the coupling coefficients can also be derived analytically. Such analysis has revealed, moreover, that this behavior is due to the singularity in the derivative of the reflection coefficient at the scan angle, which corresponds to a beam at grazing. This appears, from all the evidence available to date, to be a general property of infinite arrays which may be used as a check for numerically obtained solutions. In fact, this concept has been applied to most of the solutions obtained in this book. By extending the concept of mutual coupling to the higher-order modal coefficients, it is possible to show that the element pattern, that is, the radiation pattern of an infinite array that has only one element excited, with the rest terminated in their characteristic impedances, is proportional to the transmission coefficient as a function of scan of the active array when all elements are excited.

Finally, the effects of an anisotropic plasma on the properties of an array are examined. In spite of the anisotropy of the medium, the problem is still amenable to an exact solution by the Wiener-Hopf technique, which will be discussed in some detail. The anisotropy of the medium introduces an asymmetry in the reflection and coupling coefficients. The asymptotic behavior of the coupling coefficients, however, remains the same as in the isotropic medium. The latter finding, together with other independently derived results, indicates that the asymptotic behavior of the coupling may be a general property of all phased arrays. An extension of these results to certain planar arrays will be discussed in Chapter 7.

2. RELATION BETWEEN RECTANGULAR WAVEGUIDE ARRAYS AND EQUIVALENT PARALLEL-PLATE ARRAYS[1,2]

A rectangular waveguide array may be constructed by placing rectangular waveguides contiguous to each other along two coordinate axes. They may be arranged so that the centers of the waveguides form a rectangular grid, or the centers in one row are displaced from those in an adjacent row to form a brick layer pattern. Metallic spacers may be placed between the waveguides to adjust the interelement spacing. Figure 4.1 shows an array arranged in a rectangular grid. We denote the spacing in the x and y directions by b and

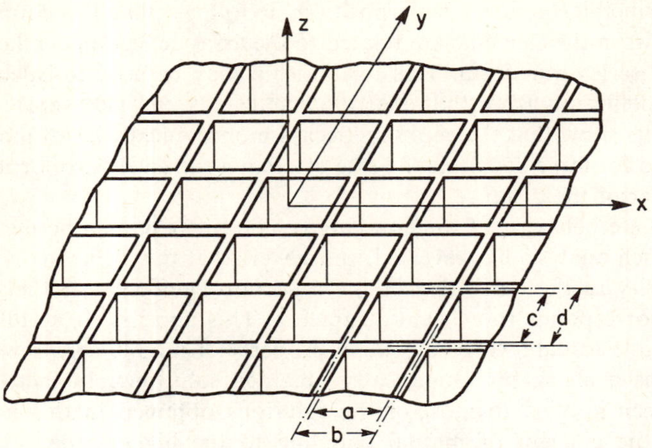

Fig. 4.1. An infinite array of rectangular waveguides.

d, respectively. The inside dimensions of the waveguides are $a \times c$.

For analytical purposes, the array is assumed infinite in extent; the justification for this assumption was discussed in Chapter 1. As a result, a Floquet type modal expansion of the fields in the region exterior to the waveguides may be effected. The formulation of the array boundary value problem leads in general to vector two-dimensional integral equations. We shall show that, by introducing idealizing assumptions, the problem can be reduced to scalar one-dimensional equations, which are considerably simpler to treat. Moreover, with additional assumptions the problem can be solved exactly.

Consider the array to be excited in the TE_{10} mode by sources which are located far away from the array aperture. These sources have equal amplitude and constant progressive phases in the x and y directions. The incident wave is assumed polarized in the y direction.

Let the array elements be given a pair of numbers (m, n) to identify their location relative to the coordinate origin. The element at the origin is denoted the (0, 0)th element. Now, in order to scan the beam to the (θ, ϕ) direction, the (m, n)th elements is excited with a voltage $e^{+j(m\psi_x+n\psi_y)}$, where ψ_x and ψ_y are the phase shifts between adjacent elements (called the steering phases) in the x and y directions, respectively. They are related to the scan angle (θ, ϕ) of the beam by

$$\begin{aligned} \psi_x &= \frac{2\pi b}{\lambda} \sin\theta \cos\phi \\ \psi_y &= \frac{2\pi d}{\lambda} \sin\theta \sin\phi \end{aligned} \tag{1}$$

When $\psi_y = 0$, implying that $\phi = 0$ or π, the beam is scanned in a plane perpendicular to the incident electric field and the array is said to be scanned in the H plane. Similarly, when $\psi_x = 0$, we find $\phi = \pm\pi/2$ and the beam is scanned in the plane containing the incident electric field. This mode is called the E plane of scanning.

One of the situations which allows the vector two-dimensional problem to be reduced to a scalar one-dimensional one occurs when the beam is scanned in the H plane and the waveguide walls perpendicular to the incident electric field are assumed to be infinitesimally thin, that is, $c = d$. Under these conditions, the waveguide walls do not have any effect on the field distributions and hence may be removed. When this is done, the rectangular waveguide array reduces to a parallel-plate waveguide array as illustrated in Fig. 4.2. The incident fields in each column of elements combine to form a

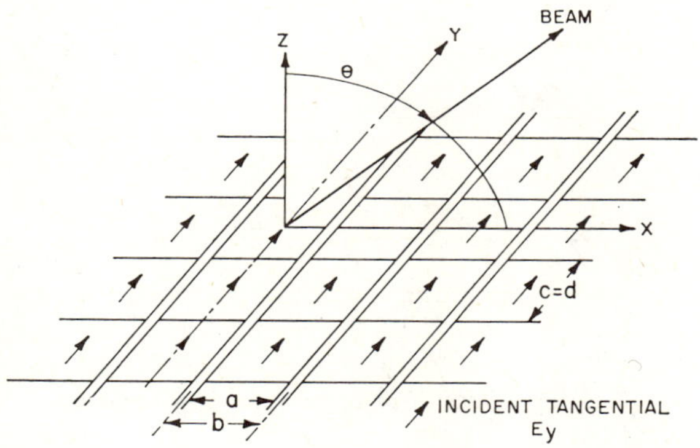

Fig. 4.2. H-plane scanning in x-z plane.

TE_1 mode in the parallel-plate region. Because there is no discontinuity along y and the incident fields do not vary with y, the scattered fields contain only TE modes with no y variation. There are only three nonvanishing field components in such modes: E_y, H_x, and H_z. The problem is thus scalar one dimensional, and its solution is much simplified. This will be discussed presently.

An alternative scan which permits a similar simplification occurs when the (m, n)th element is excited by a voltage $e^{+j(\pm m\pi + n\psi_y)}$. Because of the constant phase shift applied along the x axis, the variation of ψ_y causes the beam to be scanned in a plane parallel to the E plane. In fact, there are two radiated beams equally displaced with respect to the y–z plane. The direction of these beams may be determined from

$$\sin\theta\cos\phi = \frac{\pm\lambda}{2b} \quad \text{and} \quad \sin\theta\sin\phi = \frac{\psi_y\lambda}{2\pi d}$$

This mode of scanning is called the quasi-E-plane scan to distinguish it from the true E-plane scan. (For the true E-plane scan, the problem remains vector two dimensional. No simplification is possible.)

Since the incremental phase shift applied in the x direction is π radians per column of elements, the incident fields alternate in sign from column to column. If we assume that the waveguide walls parallel to y are infinitesimally thin, that is, $a = b$, the aggregate of the incident waves may be regarded as a

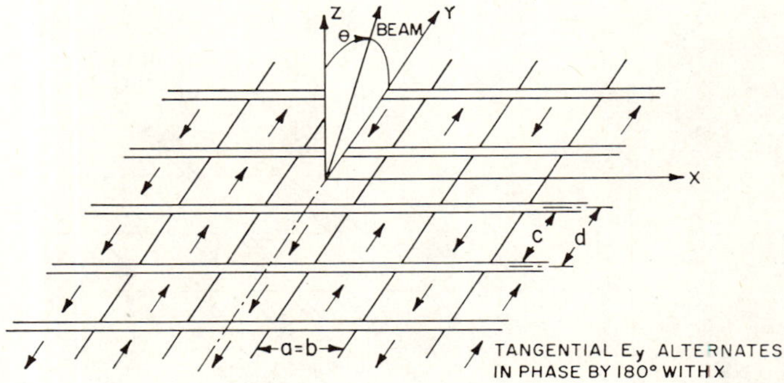

Fig. 4.3. Quasi-E-plane scanning in y-z plane.

standing wave in the x direction. These infinitesimally thin waveguide walls parallel to the electric fields are located at the nulls of the standing wave. Therefore, these walls do not perturb the field distribution and may be removed. When this is done, the rectangular waveguide array becomes a parallel-plate array as depicted in Fig. 4.3. Or, equivalently, these walls may be extended to infinity in the z direction to form linear rectangular waveguide arrays radiating into parallel-plate regions. Since in the reduced structure the incident field has no E_x component and there is no discontinuity in the x direction, the scattered field will have no E_x component. Moreover, the nonvanishing field components will have the same functional dependence on x as the incident wave. This type of wave mode behaves essentially like TM modes in parallel-plate waveguides. In fact, by letting $a \to \infty$, the fields degenerate into TM modes with nonvanishing field components H_x, E_y, and E_z. Thus, the reduced problem is again scalar one dimensional.

3. INTEGRAL EQUATIONS FOR THE QUASI-*E* AND *H* PLANES OF SCAN

Although it is possible to reduce the general integral equations derived in Chapter 2 to the special cases to be considered here, this approach does not prove very productive in view of the labor involved. Instead, we shall rederive the equations here, taking advantage of the notational simplicity for the present case to illustrate the details of the basic steps.

The formulation of the integral equations for the two planes of scan, the *H* plane and the quasi-*E* plane, follows a similar procedure. We shall therefore discuss only one of the cases, say that of the *H* plane. (see Fig. 4.2.) The results for the other case will be given directly.

3.1 *H* Plane of Scan

The integral equations may be derived to have either the tangential electric field or the magnetic field at the aperture as the unknown function. The first step in the procedure is to introduce suitable representations for the tangential field components in the regions both inside and outside the waveguides. These representations should satisfy the boundary conditions at $z = +\infty$ and $z = -\infty$. When the boundary conditions across the common aperture are imposed, the desired integral equations are obtained.

Recall now that the array is assumed to be infinite in extent, and that the elements are excited with sources which have constant amplitudes and linear progressive phases. The fields in each periodic cell are identical except for the constant phase shift which exists from element to element. It is necessary, therefore, to solve for the fields in a single cell only, say the zeroth cell at the origin.

The fields inside the waveguide are most conveniently expressed in terms of the waveguide modes. The reason is that the modal functions are solutions to the wave equation subject to the appropriate boundary conditions on the waveguide walls.

Assume now that the lowest-order mode, that is, the TE_{10} mode with unit modal current, is incident. The tangential (to z) field components inside the waveguide may be written as

$$\left.\begin{aligned}\mathscr{H}_x^-(x, z) &= (e^{-j\gamma_1 z} - Re^{j\gamma_1 z})\Phi_1(x) + \sum_{q=2}^{\infty} i_q \Phi_q(x) e^{j\gamma_q z} \\ \mathscr{E}_y^-(x, z) &= -z_1(e^{-j\gamma_1 z} + Re^{j\gamma_1 z})\Phi_1(x) + \sum_{q=2}^{\infty} z_q i_q \Phi_q(x) e^{j\gamma_q z}\end{aligned}\right\}, \quad z \leq 0 \quad (2)$$

Here,

$$\Phi_q(x) = \begin{cases} \left.\begin{matrix} \sqrt{2/a}\cos\dfrac{q\pi}{a}x, & q \text{ odd} \\ \sqrt{2/a}\sin\dfrac{q\pi}{a}x, & q \text{ even} \end{matrix}\right\}, & |x| \le \dfrac{a}{2}, \\ 0 & \dfrac{a}{2} \le |x| \le \dfrac{b}{2} \end{cases}$$

are the orthonormalized waveguide modal functions, and

$$z_q = \frac{\omega\mu_0}{\gamma_q}$$

are the modal impedances with the corresponding propagation constants:

$$\gamma_q = \begin{cases} \sqrt{k^2 - (q\pi/a)^2} & \text{for } k^2 \ge \left(\dfrac{q\pi}{a}\right)^2 \\ -j\sqrt{(q\pi/a)^2 - k^2} & \text{for } k^2 < \left(\dfrac{q\pi}{a}\right)^2 \end{cases}$$

The $\{i_q\}$ are the unknown modal coefficients (also called modal currents) yet to be determined. Notice that the first mode is singled out from the summation for convenience, because it is the only term which contains two parts; one part represents a wave traveling toward the aperture and the other part a wave reflected away from the aperture. This representation is valid everywhere inside the waveguide. In particular, at the aperture $z = 0$, we find

$$H_x^-(x) = \mathscr{H}_x^-(x, 0) = (1 - R)\Phi_1(x) + \sum_{q=2}^{\infty} i_q\Phi_q(x) \tag{3}$$

$$E_y^-(x) = \mathscr{E}_y^-(x, 0) = -z_1(1 + R)\Phi_1(x) + \sum_{q=2}^{\infty} z_q i_q \Phi_q(x) \tag{3a}$$

The orthonormality relation between the modal functions can be utilized to express the unknown modal coefficients in terms of the aperture fields. For example, multiplying the first equation by $\Phi_q(x)$ and then integrating over $(-b/2, b/2)$, we find

$$i_p = \int_{-b/2}^{b/2} \Phi_p(x')H_x^-(x')\,dx' \qquad \text{with} \quad i_1 = 1 - R \tag{4}$$

This relation can be used in (3a) to express the tangential electric field at the aperture in terms of the tangential magnetic field there. Thus

$$E_y^-(x) = -2z_1\Phi_1(x) + \int_{-b/2}^{b/2}\left\{\sum_{q=1}^{\infty} z_q\Phi_q(x)\Phi_q(x')\right\}H_x^-(x')\,dx' \tag{5}$$

In obtaining (5) the order of integration and summation is interchanged. This is permissible because the kernel

$$\left\{\sum_{q=1}^{\infty} z_q \Phi_q(x)\Phi_q(x')\right\}$$

behaves like $\ln|x - x'|$ and the integral is absolutely convergent for physically acceptable solutions.

The fields in the region outside the waveguide may be expanded in terms of the Floquet modes because of the periodic geometry and excitation. The Floquet modes for a one-dimensional geometry can be determined in the same manner as discussed in Chapter 2. They are given by

$$\Psi_m(x) = \sqrt{1/b}\, e^{j[(2m\pi+\psi)/b]x}, \quad m = 0, \pm 1, \pm 2 \tag{6}$$

with the associated modal impedances

$$Z_m = \frac{\omega\mu_0}{\Gamma_m}$$

where

$$\Gamma_m = \begin{cases} \sqrt{k^2 - [(2m\pi + \psi)/b]^2} & \text{for } k^2 \geq \left(\dfrac{2m\pi + \psi}{b}\right)^2 \\[2ex] -j\sqrt{[(2m\pi + \psi)/b]^2 - k^2} & \text{for } k^2 < \left(\dfrac{2m\pi + \psi}{b}\right)^2 \end{cases}$$

are the z-directed propagation constants. These modal functions are scan dependent, as the $\{\Psi_m\}$ are functions of the steering phase, ψ. (The subscript x is dropped from ψ_x for brevity.) They are orthonormalized over the interval $(-b/2, b/2)$ in accordance with the following relation:

$$\langle \Psi_m, \Psi_n^* \rangle = \int_{-b/2}^{b/2} \Psi_m(x)\Psi_n^*(x)\, dx = \delta_{mn} \tag{7}$$

The fields in $z \geq 0$ may now be described in terms of $\{\Psi_m\}$ by

$$\left.\begin{aligned} \mathscr{H}_x^+(x, z) &= \sum_{m=-\infty}^{\infty} I_m \Psi_m(x) e^{-j\Gamma_m z} \\ \mathscr{E}_y^+(x, z) &= -\sum_{m=-\infty}^{\infty} Z_m I_m \Psi_m(x) e^{-j\Gamma_m z} \end{aligned}\right\}, \quad z \geq 0 \tag{8}$$

where $\{I_m\}$ are the unknown modal currents to be determined. At the aperture, these expressions reduce to

$$H_x^+(x) = \mathscr{H}_x^+(x, 0) = \sum_{m=-\infty}^{\infty} I_m \Psi_m(x) \tag{9}$$

and

$$E_y^+(x) = \mathscr{E}_y^+(x, 0) = -\sum_{m=-\infty}^{\infty} Z_m I_m \Psi_m(x) \tag{9a}$$

By virtue of the orthonormality of the modal functions,

$$I_m = \int_{-b/2}^{b/2} \Psi_m^*(x') H_x^+(x') \, dx' \tag{10}$$

Using (10) in (9a), we have

$$E_y^+(x) = \int_{-b/2}^{b/2} \left\{ \sum_{m=-\infty}^{\infty} Z_m \Psi_m(x) \Psi_m^*(x') \right\} H_x^+(x') \, dx' \tag{11}$$

The boundary conditions require that the tangential field components be continuous across the aperture:

$$\left.\begin{aligned} E_y^+(x) &= E_y^-(x) \\ H_y^+(x) &= H_y^-(x) \end{aligned}\right\} \quad \text{for} \quad |x| \leq \frac{a}{2}$$

Notice that, since the tangential electric field vanishes on and inside the waveguide walls, $-a/2 \leq |x| \leq b/2$, the continuity of E_y may be extended over the entire period: $-b/2 \leq x \leq b/2$. A step discontinuity in the tangential magnetic field exists over the range $a/2 \leq |x| \leq b/2$ at the aperture because $H_y^-(x) \equiv 0$, while $H_y^+(x) \neq 0$. We may therefore define the tangential magnetic field in $|x| < b/2$ as

$$H_y(x) = \begin{cases} H_y^+(x) = H_y^-(x) & \text{for} \quad |x| \leq \dfrac{a}{2} \\ H_y^+(x) & \text{for} \quad \dfrac{a}{2} \leq |x| \leq \dfrac{b}{2} \end{cases}$$

When the continuity conditions are imposed using (5) and (11), the desired integral equation is obtained:

$$2z_1\Phi_1(x) = \int_{-b/2}^{b/2} \left\{ \sum_{q=1}^{\infty} z_q \Phi_q(x)\Phi_q(x') + \sum_{m=-\infty}^{\infty} Z_m \Psi_m(x) \Psi_m^*(x') \right\} H_x(x') \, dx' \qquad \text{for} \quad |x| \leq \frac{b}{2} \tag{12}$$

This equation is derived for a transmitting array excited by the lowest-order mode, the TE_{10} mode, of the waveguide. The excitation appears as the free term on the left. Notice that the incident wave was assumed to have a unit modal current. Since a linear system is being considered, if the incident wave has a modal current A_1, the corresponding equation is readily obtained by multiplying the left-hand side of (12) by A_1. The solution for the tangential magnetic field will thus be A_1 times the solution for a unit incident wave. The kernel, however, stays unchanged because it is determined solely by the geometry of the problem, not by the mode of excitation.

On the basis of these observations, it is easy to write the integral equations for various different situations. For example, if the array is excited by a linear combination of the first N modes with modal currents $\{A_p\}$, only the left side of (12) has to be replaced by $2\sum_{p=1}^{N} z_p A_p \Phi_p(x)$. Similarly, for a receiving array only the excitation function in (12) has to be changed to the appropriate one, such as $2Z_0\Psi_0(x)$, when the incident wave consists of the zeroth space harmonic.

The formulation of the integral equation with the aperture electric field as the unknown function follows a procedure parallel to that used for the tangential magnetic field case. Instead of expressing the unknown modal currents in terms of the tangential magnetic field as shown in (4) and (10), the modal voltages, $\{v_q\}$ and $\{V_m\}$, are derived as functions of the aperture electric field:

$$i_q = v_q y_q = y_q \int_{-a/2}^{a/2} \Phi_q(x') E_y^-(x')\, dx', \qquad 1 + R = y_1 \int_{-a/2}^{a/2} \Phi_1(x') E_y^-(x')\, dx' \tag{4a}$$

and

$$I_m = V_m Y_m = Y_m \int_{-a/2}^{a/2} \Psi_m^*(x') E_y^+(x')\, dx' \tag{10a}$$

where

$$y_q = \frac{1}{z_q} \quad \text{and} \quad Y_m = \frac{1}{Z_m}$$

These results may be used to express the tangential magnetic field in terms of the tangential aperture electric field as

$$H_x^-(x) = 2\Phi_1(x) + \int_{-a/2}^{a/2} \left\{ \sum_{q=1}^{\infty} y_q \Phi_q(x) \Phi_q(x') \right\} E_y^-(x')\, dx' \tag{5a}$$

and

$$H_x^+(x) = \int_{-a/2}^{a/2} \left\{ \sum_{m=-\infty}^{\infty} Y_m \Psi_m(x) \Psi_m^*(x') \right\} E_y^+(x')\, dx' \tag{11a}$$

Imposing the continuity conditions across the aperture, $|x| \le a/2$, yields the integral equation. Thus, under the assumption that the incident wave has a unit modal voltage instead of a unit modal current, the integral equation is given by*

$$2y_1\Phi_1(x) = \int_{-a/2}^{a/2} \left\{ \sum_{q=1}^{\infty} y_q \Phi_q(x)\Phi_q(x') + \sum_{m=-\infty}^{\infty} Y_m \Psi_m(x)\Psi_m^*(x') \right\} E_y(x')\, dx' \quad \text{for } |x| \leq \frac{a}{2} \tag{13}$$

Equations (12) and (13) may be regarded as a pair of integral equations associated with an array. A comparison between them shows that the equation for the aperture electric field has the modal admittances in the kernel, whereas the equation for the tangential magnetic field has the modal impedances in the kernel. Depending on whether TE or TM modes are involved, these modal impedances (or admittances) determine how fast these kernels converge. Another point of interest is the domain and range over which the integral equations are defined. The one for the electric field is valid only over the aperture, $|x| \leq a/2$, whereas the one for the magnetic field applies over an entire unit cell, $|x| \leq b/2$.

3.2 Quasi-*E* Plane of Scan

The derivation of the integral equations for the quasi-*E* plane of scan (Fig. 4.3), may be carried out by an analogous procedure. In fact, by referring to (12) and (13), the equations for this case may be written almost by inspection, so long as the pertinent complete sets of modal functions and the associated modal impedances are known. These quantities may be determined by using standard techniques for the analysis of the eigenmodes in cylindrical waveguides.

As was mentioned earlier, the fields which exist in the quasi-*E* plane of scan are characterized by the absence of the E_x component, and the same $\cos(\pi x/a)$ variation for all tangential field components. Although there are three nonvanishing tangential field components, only two of them, E_y and H_x, are needed in the formulation. The third one is redundant.

Since the common factor, $\cos(\pi x/a)$ does not play any role in the subsequent analysis, it will be dropped for convenience. Thus, the orthonormal

* In arriving at this result, the interchange of the order of integration and summation has been performed. Strictly speaking, this step is not permissible as the resultant integral becomes divergent in the ordinary sense. The equation is put into this form symbolically for the sake of compactness. In actual computation, this step need not be carried out so that no difficulty arises. However, there are situations in which the difficulty cannot be avoided in the usual way and special care should be exercised. Examples of such cases can be found in Chapter 8.

modal functions without this common factor are given by

$$\Phi_q'(y) = \begin{cases} \left.\begin{array}{ll} \sqrt{(2-\delta_{0q})/c}\,\cos\left(\dfrac{q\pi}{c}y\right), & q \text{ even} \\ \sqrt{(2-\delta_{0q})/c}\,\sin\left(\dfrac{q\pi}{c}y\right), & q \text{ odd} \end{array}\right\}, & |y| \le \dfrac{c}{2} \\ 0 & \text{otherwise} \end{cases} \tag{14}$$

with $q = 0, 1, 2, \ldots, \infty$. The associated modal admittances are

$$y_q' = \frac{k^2 - (\pi/a)^2}{\omega\mu_0\gamma_q'} \tag{15}$$

where

$$\gamma_q' = \begin{cases} \sqrt{k^2 - (\pi/a)^2 - (q\pi/c)^2} & \text{for } k^2 \ge \left(\dfrac{\pi}{a}\right)^2 + \left(\dfrac{q\pi}{c}\right)^2 \\ -j\sqrt{(\pi/a)^2 + (q\pi/c)^2 - k^2} & \text{for } k^2 < \left(\dfrac{\pi}{a}\right)^2 + \left(\dfrac{q\pi}{c}\right)^2 \end{cases}$$

are the propagation constants. The Floquet modes for the quasi-*E*-plane case are similar to those for the *H*-plane case with an appropriate change in notation. Thus

$$\Psi_m'(y) = \sqrt{1/d}\; e^{j[(2m\pi+\psi)/d]y}, \qquad m = 0, \pm 1, \pm 2, \ldots, \pm\infty \tag{16}$$

The modal admittances and the z-directed propagation constants of the Floquet modes are, respectively,

$$Y_m' = \frac{k^2 - (\pi/a)^2}{\omega\mu_0\Gamma_m'}$$

and

$$\Gamma_m' = \begin{cases} \sqrt{k^2 - (\pi/a)^2 - [(2m\pi+\psi)/d]^2} & \text{for } k^2 \ge \left(\dfrac{\pi}{a}\right)^2 + \left(\dfrac{2m\pi+\psi}{d}\right)^2 \\ -j\sqrt{(\pi/a)^2 + [(2m\pi+\psi)/d]^2 - k^2} & \text{for } k^2 < \left(\dfrac{\pi}{a}\right)^2 + \left(\dfrac{2m\pi+\psi}{d}\right)^2 \end{cases} \tag{17}$$

where the subscript y has been dropped from ψ_y for brevity.

Using the various expressions listed above, we can write the integral equations when the array is excited by the lowest order mode, $\Phi_0'(y)$, as

$$2z_0'\Phi_0'(y) = \int_{-d/2}^{d/2} \left\{ \sum_{q=0}^{\infty} z_q'\Phi_q'(y)\Phi_q'(y') + \sum_{m=-\infty}^{\infty} Z_m'\Psi_m'(y)\Psi_m'^{*}(y') \right\} H_x(y')\,dy' \tag{18}$$

for the tangential magnetic field, and

$$2y_0'\Phi_0'(y) = \int_{-c/2}^{c/2} \left\{ \sum_{q=0}^{\infty} y_q'\Phi_q'(y)\Phi_q'(y') + \sum_{m=-\infty}^{\infty} Y_m'\Psi_m'(y)\Psi_m'^*(y') \right\} E_y(y')\, dy' \quad (19)$$

for the tangential aperture electric field.

4. CANONICAL SOLUTIONS

The integral equations derived in Section 3 are valid for parallel plates with finite wall thickness, that is, for H-plane scan, $b > a$, and for quasi-E-plane scan, $d > c$. In fact, the integral equations for the aperture electric field remain valid even when thin irises are introduced at the array apertures, provided that the limits of integration over the apertures are reduced accordingly to the aperture region only. The irises need not be placed symmetrically inside the apertures. On the other hand, the integral equation for the tangential magnetic field is much more complex when irises are present.[22]

When waveguide walls have finite thickness, only approximate solutions are possible for these integral equations. The application of the moments method, using different sets of functions as the basis and weighting functions to obtain numerical results, will be discussed in the next chapter. We shall consider here only the special cases which are amenable to exact solutions.

Exact solutions can be obtained when all the waveguide walls are infinitesimally thin. This means that in the H-plane case $a = b$ and in the quasi-E-case $c = d$. In order to be able to solve the integral equations exactly, they are first converted into matrix equations by using *special* functions as the basis for representation and testing. We shall apply the procedure to (19) for an illustration. Similar techniques may be applied to other integral equations.

Assume that

$$E_y(y) \approx \sum_{n=0}^{N} v_n'\Phi_n'(y) \quad (20)$$

We shall let N approach infinity afterwards. As N increases, the approximation improves, and eventually, for $N \to \infty$, the above relation becomes an equality.

When (20) is substituted into (19), we find

$$2y_0'\Phi_0'(y) \approx \sum_{n=0}^{N} v_n' \left[y_n'\Phi_n'(y) + \sum_{m=-\infty}^{\infty} Y_m'\Psi_m'(y)C_{nm}^* \right] \quad (21)$$

where the intermodal coupling coefficients obtained by using (14) and (17) are given by

$$C_{nm} = \int_{-c/2}^{c/2} \Phi'_n(y)\Psi'_m(y)\,dy$$

$$= \frac{\sqrt{2-\delta_{0n}}}{d}\,\frac{2(-1)^{n/2+m}[(2m\pi+\psi)/d]}{\Gamma'^2_m - \gamma'^2_n}\begin{cases}\sin\dfrac{\psi}{2}, & n \text{ even}\\[2ex] \cos\dfrac{\psi}{2}, & n \text{ odd}\end{cases}$$

The relation

$$\Gamma'^2_m - \gamma'^2_n = \left(\frac{n\pi}{d}\right)^2 - \left(\frac{2m\pi+\psi}{d}\right)^2$$

has been used in obtaining the expression for C_{nm}. We now weight (21) with $\Psi'^*_q(y)$, $q = 0, \pm 1, \ldots$. This step yields

$$2y'_0 C^*_{0q} = \sum_{n=0}^{N} v'_n[(y'_n + Y'_q)C^*_{nq}], \quad q = 0, \pm 1, \ldots, \tag{22}$$

By substituting the expressions for y'_n, Y'_q, and C_{nq} into (22), letting $N \to \infty$, and

$$\begin{aligned} v'_0 &= 1 + v''_0 \\ v'_n &= v''_n, \quad n > 0 \end{aligned} \tag{23}$$

and then transferring the known quantities to the left side, we may rearrange (22) as

$$\sqrt{2/d}\,\frac{\sin(\psi/2)}{\gamma'_0(\Gamma'_q + \gamma'_0)} = \sum_{n=0}^{\infty} \frac{1}{\Gamma'_q - \gamma'_n} B_n, \quad q = 0, \pm 1, \pm 2, \ldots, \pm\infty \tag{24}$$

where

$$B_n = (-1)^{n/2}\frac{\sqrt{2-\delta_{0n}}}{d}\,\frac{v''_n}{\gamma'_n}\begin{cases}\sin\dfrac{\psi}{2}, & n \text{ even}\\[2ex] \cos\dfrac{\psi}{2}, & n \text{ odd}\end{cases}$$

Equations (24) constitute an infinite set of algebraic equations of a very special form. When the equations are arranged in matrix form, the (n, q)th element of the matrix is $1/(\Gamma'_q - \gamma'_n)$, that is, γ'_n always appears in the nth column, and Γ'_q in the qth row. Although there are infinitely many equations for the infinitely many unknowns, B_n, this special form of equations can be solved exactly. The details of finding a solution to such equations by the function-theoretic method are discussed in Chapter 3. It should be noted that the problem can be solved by the Wiener-Hopf technique also. This technique is discussed in Section 10, where the effects of an anisotropic plasma are considered.

The solutions for the modal coefficients are in the form of infinite products. Although all the modal coefficients can be obtained by this method, since the propagating modes are of principal interest only results for these modes are given below. Thus, for a single propagating waveguide mode, a situation most commonly employed in practical arrays, the waveguide dimensions must be such that $\sqrt{1-(\lambda/2b)^2} \leq \lambda/2d$. Under this condition, when $0 \leq \psi_y \leq (2\pi d/\lambda)\sqrt{1-(\lambda/2b)^2}$, there is a propagating mode in the free space region, whereas when $(2\pi d/\lambda)\sqrt{1-(\lambda/2b)^2} \leq \psi_y \leq \pi$ no propagating mode exists. (Because of the symmetry in the array geometry, only the range $0 \leq \psi_y \leq \pi$ need be considered.)

The reflection coefficient and transmission coefficients, respectively, are given by[1]

$$R = \begin{cases} \dfrac{\gamma_0' - \Gamma_0'}{\gamma_0' + \Gamma_0'} \exp\left\{-j2\left[G(\gamma_0') + \tan^{-1}\dfrac{\gamma_0'}{|\Gamma_{-1}'|} - \tan^{-1}\dfrac{\gamma_0'}{|\gamma_1'|}\right]\right\} \\ \qquad\qquad \text{for} \quad 0 \leq \psi_y \leq (2\pi d/\lambda)\sqrt{1-(\lambda/2b)^2} \leq \pi \\ \exp\left\{j\left[\dfrac{\pi}{2} - G(\gamma_0') - \tan^{-1}\dfrac{\gamma_0'}{|\Gamma_0'|} - \tan^{-1}\dfrac{\gamma_0'}{|\Gamma_{-1}'|} - \tan^{-1}\dfrac{\gamma_0'}{|\gamma_1'|}\right]\right\} \\ \qquad\qquad \text{for} \quad (2\pi d/\lambda)\sqrt{1-(\lambda/2b)^2} \leq \psi_y \leq \pi \end{cases} \tag{25}$$

and

$$T_0 = \frac{2\sqrt{\gamma_0'\Gamma_0'}}{\gamma_0' + \Gamma_0'} \exp\left\{j\left[G(\Gamma_0') + \tan^{-1}\frac{\Gamma_0'}{|\Gamma_{-1}'|} - \tan^{-1}\frac{\Gamma_0'}{|\gamma_1|} - G(\gamma_0') - \tan^{-1}\frac{\gamma_0'}{|\Gamma_{-1}'|} + \tan^{-1}\frac{\gamma_0'}{|\gamma_1'|}\right]\right\}$$
$$\text{for} \quad 0 \leq \psi_y \leq (2\pi d/\lambda)\sqrt{1-(\lambda/2b)^2} \leq \pi \tag{26}$$

where

$$G(x) = -\frac{bx}{\pi}\ln 2 + \tan^{-1}\frac{x}{|\Gamma_1'|} + \sum_{n=2}^{\infty}\left(\tan^{-1}\frac{x}{|\Gamma_n'|} + \tan^{-1}\frac{x}{|\Gamma_{-n}'|} - \tan^{-1}\frac{x}{|\gamma_n'|}\right)$$

The transmission coefficient is normalized so that $|T_0|^2$ is equal to the power transmitted when a unit power is supplied to each waveguide.

In the case of *H*-plane scan, a similar procedure may be applied to obtain a solution. For a single propagating mode, the waveguide dimension must be $\frac{1}{2} < b/\lambda < 1$. With this spacing, there will be one propagating mode in the region exterior to the waveguides (corresponding to one radiated beam) when $0 \leq \psi_x \leq 2\pi(1 - b/\lambda)$, and two propagating modes (corresponding to the main beam and a grating lobe) when $2\pi(1 - b/\lambda) \leq \psi_x \leq \pi$. The reflection coefficient has different expressions for the two respective scan ranges.

The two transmission coefficients are denoted by T_0 for the main beam and T_{-1} for the grating lobe.* The results are as follows:[1]

$$R = \begin{cases} \left(\dfrac{\Gamma_0 - \gamma_1}{\Gamma_0 + \gamma_1}\right) \exp\left\{2j\left[\dfrac{\pi}{2} - F(\gamma_1) - \tan^{-1}\dfrac{\gamma_1}{|\Gamma_{-1}|}\right]\right\} \\ \qquad\qquad \text{for} \quad 0 \le \psi_x \le 2\pi(1 - b/\lambda) \\ \left(\dfrac{\Gamma_0 - \gamma_1}{\Gamma_0 + \gamma_1}\right)\left(\dfrac{\gamma_1 - \Gamma_{-1}}{\gamma_1 + \Gamma_{-1}}\right) \exp\left[-j2F(\gamma_1)\right] \\ \qquad\qquad \text{for} \quad 2\pi(1 - b/\lambda) \le \psi_x \le \pi \end{cases} \tag{27}$$

$$T_0 = \begin{cases} \dfrac{\sqrt{2\gamma_1\Gamma_0}}{\Gamma_0 + \gamma_1} \exp\left\{j\left[\pi + F(\Gamma_0) + \tan^{-1}\dfrac{\Gamma_0}{|\Gamma_{-1}|} - F(\gamma_1) - \tan^{-1}\dfrac{\gamma_1}{|\Gamma_{-1}|}\right]\right\} \\ \qquad\qquad \text{for} \quad 0 \le \psi_x \le 2\pi(1 - b/\lambda) \\ \dfrac{\sqrt{2\gamma_1\Gamma_0}}{\Gamma_0 + \gamma_1}\sqrt{\left(\dfrac{\Gamma_0 + \Gamma_{-1}}{\Gamma_0 - \Gamma_{-1}}\right)\left(\dfrac{\gamma_1 - \Gamma_{-1}}{\gamma_1 + \Gamma_{-1}}\right)} \exp\left\{j[\pi + F(\Gamma_0) - F(\gamma_1)\right\} \\ \qquad\qquad \text{for} \quad 2\pi(1 - b/\lambda) \le \psi_x \le \pi \end{cases} \tag{28}$$

and

$$T_{-1} = \frac{\sqrt{2\gamma_1\Gamma_{-1}}}{\Gamma_{-1} + \gamma_1}\sqrt{\left(\frac{\Gamma_0 + \Gamma_{-1}}{\Gamma_0 - \Gamma_{-1}}\right)\left(\frac{\Gamma_0 - \gamma_1}{\Gamma_0 + \gamma_1}\right)} \exp\left\{j[\pi + F(\Gamma_{-1}) - F(\gamma_1)\right\} \qquad \text{for} \quad 2\pi(1 - b/\lambda) \le \psi_x \le \pi \tag{29}$$

where

$$F(x) = -\frac{bx}{\pi}\ln 2 + \tan^{-1}\frac{x}{|\Gamma_1|} + \sum_{m=2}^{\infty}\left(\tan^{-1}\frac{x}{|\Gamma_m|} + \tan^{-1}\frac{x}{|\Gamma_{-m}|} - \tan^{-1}\frac{x}{|\gamma_m|}\right)$$

5. NUMERICAL EXAMPLES

The solutions obtained above are exact and apply for the situation in which all the waveguide walls are infinitesimally thin. They are useful for calculating numerical data for various waveguide dimensions. With due care the numerical calculation may be carried out with a high degree of accuracy. Numerical results for the two planes of scan are depicted below to illustrate the reflection and transmission properties of the array.

The results for the quasi-E plane of scan are displayed in Fig. 4.4; some interesting characteristics are associated with these results. We recall that

* This grating lobe is in fact the Ψ_{-1} Floquet mode, which is propagating for a certain scan range.

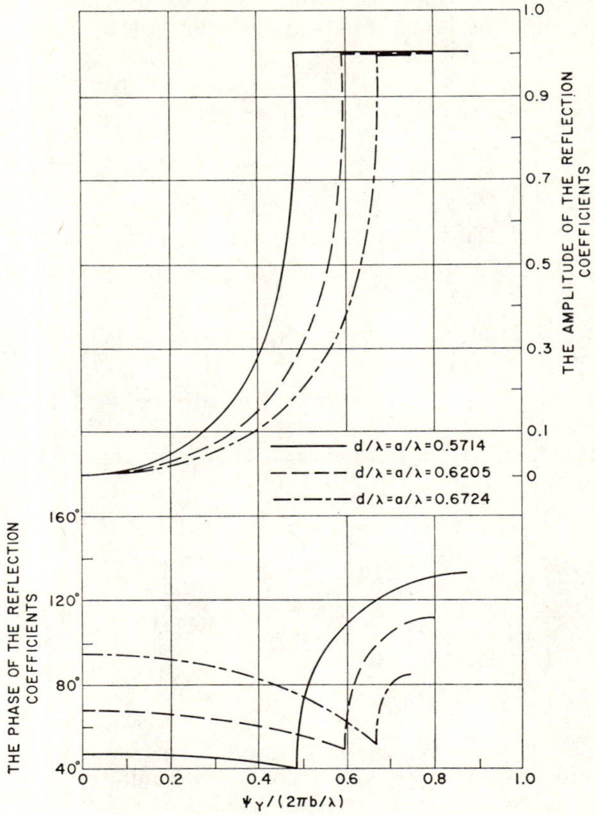

Fig. 4.4. Reflection coefficients for quasi-*E* plane.

in the quasi-*E* plane of scan the waveguide walls parallel to the incident electric field, that is, the y–z plane, do not play any role in the field distribution, so that they may be removed. (See Fig. 4.3.) When $\psi_y = 0$, each row of elements "sees" an identical environment. In consequence, exactly the same field distribution exists in each row. Since the incident electric field in each element is perpendicular to the planes $y = pd/2$, $p = 0, \pm 1, \pm 3, \pm 5, \ldots$, the scattered electric fields will remain perpendicular to these planes. Therefore the waveguide walls parallel to the x–z plane can be extended to infinity without perturbing the fields. When this is done, there is no discontinuity at the array aperture, and thus the array is perfectly matched. The reflection coefficient at broadside is therefore equal to zero, as shown in Fig. 4.4.

The waveguide dimensions chosen for these computations are such that no beam is radiated when the steering phase exceeds the value $\psi_{yc} = (2\pi d/\lambda)\sqrt{1 - (\lambda/2b)^2}$. As a result, the reflection coefficients for the range

$$(2\pi d/\lambda)\sqrt{1 - (\lambda/2b)^2} \leq \psi_y \leq \pi$$

are of unit magnitude, implying that under such conditions all the incident power will be totally reflected back toward the sources. The phase of R, the reflection coefficient, varies rapidly in this scan range.

When a total reflection takes place, the dominant mode fields interfere to form a pure standing wave inside the waveguide. At distances far removed from the aperture, the evanescent modes are sufficiently attenuated and the tangential electric field exhibits repetitive null positions. Electric conductors may be placed at these null positions without disturbing the field distributions. When this is done, the array structure may be viewed as a corrugated surface which is completely isolated from the source region. The solutions for the modal coefficients obtained for the array geometry may then be regarded as the homogeneous (source free) solution for the modified structure of a corrugated surface. Since there is no propagating mode in the free space region in such a situation, all the energy is confined to the vicinity of the corrugated surface. Hence the solution actually may be viewed as that of a surface wave propagating along the corrugated structure.[3] A calculation by the Poynting theorem shows that the power flow is directed along the direction of the phase variation. The surface wave is thus a forward wave. Since the slope of the phase variation is larger than the free space wave number, this is a slow surface wave. The conditions for the occurrence of the kind of surface wave encountered here are entirely predictable from the array geometry.

There are other situations in which the occurrence of a total reflection cannot be predicted accurately without actually solving the boundary value problem. The surface wave supported in the modified corrugated surface also has some interesting and unusual characteristics related to phased arrays. These various aspects will be discussed in more detail in Chapters 5–8.

Figures 4.5 and 4.6 show the results for the reflection and power transmission coefficients in the H plane of scan. In this plane of scan, because the element spacing $b/\lambda > \frac{1}{2}$, a grating lobe in addition to the main beam will be radiated when the steering phase falls in the range $2\pi(1 - b/\lambda) \leq \psi_x \leq \pi$. The reflection coefficient decreases in magnitude as the beam is scanned away from broadside. In the region where there are two radiated beams, the reflection coefficient rapidly diminishes in amplitude with increasing values of the steering phase, implying that two beams together make the impedance better matched to that of the waveguide. The case of $\psi_x = 180°$ is the same as $\psi_y = 0$ in the quasi-E plane of scan, and the array is perfectly matched.

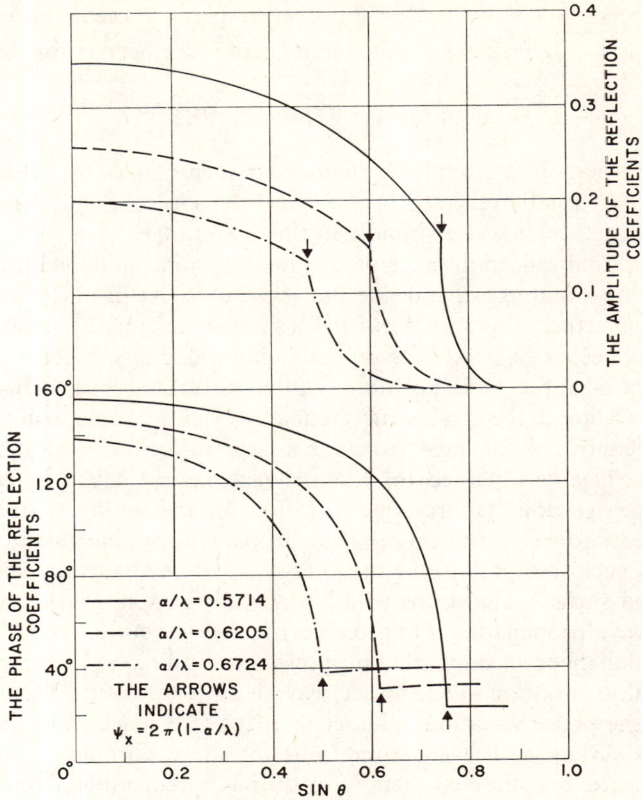

Fig. 4.5. Reflection coefficients for *H*-plane scan.

All the incident power is radiated. The two radiated beams are equally disposed with respect to the broadside direction, and each shares one half of the incident power. This fact is illustrated in Fig. 4.6, in which the variation of the power transmission coefficients of the two beams is shown.

As the steering phases are varied to scan the beam of an array, there are angles of transition marking the appearance or disappearance of a beam (or grating lobe). In the quasi-*E*-plane case considered above, a pair of beams are radiated which are equally displaced relative to the y–z plane at scan angles $\theta = \sin^{-1}(\lambda/2b)$, $\phi = 0$, and $\theta = \sin^{-1}(\lambda/2b)$, $\phi = \pi$, when $\psi_y = 0$. As ψ_y is increased, these beams are scanned along planes parallel to the y–z plane away from the broadside until the steering phase, ψ_y, reaches the critical value $(2\pi d/\lambda)\sqrt{1-(\lambda/2b)^2}$, when these beams disappear. On the

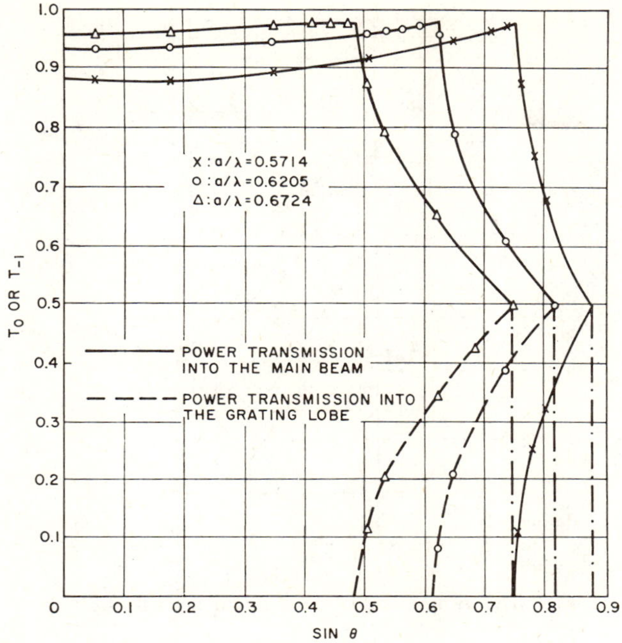

Fig. 4.6. Power transmission coefficients for *H*-plane scan.

other hand, the steering phase $\psi_{xc} = 2\pi(1 - b/\lambda)$ in the H plane of scan marks the transition at which a grating lobe emerges. A beam is at grazing, that is, the beam is parallel to the array face, when the steering phase is exactly at these critical values.

The appearance or disappearance of these beams is manifested by a sharp discontinuity in the slope of the reflection and transmission coefficients. In fact, the first derivative of the reflection coefficients is singular at these angles. It will be shown presently that the order of this singularity determines the asymptotic behavior of the coupling coefficients between elements which are separated far apart.

6. MUTUAL COUPLING

So far, the array has been analyzed under the assumption that all the elements are excited with sources of uniform amplitudes and linear phase taper so that we could study the array properties when the beam is scanned. The effects of mutual coupling are implicitly incorporated in the formulation.

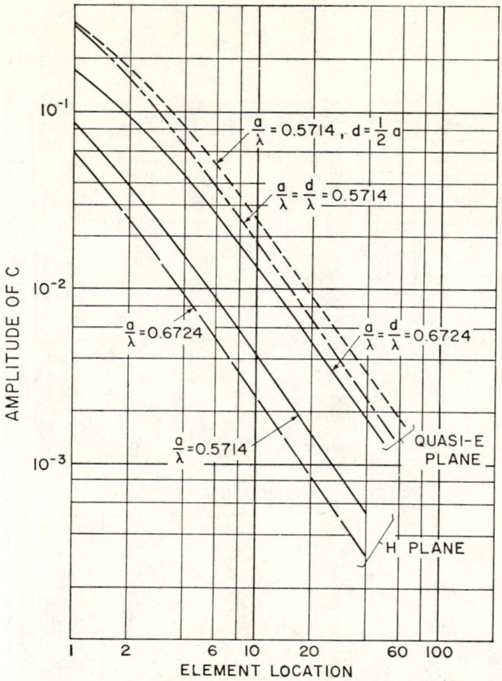

Fig. 4.7. Amplitudes of the coupling coefficients.

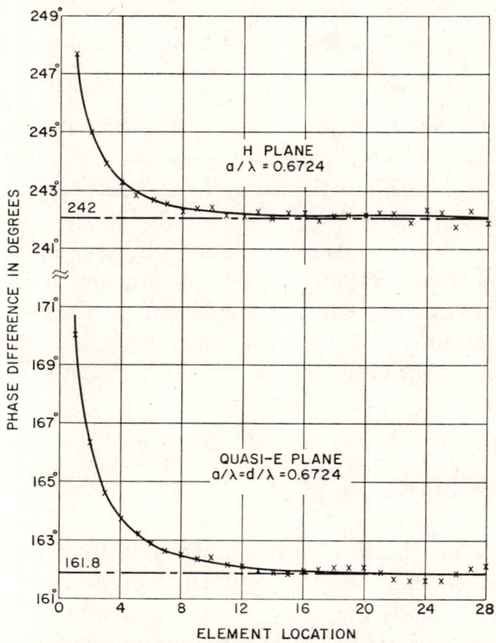

Fig. 4.8. Phase difference between successive coupling coefficients.

The array characteristics can also be investigated by exciting one element at a time. This method permits the determination of the element patterns and the mutual coupling between the elements directly.[4,5,6] The coupling coefficients can be related to the reflection coefficient of a scanned array through the principle of linear superposition. The relation becomes a Fourier series transform in infinite arrays and is useful as an analytical tool to calculate the coupling coefficients from the reflection coefficient.

Let $\mathscr{C}_{mn}$ denote the coupling (scattering) coefficient into the mth element when the nth one is excited alone by unit amplitude while the rest of the array elements are terminated in the characteristic impedance of the waveguides. Here the coupling refers to the interaction between the fundamental mode of the waveguides. Later on, we shall discuss the coupling into the higher-order modes.

As was shown in Chapter 1, the active reflection coefficient of the mth element, R_m, of a uniformly illuminated array with a linear phase taper is

$$R_m = \sum_n \mathscr{C}_{mn} e^{-j(n-m)\psi} \tag{30}$$

In an infinite array, the reflection coefficients in all the elements are identical. Without loss of generality, a reference element, denoted the zeroth element, may be chosen for consideration. Dropping the subscript 0 for convenience, we can rewrite (30) as

$$R = \sum_{n=-\infty}^{\infty} \mathscr{C}_n e^{-jn\psi} \tag{30a}$$

where $\mathscr{C}_n$ is the coupling coefficient between the zeroth element and the nth element.

Equation (30a) may be regarded as a Fourier series expansion of the reflection coefficient in which $\mathscr{C}_n$ is the nth Fourier coefficient. It follows immediately that

$$\mathscr{C}_n = \frac{1}{2\pi} \int_{-\pi}^{\pi} R(\psi) e^{-jn\psi}\, d\psi \tag{31}$$

Thus, if the reflection coefficient is determined over a basic scan range, $-\pi \leq \psi \leq \pi$, the coupling coefficients may be calculated by a Fourier analysis. This approach is particularly useful when closed form solutions for the reflection coefficients are available.

Numerical results for the coupling coefficients calculated from (31), using (25) for the quasi-E plane of scan and (27) for the H plane of scan, have been obtained. The results of such computations are shown in Figs. 4.7 and 4.8.

Some general characteristics of the mutual coupling are evident from the numerical data. The coupling decreases monotonically with both increasing element spacing in terms of the wavelength (a/λ in H-plane scan and d/λ in quasi-E-plane scan) and increasing separation between the waveguides. For waveguides which are far removed from the excited element, say in the order of five or more elements, the decay of coupling coefficients reaches an asymptotic region in which the coupling variation with element separation, when plotted in log-log scale, approaches a straight line with slope equal to $-\frac{3}{2}$, that is, $|\mathscr{C}_n| \sim (na)^{-3/2}$. This is true for all a/λ and d/λ values and for both the H and quasi-E planes of scan. The differences between the phases of the coupling coefficients to adjacent elements approach asymptotically to free space propagation values. The asymptotic behavior of coupling coefficients is related to a certain property of the array reflection coefficient, as we shall discuss in the next section.[7]

7. ASYMPTOTIC BEHAVIOR OF THE COUPLING COEFFICIENTS

As shown in (31), the coupling coefficient between the pth element and the zeroth element is the pth Fourier coefficient of the reflection coefficient. From the theory of Fourier analysis, it is known that the behavior of the Fourier coefficients of a function is intimately related to the continuity and differentiability of the function. It was pointed out earlier that the reflection coefficient of an array has a singularity in the first derivative, occurring at the steering phase when a beam is at grazing. The order of this singularity determines the rate of decay of the coupling coefficients. We shall analyze the quasi-E-plane case as an illustration. The same procedure has been applied to investigate the H-plane case, and the same results have been obtained.

The reflection coefficient, R, is a continuous function of the steering phase, ψ. This can be verified by examining expression (25). It is also evident in the numerical results shown in Fig. 4.4. To see that the first derivative of R has a singularity and to determine the order of the singularity, we write R in polar form as

$$R(\psi) = M(\psi)e^{jf(\psi)} \tag{32}$$

Differentiating with respect to ψ yields

$$\frac{dR(\psi)}{d\psi} = \left[\frac{dM(\psi)}{d\psi} + jM(\psi)\frac{df(\psi)}{d\psi}\right]e^{jf(\psi)} \tag{33}$$

Since the reflection coefficient R has different expressions for different scan ranges, we must use the appropriate one for each range of ψ. Thus, for

$0 \leq \psi \leq k'd$, where $k'^2 = k^2 - (\pi/b)^2$, we have

$$M(\psi) = \frac{\gamma_0' - \Gamma_0'}{\gamma_0' + \Gamma_0'}$$

$$f(\psi) = f_1(\psi) = -2\left[G(\gamma_0') + \tan^{-1}\frac{\gamma_0'}{|\Gamma_{-1}'|} - \tan^{-1}\frac{\gamma_0'}{|\gamma_1'|}\right] \tag{34}$$

$$\frac{dM(\psi)}{d\psi} = \frac{2\gamma_0'\psi}{d^2\Gamma_0'(\gamma_0'^2 + \Gamma_0'^2)}$$

and

$$\frac{df_1(\psi)}{d\psi} = \sum_{n=-\infty}^{\infty}{}' \frac{2k'}{|\Gamma_n'|\,(2n\pi + \psi)}$$

where the symbol Σ' implies that the summation is to exclude the zeroth term. For $k'd \leq \psi \leq \pi$,

$$M(\psi) = 1$$

and

$$f(\psi) = \pi - 2\tan^{-1}\frac{\gamma'}{|\Gamma_0'|} + f_1(\psi)$$

so that

$$\frac{dM(\psi)}{d\psi} = 0$$

$$\frac{df(\psi)}{d\psi} = \frac{2k'}{|\Gamma_0'|\,\psi} + \frac{df_1(\psi)}{d\psi} \tag{35}$$

Only the propagation constants pertinent to the Floquet modes depend on the steering phase, ψ. Among these propagation constants, only the zeroth-order one, Γ_0', can become zero as it changes from propagating to nonpropagating for the element spacing $(d/\lambda)\sqrt{1 - (\lambda/2b)^2} \leq \frac{1}{2}$. It is thus seen that when (34) and (35) are substituted into (33), $dR(\psi)/d\psi$ becomes singular as $\psi \to k'd$ because of the appearance of Γ_0' in the denominator. The singularity occurs in the amplitude function when ψ approaches $k'd$ from below, while the singularity occurs in the phase function when ψ approaches $k'd$ from above. That the singularity appears in such a fashion is characteristic of the situation in which a beam disappears from the real space. When a beam emerges into the real space as the steering phase is increased, as in the case of H-plane scan, the singularity of $dR/d\psi$ appears in reversed order. This observation applies, in fact, quite generally, as we shall see later when the properties of other type of arrays are discussed. (See Chapter 7.)

The singularity is of the form

$$\frac{dR}{d\psi} \sim \frac{1}{\Gamma_0'} = \frac{1}{\sqrt{(\psi/d)^2 - k'^2}}$$

where $\Gamma_0' \to 0$ as $\psi \to k'd$.

It is this behavior of $dR/d\psi$ which determines the asymptotic decay of the coupling coefficients. To demonstrate this, the integral in (31) is integrated by parts, taking into account the fact that $R(\psi)$ is an even function of ψ, to obtain

$$\begin{aligned}\mathscr{C}_p &= \frac{1}{2\pi}\int_{-\pi}^{\pi} R(\psi)e^{-jp\psi}\, d\psi = \frac{1}{\pi}\int_0^{\pi} R(\psi)\cos p\psi\, d\psi \\ &= -\frac{1}{\pi p}\left[\int_0^{k'd} \frac{dR(\psi)}{d\psi}\sin p\psi\, d\psi + \int_{k'd}^{\pi} \frac{dR(\psi)}{d\psi}\sin p\psi\, d\psi\right]\end{aligned}$$

Since $M(\psi)$ and $df(\psi)/d\psi$ are continuous functions of ψ in $0 \leq \psi \leq k'd$, while $M(\psi)$, $dM(\psi)/d\psi$, and $df_1(\psi)/d\psi$ are continuous in $k'd \leq \psi \leq \pi$, for large p,

$$\mathscr{C}_p = -\frac{1}{\pi p}\left[\int_0^{k'd} \frac{dM(\psi)}{d\psi} e^{jf_1(\psi)} \sin p\psi\, d\psi + j\int_{k'd}^{\pi} \frac{2k'}{\Gamma_0'\psi} e^{jf_1(\psi)} \sin p\psi\, d\psi\right] + 0(p^{-2}) \quad (36)$$

[The neglected terms in (36) are of bounded variation because they are continuous. Hence their contribution to $\mathscr{C}_p$, for large p, is at least as fast as the indicated $0(p^{-2})$ for these terms.[8]] The first integral in (36) may be rearranged as

$$I_1 = \int_0^{k'd} \frac{1}{|\Gamma_0'|}\left[\frac{2\,|\gamma_0'|\,\psi}{d^2(\gamma_0' + \Gamma_0')^2} e^{jf_1(\psi)} - \frac{2}{d} e^{jf_1(k'd)}\right] \sin p\psi\, d\psi + \frac{2}{d} e^{jf_1(k'd)} \int_0^{k'd} \frac{1}{|\Gamma_0'|} \sin p\psi\, d\psi \quad (37)$$

This isolates the singular portion of the integral in the second term. The integrand in the first term of (37) is now continuous and, by the same reasoning as above, contributes terms of $0(p^{-2})$. Thus,

$$\begin{aligned}I_1 &= \frac{2}{d} e^{jf_1(k'd)} \int_0^{k'd} \frac{1}{|\Gamma_0'|} \sin p\psi\, d\psi + 0(p^{-1}) \\ &= \pi e^{jf_1(k'd)} S_0(k'pd) + 0(p^{-1})\end{aligned}$$

where

$$S_0(k'pd) = \frac{2}{\pi}\int_0^1 \frac{\sin(k'pd\, x)}{\sqrt{1 - x^2}}\, dx$$

The second integral of (36) may be handled similarly. Thus

$$
\begin{aligned}
I_2 &= \int_{k'd}^{\pi} \frac{2k'}{\Gamma_0'\psi} e^{jf(\psi)} \sin p\psi \, d\psi \\
&= \frac{2}{d} e^{jf_1(k'd)} \int_{k'd}^{\pi} \frac{1}{\Gamma_0'} \sin p\psi \, d\psi + 0(p^{-1}) \\
&= \frac{2}{d} e^{jf_1(k'd)} \int_{k'd}^{\infty} \frac{1}{\Gamma_0'} \sin p\psi \, d\psi + 0(p^{-1}) \\
&= j\pi e^{jf_1(k'd)} J_0(k'pd) + 0(p^{-1})
\end{aligned}
$$

The coupling coefficient, $\mathscr{C}_p$, for large p, therefore, may be written as

$$
\mathscr{C}_p = -\frac{1}{p} e^{jf_1(k'd)}[S_0(k'pd) + jJ_0(k'pd)] + 0(p^{-2}) \tag{38}
$$

Introducing the asymptotic expansion for the functions $S_0(k'pd)$ and $J_0(k'pd)$ for large p,[9] we finally obtain

$$
\mathscr{C}_p = -j\sqrt{2/\pi}\,(k'd)e^{j[f_1(k'd)-\pi/4]} \frac{e^{-jk'pd}}{(k'pd)^{3/2}} + 0(p^{-2}) \tag{39}
$$

This result indicates that for large separations the magnitude of the fundamental mode coupling coefficient is inversely proportional to the $\frac{3}{2}$ power of the separation pd, and the phase of the coupling coefficient varies in accordance with free space propagation.

The above analysis may be extended to investigate the asymptotic behavior of the coupling in the higher-order modes. Analysis for the coupling between the fundamental mode in the excited zeroth element and the mth mode of the pth element shows that only the factor $(\gamma_m' - \Gamma_0')/(\gamma_0' - \Gamma_0')$ contributes to the asymptotic behavior. Therefore, the functional dependence on element separation is similar to that for the fundamental mode. Since all the modal coefficients have the same asymptotic behavior, the aperture field must have the same asymptotic dependence on distance.

The asymptotic dependence on distance of the coupling coefficients seems to be universal for all phased arrays. The result remains valid even when the array is immersed in an anisotropic plasma, as will be discussed in Section 10. Moreover, the asymptotic behavior is similar to the behavior of the fields which exist at the interface between a lossy dielectric and free space due to a magnetic line source.[10] This similarity suggests that the array face may be treated as a lossy surface.

The generality of this property suggests, moreover, that by geometrical extrapolation the asymptotic decay of the coupling coefficients should be proportional to e^{-jkr}/r^2 for planar arrays, where r denotes the distance from the excited element. An extension of this calculation to a broad class of planar arrays will be carried out in Chapter 7.

8. ELEMENT PATTERNS AND TRANSMISSION COEFFICIENTS IN INFINITE ARRAYS

The reflection coefficient of an infinite phased array was shown to be related to the coupling coefficients between the elements by the Fourier series transform [see (30a)]. This is the result of linear superposition. The same principle may be applied to show that the element pattern of an infinite array is proportional to the transmission coefficient of the phased array.[11–14] The element pattern refers to the radiation pattern of an array with a single excited element, the other elements being terminated in the characteristic impedance of the waveguides. This relation is quite general and holds for all arrays. For clarity, however, we will provide here a proof for only the scalar one-dimensional problem. The extension to planar arrays of arbitrary antenna elements is straightforward.

Consider the H-plane case as an example (see Fig. 4.2). There is only one tangential electric field component in this plane of scan, that is,

$$\mathbf{E}(x, 0) = \hat{y}E_y(x)$$

The aperture fields for both a single-element excitation and the whole array excitation with uniform amplitudes and linear phase taper are needed in the analysis. The aperture field of the latter situation is scan dependent and may be written as

$$\begin{aligned} E_y(x, \psi) &= \Phi_1(x) + \sum_{n=1}^{\infty} v_n(\psi)\Phi_n(x) \\ &= \sum_{m=-\infty}^{\infty} V_m(\psi)\Psi_m(x) \end{aligned} \tag{40}$$

The dependence on scan is explicitly exhibited in the modal voltages. Like the derivation of the integral equations, the unknown modal coefficients, v_n and V_m, may be expressed in terms of the aperture field as

$$\begin{aligned} \delta_{1n} + v_n(\psi) &= \int_{-a/2}^{a/2} \Phi_n(x)E_y(x, \psi)\, dx \\ V_m(\psi) &= \int_{-b/2}^{b/2} \psi_m^*(x)E_y(x, \psi)\, dx \end{aligned} \tag{41}$$

Consider next the aperture field in the pth waveguide, $E_y{}^p(x)$, when only the zeroth waveguide is excited. Let $\mathscr{C}_n{}^p$ be the coupling into the nth mode of the zeroth waveguide when the pth waveguide is excited in the fundamental mode. The aperture field $E_y^p(x)$ is then the sum of all the pertinent modes:

$$E_y^p(x) = \delta_{0p}\Phi_1(x) + \sum_{n=1}^{\infty} \mathscr{C}_n^{-p}\Phi_n(x) \tag{42}$$

The first term on the right-hand side accounts for the incident wave in the zeroth element, the excited waveguide. Notice that $\mathscr{C}_n^{-p}$ instead of $\mathscr{C}_n^{p}$ is used, because the coupling from the zeroth element to the pth element is needed here, and $\mathscr{C}_n^{p}$ is not necessarily equal to $\mathscr{C}_n^{-p}$. The coupling coefficient $\mathscr{C}_n^{-p}$ is related to the modal coefficients $v_n(\psi)$ of (40) and (41) by linear superposition. Thus,

$$v_n(\psi) = \sum_{p=-\infty}^{\infty} \mathscr{C}_n^{-p} e^{jp\psi} \tag{43}$$

and hence

$$\mathscr{C}_n^{-p} = \frac{1}{2\pi}\int_{-\pi}^{\pi} v_n(\psi)e^{-jp\psi}\, d\psi$$

Substituting this relation into (42) and then using (40) yields

$$E_y^p(x) = \frac{1}{2\pi}\int_{\pi}^{\pi} E_y(x, \psi)e^{-jp\psi}\, d\psi \tag{44}$$

Equation (44) indicates that the aperture field at the pth waveguide for the single-element excitation of the zeroth element is equal to the pth Fourier component of the aperture field when the entire array is excited. By using (40), (44) may also be written as

$$E_y^p(x) = \sum_{m=-\infty}^{\infty} \Psi_m(x)\frac{1}{2\pi}\int_{-\pi}^{\pi} V_m(\psi)e^{-jp\psi}\, d\psi \tag{45}$$

which is more convenient for later use.

It is known that the radiation pattern of an aperture type antenna is uniquely related to its aperture field distribution. In particular, the far-zone field at distance r due to a single excited element for the polarization under consideration is given by[15]

$$E_y^f(r, \theta) = \left(\frac{jk\cos^2\theta}{2\pi r}\right)^{1/2}\int_{-\infty}^{\infty} E_y^s(x')e^{j\psi x'/b}\, dx' \tag{46}$$

where $\psi = kb\sin\theta$, and $E_y^s(x')$ is the aperture field over the entire range, $-\infty < x' < \infty$, when the zeroth element is excited. Since

$$E_y^s(x') = \sum_{p=-\infty}^{\infty} E_y^p$$

(46) may be written as

$$E_y^f(r, \theta) = \left(\frac{jk\cos^2\theta}{2\pi r}\right)^{1/2} \sum_{p=-\infty}^{\infty} \int_{-b/2}^{b/2} E_y^p(x) e^{j(\psi x/b + p\psi)}\, dx \tag{47}$$

where the exponential factor in the integrand accounts for the phase difference due to the difference in location of the apertures. Substituting for $\Psi_m(x)$ from (6) and for E_y^p from (45) and rearranging, we may write (47) as

$$E_y^f(r, \theta) = \left(\frac{jk\cos^2\theta}{2\pi r}\right)^{1/2} \sum_{p=-\infty}^{\infty} \sum_{m=-\infty}^{\infty} \int_{-b/2}^{b/2} dx \sqrt{1/b}\, e^{j[(2m\pi/b)x]} \times \frac{1}{2\pi} \int_{-\pi}^{\pi} d\psi'\, V_m(\psi') e^{j(p+x/b)(\psi-\psi')} \tag{48}$$

By applying the following identity:

$$\sum_{p=-\infty}^{\infty} e^{jp(\psi-\psi')} = 2\pi\delta(\psi - \psi')$$

(48) finally becomes

$$E_y^f(r, \theta) = \left(\frac{jkb\cos^2\theta}{2\pi r}\right)^{1/2} V_0(\psi) \tag{49}$$

where

$$\psi = kb\sin\theta$$

It is well known that the far-zone magnetic field is related to the electric field by the free space admittance, $\eta_0 = \sqrt{\varepsilon_0/\mu_0}$, that is,

$$H_\theta^f(r, \theta) = \eta_0 E_y^f(r, \theta) \qquad \text{for} \quad kr \gg 1$$

The power density radiated in the θ direction for a unit power incident in the zeroth element is then

$$\mathscr{S}(\theta) = r\,\frac{E_y^f H_\theta^{f*}}{y_1} = \frac{k}{\gamma_1}\left(\frac{kb\cos^2\theta}{2\pi}\right) |V_0(\psi)|^2 \tag{50}$$

On the other hand, the transmission coefficient of the phased array is related to the modal voltage of the propagating mode by

$$T(\psi) = \sqrt{\frac{\Gamma_0}{\gamma_1}}\, V_0(\psi) \tag{51}$$

The quantity $|T(\psi)|^2$ is the power radiated in the direction normal to the array aperture for a unit power incident in each waveguide.

A comparison between (50) and (51) shows that, with $\psi = kb \sin \theta$,

$$\mathscr{S}(\theta) = |T(kb \sin \theta)|^2 \frac{b}{\lambda} \cos \theta \tag{52}$$

This is a relation for the power density patterns. The field pattern is therefore proportional to the transmission coefficient, as stated.

9. FINITE EXCITATION IN AN INFINITE ARRAY ENVIRONMENT

Infinite arrays are good approximate models for large arrays, and yet are much easier to analyze. Since all practical arrays are finite, two important questions arise with regard to the approximation. The first is, When can the approximation be applied, that is, how many elements are needed to surround a given element in order for it to behave approximately as an element in an infinite array environment? The next question concerns the edge elements: How does the behavior of the edge elements affect the performance of an array? This second question is of particular interest when small arrays are considered.

The above questions and others can best be answered at present by directly comparing the results for finite arrays with those of the corresponding infinite arrays. The solutions to finite arrays depend on how the arrays are terminated. By using the Fourier series transform relationship, we have calculated the mutual coupling between the elements in an infinite array from the reflection coefficient. Using this coupling information, we shall now determine the reflection which takes place in each waveguide when only a finite number of elements are excited in a passive infinite array environment. This approach yields information on the match characteristics of a particular finite array. In Chapter 8, we shall discuss methods of determining the radiation patterns and reflection coefficients of finite arrays under different environments. In particular, analysis and numerical data for finite arrays terminated in an infinite ground plane and in a corrugated surface will be given.

Although there is no restriction on the arrangement of a finite number of excited elements, we shall confine ourselves to two special cases which are relevant to the infinite array approximation. We first consider the simulation of a large array environment, using the square waveguide array of Fig. 4.1. Only the H plane and quasi-E plane of scan are considered. The reflection coefficient of the zeroth column of elements, when it is flanked by M active columns of elements on one side and N active columns of elements on the other side, phased for ψ scanning direction, is given by

$$R(M, N) = \sum_{m=-M}^{N} \mathscr{C}_m e^{+jm\psi} \tag{53}$$

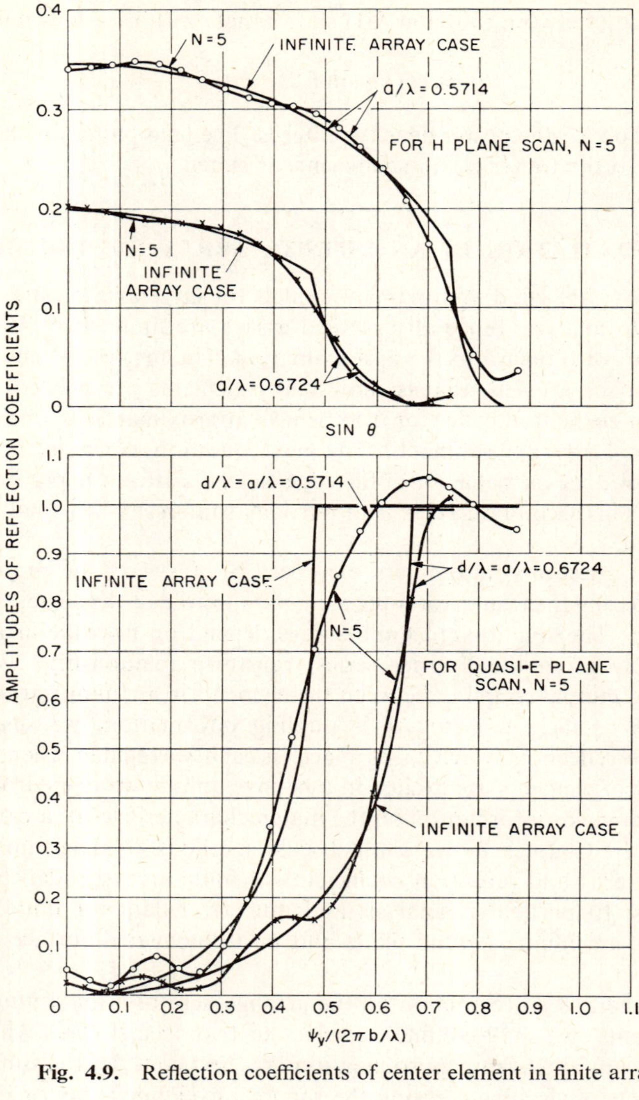

Fig. 4.9. Reflection coefficients of center element in finite array.

Strictly speaking, (53) gives the reflection coefficient of the zeroth column of elements in an array of $M + N + 1$ active columns of elements flanked by infinitely many columns of passive elements. However, both experimental and analytical studies of mutual coupling indicate that the coupling between two elements embedded in an array environment is rather independent of the location of the pair, provided that neither element is at the edge of the array.[16,17] This observation suggests that the calculation based on (53)

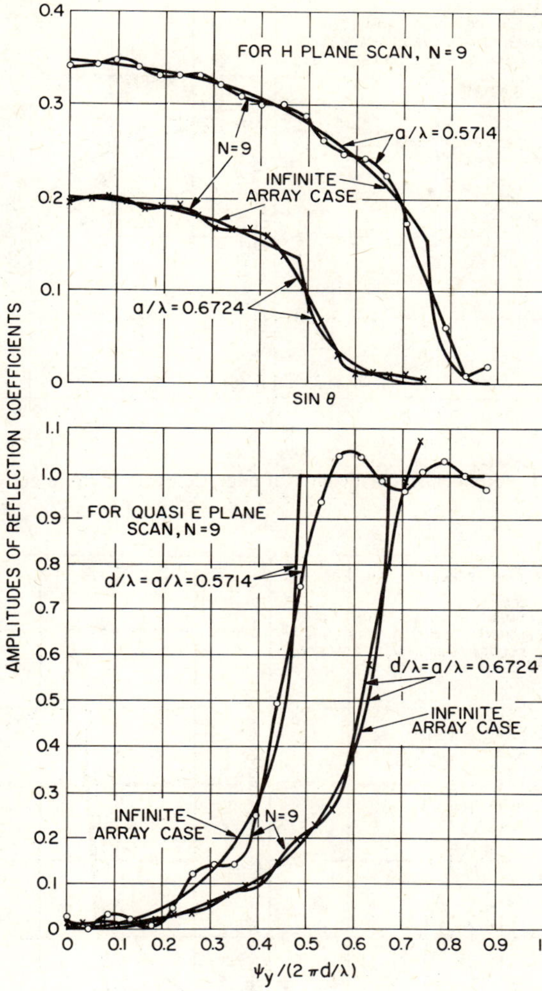

Fig. 4.10. Reflection coefficients of center element in finite array.

should be a good approximation to the finite array case if at least several columns of dummy elements are provided. Numerical results for two cases, one with $M = N = 5$ and one with $M = N = 9$, are shown in Figs. 4.9 and 4.10.

The calculation of the finite array reflection coefficients actually is equivalent to taking the Fourier partial sum of the reflection coefficient of an

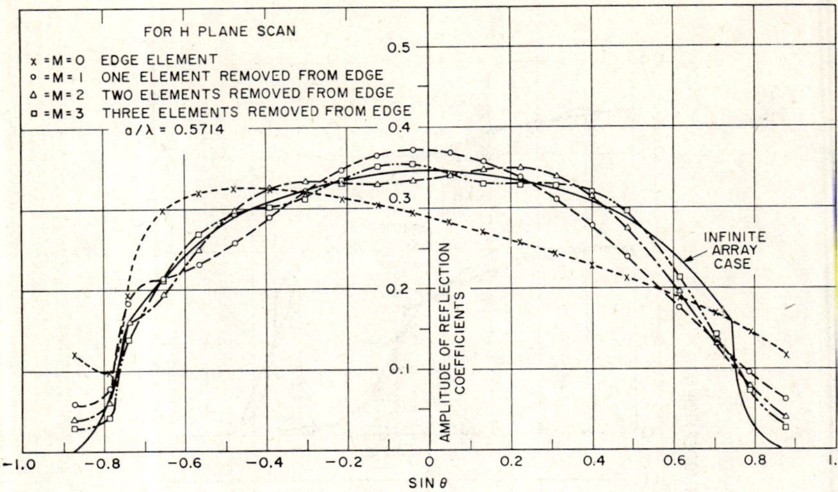

Fig. 4.11. Reflection coefficients of edge elements in finite active array.

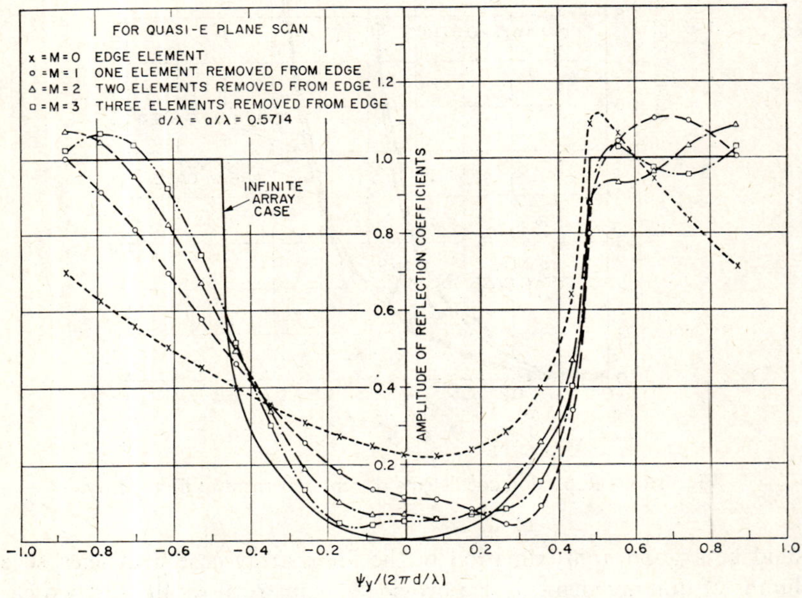

Fig. 4.12. Reflection coefficients of edge elements in finite active array.

infinite array. The rate of convergence of the partial sums is governed by the properties of the function to be approximated and depends strongly on the angle of scan. In the region where the reflection coefficient is smooth and varies slowly, the convergence is fairly rapid. On the other hand, in the region where the reflection coefficient has rapid variations, such as in the vicinity of the emergence or disappearance of a beam, the convergence is usually slow. Generally speaking, for a centrally located element in the finite array ($M = N$), as can be seen from Figs. 4.9 and 4.10, the results for $N = 5$ reproduce the infinite array results very well when the beam is not too close to the grazing position. Increasing the number of elements makes the approximation better for a wider scan range. This means that the accuracy of the simulation of a large array environment depends on the range of scan. In practical applications grating lobes are undesirable, and the scanning is usually limited to many degrees from the grating lobe formation angles. Therefore the use of five dummy elements on each side of the array should be sufficient to produce an infinite array environment. It should be noted, however, that in arrays with dielectric plug loading or dielectric sheath covering there are conditions under which certain resonance characteristics appear in the array behavior. In such situations more elements are needed to simulate a large array environment. The reader is referred to Chapter 8 for further discussion on this subject.

The coupling coefficients may be used to evaluate the edge effects in an array also. In a finite sized array, there are two edges, which may interact with each other if they are not sufficiently separated. Thus, a semi-infinite array should be considered if the effects of a single edge are to be isolated. This may be achieved by choosing a large value for N in (53), while M is set equal to 0, 1, 2, etc. This will then yield the reflection coefficient of an element located at the very edge, one element removed from the edge, etc., in a semi-infinite active array.

Since a maximum value of 40 used for M and N in (53) proved to be an excellent approximation to an infinite array situation for most angles, ψ, N is set equal to 40 in the semi-infinite array case. The results for $M = 0$ through 3 are compared with the results for infinite arrays in Figs. 4.11 and 4.12. Because of the asymmetry in the distribution of active elements around the zeroth (reference) element, scanning in both positive and negative directions has to be considered. The reflection coefficients for $M = 0$ and $M = 1$ show a considerable departure from the infinite array case. This departure diminishes as M is increased; it diminishes faster in the H-plane than in the quasi-E-plane case. Whereas in a finite array approximately five elements are required on both sides of a given element to simulate a large array environment, in a semi-infinite array an element need be removed from the edge by only about half of that number for an equivalent simulation of an infinite array. The difference is due, apparently, to the fact that in the latter case the effects of only one edge need be considered.

10. EFFECTS OF AN ANISOTROPIC PLASMA

In this section, we consider the properties of an infinite array which is immersed in an infinite anisotropic plasma.[20,21] There are two reasons for examining such a model. First, like the cases treated earlier in this chapter, the problem involves one of the few isolated situations which are amenable to exact solutions by the Wiener-Hopf technique. This is of interest in itself. Second, the closed form solution can be used to calculate the mutual coupling between the elements. Of particular interest is the asymptotic behavior of the coupling coefficients into elements which are located far away from the single excited element. We show that, because of the anisotropy of the medium, the reflection coefficient becomes asymmetric with respect to the scan angle. Consequently, the coupling into the elements on one side of an excited element is different from the coupling into the elements on the opposite side of the excited element, although the array geometry possesses reflection symmetry. On the other hand, the asymptotic behavior of the coupling coefficients which holds for an isotropic medium remains the same when the array is immersed in an anisotropic plasma.

Figure 4.13 illustrates the geometry of the problem. A thin-walled parallel-plate array is placed in an infinite plasma medium. A uniform static magnetic field oriented in the y direction is applied so that the medium is anisotropic. The plasma may be characterized by a dyadic permittivity expressible as[18]

$$\bar{\bar{\varepsilon}} = \begin{bmatrix} \varepsilon_1 & 0 & j\varepsilon_2 \\ 0 & \varepsilon_3 & 0 \\ -j\varepsilon_2 & 0 & \varepsilon_1 \end{bmatrix} \tag{54}$$

where

$$\begin{aligned} \varepsilon_1 &= 1 - \left(\frac{\omega_p}{\omega}\right)^2 \left[1 - \left(\frac{\omega_b}{\omega}\right)^2\right]^{-2} \\ \varepsilon_2 &= \left(\frac{\omega_p}{\omega}\right)\left(\frac{\omega}{\omega_b} - \frac{\omega_b}{\omega}\right)^{-1} \\ \varepsilon_3 &= 1 - \left(\frac{\omega_p}{\omega}\right)^2 \end{aligned} \tag{55}$$

with ω_p being the plasma frequency and ω_b the cyclotron frequency. We assume that the fields are independent of y. The consequence of this

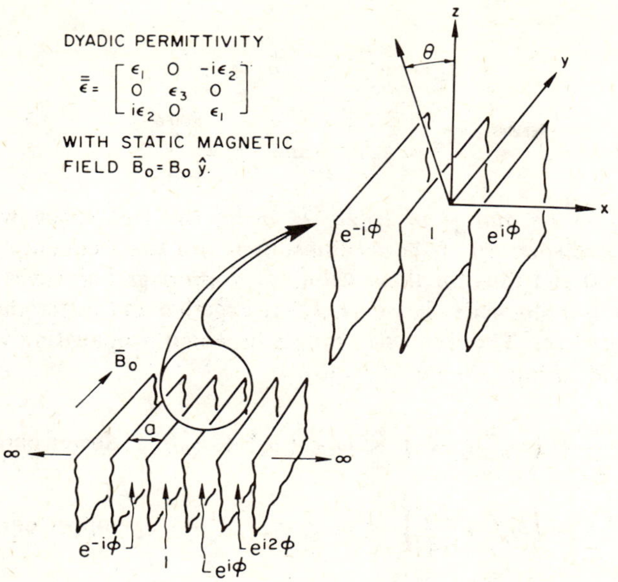

Fig. 4.13. An infinite array immersed in an anisotropic plasma.

assumption, together with the choice of the direction of the static magnetic field, is that the fields may still be decomposed into TE and TM modes with respect to z. This can be readily seen from the Maxwell equations. The TE mode is not influenced by the static magnetic field. The presence of the plasma is manifest only in a uniform change of the dielectric constant of the medium. (The dielectric constant is a function of frequency, however.) The solution of this case may be carried out in the same way as if the array were in free space, and thus the analysis of Section 4 applies.

For the TM modes, there are three nonvanishing field components, H_y, E_x, and E_z. They are related by

$$
\begin{aligned}
E_x &= \left(q_2 \frac{\partial}{\partial x} - q_1 \frac{\partial}{\partial z}\right) H_y \\
E_z &= \left(q_1 \frac{\partial}{\partial x} + q_2 \frac{\partial}{\partial z}\right) H_y \qquad (56) \\
&\left(\frac{\partial^2}{\partial x^2} + \frac{\partial^2}{\partial z^2} + k_e^2\right) H_y = 0
\end{aligned}
$$

where

$$q_1 = \frac{-j\varepsilon_1}{\omega\varepsilon_0\varepsilon}, \qquad q_2 = \frac{-\varepsilon_2}{\omega\varepsilon_0\varepsilon}$$
$$k_e^2 = k_0^2\varepsilon_e, \quad \text{and} \quad \varepsilon_e = \frac{\varepsilon}{\varepsilon_1} \tag{57}$$

with $\varepsilon = \varepsilon_1^2 - \varepsilon_2^2$ and $k_0 = 2\pi/\lambda_0$, λ_0 being the free space wavelength. Owing to the dispersion of the medium, there are two frequency regions in which $k_e^2 > 0$ and thus for these values of ω propagating waves exist. For the remaining frequencies, however, k_e^2 is negative and hence there are no propagating waves. The frequency ranges in which propagating waves exist may be found, using (55) and (57), to be

$$-\frac{\omega_b}{2} + \left(\omega_p^2 + \frac{\omega_b^2}{4}\right)^{1/2} < \omega < (\omega_p^2 + \omega_b^2)^{1/2}, \quad \text{lower band}$$
$$\frac{\omega_b}{2} + \left(\omega_p^2 + \frac{\omega_b^2}{4}\right)^{1/2} < \omega < \infty, \quad \text{upper band} \tag{58}$$

The wave modes supported by a waveguide filled with an anisotropic plasma differ from those of an empty waveguide. They may be determined

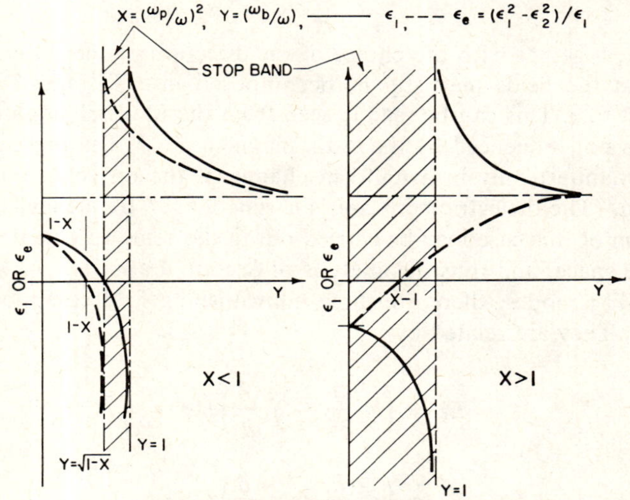

	LOWEST ORDER MODE PROPAGATES WHEN
$X<1$	$0 \le Y < \sqrt{1-X}$ OR $1<Y$
$X>1$	$1 < Y$

Fig. 4.14. Frequency bands in which incident wave is propagating.

from (56) by using the separation of variable technique with the appropriate boundary conditions. The fields of the lowest-order mode are given by

$$
\begin{aligned}
H_y &= e^{\mp\alpha_0 x \mp j\gamma_0 z} \\
E_x &= \frac{\gamma_0}{\omega\varepsilon_0\varepsilon_1}\, e^{\mp\alpha_0 x \mp j\gamma_0 z}
\end{aligned}
\tag{59}
$$

where $\gamma_0 = k_0\sqrt{\varepsilon_1}$ and $\alpha_0 = k_0\varepsilon_2/\sqrt{\varepsilon_1}$. This is a TEM mode because of the absence of the field component in the direction of propagation. It differs from an ordinary TEM mode in several respects, however. First, there is an exponential variation of the field across the waveguide cross section. This exponential variation, moreover, depends on the direction of propagation. Furthermore, the mode propagates only when γ_0 is real, that is, $\varepsilon_1 > 0$. The conditions for the incident wave to be propagating are shown in Fig. 4.14, in which $Y = \omega_b/\omega$ is varied with $X = (\omega_p/\omega)^2$ chosen as a parameter.

10.1 Solution by the Wiener-Hopf Technique*

It is assumed that the array is excited in this lowest TEM mode with a uniform amplitude and a linear phase taper. [We choose the upper sign in (59) for the wave propagating in the z direction.] The problem thus posed may be solved exactly by using the Wiener-Hopf technique.[19,20] Let $f(x, z)$ and $g(x, y)$ represent, for notational convenience, the scattered H_y and E_z. These scattered fields must satisfy the wave equation, (56). Thus,

$$
\begin{aligned}
g(x, z) &= \left(q_1\frac{\partial}{\partial x} + q_2\frac{\partial}{\partial z}\right) f(x, z) \\
\left(\frac{\partial^2}{\partial x^2} + \frac{\partial^2}{\partial z^2} + k_e^{\,2}\right) f(x, z) &= 0
\end{aligned}
\tag{60}
$$

Let the Fourier integral of functions $f(x, z)$ be defined by

$$
F(x, \gamma) = \left(\frac{1}{2\pi}\right)^{1/2} \int_{-\infty}^{\infty} f(x, z) e^{j\gamma z}\, dz
$$

Taking the Fourier transform of (60), we find that

$$
\begin{aligned}
G(x, \gamma) &= \left(q_1\frac{\partial}{\partial x} - j\gamma q_2\right) F(x, \gamma) \\
\left(\frac{\partial^2}{\partial x^2} - w^2\right) F(x, \gamma) &= 0
\end{aligned}
\tag{61}
$$

where $w^2 = \gamma^2 - k_e^{\,2}$.

* The reader who is not interested in the details of the application of the Wiener-Hopf technique may proceed directly to p. 161 for a presentation and discussion of the results of the analysis.

To ensure the convergence of the Fourier transforms and to facilitate the analysis, the medium is assumed to have a slight loss. Thus, we define $k_0 = k_0' - jk_0''$. This is an essential step in the Wiener-Hopf technique. The loss will ultimately be set equal to zero when the analysis is completed.

A general solution of the second equation in (61) may be written as

$$F(x, \gamma) = A(\gamma) \sinh wx + B(\gamma) \cosh wx \tag{62a}$$

where $A(\gamma)$ and $B(\gamma)$ are unknowns yet to be determined. From (61) and (62a) we find

$$G(x, \gamma) = [-jq_2\gamma A(\gamma) + q_1 w B(\gamma)] \sinh wx + [q_1 w A(\gamma) - jq_2\gamma B(\gamma)] \cosh wx \tag{62b}$$

The next step is to apply the boundary conditions to determine the unknowns, $A(\gamma)$ and $B(\gamma)$. For a periodic array with a uniform illumination and a linear phase taper, the scattered fields must satisfy the Floquet periodic boundary conditions, given by

$$\begin{aligned} F(x + md, \gamma) &= F(x, \gamma)e^{jm\psi} \\ G(x + md, \gamma) &= G(x, \gamma)e^{jm\psi} \end{aligned} \tag{63}$$

Applying these conditions to (62b) with $x = 0$, $m = 1$ gives

$$\begin{aligned} [-jq_2\gamma A(\gamma) + q_1 w B(\gamma)] \sinh wd + [q_1 w A(\gamma) - jq_2\gamma B(\gamma)] \cosh wd \\ = e^{j\psi}[q_1 w A(\gamma) - jq_2\gamma B(\gamma)] \end{aligned}$$

This relation may be used to express $B(\gamma)$ in terms of $A(\gamma)$ as

$$B(\gamma) = \frac{q_1 w(e^{j\psi} - \cosh wd) + jq_2\gamma \sinh wd}{q_1 w \sinh wd + jq_2\gamma(e^{j\psi} - \cosh wd)} A(\gamma)$$

This result may be substituted into (62a) and (62b) to eliminate one of the unknowns, say $B(\gamma)$. After some manipulation, we find

$$F(x, \gamma) = \frac{\begin{aligned} A(\gamma)\{q_1 w[e^{j\psi} \cosh wx - \cosh w(d - x)] \\ + jq_2\gamma[e^{j\psi} \sinh wx + \sinh w(d - x)]\} \end{aligned}}{q_1 w \sinh wd + jq_2\gamma(e^{j\psi} - \cosh wd)} \tag{64a}$$

and

$$G(x, \gamma) = A(\gamma)(q_1^2 w^2 + q_2^2\gamma^2) \frac{e^{j\psi} \sinh wx + \sinh w(d - x)}{q_1 w \sinh wd + jq_2\gamma(e^{j\psi} - \cosh wd)} \tag{64b}$$

The other boundary conditions require that the tangential electric field vanish on the conducting plane and that the tangential magnetic fields be continuous in the region $z \geq 0$. This means that

$$g(x, z) = 0 \qquad \text{for} \quad x = md, \quad z < 0 \tag{65}$$

and

$$H_y^i(x = md^+, z) + f(x = md^+, z) = H_y^i(x = md^-, z) + f(x = md^-, z) \quad \text{for} \quad z \geq 0$$

At this stage, it is desirable to introduce the following standard notations commonly used in the Wiener-Hopf procedure:[19]

$$F_+(x, \gamma) = \left(\frac{1}{2\pi}\right)^{1/2} \int_0^{\infty} f(x, z)e^{j\gamma z}\, dz$$

and

$$F_-(x, \gamma) = \left(\frac{1}{2\pi}\right)^{1/2} \int_{-\infty}^{0} f(x, z)e^{j\gamma z}\, dz$$

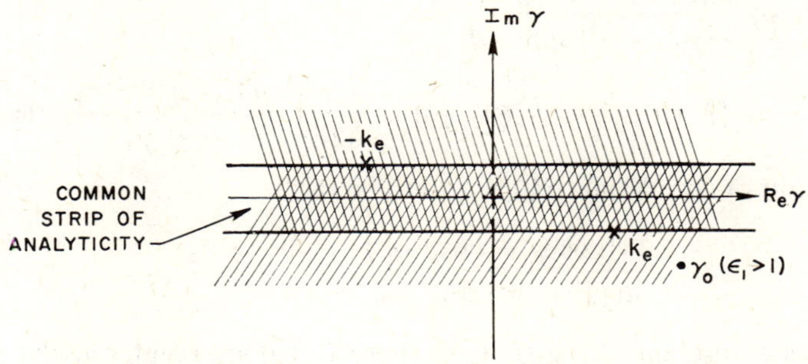

Fig. 4.15. Complex γ plane.

The subscripts + and − denote that these expressions are analytic in the appropriate upper and lower halves of the complex γ plane, respectively, as shown in Fig. 4.15. This is a result of the Fourier transform theory.[19] Taking the Fourier transform of the boundary conditions as given by (65) for $m = 0$ yields

$$G_-(0, \gamma) = 0$$

and (66)

$$[F_+^i(0^+, \gamma) + F_+(0^+, \gamma)] - [F_+^i(0^-, \gamma) + F_+(0^-, \gamma)] = 0$$

where F_+^i is the transform of the incident magnetic field and is given by

$$F_+^i(x, \gamma) = \left(\frac{1}{2\pi}\right)^{1/2} \int_0^{\infty} (e^{-\alpha_0 x - j\gamma_0 z})e^{j\gamma z}\, dz$$

$$= \left(\frac{1}{2\pi}\right)^{1/2} e^{-\alpha_0 x} \frac{j}{\gamma - \gamma_0}, \quad \text{Im}\,(\gamma - \gamma_0) > 0$$

Using these results, we find that

$$A(\gamma) = G_+(0, \gamma) \frac{q_1 w \sinh wd + jq_2\gamma(e^{j\psi} - \cosh wd)}{(q_1^2 w^2 + q_2^2\gamma^2) \sinh wd} \tag{67a}$$

and

$$[F_-(0^+, \gamma) - F_-(0^-, \gamma)] - \left(\frac{1}{2\pi}\right)^{1/2} \frac{j(1 - e^{-j\psi - \alpha_0 d})}{\gamma - \gamma_0}$$
$$= A(\gamma) \frac{2q_1 w(\cos\psi - \cosh wd)}{q_1 w \sinh wd + jq_2\gamma(e^{j\psi} - \cosh wd)} \tag{67b}$$

After slight rearrangement, combination of (67a) and (67b) yields

$$\frac{q_1 d}{q_1^2 + q_2^2} G_+(0, \gamma) \frac{2w(\cos\psi - \cosh wd)}{d(\gamma_0^2 - \gamma^2) \sinh wd}$$
$$= [F_-(0^+, \gamma) - F_-(0^-, \gamma)] - \left(\frac{1}{2\pi}\right)^{1/2} \frac{j(1 - e^{-j\psi - \alpha_0 d})}{\gamma - \gamma_0} \tag{68}$$

This is an equation of the Wiener-Hopf type. The term

$$\frac{2w(\cos\psi - \cosh wd)}{d(\gamma_0^2 - \gamma^2) \sinh wd}$$

may be written as a quotient: $L_+(\gamma)/L_-(\gamma)$, where $L_\pm(\gamma)$ are regular analytic in the upper and the lower halves, respectively, of the complex γ plane. The detail of this factorization is given in the appendix. Multiplying the equation throughout by $L_-(\gamma)$ gives

$$\frac{q_1 d}{q_1^2 + q_2^2} G_+(0, \gamma) L_+(\gamma) + \left(\frac{1}{2\pi}\right)^{1/2} \frac{j(1 - e^{-j\psi - \alpha_0 d})}{\gamma - \gamma_0} L_-(\gamma_0)$$
$$= [F_-(0^+, \gamma) - F_-(0^-, \gamma)]L_-(\gamma) - \left(\frac{1}{2\pi}\right)^{1/2} \frac{j(1 - e^{-j\psi - \alpha_0 d})}{\gamma - \gamma_0} [L_-(\gamma) - L_-(\gamma_0)] \tag{69}$$

where the second term on the left side of (69) has been added on both sides of the equation. This is done to regularize the second term on the right-hand side, which without the added term would have a pole at $\gamma = \gamma_0 = k_0'\sqrt{\varepsilon_1} - jk_0''\sqrt{\varepsilon_1}$. Now, the left side is regular analytic in the upper half plane, while the right side is regular analytic in the lower half plane. Both sides have a common strip of analyticity. Therefore, each side is the analytic continuation of the other, and taken together they form an entire function. Since it can be shown that each side approaches zero as $|\gamma| \to \infty$, by a theorem due to

Liouville,[19] each side must be equal to the constant zero. Thus, we find

$$G_+(0, \gamma) = -\left(\frac{1}{2\pi}\right)^{1/2} \frac{q_1^{\,2} + q_2^{\,2}}{q_1 d} \frac{j(1 - e^{-j\psi - \alpha_0 d})}{\gamma - \gamma_0} \frac{L_-(\gamma_0)}{L_+(\gamma)} \tag{70}$$

and by (67a)

$$A(\gamma) = \left(\frac{1}{2\pi}\right)^{1/2} \frac{j(1 - e^{-j\psi - \alpha_0 d})}{q_1 d(\gamma - \gamma_0)^2(\gamma + \gamma_0)} \frac{L_-(\gamma_0)}{L_+(\gamma)} \times \frac{q_1 w \sinh wd + jq_2\gamma(e^{j\psi} - \cosh wd)}{\sinh wd} \tag{71}$$

Using (71) in (64a) and (64b), we determine the expression for $F(x, \gamma)$ and $G(x, \gamma)$. The fields may then be calculated by the inverse transforms of these functions as

$$\begin{matrix} f(x, z) \\ g(x, z) \end{matrix} = \left(\frac{1}{2\pi}\right)^{1/2} \int_{-\infty}^{\infty} \begin{matrix} F(x, \gamma) \\ G(x, \gamma) \end{matrix} e^{-j\gamma z}\, d\gamma \tag{72}$$

In evaluating these integrals it is convenient to divide the space into two regions: the region inside and the region outside the parallel plates. For the region inside the plates, $z \leq 0$, we may close the contour of integration in the upper half plane by a semicircle of large radius, while for the region outside the plates, $z \geq 0$, we may close the contour in the lower half plane (see Fig. 4.15). The radius may then be allowed to grow to infinity. Since it can be shown that the integral along the infinite semicircle vanishes, and that there are only simple poles in both the upper and lower half planes, the integral of (72) is then equal to $\pm 2\pi i$ times the residues contributed from the poles in the respective half plane. An examination of the expression for $F(x, \gamma)$ and $G(x, \gamma)$ will show that in the upper half plane the poles are located at $\gamma = -\gamma_n$, $n = 0, 1, 2, \ldots$. The poles in the lower half planes are situated at $\gamma = \gamma_0$ and Γ_n, for $n = 0, \pm 1, \pm 2$. Although the contour integration yields all the terms contributed by the pole singularities, only those corresponding to the propagating modes are of interest. Under the assumption that $d/\lambda_0 < \frac{1}{2}$, a single propagating scattered field exists in $z \leq 0$, which is found by a standard residue evaluation at $\gamma = -\gamma_0$. Thus,

$$f(x, z) \approx R(\psi)e^{\alpha_0 x + j\gamma_0 z} \qquad \text{for} \quad z \leq 0$$

where

$$R(\psi) = \frac{\sinh \frac{1}{2}(j\psi - \alpha_0 d)}{\sinh \frac{1}{2}(j\psi + \alpha_0 d)} \left(\frac{\gamma_0 + \Gamma_0}{\gamma_0 \Gamma_0}\right) \times \prod_{n=1}^{\infty} \left(\frac{\gamma_n + \gamma_0}{\gamma_n - \gamma_0}\right) \left(\frac{\Gamma_n - \gamma_0}{\Gamma_n + \gamma_0}\right) \left(\frac{\Gamma_{-n} - \gamma_0}{\Gamma_{-n} + \gamma_0}\right) e^{j(2\gamma_0 d/\pi)\ln 2} \tag{73a}$$

is the reflection coefficient, and

$$\Gamma_n = \sqrt{k_e^{\,2} - [(2n\pi + \psi)/d]^2} \quad \text{and} \quad \gamma_n = \sqrt{k_e^{\,2} - (n\pi/d)^2}$$

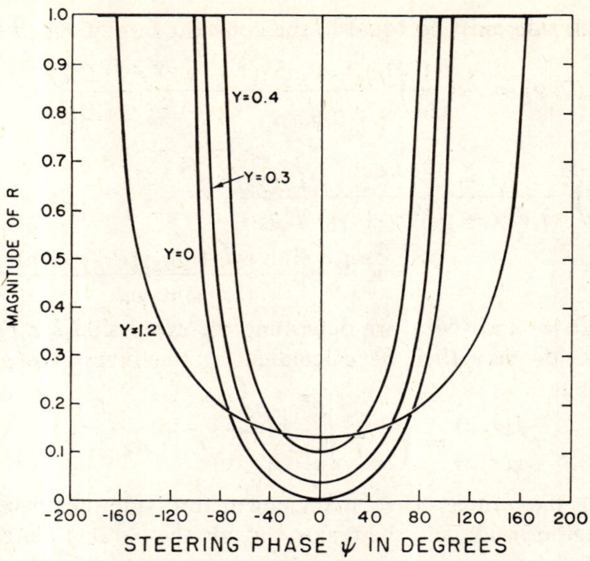

Fig. 4.16*a*. Reflection coefficient of an array in an anisotropic plasma; $X = 0.49$, $d/\lambda_0 = 0.4$.

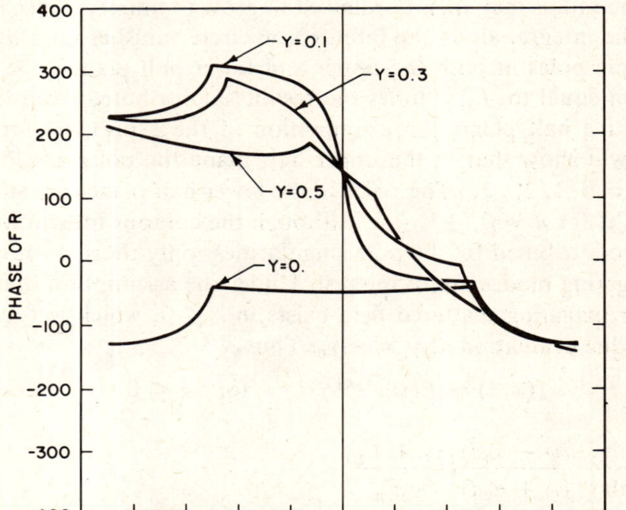

Fig. 4.16*b*. Reflection coefficient of an array in an anisotropic plasma; $X = 0.49$, $d/\lambda_0 = 0.4$.

are the propagation constants of the nth mode in the region exterior and the region interior to the waveguides, respectively. The field in the region $z \geq 0$ is found to be

$$f(x, z) \approx T(\psi)e^{-j\Gamma_0 x} \tag{73b}$$

where

$$T(\psi) = 2j \sinh \tfrac{1}{2}(j\psi + \alpha_0 d)\, \frac{q_1\psi/d + jq_2\Gamma_0}{q_1\psi/d}\left(\frac{\gamma_0}{\gamma_0 + \Gamma_0}\right)\left(\frac{\sin\psi}{\cosh\alpha_0 d - \cos\psi}\right)$$

$$\times \prod_{n=1}^{\infty}\left(\frac{\gamma_n + \gamma_0}{\gamma_n - \gamma_0}\right)\left(\frac{\Gamma_n - \gamma_0}{\Gamma_n + \gamma_0}\right)\left(\frac{\Gamma_{-n} - \gamma_0}{\Gamma_{-n} + \gamma_0}\right)e^{j[(\Gamma_0 - \gamma_0)d/\pi]\ln 2} \tag{73c}$$

is the transmission coefficient.

Notice that in (73a), because of the term $\sinh \frac{1}{2}(j\psi - \alpha_0 d)/\sinh \frac{1}{2}(j\psi + \alpha_0 d)$, which arises as a result of the anisotropy of the medium, the reflection coefficient is neither an even nor an odd function of the steering phase, ψ, even though there exists a reflection symmetry in the array geometry. The asymmetry occurs only in the phases of R, however.

In addition to the dependence on the element spacing and the scan angle, the reflection coefficients are functions of the properties of the medium also. Figures 4.16a and 4.16b present the reflection coefficients versus ψ for $d/\lambda_0 = 0.4$ and $X = (\omega_p/\omega)^2 = 0.49$ with $Y = \omega_b/\omega$ kept as a parameter. Since the values of Y affect the effective propagation constant, k_e, and consequently the effective element spacing, d/λ_e, the ranges of the steering phase, ψ, for which a beam is radiated vary as Y is changed. Another interesting point is that, unlike the results for this polarization in the isotropic medium, the reflection coefficient here is not equal to zero in the broadside direction.

10.2 Coupling Analysis

As was discussed in Section 6, the reflection coefficient in an element may be regarded as a linear superposition of the coupling from all the elements. Notice that $\mathscr{C}_n$ in (30a) should be interpreted as the coupling from the nth element to the zeroth element instead of the other way around. The reason is that the coupling from the nth element to the zeroth element is not equal to the coupling from the zeroth element to the nth element, as the ordinary reciprocity relation does not hold in an anisotropic medium.* However, the

* These two couplings will be equal to each other if, in determining the coupling from the zeroth element to the nth one, the array is immersed in a transposed medium, that is, a medium in which the direction of the static magnetic field is reversed. This, in the absence of losses, is equivalent to reversing the sign of the off-diagonal terms of the dyadic permittivity.

coupling from the nth element to the zeroth element is equal to the coupling from the zeroth element to the $(-n)$th element, because of the translational invariance of the array geometry.

Equation (31) may be utilized to calculate the coupling coefficients from the reflection coefficient. The closed form solution of (73a) is particularly useful for this purpose. As the above discussion indicates, the coupling from the zeroth element to the nth element is not equal to the coupling to the $(-n)$th element. How different these couplings are depends on the anisotropy of the medium. Calculations for $X = 0.49$ and $d/\lambda_0 = 0.4$ are shown in Fig. 4.17.

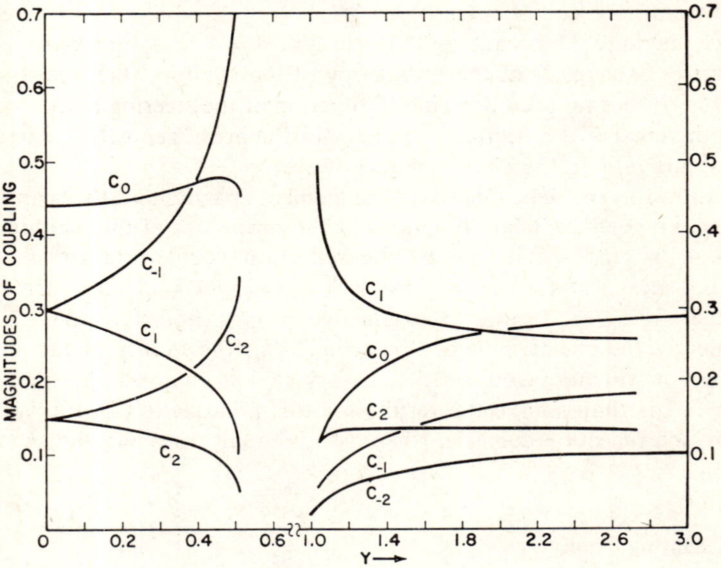

Fig. 4.17. Near-in couplings as a function of Y.

For $Y = 0$, corresponding to the isotropic case, $\mathscr{C}_n$ and $\mathscr{C}_{-n}$ are equal. They diverge from each other, however, as Y is increased. Similarly, for very large values of Y such that the medium becomes isotropic, these couplings converge to the same values again.

Figures 4.18a and 4.18b show the coupling coefficients as functions of the element location for $X = 0.49$ and $Y = 0.5$. Aside from the fact that the coupling to the elements in the positive x direction differs from the coupling to the elements in the negative x direction, the coupling coefficients display a monotonic decay with increasing element separation such as was found in

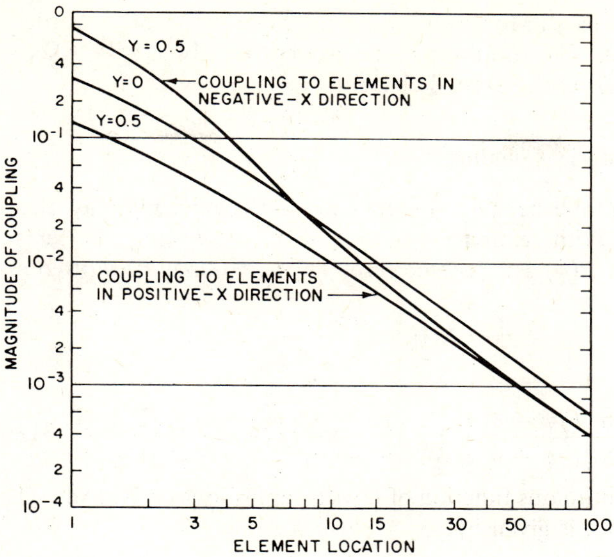

Fig. 4.18*a*. Coupling coefficient as a function of element location; $X = 0.49$, $d/\lambda_0 = 0.4$.

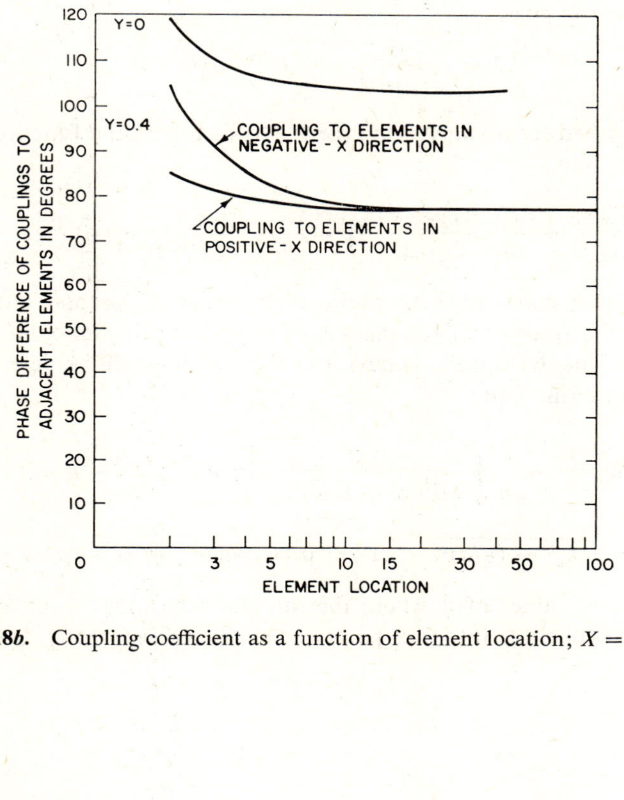

Fig. 4.18*b*. Coupling coefficient as a function of element location; $X = 0.49$, $d/\lambda_0 = 0.4$.

the isotropic case. Moreover, for large separations of the order of ten or more elements, the coupling decays asymptotically as $r^{-3/2}$, where r denotes the distance from the excited element.

10.3 Asymptotic Coupling

The asymptotic behavior of the coupling coefficients may also be shown analytically. With reference to (73a), if $d/\lambda_e < \frac{1}{2}$, only the term Γ_0 in the expression of $R(\psi)$ can become zero. Thus $R(\psi)$ may be written as

$$R(\psi) = \left(\frac{\gamma_0 - \Gamma_0}{\gamma_0 + \Gamma_0}\right) U(\psi) \tag{74}$$

where

$$U(\psi) = \frac{\sinh \frac{1}{2}(j\psi - \alpha_0 d)}{\sinh \frac{1}{2}(j\psi + \alpha_0 d)} \prod_{n=1}^{\infty} \left(\frac{\gamma_n + \gamma_0}{\gamma_n - \gamma_0}\right) \left(\frac{\Gamma_n - \gamma_0}{\Gamma_n + \gamma_0}\right) \left(\frac{\Gamma_{-n} - \gamma_0}{\Gamma_{-n} + \gamma_0}\right) e^{j(2\gamma_0 d/\pi) \ln 2}$$

which is a continuous function of ψ with continuous derivatives. The coupling coefficient, $\mathscr{C}_n$, is given by

$$\mathscr{C}_n = \frac{1}{2\pi} \int_{-\pi}^{\pi} R(\psi) e^{-jn\psi}\, d\psi = \frac{1}{2\pi} \int_{-\pi}^{\pi} \left(\frac{\gamma_0 - \Gamma_0}{\gamma_0 + \Gamma_0}\right) U(\psi) e^{-jn\psi}\, d\psi \tag{75}$$

Integration by parts gives

$$2\pi\mathscr{C}_n = \frac{1}{jn} \int_{-\pi}^{\pi} \frac{dR}{d\psi} e^{-jn\psi}\, d\psi \tag{76}$$

where the integrated term vanishes because $R(\psi)$ is a periodic function of ψ with period 2π. Now,

$$\frac{dR(\psi)}{d\psi} = \frac{dU(\psi)}{d\psi}\left(\frac{\gamma_0 - \Gamma_0}{\gamma_0 + \Gamma_0}\right) - \frac{2\gamma_0\psi}{d^2(\gamma_0 + \Gamma_0)^2\Gamma_0} U(\psi) \tag{77}$$

The first term is a continuous function of ψ, while the second term has singularities at the zeros of Γ_0 (as this factor appears in the denominator of the expression). The asymptotic behavior of the coupling will be determined from the second term. Thus

$$\lim_{n\to\infty} 2\pi\mathscr{C}_n = -\frac{1}{jn} \int_{-\pi}^{\pi} \frac{2\gamma_0\psi}{d^2(\gamma_0 + \Gamma_0)^2\Gamma_0} U(\psi) e^{-jn\psi}\, d\psi + 0(n^{-2}) \tag{78}$$

Since

$$\Gamma_0 = \sqrt{k_e^2 - (\psi/d)^2}, \qquad \Gamma_0 = 0 \qquad \text{when} \quad \psi = \pm k_e d$$

These are the two values of ψ where the integrand becomes singular. The integral may be broken up into four parts and evaluated separately. Consider the following integral first:

$$\Lambda_1 = \int_{-\pi}^{-k_e d} \frac{2\gamma_0 \psi}{d^2(\gamma_0 + \Gamma_0)^2 \Gamma_0} U(\psi) e^{-jn\psi}\, d\psi$$

$$= -\int_{k_e d}^{\pi} \frac{2\gamma_0 \psi}{d^2(\gamma_0 + \Gamma_0)^2 \Gamma_0} U(-\psi) e^{jn\psi}\, d\psi$$

By following the procedure used in (37) to isolate the singular point in the integrand, we find that

$$\Lambda_1 = j \frac{2k_e}{\gamma_0} U(-k_e d) \left(\frac{\pi}{2nk_e d}\right)^{1/2} e^{j(nk_e d + \pi/4)} + 0(n^{-1})$$

Similarly, we obtain

$$\Lambda_2 = \int_{-k_e d}^{0} \frac{2\gamma_0 \psi}{d^2(\gamma_0 + \Gamma_0)^2 \Gamma_0} U(\psi) e^{-jn\psi}\, d\psi$$

$$= -\int_{0}^{k_e d} \frac{2\gamma_0 \psi}{d^2(\gamma_0 + \Gamma_0)^2 \Gamma_0} U(-\psi) e^{jn\psi}\, d\psi$$

$$= j \frac{2k_e}{\gamma_0} U(-k_e d) \left(\frac{\pi}{2nk_e d}\right)^{1/2} e^{j(nk_e d + \pi/4)} + 0(n^{-1})$$

The results show that Λ_1 and Λ_2 make the same contribution to the asymptotic behavior of the coupling. Continuing the process, we may evaluate the integrals from 0 to $k_e d$ and from $k_e d$ to π to determine the contribution from the other singularity at $\psi = k_e d$. It turns out that the singularity approached from below gives rise to the same asymptotic dependence as is contributed by the singularity approached from above, but with an opposite sign. Therefore, the contribution from $\psi = k_e d$ is of a higher order, $0(n^{-1})$.

Adding up the results, we find that

$$\lim_{n \to \infty} \mathscr{C}_n = \frac{k_e}{\gamma_0} U(-k_e d) \left(\frac{2}{\pi k_e d}\right)^{1/2} \frac{1}{n^{3/2}} e^{j(nk_e d + \pi/4)} + 0(n^{-2})$$

that is, asymptotically, the coupling decays as $n^{-3/2}$ for large n. The asymptotic phase difference of the coupling into two adjacent elements is equal to $k_e d$, the electrical length by free propagation. This asymptotic behavior is identical to the result for an isotropic medium, as discussed in Section 6. The fact that the anisotropy of the medium does not alter the asymptotic dependence of the coupling lends strong support to the conjecture that the asymptotic behavior may be a very general property of phased array antennas. An extension to a broad class of planar arrays will be carried out in Chapter 7 with the same conclusion.

APPENDIX WIENER-HOPF FACTORIZATION

The factorization of the expression

$$\frac{2w\,(\cos\psi - \cosh wd)}{d(\gamma_0^2 - \gamma^2)\sinh wd} \tag{A.1}$$

may be carried out as follows. Since

$$\cos\psi - \cosh wd = 2\sin\tfrac{1}{2}(\psi - jwd)\sin\tfrac{1}{2}(\psi + jwd)$$

we can write the sine functions in terms of infinite products and obtain

$$\begin{aligned}
&\cos\psi - \cosh wd \\
&= 2\cdot\tfrac{1}{2}(\psi - jwd)\prod_{n=1}^{\infty}\left[1 - \frac{(\psi - jwd)^2}{4n^2\pi^2}\right]\cdot\tfrac{1}{2}(\psi + jwd)\prod_{n=1}^{\infty}\left[1 - \frac{(\psi + jwd)^2}{4n^2\pi^2}\right] \\
&= \frac{d^2}{2}\left[\left(\frac{\psi}{d}\right)^2 + \gamma^2 - k_e^2\right] \\
&\quad\times\prod_{n=1}^{\infty}\left(\frac{d}{2n\pi}\right)^2\left[\left(\frac{2n\pi - \psi}{d}\right)^2 + \gamma^2 - k_e^2\right]\left(\frac{d}{2n\pi}\right)^2\left[\left(\frac{2n\pi + \psi}{d}\right)^2 + \gamma^2 - k_e^2\right]
\end{aligned} \tag{A.2}$$

We define

$$\Gamma_n^2 = k_e^2 - \left(\frac{2n\pi + \psi}{d}\right)^2 \tag{A.3}$$

The above results may then be written as

$$\begin{aligned}
&\frac{2}{d^2}(\cos\psi - \cosh wd) \\
&\qquad = (\gamma^2 - \Gamma_0^2)\prod_{n=1}^{\infty}\left(\frac{d}{2n\pi}\right)^2(\gamma^2 - \Gamma_{-n}^2)\left(\frac{d}{2n\pi}\right)^2(\gamma^2 - \Gamma_n^2)
\end{aligned} \tag{A.4}$$

A similar expansion may be applied to sinh wd to yield

$$\sinh wd = wd\prod_{n=1}^{\infty}\left(\frac{d}{n\pi}\right)^2(\gamma^2 - \gamma_n^2) \tag{A.5}$$

where

$$\gamma_n^2 = k_e^2 - \left(\frac{n\pi}{d}\right)^2 \tag{A.6}$$

Combining (A.4) and (A.5), we may write (A.1) as

$$\frac{2w(\cos\psi - \cosh wd)}{d(\gamma_0^2 - \gamma^2)\sinh wd} = \left(\frac{\gamma^2 - \Gamma_0^2}{\gamma_0^2 - \gamma^2}\right)\prod_{n=1}^{\infty}\frac{(d/2n\pi)^2(\gamma^2 - \Gamma_n^2)(d/2n\pi)^2(\gamma^2 - \Gamma_{-n}^2)}{(d/n\pi)^2(\gamma^2 - \gamma_n^2)} \quad \text{(A.7)}$$

Since the medium is assumed slightly lossy, $k_e = k_e' - jk_e''$. Moreover, to ensure convergence of the inverse Fourier integrals in (72), it is necessary to introduce proper branch cuts into the complex γ plane. Consequently, it is found that both γ_n and Γ_n have negative imaginary parts. Thus, the expression

$$L_+(\gamma) = j\left(\frac{\gamma - \Gamma_0}{\gamma_0 - \gamma}\right)\prod_{n=1}^{\infty}\frac{(d/2n\pi)^2(\gamma - \Gamma_n)(\gamma - \Gamma_{-n})}{(jd/n\pi)(\gamma - \gamma_n)}\, e^{j(\gamma d/\pi)\ln 2}$$

is regular analytic in the upper half plane,

$$\frac{1}{L_-(\gamma)} = -j\left(\frac{\gamma + \Gamma_0}{\gamma_0 + \gamma}\right)\prod_{n=1}^{\infty}\frac{(d/2n\pi)^2(\gamma + \Gamma_n)(\gamma + \Gamma_{-n})}{(-jd/n\pi)(\gamma + \gamma_n)}\, e^{-j(\gamma d/\pi)\ln 2}$$

is regular analytic in the lower half plane, and

$$\frac{2w(\cos\psi - \cosh wd)}{d(\gamma_0^2 - \gamma^2)\sinh wd} = \frac{L_+(\gamma)}{L_-(\gamma)}$$

The exponential factors in $L_+(\gamma)$ and $L_-(\gamma)$ are introduced in order to render the infinite product covergent. Notice also that $L_+(\gamma) = 1/L_-(-\gamma)$.

REFERENCES

1. C. P. Wu and V. Galindo, "Properties of a Phased Array of Rectangular Waveguides with Thin Walls," *IEEE Trans. Antennas and Propagation*, Vol. AP-14, pp. 163–172, March 1966.
2. V. Galindo and C. P. Wu, "Numerical Solutions for an Infinite Phased Array of Rectangular Waveguides with Thick Walls," *IEEE Trans. Antennas and Propagation*, Vol. AP-14, pp. 149–158, March 1966.
3. R. E. Collin, *Field Theory of Guided Waves*, McGraw-Hill Book Company, New York, 1960, pp. 465–468.
4. J. Blass and S. J. Rabinowitz, "Mutual Coupling in Two Dimensional Arrays," 1957 *IRE WESCON Convention Record*, Part 1, pp. 134–150.
5. P. W. Hannan, "The Element-Gain Paradox for a Phased-Array Antenna," *IEEE Trans. Antennas and Propagation*, Vol. AP-12, pp. 423–433, July 1964.
6. N. Amitay, J. S. Cook, R. G. Pecina, and C. P. Wu, "Mutual Coupling and Matching Conditions in Large Planar Phased Arrays," *1964 PTGAP International Symposium Program and Digest*, pp. 150–156.

7. V. Galindo and C. P. Wu, "Asymptotic Behavior of the Coupling Coefficients for an Infinite Array of Thin-Walled Rectangular Waveguides," *IEEE Trans. Antennas and Propagation*, Vol. AP-14, pp. 248–249, March 1966.

8. E. C. Titchmarsh, *The Theory of Functions*, 2nd ed., Oxford University Press, London, 1939, pp. 426–427.

9. P. M. Morse and H. Feshback, *Methods of Theoretical Physics*, Vol. II, McGraw-Hill Book Company, New York, 1953, p. 1324.

10. J. R. Wait, *Electromagnetic Waves in Stratified Media*, Pergamon Press, New York, 1962, pp. 25–35.

11. V. Galindo and C. P. Wu, "The Relation Between the Far-Zone Pattern of the Singly Excited Element and the Transmission Coefficient of the Principal Lobe in an Infinite Array," *IEEE Trans. Antennas and Propagation*, Vol. AP-14, pp. 397–398, May 1966.

12. E. V. Baklanov, "Radiation of Electromagnetic Waves from a System of Semi-infinite Plates," *Soviet Phys. Doklady*, Vol. 8, pp. 1100–1102, May 1964.

13. D. S. Karjala and R. Mittra, "Radiation from Some Periodic Structures Excited by a Waveguide," *Electronics Letters*, Vol. 1, pp. 111–112, June 1965.

14. E. C. Du Fort, "Finite Scattering Matrix for an Infinite Antenna Array," *Radio Science*, Vol. 2 (New Series) pp. 19–27, January 1967.

15. S. Silver (ed.), *Microwave Antenna Theory and Design*, M.I.T. Radiation Laboratory Series, Vol. 12, McGraw-Hill Book Company, New York, 1949, p. 89.

16. S. W. Lee, "Radiation from an Infinite Aperiodic Array of Parallel-Plate Waveguides," *IEEE Trans. Antennas and Propagation*, Vol. AP-15, pp. 598–606, September 1967.

17. C. P. Wu, "Analysis of Finite Parallel-Plate Waveguide Arrays," *IEEE International G-AP Symposium Digest*, pp. 124–133, 1968, and *IEEE Trans. Antennas and Propagation*, pp. 328–334, May 1970.

18. K. G. Budden, *Radio Waves in the Ionosphere*, Cambridge University Press, London, 1961.

19. N. Noble, *Methods Based on the Wiener-Hopf Technique*, Pergamon Press, New York, 1965.

20. C. P. Wu, "Diffraction of a Plane Electromagnetic Wave by an Infinite Set of Parallel Metallic Plates in an Anisotropic Plasma," *Can. J. Phys.*, Vol. 45, pp. 1911–1923, May 1967.

21. C. P. Wu and V. Galindo, "Radiation from an Infinite Phased Array of Parallel-Plate Waveguides in an Anisotropic Plasma," *IEEE International G-AP Symposium Digest*, pp. 98–101, 1966.

22. J. Schwinger and D. Saxon, *Discontinuities in Waveguides—Notes on Lectures by J. Schwinger*, Gordon and Breach, New York, 1968, Chapter V.

5. Arrays of Rectangular Waveguides

In Chapter 4 we saw examples of essentially the only exact solutions to phased array problems that are available. Although much information concerning this type of array can be derived from the exact solution, the limitations are very severe. For example, to determine the effect of waveguide wall thickness or dielectric covers or plugs in the parallel-plate arrays, we must resort to approximate methods such as those discussed in Chapter 3. Furthermore, when a planar or two-dimensional array of waveguides is scanned to an arbitrary pointing direction, the relevant integral equation is vector two dimensional (Chapter 2) and, as a result, cannot be solved exactly.

In this chapter, therefore, we will apply some of the approximation techniques discussed earlier to solve the integral equations appropriate to thick-walled parallel-plate waveguides and planar arrays of rectangular waveguides. In addition to presenting results from the applied numerical analysis, we will discuss the very significant problem of verifying the validity of these results by various tests.

The approximate solutions to the thick-walled parallel-plate array and also the planar array of rectangular waveguides illustrate *new* phenomena that did not occur in the linear array of thin-walled parallel plates. In particular, an unusual scan-dependent resonance phenomenon will be discussed.

1. NUMERICAL METHODS AND RESULTS FOR *E* AND *H* PLANES OF SCAN

The thin-walled parallel-plate waveguide phased array with a homogeneous dielectric in the waveguides and free space region was solved exactly in Chapter 4. The problem, put in the form of a *scalar one-dimensional* integral equation of the first kind, was derived from two particular scan planes of a rectangular array of rectangular waveguides.

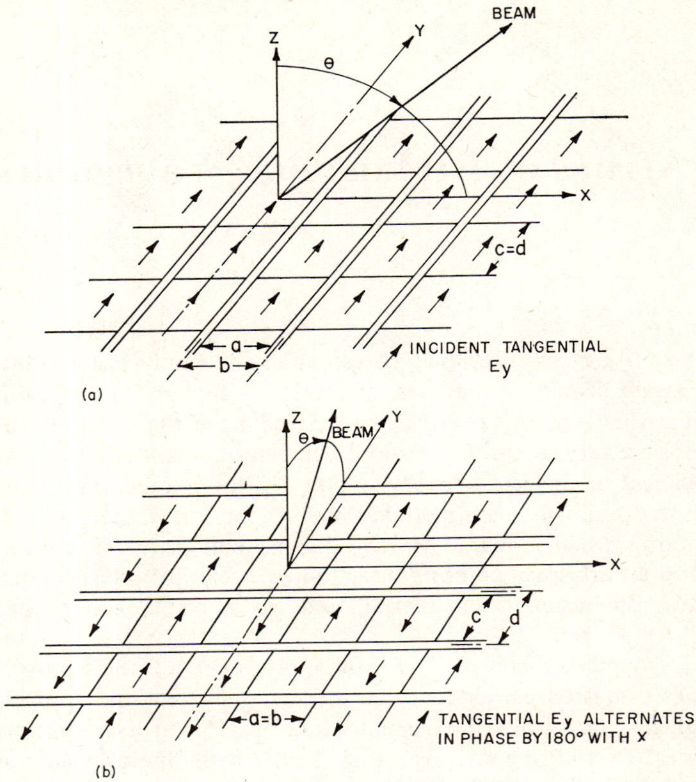

Fig. 5.1. Infinite array geometry. (*a*) *H*-plane scanning in x-z plane. (*b*) Quasi-*E*-plane scanning in y-z plane.

The geometry of this array is shown in Fig. 5.1, in which the *H* plane and quasi-*E* planes of scan are illustrated. The quasi-*E* plane of scan becomes an *E*-plane scan of a parallel-plate array as the dimension a becomes infinitely large. When the waveguide walls are thin, $a = b$ and $c = d$, the appropriate integral equation is scalar one dimensional and can be solved exactly as discussed in Chapter 4. If, however, we allow the waveguide walls to become *thick* only in the plane of scan (see Fig. 5.1), we can again reduce the problem to a scalar one-dimensional integral equation. However, we cannot reduce the integral equation to a form suitable for an *exact* solution, and therefore we will consider various approximate solutions.

Finally, it may be pointed out here that an inhomogeneous or stratified dielectric filling of the waveguides and free space regions will also permit a scalar one-dimensional integral equation to be formulated for the structure in Fig. 5.1. The effects of dielectric stratification will be discussed in Chapter 6.

1.1 The Scalar Integral Equations

Integral equations for the unknown tangential electric or magnetic fields in the aperture ($z = 0$ in Fig. 5.1) were derived in Chapter 2. In Chapter 4, more specific forms were derived for the structure in Fig. 5.1.

For the H-plane scan, we will solve the integral equation for the unknown tangential magnetic field, H_x, in the aperture. As we know from Chapter 2, the only field components that exist for this plane of scan are the $\mathscr{H}_x$, $\mathscr{H}_z$, and $\mathscr{E}_y$ components. We can represent the $\mathscr{H}_x$ field, interior to the waveguides, as

$$\mathscr{H}_x = \left[(e^{-j\beta_1 z} - Re^{+j\beta_1 z})\phi_1(x) + \sum_{n=2}^{\infty} i_n e^{+j\beta_n z}\phi_n(x)\right] e^{-jk\,(\sin\theta)\,x_p} \qquad (1)$$

where R is the scattering matrix reflection coefficient. In this expression, θ denotes the scan angle, and the factor $e^{-jk\,(\sin\theta)\,x_p}$, by letting

$$x_p = pb \qquad \text{where} \quad p = 0, \pm 1, \pm 2, \ldots \qquad (2)$$

designates the phase of any particular waveguide relative to the others in the plane of scan ($p = 0$ designates the guide at $x = 0$). Because of the periodicity of the fields, it is sufficient to consider only $p = 0$.

The incident mode $\phi_1 e^{-j\beta_1 z}$ is assigned a unit amplitude. The interior normal modal functions for rectangular waveguides are given as[1]

$$\phi_n = \sqrt{\frac{2}{a}}\,\begin{matrix}\cos\\ \sin\end{matrix}\left(\frac{n\pi}{a}x\right) \qquad \text{for} \quad \begin{pmatrix} n \text{ odd} \\ n \text{ even}\end{pmatrix} \qquad \text{for} \quad |x| \le \frac{a}{2} \qquad (3)$$

and

$$\phi_n = 0 \qquad \text{for} \quad \left(\frac{a}{2} < |x| < \frac{b}{2}\right) \qquad (4)$$

The propagation constants, β_n, are given as

$$\beta_n = \begin{Bmatrix} \sqrt{k^2 - (n\pi/a)^2} \\ -j\sqrt{(n\pi/a)^2 - k^2} \end{Bmatrix} \qquad \text{for} \quad k^2 \gtrless \left(\frac{n\pi}{a}\right)^2 \qquad (5)$$

For

$$\frac{n\pi}{a} > k = \frac{2\pi}{\lambda}$$

ϕ_n becomes evanescent.

The coefficients i_n are found by applying the orthonormality relation between the ϕ_n's so that

$$i_n = \int_{-b/2}^{b/2} \phi_n H_x\, dx \qquad \text{where} \quad H_x = \mathscr{H}_x(x, z = 0) \qquad (6)$$

In the same way, R can be found directly from the aperture field by

$$1 - R = \int_{-b/2}^{b/2} \phi_1 H_x \, dx \tag{7}$$

The field $\mathscr{E}_y$ is found from $\mathscr{H}_x$ and the modal impedances,

$$z_n = \frac{\omega\mu}{\sqrt{k^2 - (n\pi/a)^2}} \tag{8}$$

as

$$\mathscr{E}_y = -\left[(e^{-j\beta_1 z} + Re^{+j\beta_1 z})z_1\phi_1(x) - \sum_2^\infty i_n z_n e^{+j\beta_n z}\phi_n(x)\right], \quad p = 0 \tag{9}$$

In the exterior region, $z \geq 0$, by Floquet's theorem, we represent $\mathscr{E}_y$ and $\mathscr{H}_x$ in terms of the exterior orthonormal modes and modal impedances:

$$\Psi_m = \sqrt{1/b}\, e^{j[(2\pi m/b) - k \sin\theta]x} \tag{10}$$

and

$$Z_m = \frac{\omega\mu}{\Gamma_m}$$

where

$$\Gamma_m = \sqrt{k^2 - [(2\pi m/b) - k\sin\theta]^2} \tag{11}$$

For $\mathscr{H}_x$ we find

$$\mathscr{H}_x = \sum_{m=-\infty}^{\infty} I_m e^{-j\Gamma_m z}\Psi_m \tag{12}$$

By the orthonormality of the Ψ_m we find

$$I_m = \int_{-b/2}^{b/2} \Psi_m^* H_x \, dx \tag{13}$$

where $H_x = \mathscr{H}_x(x, z = 0)$ is the tangential aperture field. For $\mathscr{E}_y$ we obtain, as before,

$$\mathscr{E}_y = -\sum_{m=-\infty}^{\infty} I_m Z_m e^{-\Gamma_m z}\Psi_m \tag{14}$$

The continuity of H_x and E_y across the boundary at $z = 0$ gives directly

$$H_x = (1 - R)\phi_1 + \sum_{n=2}^{\infty}\phi_n i_n = \sum_{m=-\infty}^{\infty}\Psi_m I_m \tag{15}$$

and

$$-E_y = (1 + R)z_1\phi_1 - \sum_{n=2}^{\infty} z_n\phi_n i_n = \sum_{m=-\infty}^{\infty} Z_m\Psi_m I_m \tag{16}$$

Manipulation of (15) and (16) permits the formulation of Fredholm integral equations of either the first or second kind and a variational expression for R. With i_n and I_m expressed as in (6) and (13), the unknown field in each of these equations is the aperture field, H_x. It is possible to write precisely similar equations for the H-plane scan case with E_y as the unknown variable. In this case, (15) and (16) appear the same, with E_y and H_x reversing positions and the modal impedances, z_n, being replaced by the modal admittances, $y_n = 1/z_n$. However, the resultant integral equations have kernels with infinite series dyadic representations that converge much more slowly than the representations for the kernels of the integral equations for H_x. The reverse is true, however, for the quasi-E-plane scan case, for which, therefore, we will obtain equations with E_y as the unknown.

By manipulation of (7) and (16), we readily develop an integral equation of the first kind for H_x as

$$2z_1\phi_1(x) = \int_{-b/2}^{b/2}\left[\sum_{n=1}^{\infty} z_n\phi_n(x)\phi_n(x') + \sum_{-\infty}^{\infty} Z_m\Psi_m(x)\Psi_m^*(x')\right]H_x(x')\,dx' \quad (17)$$

The derivation of the integral equation of the second kind is straightforward. The free term on the left-hand side of (17) denotes the incident field, whether from the interior or exterior region. The reversal of integration and summation, in deriving (17) from (16), was discussed in Chapter 2.

In the quasi-E-plane scan case we may represent E_y and H_x in terms of the interior mode functions,[1]

$$\phi_n' = -\sqrt{\frac{\varepsilon_n}{c}}\,\begin{matrix}\cos\\ \sin\end{matrix}\left(\frac{n\pi}{c}\,y\right) \qquad \text{for} \quad \begin{pmatrix} n \text{ even} \\ n \text{ odd}\end{pmatrix} \quad \text{and} \quad y \le \frac{c}{2} \qquad (18)$$

where a dependence, $\cos(\pi/b)x$, common to all tangential fields has been dropped, and where $\varepsilon_0 = 1$, $\varepsilon_n = 2$ for $n > 0$, and $\phi_n' = 0$ for $|y| > (c/2)$, and the exterior mode functions Ψ_m as given in (10) with b replaced by d and x replaced by y. An integral equation for E_y is found as before with the following result:

$$2y_0\phi_0'(y) = \int_{-c/2}^{c/2}\left[\sum_{0}^{\infty} y_n\phi_n'(y)\phi_n'(y') + \sum_{-\infty}^{\infty} Y_m\Psi_m(y)\Psi_m^*(y')\right]E_y(y')\,dy' \quad (19)$$

with the modal admittances

$$y_n = \frac{k^2 - (\pi/b)^2}{\omega\mu\sqrt{k^2 - (\pi/b)^2 - (n\pi/c)^2}}$$

$$Y_m = \frac{k^2 - (\pi/b)^2}{\omega\mu\sqrt{k^2 - (\pi/b)^2 - [(2\pi m/d) - k\sin\theta]^2}} \qquad (20)$$

The integral equations, (17) and (19), are scalar one dimensional and of the same form as the integral equations for the thin-walled parallel plates (Chapter 4). However, these equations cannot be solved *exactly* by the same techniques used earlier or, in fact, by any known method. As a result, we will consider various approximate methods of solution next.

1.2 Approximate Solutions and Verification of Accuracy

A useful method for solving (17) and (19) approximately is the Galerkin or Ritz method, as discussed in Chapter 3. This approach was first employed by Primich[2] for a solution of the problem with $\theta = 0$, the broadside radiation case. Primich employed a variational technique directly, as opposed to solving the integral equation by a moments technique (see Chapters 2, 3, and Ref. 3). The broadside solution, which is valuable for the design of arrays, was extended to include all scan angles θ (including imaginary space) by Galindo and Wu.[4] The results have proved useful for the understanding and design of phased arrays.

The results of Primich and those obtained by Galindo and Wu for $\theta = 0$ were essentially identical. These independently obtained results thus proved to be a valuable check of the accuracy of the approximate solutions. Another independent and mathematically elegant approach to the problem for $\theta = 0$ was taken by Lee.[5] His results also checked very well with those of Primich and those of Galindo and Wu, although he used neither a variational nor a moments method.

Lee's approach is essentially a *perturbation* type solution of the original thin-wall solution, although it yields quite accurate results, with moderate effort, for substantial wall thicknesses. It will be recalled (Chapter 4) that a function-theoretic solution of the thin-wall problem is possible because the *infinite* linear set of equations, (24) of Chapter 4, derivable from the integral equation, could be put into the form

$$\sum_{n=1}^{\infty} \frac{1}{\Gamma_m - \beta_n} B_n = \frac{1}{\Gamma_m + \beta_1}, \qquad m = 0, \pm 1, \pm 2, \ldots \tag{21}$$

where B_n is related to i_n in (1) and (6). With a moderate wall thickness proportional to perturbation parameter δ_m (see Ref. 5), it is possible to obtain a form similar to (21):

$$\sum_{n=1}^{\infty} \left(\frac{1}{\Gamma_m - \beta_n} - \frac{\delta_m}{\Gamma_m + \beta_n} \right) B_n = \left(\frac{1}{\Gamma_m + \beta_1} - \frac{\delta_m}{\Gamma_m - \beta_1} \right), \qquad m = 0, \pm 1, \pm 2, \ldots \tag{22}$$

This *perturbed* set of equations can now be solved approximately by a *modified*[6] residue calculus or function-theoretic technique.

A third independent mathematical approach to the thick-wall parallel-plate problem was taken by Van Blaricum and Mittra.[13] Their approach was also, essentially, a perturbation analysis, expressed in terms of a Neumann series (see Chapter 3) in this case. Two particular features distinguish their approach, which they designate as the "generalized scattering matrix" approach.* Their method is applicable for all scan angles of the parallel plates, and Pace[14] has proved that the Neumann series will converge. The starting point or initial approximation in this Neumann series involves known solutions to two "auxiliary" problems, as shown in Fig. 5.2. One problem, designated

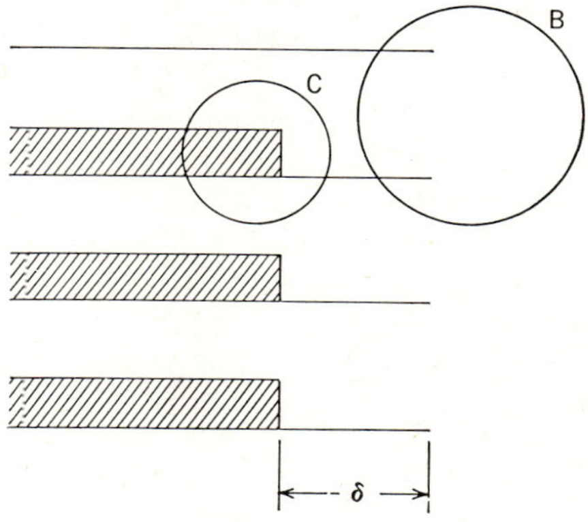

Fig. 5.2. Auxiliary generalized scattering matrix problem for thick parallel plates.

as *B* in Fig. 5.2, is the exact residue calculus solution for the thin-wall problem. The other problem, designated as *C* in the figure, is any approximate solution to the step in a waveguide problem. Van Blaricum and Mittra used the *modified* residue calculus technique for this approximate solution. When $\delta \to 0$ in Fig. 5.2, the composite auxiliary solutions in the Neumann series give the thick-wall parallel-plate solution, in the *E* or *H* plane of scan. The numerical solutions of Van Blaricum and Mittra were the same as those of Galindo and Wu, which will be presented later.

The first step in obtaining a Galerkin or moments type solution to (17) or (19), it will be recalled from Chapter 3, is to represent the unknown field in some basis of functions which are *complete* over the prescribed interval. For example, in (19), we may choose to represent the unknown tangential electric field in terms of the basis of $\phi_m'(y)$ functions, which we know are complete in the interval $-b/2$ to $b/2$. The choice of basis functions will of course determine the *rate* of convergence of the representation. If the choice of basis is the *orthonormal* set $\{\eta_n(y){:}n = 0, 1, 2, \ldots\}$ for $E_y(y)$ in (19), so that

$$E_y(y) = \sum_{n=0}^{\infty} v_{\eta_n}\eta_n(y) \tag{23}$$

* The "generalized" scattering matrix of Van Blaricum and Mittra includes evanescent as well as propagating modes.

then, assuming that moments of (19) are taken in the same basis, a measure of the accuracy of the approximate moments solution is determined by the *convergence* of the approximate solution when we approximate

$$E_y(y) \approx \sum_{n=0}^{N} v'_{\eta_n} \eta_n(y) \tag{24}$$

The coefficients v'_{η_n} should, in the limit of $N \to \infty$, become equal to the v_{η_n}. Some examples of this convergence with increasing N will be illustrated later for a particular basis.

The rate of convergence of the series (23) and (24) will depend not only on the choice of basis, $\eta_n(y)$, but also on the functional form of the field function, $E_y(y)$. The asymptotic behavior, for example, of the coefficients v_{η_n} for large n in (23) will depend on the highest-order singularity of the field function.[7] This singular behavior of E_y is known in the case of parallel-plate or rectangular waveguide arrays from a knowledge of the behavior of the field near the edges at $y = \pm b/2$. It is well known[8] that in the vicinity of such edges the normal component of the electric field varies as

$$E_y(y) \sim \frac{1}{(y \mp b/2)^{1/2}}$$

where $(y \to \pm b/2)$ for a thin wall or knife edge, and

$$E_y(y) \sim \frac{1}{(y \mp b/2)^{1/3}}$$

where $(y \to \pm b/2)$ for a right angle (90°) edge (thick-wall case). In the thin-wall case, for example, the representation of E_y in terms of the $\phi'_n(y)$,

$$E_y(y) = \sum_{n=0}^{\infty} v_n \phi'_n(y) \tag{25}$$

will have coefficients which decay asymptotically[8] as

$$v_n \to \frac{K}{n^{1/2}} (n \to \infty) \tag{26}$$

where K is a constant independent of n.

As we see from (25) and (26), an a priori knowledge of the singular behavior of the aperture field (or its derivatives, if the field is continuous) will determine the asymptotic behavior of the coefficients in any given representation. This knowledge of the field singularities and consequently of the asymptotic coefficient behavior can be incorporated into a moments or Galerkin type solution by a method sometimes referred to as the asymptotic anticipation method.[10] For example, (25) can be written approximately as

$$E_y(y) \approx \sum_{n=0}^{N} v'_n \phi'_n(y) + K' \sum_{N+1}^{\infty} \frac{1}{n^{1/2}} \phi'_n(y) \tag{27}$$

We note that

$$v'_n \to v_n, \qquad K' \to K \quad \text{in the limit as } N \to \infty \tag{28}$$

In (27) we have $N + 2$ unknown coefficients which can be determined by the moments method applied to the integral equation, (19). For an equivalent number of unknown constants we could have used the approximation

$$E_y(y) \approx \sum_{n=0}^{N+1} v''_n \phi_n(y) \tag{29}$$

This approximation, however, will not contain the edge singularity of the field, and inclusion of the singularity leads, in general, to more accurate results.[10]

It should be noted that the function

$$\sum_{N+1}^{\infty} \frac{1}{n^{1/2}} \phi'_n(y)$$

is *orthogonal* to the ϕ'_n $(n \leq N)$. Very often, this summation, which is singular for a given value or values of y, can be summed in closed form when the sum is extended from $n = 0$ to $n = \infty$. An example of this is the *quasi-static* approximation to the kernels of either (17) or (19). From the discussion in Chapter 3, it is clear that the process of substituting an approximate E_y such as (27) or (29) into (19) and then taking moments is equivalent to approximating the kernel of (19) in the subspace spanned by the finite basis[11] $\{\phi'_n : n = 0, \ldots, N\}$. The quasi-static approximation of the integral equation kernel is a method by which the singularity of the kernel function $K(y, y')$ at $y = y'$ is extracted in closed form and has this similarity to the asymptotic anticipation method just discussed. We will illustrate the quasi-static form of the kernel of (19) for the special *thin-wall* case, $c = d$ (see Fig. 5.1*b*).

Consider, first, the part of the kernel of (19) arising from the interior waveguide region:

$$K_i(y, y') = \sum_{n=0}^{\infty} y_n \phi'_n(y) \phi'_n(y') \tag{30}$$

where we represent the ϕ'_n by

$$\phi'_n(y) = \sqrt{\varepsilon_n/c} \cos \frac{n\pi}{c} y \tag{30a}$$

This function is the same as that in (18) except for a shift in coordinates (see Fig. 5.3) of $y = 0$ to the edge of the parallel-plate waveguide. Furthermore,

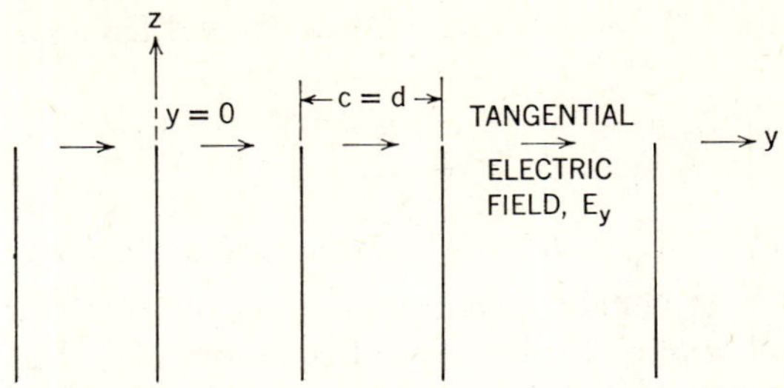

Fig. 5.3. Coordinate shift.

for convenience, we will consider an E-plane scan for parallel plates by letting $a = b = \infty$. Hence the interior modal admittances become

$$y_n = \frac{\omega\varepsilon}{\beta_n} = \frac{\omega\varepsilon}{\sqrt{k^2 - (n\pi/c)^2}}$$

In the quasi-static approximation to series (30), we allow

$$\beta_n \to -j\frac{n\pi}{c} \qquad \text{for} \quad \text{large } n > N \tag{31}$$

The name quasi-static arises from the approximation that

$$k^2 = \left(\frac{2\pi}{\lambda}\right)^2 \ll \left(\frac{n\pi}{c}\right)^2$$

a condition also possible for *large* λ. As a result of (31) we can write

$$K_i(y, y') \approx \frac{\omega\varepsilon}{c}\frac{1}{\beta_0} + \frac{2\omega\varepsilon}{c}\sum_{n=1}^{N}\left[\frac{1}{\beta_n} - \frac{1}{-j(n\pi/c)}\right]\cos\frac{n\pi}{c}y\cos\frac{n\pi}{c}y'$$
$$+ \frac{2\omega\varepsilon}{c}\sum_{n=1}^{\infty}\frac{1}{-j(n\pi/c)}\cos\frac{n\pi}{c}y\cos\frac{n\pi}{c}y' \tag{32}$$

where we observe that the coefficients of

$$\frac{2\omega\varepsilon}{c}\sum_{n=1}^{\infty}\left[\frac{1}{\beta_n} - \frac{1}{-j(n\pi/c)}\right]\cos\frac{n\pi}{c}y\cos\frac{n\pi}{c}y'$$

(with $N \to \infty$) decay as n^{-3}. Consequently this series is not singular[7] for any y or y'. The remaining series of (32) can be summed or put in closed form:[12]

$$\frac{2\omega\varepsilon}{c}\sum_{n=1}^{\infty}\frac{1}{-j(n\pi/c)}\cos\frac{n\pi}{c}y\cos\frac{n\pi}{c}y' = -\frac{\omega\varepsilon}{\pi}j\ln\left(2\left|\cos\frac{\pi y}{c} - \cos\frac{\pi y'}{c}\right|\right) \tag{33}$$

It should be noted that the logarithmic singular part of $K_i(y, y')$ is contained in (33), and also that the singular part of $K_i(y, y')$ is pure imaginary.

The exterior dyadic sum of the kernel of (19) can also be treated in the same manner. After some manipulation we arrive at the representation

$$
\begin{aligned}
K_e(y, y') &= \sum_{m=-\infty}^{\infty} Y_m \Psi_m(y) \Psi_m^*(y') \\
&= \frac{\omega\varepsilon}{c}\left\{\frac{1}{\beta_0'} e^{-jT_y(y-y')} + j\frac{c}{\pi} \ln\left[2 \sin\frac{4\pi}{c}(y-y')\right] e^{-jT_y(y-y')}\right. \\
&\quad \left. + \sum_{-\infty}^{\infty}{}^{(0)} \left(\frac{1}{\beta_m'} + j\frac{c}{2\,|m|\,\pi}\right) e^{j[(2m\pi/c)+T_y](y-y')}\right\}
\end{aligned} \tag{34}
$$

where

$$T_y \equiv k \sin \theta \tag{35}$$

and where, from (20), with $b \to \infty$, we note

$$Y_m = \frac{\omega\varepsilon}{\beta_m'} = \frac{\omega\varepsilon}{\sqrt{k^2 - [(2\pi m/d) - T_y]^2}} \tag{36}$$

The notation $\sum^{(0)}$ denotes omission of the $m = 0$ term in the sum. Again, we have placed the logarithmic singularity of $K_e(y, y')$ in closed form in the term

$$j\frac{c}{\pi} \ln\left[2 \sin\frac{4\pi}{c}(y-y')\right] e^{-jT_y(y-y')}$$

and we note that the singularity is purely imaginary here also. Similar results are obtainable for the thick-wall quasi-E-plane and H-plane scan problems also.

The above approximation (if $N < \infty$) or transformation (if $N = \infty$) of the kernel of (19), $K_i(y, y') + K_e(y, y')$, is applicable, of course, to a moments of Galerkin type approximation of (19) in any basis of functions. We have stated that, for a given order of approximation (the number of approximating functions N), different convergence rates will lead to different accuracies in solution. It is therefore reasonable to suppose that, if several *different bases* in the moments type solution yield at least the same accuracy for the solution, then all the solutions are accurate to this degree. The key words in this statement are *different bases*. If the bases are *very* different, as, for example, between a pulse or Riemann type integration and a *smooth* function basis such as the $\phi_n'(y)$, one may usually expect solutions that are the same (within a prescribed accuracy) in both bases to be, in fact, accurate solutions. Such comparisons have been made, and some results will be presented in Chapter 8.

In regard to a moments type solution of (17) or (19), we have discussed the reduction of these equations to a matrix equation of finite order in the form

$$f = KE \tag{37}$$

where f and E are column matrices and K is a square matrix. The column matrix E is to be determined by inverting K. The invertability or *conditioning** of K is determined to a great extent by the *diagonal singularity* in the imaginary part of the kernel of (17) or (19). A strongly diagonal matrix is well conditioned. The singularity in the kernels of (17) and (19) will lead, for many bases,† to a strong diagonal in only the imaginary part of K in (37). Hence we must use care in evaluating K^{-1}.

In the matrix manipulation of K to solve (37) it is often convenient to perform all *real* arithmetic steps, as opposed to using complex arithmetic. This proves convenient when carrying extra digits for high accuracy (e.g., double precision in a computer[15]), but would be incompatible or difficult while simultaneously performing complex arithmetic.[15] Let

$$\begin{aligned} E &= E_r + jE_i \\ K &= K_r + jK_i \end{aligned} \tag{38}$$

Breaking (37) into real and imaginary parts gives

$$\begin{aligned} K_rE_r - K_iE_i &= f_r \\ K_iE_r + K_rE_i &= f_i \end{aligned} \tag{38a}$$

A straightforward manipulation leads to the solution

$$E_r + jE_i = (K_r^{-1}K_i + K_i^{-1}K_r)^{-1}[(K_i^{-1} - jK_r^{-1})f_r + (K_i^{-1} + jK_r^{-1})f_i] \tag{39}$$

Because K_r is poorly conditioned, however, we may have extreme difficulty in obtaining an accurate K_r^{-1}. An easy way to avoid this problem is to make a simple conversion to the matrices $(K_r + K_i)$ and $(K_r - K_i)$ instead of K_r and K_i. This can be done by adding and subtracting the two equations in (38a). Since K_i is contained in both of these matrices, $(K_r + K_i)^{-1}$ and $(K_r - K_i)^{-1}$ can easily be calculated with accuracy. The solutions for E_r and E_i then become

$$E_r + jE_i = (1 + j)(K_r + K_i)^{-1}f + (1 - j)(K_r - K_i)^{-1}f$$

The significance of the singularity in (17) and (19) to the numerical process of computing an approximate solution is thus illustrated. Furthermore, the existence of solutions to (17) and (19), in the exact sense, is dependent on the singularity of the kernels of these equations.

* A square matrix, K, is considered *well* conditioned if, to a prescribed *number of digits*, the arithmetic of inversion to K^{-1} leads to an accurate K^{-1}. Actually, a poorly conditioned matrix will have a vanishing determinant up to a prescribed number of retained digits. Hence it will not invert unless more than this minimum accuracy is retained in the arithmetic.

† This is particularly true when using a pulse basis [Riemann integration of (17) or (19), or equivalently trapezoidal or Simpson type integration].

One of the basic functions used to solve (17) and (19) approximately was the "pulse" basis. The interval of integration, in (19) for example, is subdivided into N subintervals of not necessarily equal length, as shown in Fig. 5.4. Each vertical pulse is chosen as an orthonormal function of y [each orthonormal pulse is multiplied by a coefficient approximately equal to $F(y)$]. The tops of the pulses may be connected by the zero-order flat top shown in

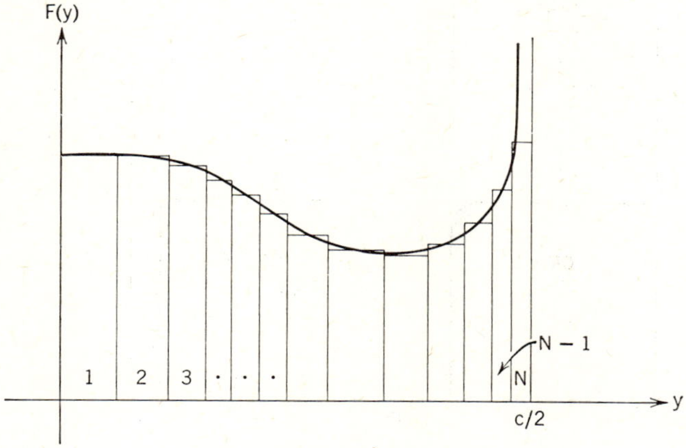

Fig. 5.4. Pulse basis integration.

Fig. 5.4, the first-order trapezoidal integration, the second-order Simpson's integration, and so on. A higher degree of smoothness of the approximating function for a given number of intervals, N, generally gives better results, although it is questionable whether equally better results, for the same effort, could not be obtained by increasing N. This integration or moments basis is essentially equivalent to the *point* matching method often referred to in the literature.

A factor of some significance is the convergence properties of this basis for large N. If the integrated functions, such as $F(y)$ in Fig. 5.4 or H_x and E_y in (17) and (19), respectively, have a singularity in the interval (including the boundary), then it is well known that the Riemann (or Simpson, etc.) integrations do *not* converge for large N. This need not dissuade us, however, from using the pulse basis, since, for a *given* N, there does exist an extended *complete* and convergent basis. That is, if $\{p_n(y):n = 1, \ldots, N\}$ denotes the N pulse functions, it is possible to complete this basis with an infinite set of functions $\{q_n(y):n = N + 1, \ldots, \infty\}$, not necessarily pulse type functions. Hence our earlier discussions on covergence apply, in this sense, to the

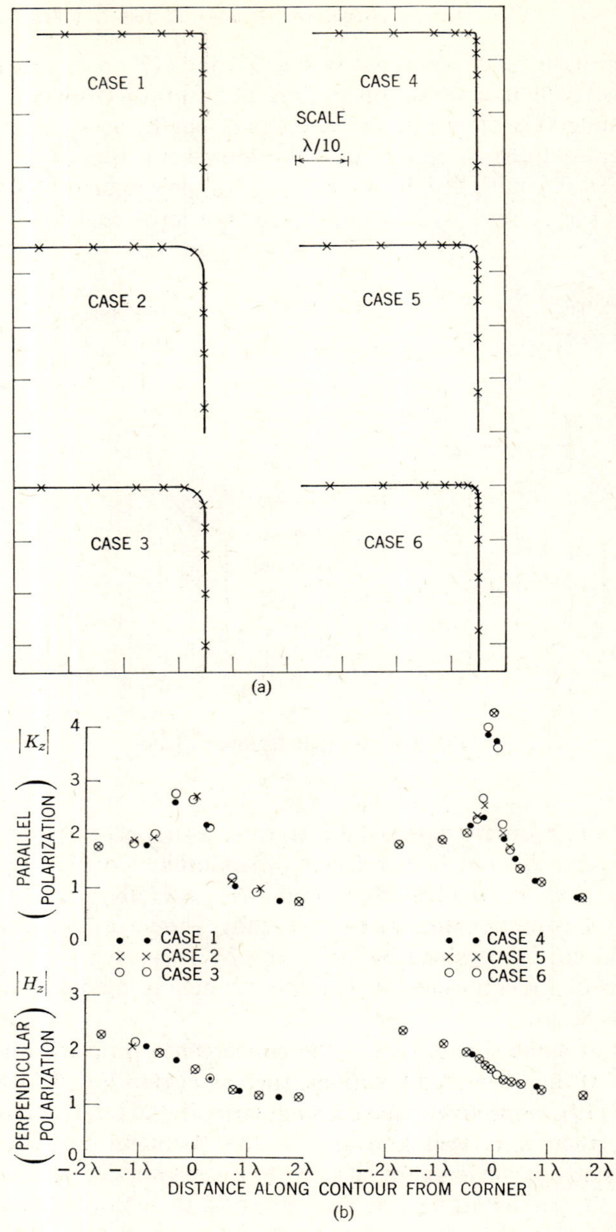

Fig. 5.5. (*a*) Mesh points near a corner of a scattering rectangle; $\lambda/10$ intervals taken away from corners. Case 1: $\lambda/20$ interval at corner; case 2: $\lambda/20$ radius, $\lambda/20$ intervals; case 3: $\lambda/20$ radius, 3 intervals at corner; cases 4, 5, 6: same as 1–3 except for $\lambda/40$ intervals at corner and $\lambda/40$ radius. (*b*) Currents near a corner of a scattering rectangle.

pulse functions as well. Nevertheless, we may (and this is common practice) refer to the convergence of a solution with increasing N when using these functions, since this terminology is appropriate for most other basis functions, such as the $\phi'_n(y)$ and $\Psi'_m(y)$.

One may expect, in using the pulse basis, that it is desirable to use closely spaced pulses or smaller intervals where the approximated function, $F(y)$ in Fig. 5.4, varies more sharply. This would be particularly true in the vicinity of a singularity, which for E_y and H_x occurs at the waveguide corners. Alternatively, the corner and the singularity may be removed by "rounding" the corner in a practical and realistic way. A study of these parameters on a corner was made,[16] and the results are shown in Fig. 5.5 for a plane wave normally incident on one face of the corner. In one case the wave is polarized parallel to the edge, and in another the wave is polarized perpendicular to the edge. The electric ($\sim H_z$) and magnetic ($= K_z$) currents found by the point matching methods indicated are shown. The more closely spaced points in cases 4, 5, and 6 clearly result in a more sharply determined peak, as illustrated. However, it is significant that the forward scattered field determined from these currents is not significantly altered by the changed spacing (or corner rounding). For the phased array, this is equivalent to saying that the edge singularity is not significant in determining the reflection coefficient and other scattering parameters. This may be expected, since, from (7), we see that R is determined by averaging H_x [for the H-plane scan equation, (17)] with a *smooth* function ϕ_1:

$$1 - R = \int_{-b/2}^{b/2} \phi_1 H_x \, dx$$

The singularity, of course, contributes significantly to the coefficients of higher order, ϕ_n, and decisively to the coefficients of asymptotic ϕ_n.

1.3 Numerical Results for *H*-Plane and Quasi-*E*-Plane Scanning

The several techniques for obtaining and verifying the accuracy of approximate solutions to (17) and (19) were applied to the rectangular waveguide arrays in Figs. 5.6*a* and 5.6*b*. Some of the results for *H*-plane and quasi-*E*-plane scanning will be presented in this section. They illustrate the importance of even a moderate waveguide wall thickness in regard to the scattering characteristics of wide angle scanning phased arrays. In addition, the near-zone and aperture ($z = 0$) tangential fields as a function of scan angle will be presented. Since it is the aperture tangential field that is approximated in a moments or variational type of approximate solution, these field plots should prove particularly interesting.

Fig. 5.6*a*. Photograph of 19 × 19 element rectangular waveguide array. Front end—thick-walled radiating aperture.

Fig. 5.6*b*. Rear end—matched loads and coupling measurements.

The variational expression (V.E.) for the input impedance or admittance can be readily derived from the integral equation, as discussed in Chapter 2. For example, in the *H*-plane scan case we obtain

$$\left(\frac{1+R}{1-R}\right)z_1\left(\int_{-b/2}^{b/2}\phi_1 H_x\,dx\right)^2 = \sum_{n=2}^{\infty} z_n(-1)^{n+1}\left(\int_{-b/2}^{b/2}\phi_n H_x\,dx\right)^2 + \sum_{-\infty}^{\infty} Z_m\left(\int_{-b/2}^{b/2}\Psi_m^* H_x\,dx\right)^2 \quad (40)$$

[see (3)–(10) for the definition of ϕ_n, Ψ_m, etc.] as a variational expression for *R*. It was shown in Chapter 3 that, if the *same* approximation for H_x is used in (40) as is used in a moments solution of (17), both (17) and (40) are stationary or variational and the same value for *R* and H_x will be obtained. If, however, moments of (17) are taken in a *different* basis from that used to approximate H_x, the V.E. (40) may be expected to give *different* (and probably better) results.

In order to test the validity of this concept, solutions of (17) were obtained by using *N* zero-order (flat top) pulse functions, $p_n(x)$, as illustrated in Fig. 5.4, to approximate H_x:

$$H_x(x) = \sum_{n=1}^{N} a_n p_n(x) \quad (41)$$

Moments of (17) were then taken on a point matching basis, that is, the left-hand term, $2z_1\phi_1(x)$, was set equal to the right-hand part of (17) at *N* points (at the pulse centers). Hence, in effect, a *mixed* basis was used for the moments type solution. Finally, the resulting solution, (41), was evaluated at the pulse centers, x_n:

$$\text{Approximate } H_x(x_n) = \sum_{n=1}^{N} a_n p_n(x_n) \quad (42)$$

for the purpose of performing a Simpson or second-order integration in (40). We note that, by the very nature of (40), a *non*mixed basis is implied. We thus expected and obtained *different* and generally better solutions by using the V.E., (40), than we obtained by solving (17) by means of mixed bases.

Some of the results obtained in this manner are illustrated in Table 5.1. The values in this table are for the thin-wall case with $b = a$ and $b/\lambda = 0.5714$, *H*-plane scan. The scan angles range from near broadside (2.87°) to near an angle (60°) wherein the main beam and grating lobe are equally disposed on either side of broadside (for exactly at this angle a perfect match, $R = 0$, occurs; see the results of Chapter 4). For two of the angles shown (2.87° and 60°), the convergence of the solution for *R* with increasing *N* is also illustrated. It was found that $N = 30$ gave *uniformly* accurate results for *R*. The exact (Chapter 4) solution is used as the basis of comparison for the solution

Table 5.1

θ	N	Directly from Integral Equation	V.E.	Thin Wall Exact Solution
2.87°	15	$R = (0.349, 158.4°)$	(0.346, 157.5°)	
2.87°	20	$R = (0.348, 157.8°)$	(0.346, 157.1°)	
2.87°	25	$R = (0.348, 157.2°)$	(0.347, 156.8°)	(0.347, 155.9°)
2.87°	30	$R = (0.347, 156.9°)$	(0.347, 156.5°)	
20°	30	$R = (0.320, 149.4°)$	(0.320, 149.0°)	(0.320, 148.4°)
40°	30	$R = (0.226, 113.4°)$	(0.225, 112.9°)	(0.226, 112.3°)
51°	30	$R = (0.457 \times 10^{-1}, 25.05°)$	$(0.461 \times 10^{-1}, 24.40°)$	$(0.462 \times 10^{-1}, 23.86°)$
60°	15	$R = (0.164 \times 10^{-2}, 163.7°)$	$(0.299 \times 10^{-3}, 31.3°)$	
60°	20	$R = (0.786 \times 10^{-3}, 169.7°)$	$(0.287 \times 10^{-3}, 26.6°)$	
60°	25	$R = (0.422 \times 10^{-3}, 162.0°)$	$(0.286 \times 10^{-3}, 25.6°)$	$(0.281 \times 10^{-3}, 24.0°)$
60°	30	$R = (0.254 \times 10^{-3}, 141.9°)$	$(0.284 \times 10^{-3}, 24.9°)$	

obtained directly from the integral equation and that obtained by using the integral equation solution in the V.E., (40). For the large values of N illustrated, the improvement obtained by using the V.E. is not evident except near $\theta = 60^\circ$, where the absolute errors are negligible but the percentage errors substantial. Here (as for smaller N) the V.E. produces improvement.

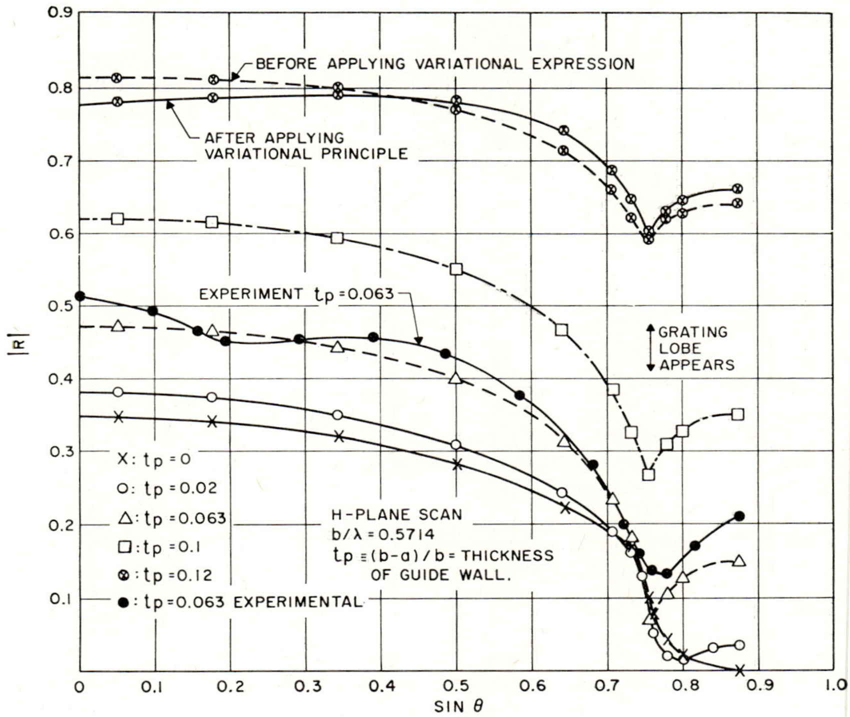

Fig. 5.7. $|R|$ vs. scan angle.

Numerical results for R, in regard to amplitude and phase, are presented in Figs. 5.7–5.17 for both the H-plane and quasi-E-plane cases. The parameters involved are scan angle, θ, wall thickness in the plane of scan, and wavelength. The integral equations, (17) and (19), were solved numerically for $N = 30$ in all cases. The wavelengths were chosen so that the variation of $R(\theta)$ over a wide frequency band could be examined while retaining a reasonably practical scan range.

The results for $|R|$ in the H-plane scan for $b/\lambda = 0.5714$ are illustrated in Fig. 5.7. The results for R in all the H-plane scan cases were corrected by application of the V.E., (40); in general, the improvement in R that resulted was less than 1 or 2%. One of the worst cases is illustrated in Fig. 5.7 before and after applying the V.E. The abrupt discontinuity in the curves occurs at the angle of emergence of a grating lobe. The significance of this phenomenon in connection with the asymptotic behavior of the mutual coupling was discussed in Chapter 4.

An experimental comparison is illustrated for thickness $t_p = (b - a)/b = 0.063$. (Although in the experiment $d - c = b - c$ rather than $d = c$, the effect of the thickness in the E plane did not appreciably affect the H-plane results.) The deviation between the experimental and theoretical curves seems to be due largely to the finite size of the experimental array in the plane of scan. Photographs of the 19 × 19 array used in the experiment are shown in Fig. 5.6. In Fig. 5.6a, the front face ($z = 0$) of the array using square waveguide elements is shown. The matched terminations of each element are shown in Fig. 5.6b. Actual determination of R was made by measuring the element-to-element coupling coefficients and then using the

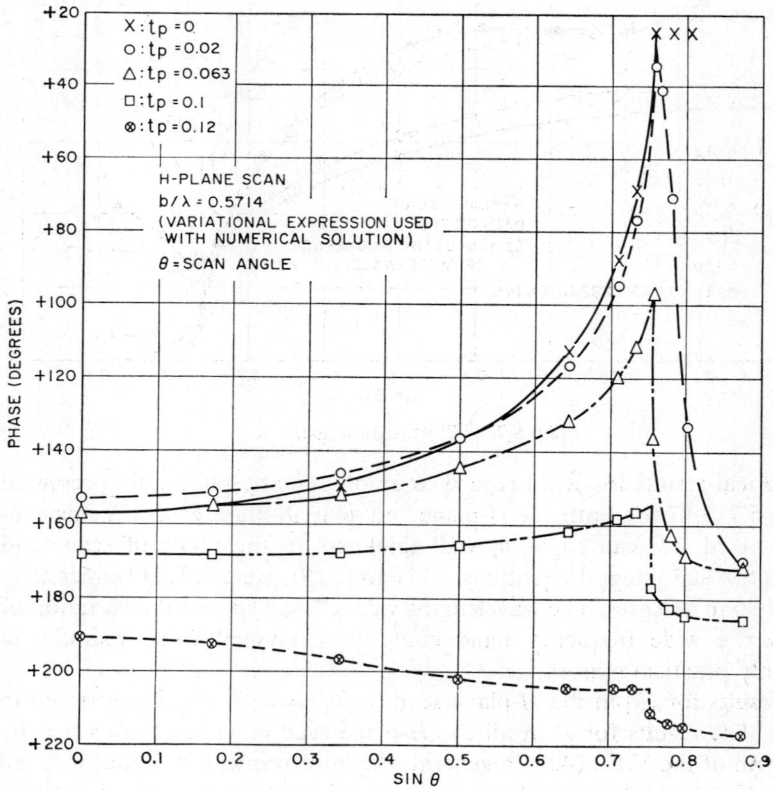

Fig. 5.8. Phase of R vs. scan angle.

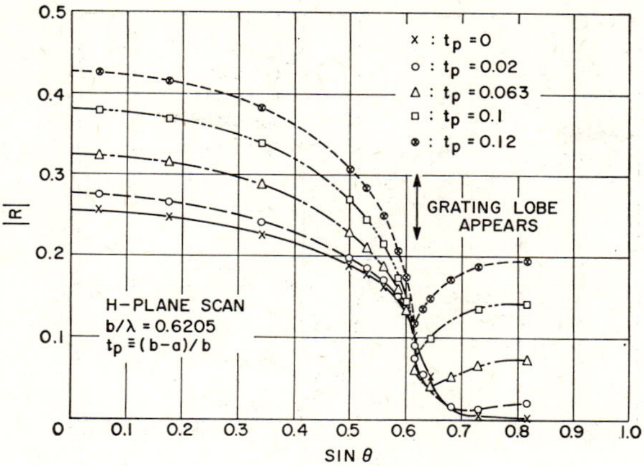

Fig. 5.9. $|R|$ vs. scan angle.

Fourier series relationship between R and the coupling coefficients to compute R. In measuring the coupling coefficients, the "central" waveguide with a protruding flange (Fig. 5.6*b*) was excited, and the incoming signal to the guide under consideration was fed through a matched waveguide-coaxial coupler to a bridge.

The effect of wall thickness is considerable here. The results for the phase of R, and further results for $|R|$, for two different wavelengths, are illustrated in Figs. 5.8–5.12 for *H*-plane scanning. The magnitude of R decreases with decreasing wavelength, particularly for the larger thicknesses. In each case

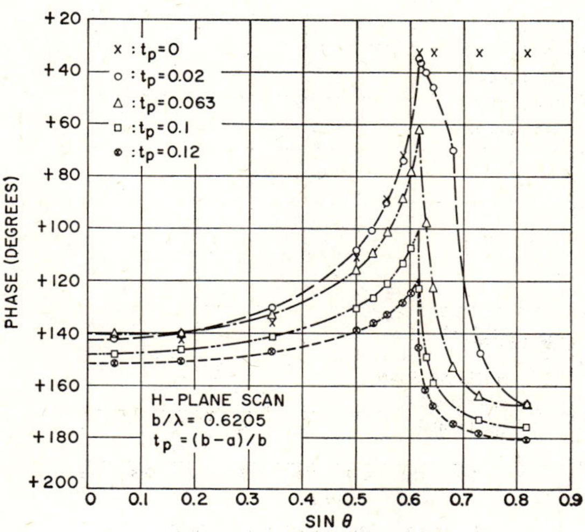

Fig. 5.10. Phase of R vs. sin θ.

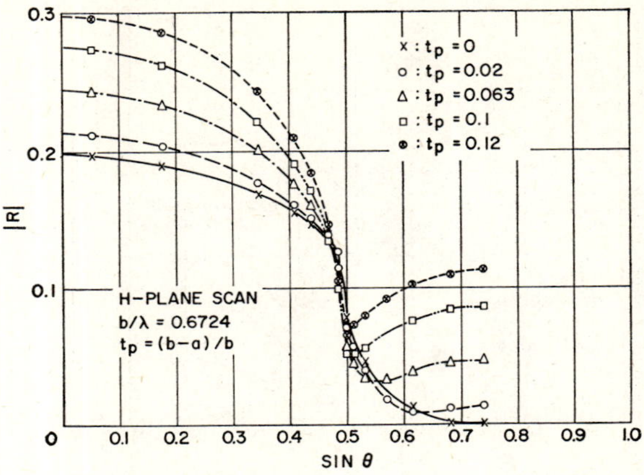

Fig. 5.11. $|R|$ vs. $\sin\theta$.

the curves are plotted to a value of $\sin\theta$ for which the main lobe and the grating lobe are symmetric about broadside. In terms of steering phase, the angle corresponds to $\psi_x = \pi$. Waveguide simulation is possible at this angle. In fact, the mismatch, R, may be viewed as a discontinuity in a waveguide due to a change in waveguide cross section corresponding to wall thickness ($R = 0$ for $t_p = 0$). The agreement between the numerical result and canonical exact results for $t_p = 0$ was consistently as good as could possibly be plotted in these graphs.

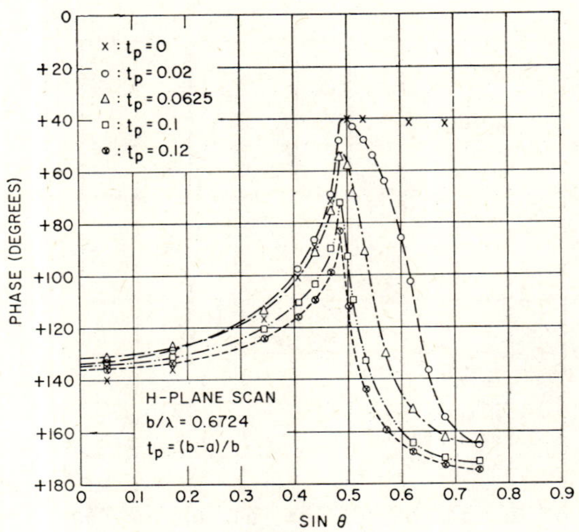

Fig. 5.12. Phase of R vs. $\sin\theta$.

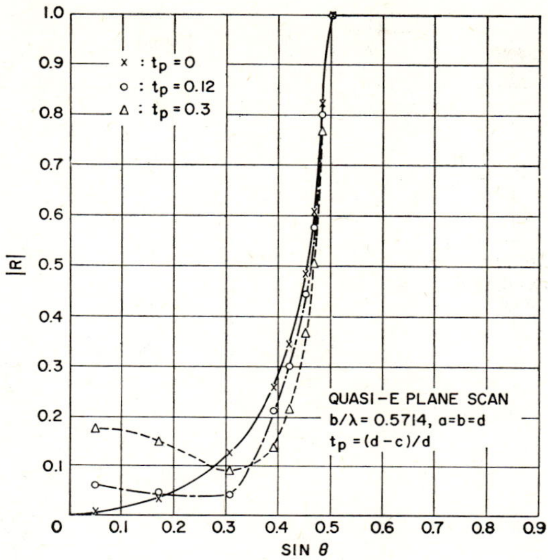

Fig. 5.13. $|R|$ vs. $\sin \theta$.

Computations for the quasi-*E*-plane scan are presented in Figs. 5.13–5.17. Comparison of the exact thin-wall solution for R and the numerical solution of (19) with $N = 30$ showed that they agreed* to three significant figures for all parameter variations (θ and λ). The results for the phase of R agreed to within 1.5° for all parameter variations. For this reason it was not considered necessary to use a V.E. to improve the values of R computed from the analog of (7).

A typical result for $|R|$ versus θ is shown in Fig. 5.13. Although the effect of thickness is great, $|R|$ is generally very small except near $\theta_{\text{cut-off}}$, the value of θ for which total reflection occurs:

$$k^2 - \left(\frac{\pi}{b}\right)^2 = (k \sin \theta_{\text{c-o}})^2$$

For this reason, the quasi-*E*-plane results are better illustrated with wall thickness, $t_p = (d - c)/d$, as the independent parameter.

In Fig. 5.14, the results for a single waveguide aperture in a solid ground plane are presented. The effect of the coupling is thus illustrated by comparison with the $\theta = 10°$ curve (this result is nearly the same as for $\theta = 0°$). The results for $|R|$ versus t_p are shown in Figs. 5.14–5.17, which indicate

* The reflection coefficient, R, for the quasi-*E* data is referred to the magnetic field (take $-R$ for reference to the electric field).

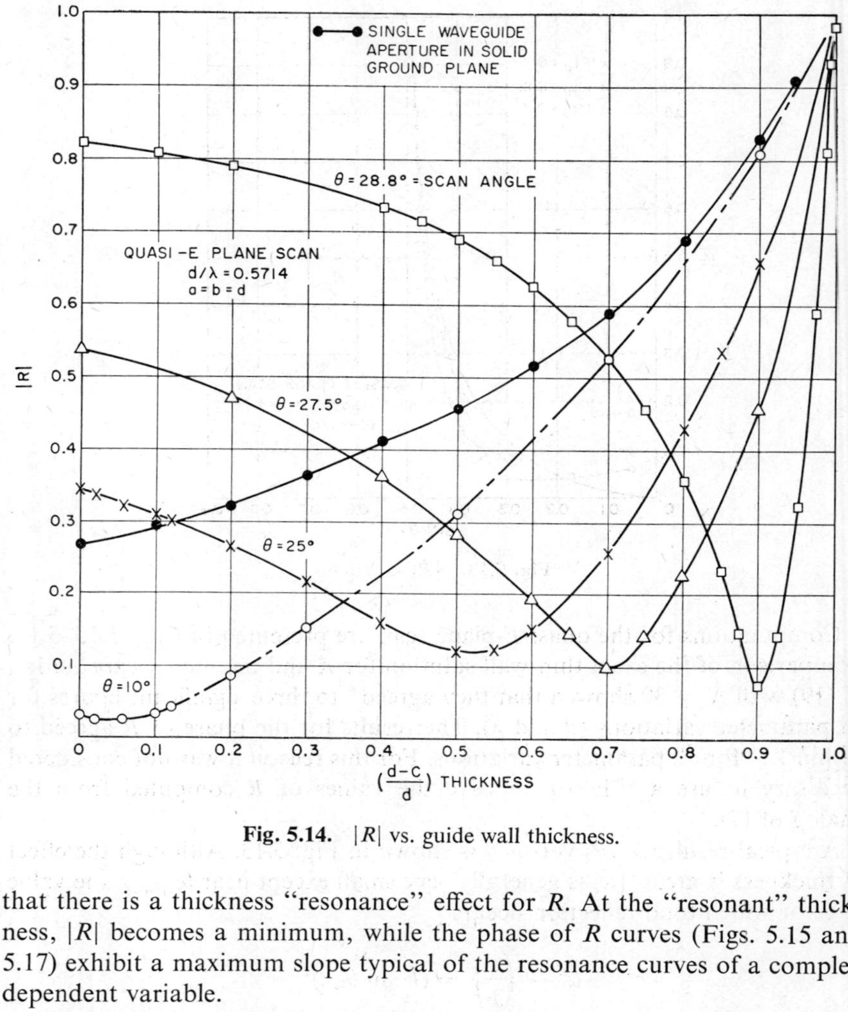

Fig. 5.14. $|R|$ vs. guide wall thickness.

that there is a thickness "resonance" effect for R. At the "resonant" thickness, $|R|$ becomes a minimum, while the phase of R curves (Figs. 5.15 and 5.17) exhibit a maximum slope typical of the resonance curves of a complex dependent variable.

A particularly interesting "resonance" effect is illustrated in Figs. 5.18*a* and 5.18*b* for two different wavelengths. As depicted in the inset to Fig. 5.18*a*, the results shown in these figures are for thick-wall parallel-plate *E*-plane scanning ($a = b = \infty$). We note that two *radiating* lobes are possible here because the thick waveguide walls allow the element spacing to be made larger than a half wavelength without multimoding in the waveguide. This resonance of $|R|$, which occurs when the scan angle is at exactly grazing for the grating lobe, is not present in the thin-wall computations. Although these large peaks in $|R|$ are serious mismatches, they are not surface waves by

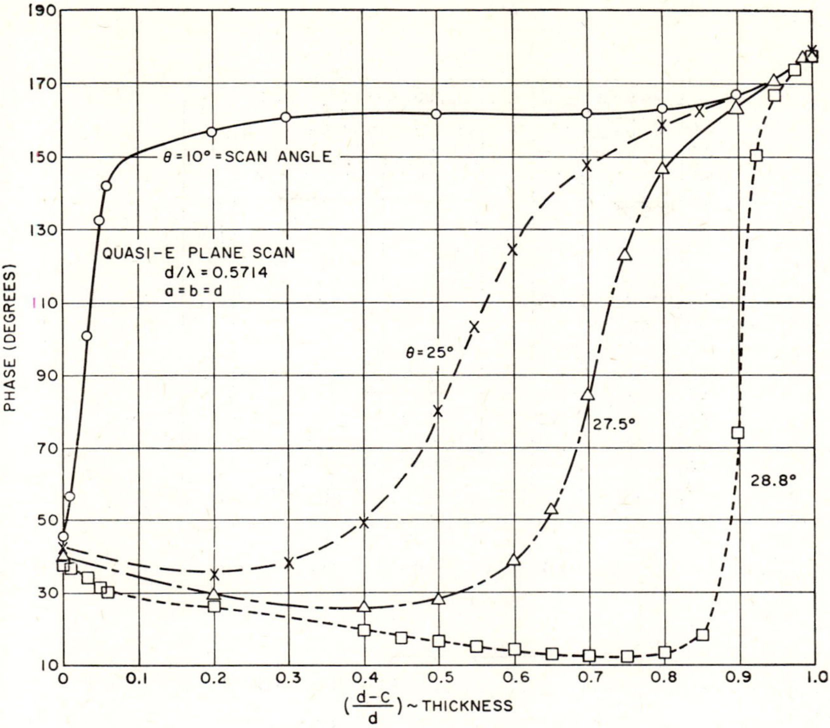

Fig. 5.15. Phase of R vs. guide wall thickness.

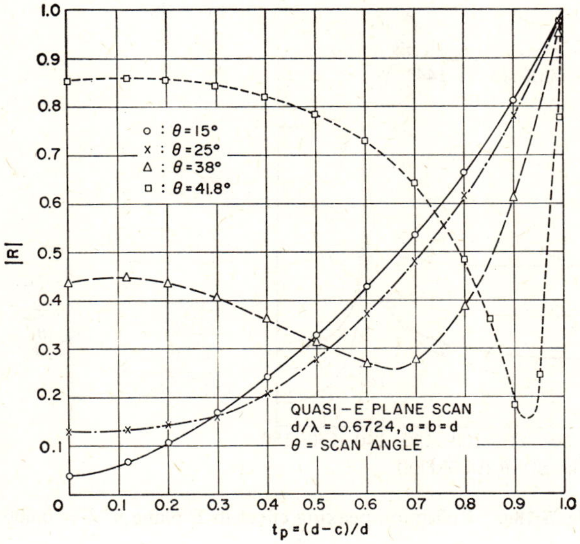

Fig. 5.16. $|R|$ vs. t_p.

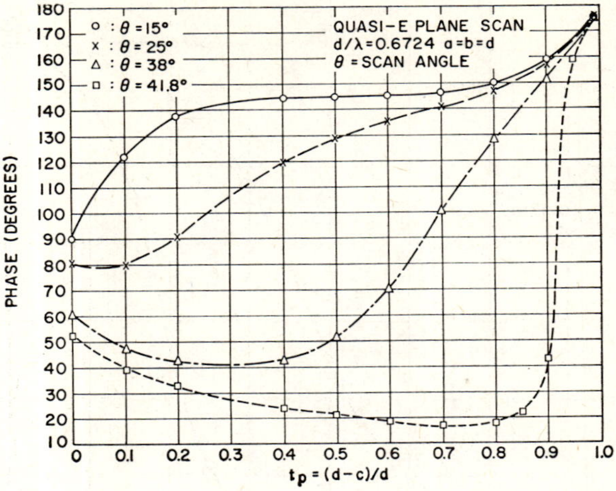

Fig. 5.17. Phase of R vs. t_p.

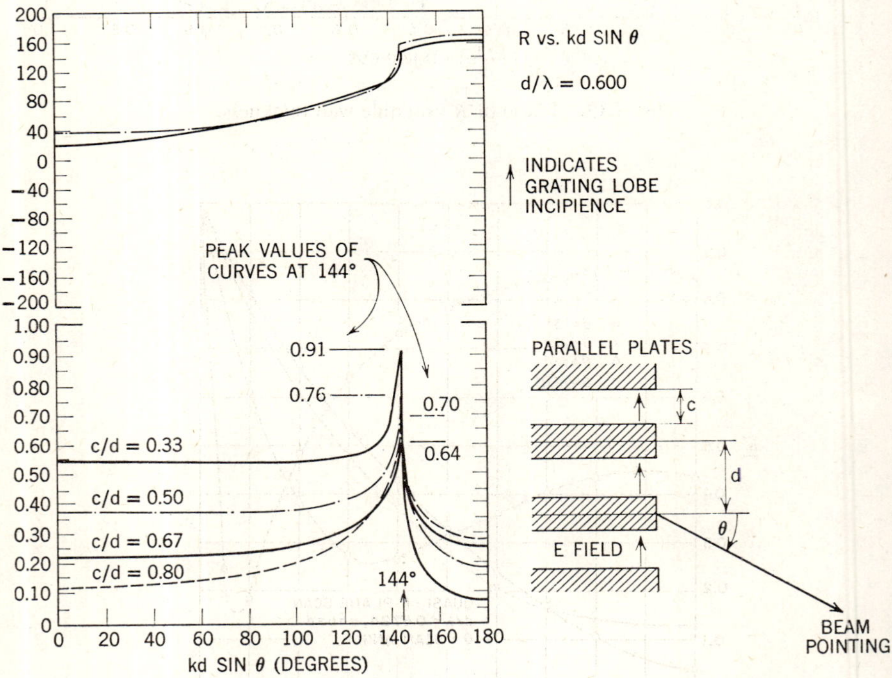

Fig. 5.18*a*. Reflection peaking effect in E plane, $d/\lambda = 0.600$.

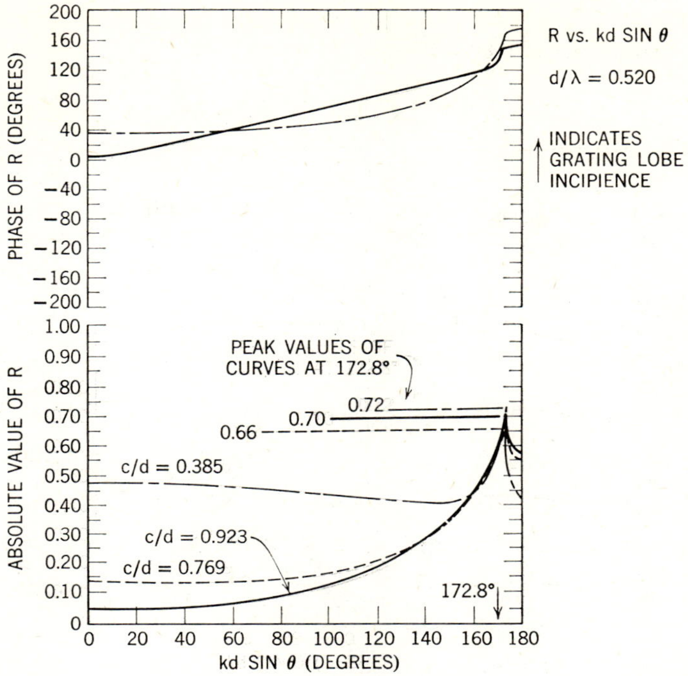

Fig. 5.18*b*. Reflection peaking effect in E plane, $d/\lambda = 0.520$.

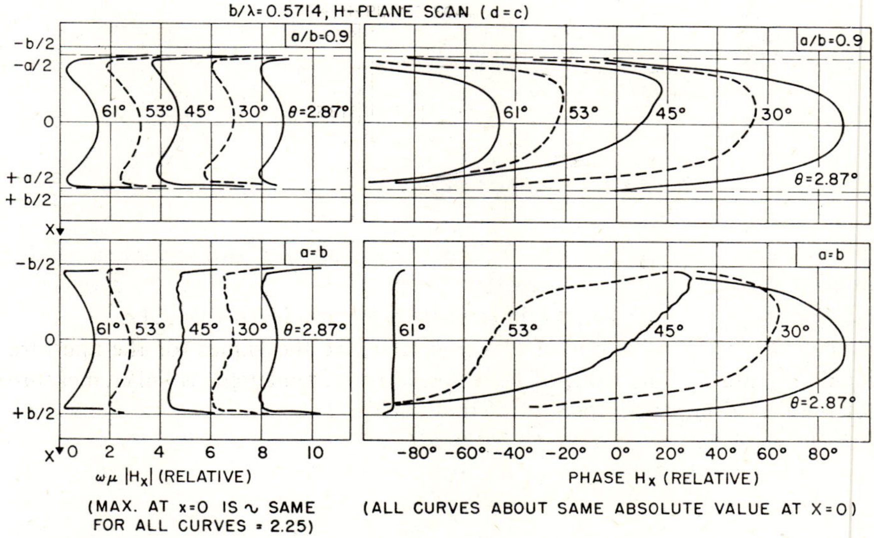

Fig. 5.19. Aperture distribution variation with scan angle and thickness.

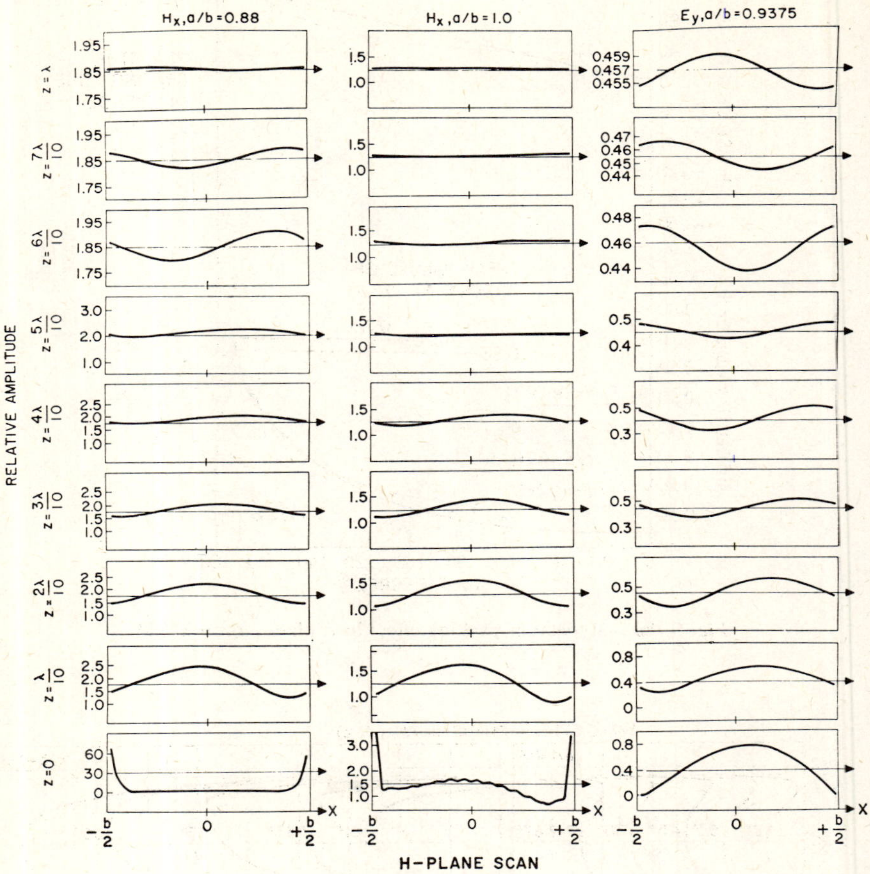

Fig. 5.20*a*. Near field amplitude $\theta = 30°$, $b/\lambda = 0.5714$.

reason of the fact that $|R| < 1.0$. The actual maximum of the peak generally increases with smaller c/d (greater thickness), although this result depends somewhat on wavelength, as illustrated by the results in Fig. 5.19.

The numerical solutions of (19) and (20) give the values for the aperture fields H_x and E_y. The value of R, it should be remembered, is only a measure of the first Fourier component of these fields. The total aperture fields, H_x, for the H-plane scan are illustrated in Fig. 5.19 for two thicknesses, zero and $(b - a)/b = 0.1$. A grating lobe exists for the $\theta = 53°$ and $61°$ curves (approximately equal symmetric lobes at $61°$). It should be observed that the effect of thickness is quite marked at the higher scan angles. Note also the sharper singularity for H_x at $|x| = a/2$ for the thin-wall case. Theoretically,[8]

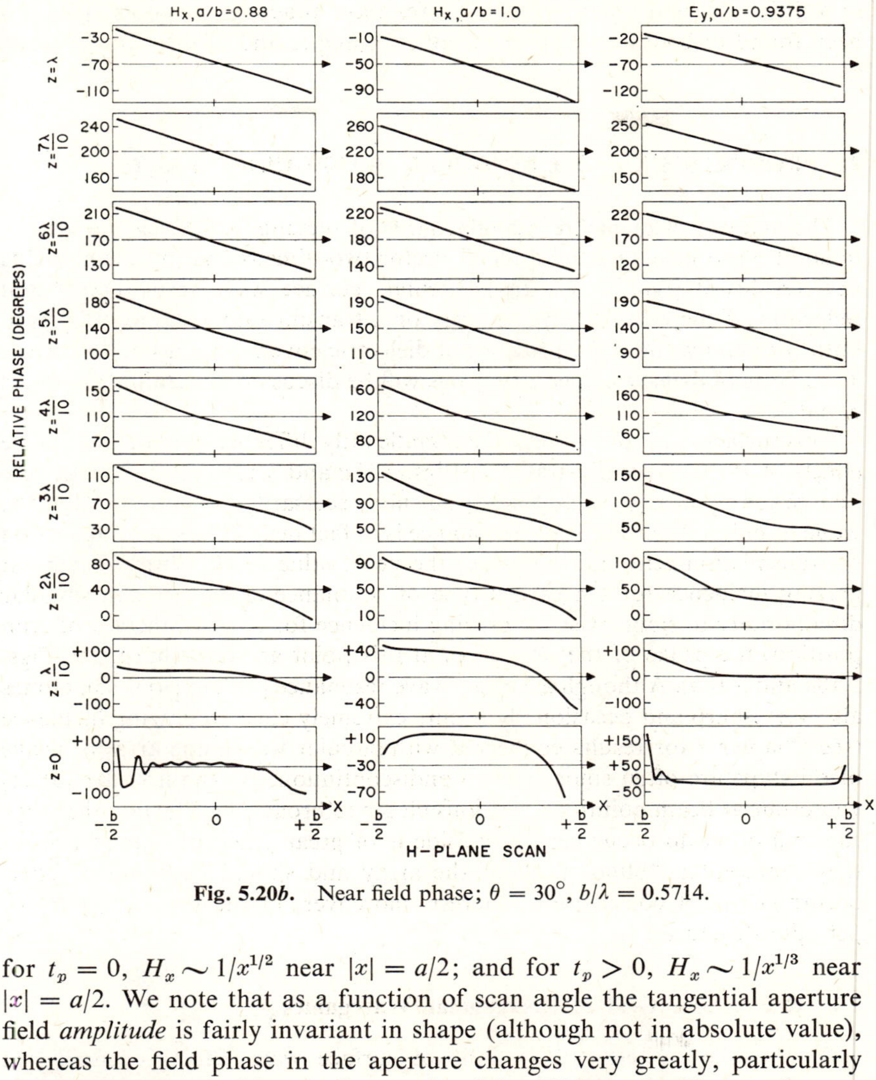

Fig. 5.20*b*. Near field phase; $\theta = 30°$, $b/\lambda = 0.5714$.

for $t_p = 0$, $H_x \sim 1/x^{1/2}$ near $|x| = a/2$; and for $t_p > 0$, $H_x \sim 1/x^{1/3}$ near $|x| = a/2$. We note that as a function of scan angle the tangential aperture field *amplitude* is fairly invariant in shape (although not in absolute value), whereas the field phase in the aperture changes very greatly, particularly near the grazing incidence of a grating lobe. This is significant when an initial approximation is used in a V.E. or iterative type method.[17]

Once the aperture fields are obtained, it is relatively simple to compute the total fields, $\mathscr{H}_x^R$ and $\mathscr{E}_y^R$ for $z > 0$. This was done for the *H*-plane scan case, and the transition from the aperture to far field is illustrated in Fig. 5.20 for both amplitude and phase and for $\mathscr{E}_y$ and $\mathscr{H}_x$ at two wall thicknesses. The far field of the infinite phased array (which is a plane wave) is reached in a little less than a half wavelength for most practical purposes. This has been found to be generally true for all scan angles and in the quasi-E plane as well.

2. PROPERTIES OF RECTANGULAR WAVEGUIDE ARRAYS

The analysis of planar rectangular grids of rectangular waveguide arrays, in addition to being a more difficult vector two-dimensional problem, yields new results of significance. In particular, surface wave resonances (total reflection of the exciting array energy or a transmission zero) are observed both with and without the presence of dielectric covers or plugs on the array. The effects of dielectric covers or plugs will be discussed in detail in Chapters 6 and 7.

This surface wave phenomenon is significantly different from the resonance effects of R that were illustrated in Figs. 5.18*a* and 5.18*b*. Those effects were manifested by a large value for $|R|$ but not necessarily, or even usually, the value of unity. A surface wave resonance is in fact defined by a *total* reflection or transmission zero. Furthermore, the peak value of $|R|$ always occurs at grazing incidence for the earlier type of resonance effect. Since a singular discontinuity in slope exists at grazing incidence for R as a function of scan position, the peaks of the $|R|$ curves at this point are very sharp (see Figs. 5.18*a* and 5.18*b*). Although a surface wave resonance peak in $|R|$ is sometimes also very sharp and occasionally occurs extremely close to grazing incidence (see Chapter 7 for results connected with circular waveguide arrays), peaks of this type are often smooth and nondiscontinuous (in the first derivative) and occur at beam pointing directions closer to broadside. The fact that they can and often do occur near broadside is of great practical importance, as they represent a "blind spot" of the array and should be avoided. Their occurrence is of considerable interest, moreover, in the general theory of periodic structures.

2.1 Rectangular Grid of Rectangular Waveguides

One of the earliest observations of surface wave effects occurring in phased arrays *without* dielectric covers was made from measurements[18] on the array depicted in Fig. 5.6. This array is a square grid of square waveguides such as are illustrated in Fig. 5.1, but in this case the waveguide walls are thick in both planes of scan (Fig. 5.21).

The spacing of the elements of this array was greater than a half wavelength in order to permit the TE_{10} mode to propagate. As a result of this spacing, an incipient or grazing *second lobe* is radiated when the principal radiated

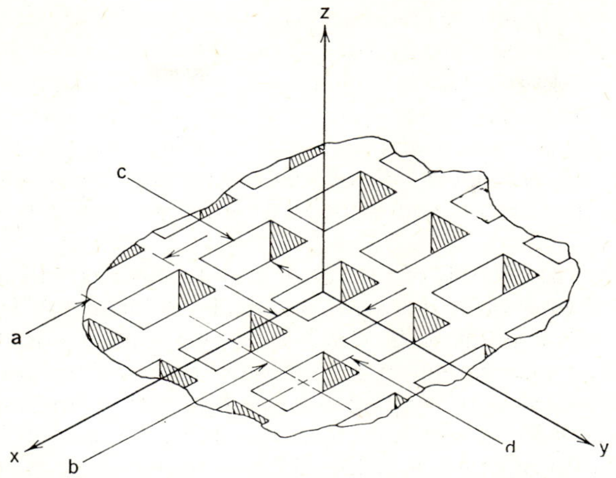

Fig. 5.21. Rectangular grid of rectangular waveguides.

lobe is pointed at $\theta = 29.1°$. The patterns depicted in Fig. 5.22 are the element radiation patterns of a centrally located element in the array. These patterns are therefore proportional to the transmission coefficient[19] of the element in an infinite excited (active) array to the extent that the number of elements illustrated approximates the infinite array environment for the central element. Comparisons with infinite array results will be made shortly.

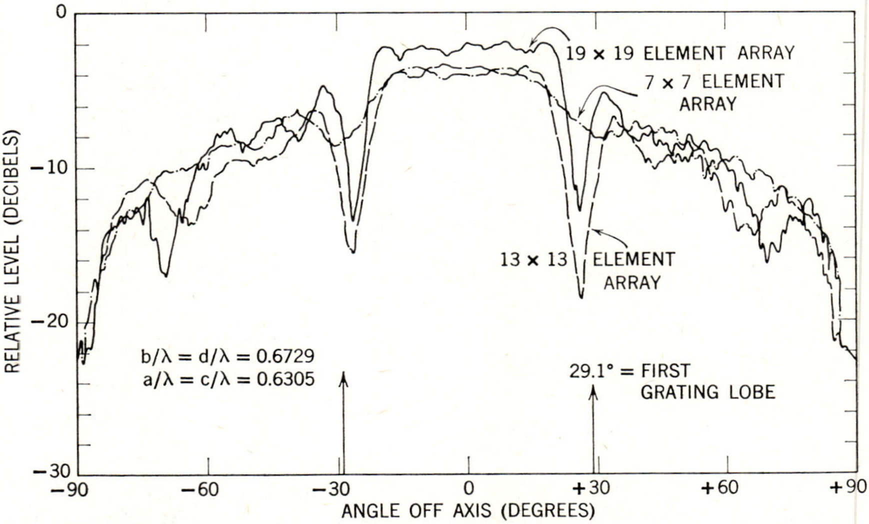

Fig. 5.22. *E*-plane radiation patterns of the center element in arrays of different sizes.

The significant feature of these results is that the sharp dip in the radiation pattern (peak in reflection) occurs at an angle closer to broadside than grating lobe incipience. Hence this dip is *not* the same phenomenon as the peaking effect discussed earlier. The fact that the dip does not go to zero is due to the finiteness of the array. Analytical results for the corresponding infinite array indicate that the dips do go to zero at these same scan angles, and therefore they are a manifestation of the forced surface wave phenomenon.

A moments type analysis of the vector two-dimensional integral equation (see Chapters 2 and 3) for this array was made by Diamond.[20] The integral equation was solved in the basis of interior or waveguide modes by using the dominant mode plus seventeen higher-order modes. This resulted in an equivalent 18 × 18 matrix equation. The principal contributions to the solutions were made, however, by the TE_{11}, TM_{11}, TE_{12}, and TM_{12} modes.

The analytical results given in Fig. 5.23 show clearly that a *zero* in radiation transmission (or total reflection: $|R| = 1.0$) occurs at an angle very near that indicated by the experiment. This angle is definitely *inside* (toward

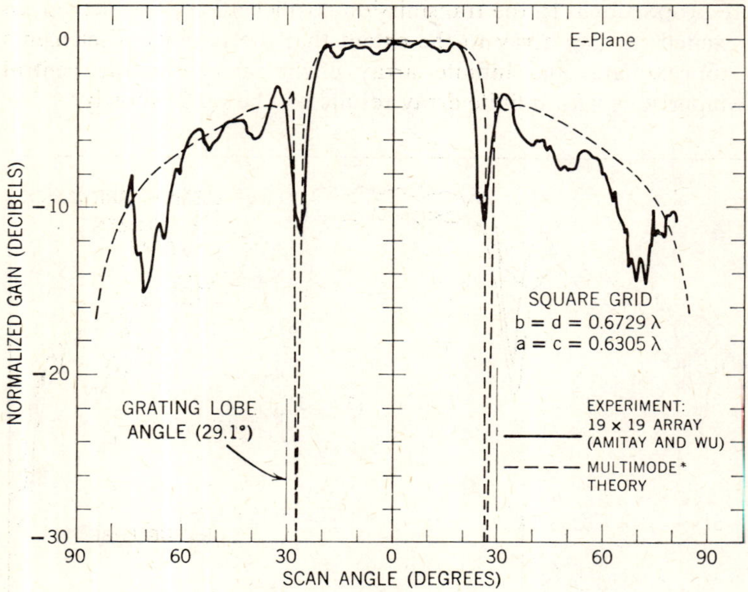

Fig. 5.23. Experimental and analytical radiation patterns—no dielectrics.
* Dominant plus 17 higher-order modes. Principal contribution due to TE-TM_{11} and TE-TM_{12} higher modes.

broadside) the grating lobe angle. This result verifies the possible implications of the experiment that a surface wave can be excited on the array without the presence of dielectric materials, such as covers or plugs.

The situation of loading the waveguides with dielectric material may be analyzed by the integral equation method. The treatment of such cases is fully discussed in Chapters 6 and 7. We shall give some results to show the

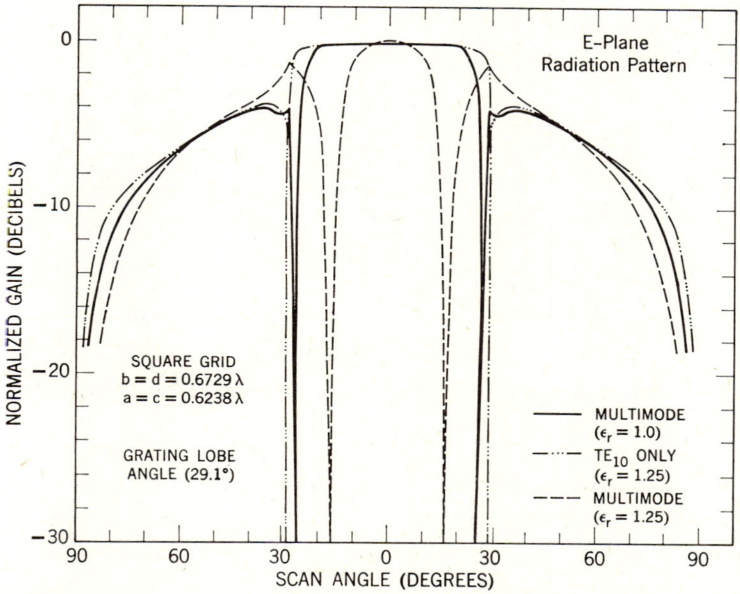

Fig. 5.24. Effect of dielectric loading on square grid radiation pattern; $\varepsilon_r = 1.25$.

effects of loading the waveguides with dielectric material on the resonance phenomenon. One would not necessarily expect the presence of the sharp nulls, the surface wave resonances, to vanish when dielectric loading is introduced into the waveguides. In fact, as is illustrated in Figs. 5.24 and 5.25, the increasing dielectric constant, varying from $\varepsilon_r = 1.0$ to $\varepsilon_r = 1.70$, causes the position of the nulls to move closer to broadside and, for most practical considerations, into a more seriously deleterious position. In Fig. 5.25, we should note that the grid spacing has been changed. In both Figs. 5.24 and 5.25, the waveguide dimensions were chosen so as to accommodate the higher dielectric constants without serious overmoding.

Another interesting aspect is shown in Figs. 5.24 and 5.25, wherein a comparison is made between the results obtained when the integral equation is solved with only one waveguide mode, the propagating TE_{10} mode, and

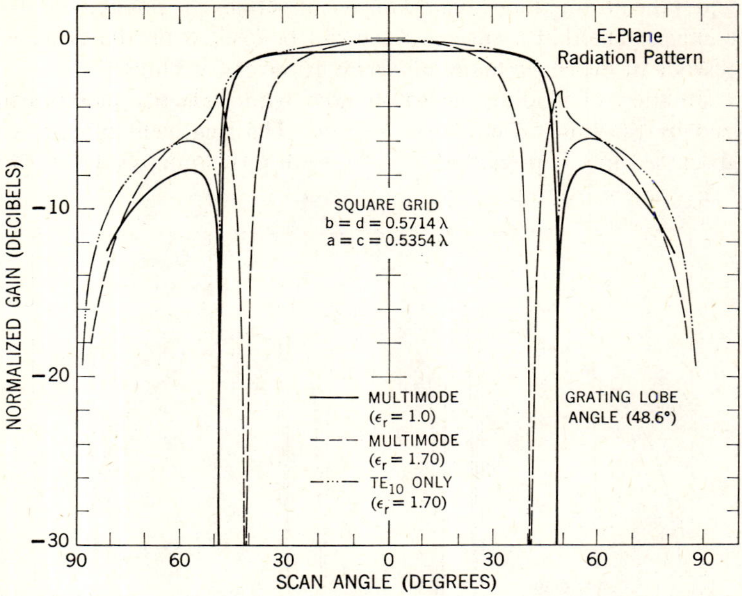

Fig. 5.25. Effect of dielectric loading on square grid radiation pattern; $\varepsilon_r = 1.70$.

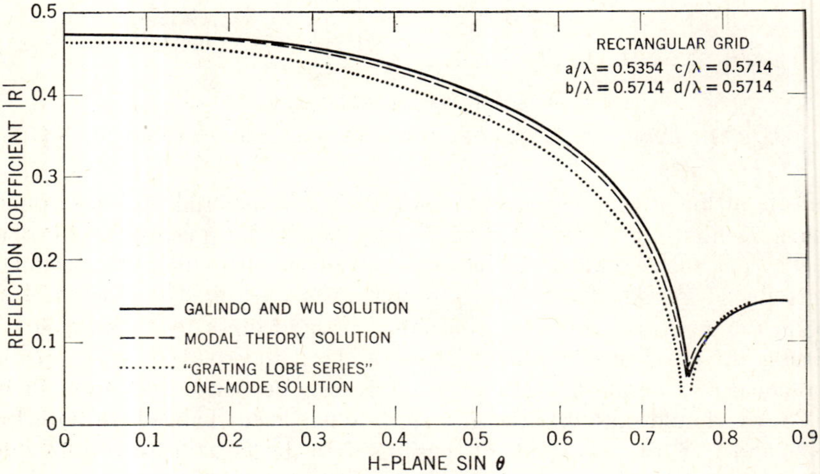

Fig. 5.26. Comparison of various basis solutions when no surface wave occurs.

those obtained when it is solved with the eighteen modes mentioned earlier (the TE_{10} plus seventeen higher-order modes). The particular result for the input impedance to the array element that is obtained when only one waveguide mode is used in the analysis is frequently called the grating lobe series. We see that the one-mode approximation is, in general, quite poor. In fact, such solutions fail to detect the very significant surface wave resonance.

It should be noted, however, that one-mode approximations are not always so singularly poor, as Farrell and Kuhn[21] have illustrated, when no surface wave resonances appear. As shown in Fig. 5.26, for a rectangular grid of rectangular waveguides in the *H*-plane scan (the walls are thin in the

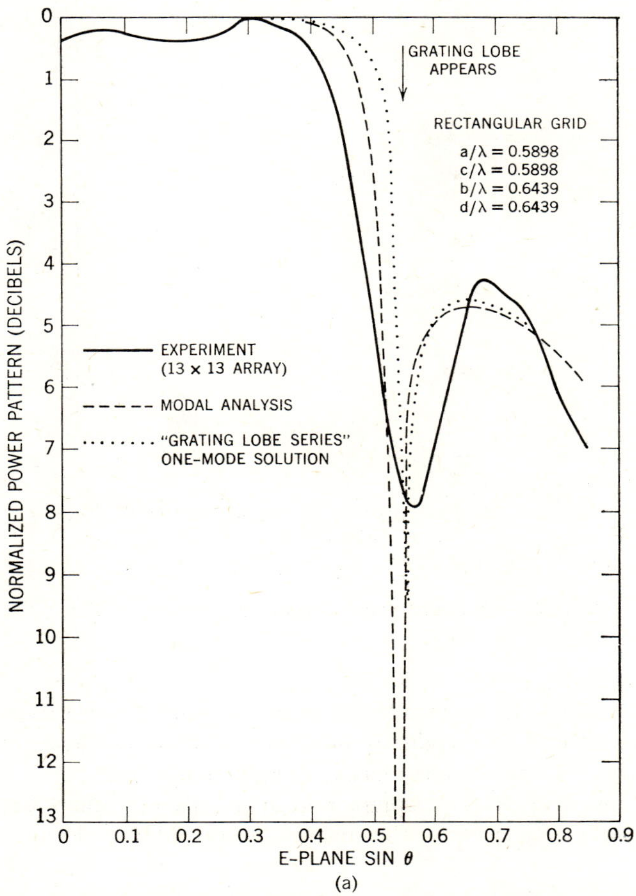

Fig. 5.27*a*. Comparison of one mode and multimode analysis of square grid of square waveguides; *E* plane.

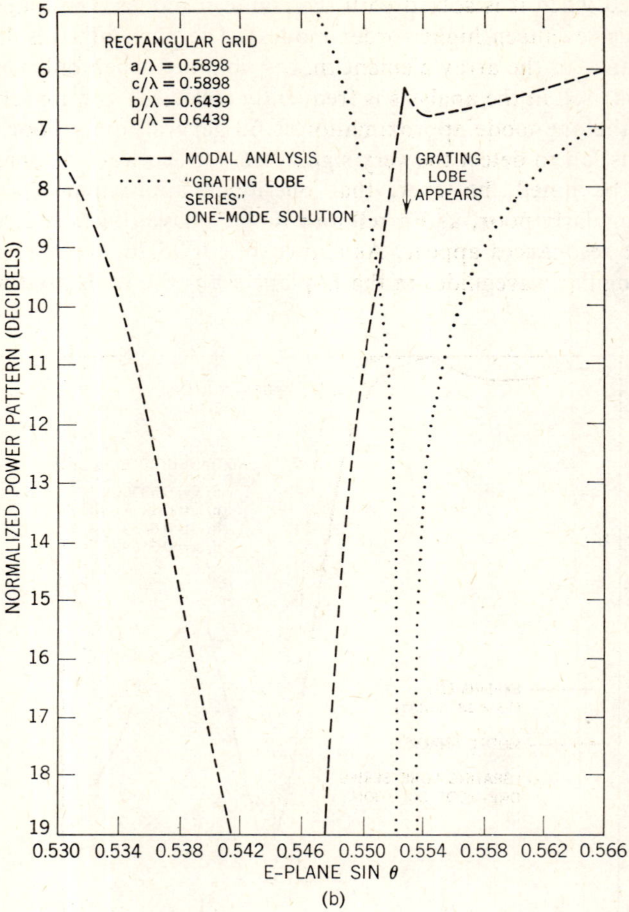

Fig. 5.27*b*. Expanded graph of comparison in Fig. 5.27*a*.

y or E-plane direction), the results obtained by using a pulse basis (thirty pulses, equally spaced; see Galindo and Wu[4]), a multiple waveguide mode basis, and a one waveguide mode basis, all agree fairly well.

In the E-plane scan for the square grid array, Farrell and Kuhn found significant discrepancy between the one-mode and multimode analyses, as shown in Fig. 5.27. In Fig. 5.27*a* we observe that the one-mode solution predicts a dip at the angle in which a grating lobe appears. The multimode analysis, on the other hand, predicts a *zero* closer to broadside, as Diamond

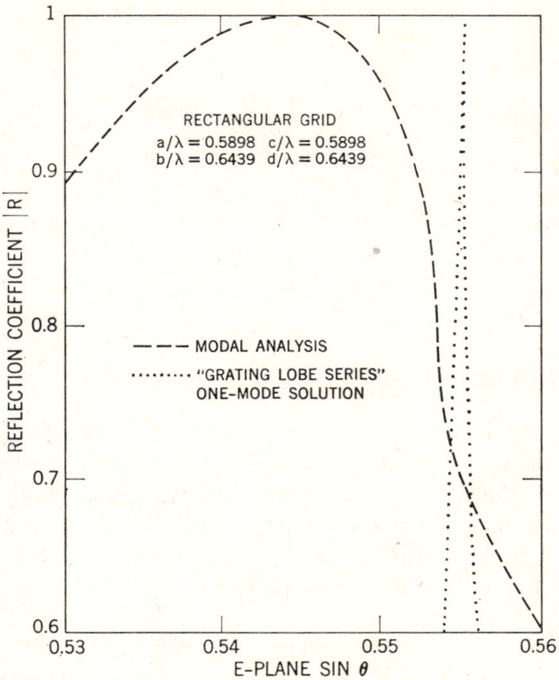

Fig. 5.28. Smooth surface wave resonance curve resulting from multimode analysis.

also found. The expanded graph in Fig. 5.27*b* shows this significant discrepancy more clearly. In fact, if we examine a plot of the reflection coefficient magnitude, $|R|$, versus $\sin\theta$ (the scan angle in the E plane), we see in Fig. 5.28 that the one-mode analysis leads to a prediction of a "peaking" effect. The multimode analysis, however, leads to a smooth surface wave resonance curve, continuous in the first and second derivatives.

2.2 Triangular Grid Arrays of Rectangular Waveguides

A triangular or brick grid of array elements has certain advantages over a square or rectangular grid. Primarily, the former arrangement is the most suitable one whenever a circular and grating lobe free scan region (in the T_x–T_y plane) is required. For such a scan region the number of elements needed per unit area is minimized[23] by the triangular arrangement. Although no fundamentally new phenomena are observed, this grid arrangement leads to different impedance and radiation characteristics for the array.

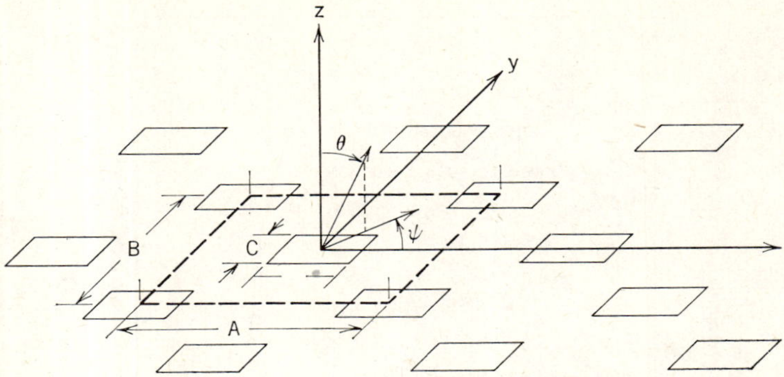

Fig. 5.29. Array geometry for triangular grid. Triangular grid angle.

The geometry of a triangular grid of rectangular waveguides is illustrated in Fig. 5.29. If the dimensions $A = B$ so that the grid is a 45° triangular array, we may view the array as a square grid one with *tilted* (at 45°) rectangular waveguide elements. With this point of view, the derivation of the appropriate integral equation for the array could proceed in the same manner as discussed in Chapter 2. Otherwise, for the *triangular* geometry shown in

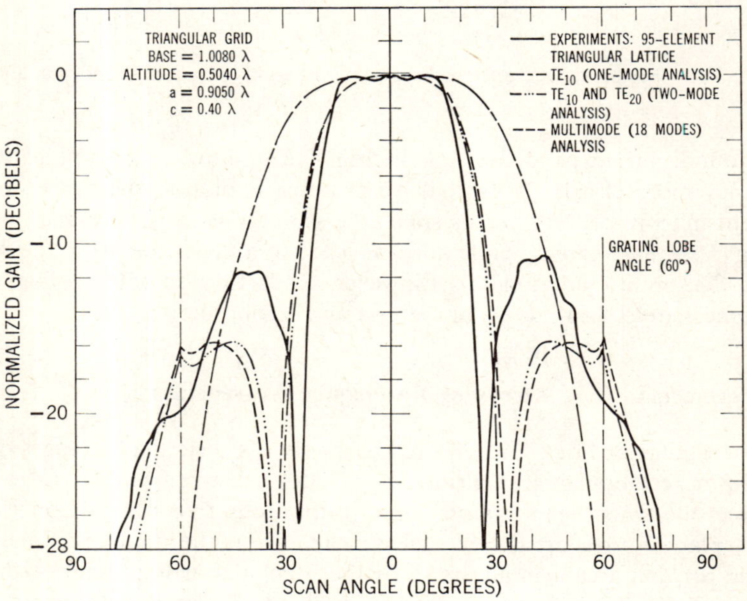

Fig. 5.30. Triangular or brick grid of rectangular waveguides; *H*-plane radiation pattern.

Fig. 5.29, it is convenient to view the array as a *skewed* element one and proceed to derive the integral equation as shown in a somewhat more general manner in Chapter 7. The resulting integral equation, in either case, is of the same essential mathematical form, and the moments type analysis proceeds in the same manner for each formulation.

Some results obtained by Diamond[20] for the array shown in Fig. 5.29, in the *H*-plane scan, are illustrated in Fig. 5.30. The radiation patterns of a single element are compared for a small 95-element experimental array and several orders of matrix approximation to the integral equation (using a waveguide mode basis). The 95-element experimental array is quite small (approximately 9 elements on a side); nevertheless it clearly shows the surface wave resonance phenomenon occurring substantially inside the grating lobe incipience angle ($\theta = 60°$). The one-mode approximation

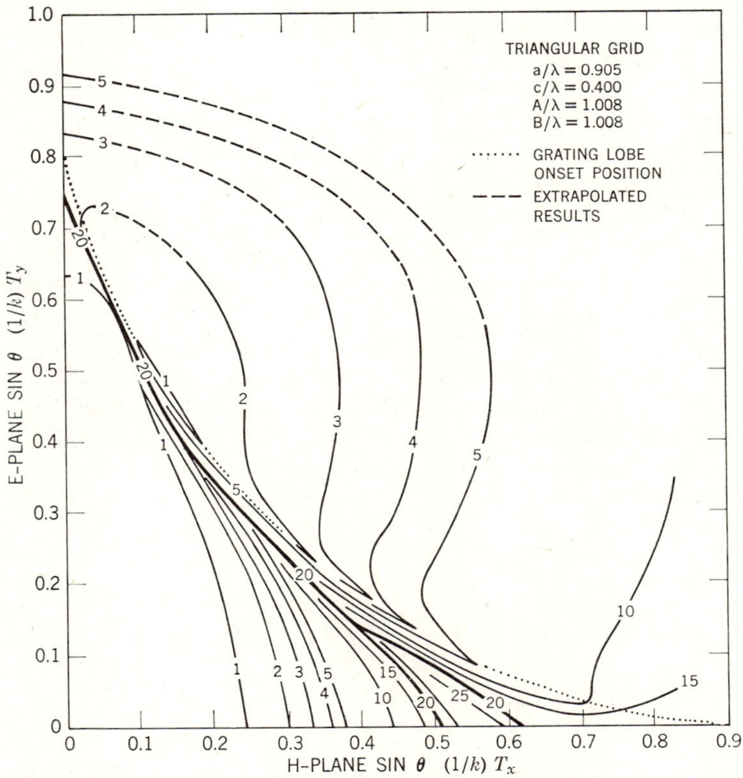

Fig. 5.31. Contour map (in decibels) of the radiation pattern of a triangular grid.

(grating lobe series) misses the surface wave entirely. The two-mode (TE_{10} and TE_{20}) approximation is seen to reveal the surface wave zero and to give a reasonable approximation to the multimode (18-mode) analysis. The discontinuity in slope at grating lobe incidence is clearly manifested in the theoretical results.

In Fig. 5.31, we see a contour plot of the radiation pattern (in decibels) of the same 45° triangular grid.[21] The plot is in a somewhat expanded (T_x, T_y) space. Since $T_x = k \sin\theta \cos\phi$ and $T_y = k \sin\theta \sin\phi$, the scales in Fig. 5.31 are simply $(1/k)T_x$ for the abscissa and $(1/k)T_y$ for the ordinate. The deep broad null along the H-plane (T_x) axis compares with results shown in Fig. 5.30. A sharper null is seen to exist along the E-plane (T_y) axis; this occurs often in planar arrays (see Chapter 7). A significant and curious aspect of this contour plot is that a null or very low minimum appears to occur along a *continuous line* inside the grating lobe incipient line (the dotted line). This may be a result of the fact that this contour plot is for a single polarization (E_y) only. Other much more extensive results for planar circular waveguide arrays (Chapter 7) indicate that these nulls are generally *isolated points* in the (T_x, T_y) plane.

2.3 Surface Wave Propagation and Related Structures

Some discussion of the origin and validity of the terms surface wave and surface wave resonance, which are used to describe the transmission nulls of the array, will be helpful.

For the sake of clarity, we will consider the parallel-plate E-plane scan problem (Fig. 5.1b with $a = b = \infty$). Extending our conclusions to the vector planar array case is straightforward by reference to the material in Chapter 2. The structure under consideration may or may not contain dielectric materials such as sheaths or plugs (Chapter 6 discusses specifically the effects of these materials). The important aspect of this entire class of problems that we wish to consider is that the fields in the *free space* region are generally represented by a Floquet type expansion, for example,

$$\mathscr{E}_y(y, z) = \sum_{m=-\infty}^{\infty} V_m e^{-j\Gamma_m z} e^{-j[(\psi - 2m\pi)/d]y} \tag{43}$$

where the z-directed propagation constant is given by

$$\Gamma_m^2 = k^2 - \left(\frac{\psi - 2m\pi}{d}\right)^2 \tag{44}$$

and ψ is the phase shift between the excitation of two adjacent elements of the typical or symbolic array shown in Fig. 5.32.

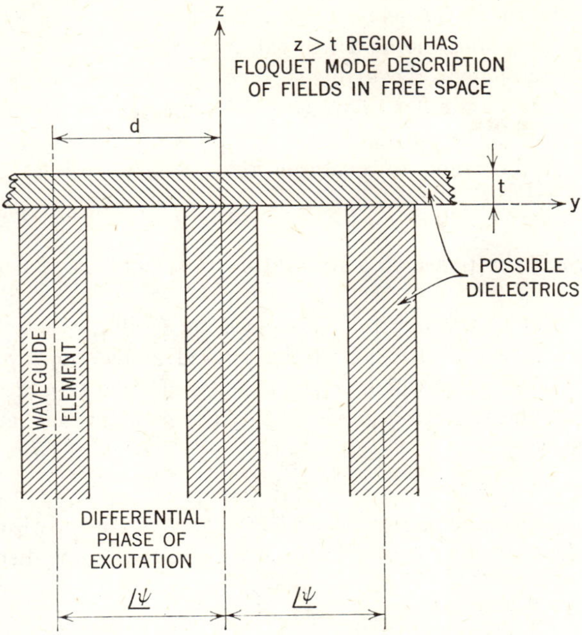

Fig. 5.32. Typical array.

In the usual mode of operation of a typical array, the steering phase, ψ, is selected so that the $m = 0$ mode is the only radiated beam:

$$\Gamma_0 = |\Gamma_0|, \qquad \Gamma_m = -j\,|\Gamma_m|, \quad m \neq 0 \tag{45}$$

In this case, all the modes for which $m \neq 0$ are evanescent and decay in the $+z$ direction as $e^{-|\Gamma_m|z}$. We have observed in this chapter that under certain circumstances the transmitted or radiation field has a zero in this visible region (or, equivalently, the incident waveguide mode is totally reflected). In this circumstance we have

$$V_0 = 0 \tag{45a}$$

in (43), and all modes with nonzero coefficients have an exponential decay with respect to the $+z$ direction. Each of the remaining modes ($m \neq 0$) has a *greater* phase variation than free space in the y direction:

$$\left|\left(\frac{\psi - 2m\pi}{d}\right)\right| > k, \quad m = 0 \tag{46}$$

Each of the remaining waves therefore is a *slow wave* with *exponential* decay away from the array and can be characterized or termed a surface wave.

Far from the array surface ($z \gg 0$), the $m = +1$ term will eventually dominate if $\psi > 0$, and the field will exhibit a *forward* $(+y)$ *phase velocity*. The phase velocity near the array surface will depend on which, if any, Floquet mode predominates. This will, of course, depend in turn on the particular array and the particular steering phase, ψ, which results in a zero transmission. In the event, for example, that

$$|V_{+1}| \gg V_m \quad (\text{all } m) \tag{47}$$

near the surface, the phase velocity is still "*slow*" but has a *backward* $(-y)$ direction.

Another useful viewpoint, first advanced by Collin,[22] is to consider the *equivalent surface wave structure* of the phased array, appropriate at only that steering phase, ψ, at which zero transmission or total reflection occurs.

Suppose that the reflection coefficient of the incident waveguide mode is determined to be

$$R = e^{-j\phi} \tag{48}$$

where the phase is referred to $z = 0$ (Fig. 5.32). It is possible to place a conducting plane or *short circuit* in the waveguides at some distance

$$z = -S = \frac{\phi - n\pi}{\gamma_0} \tag{49}$$

where γ_0 is the waveguide mode propagation constant (see Fig. 5.33). If we choose n sufficiently large in (49), the higher-order evanescent waveguide modes generated at the aperture, $z = 0$, will be decayed to a negligible value at $z = -S$. Under these conditions it is easy to see that the same electromagnetic field distribution that existed without the short circuit present will satisfy the boundary conditions for the equivalent surface wave "corrugated" structure shown in Fig. 5.33. Since no sources exist in this case, this field solution is termed a *homogeneous* solution of Maxwell's equations for the surface wave structure and is one of several possible discrete solutions to this problem.

Regardless of the interpretation, the term surface wave or surface wave resonance is seen to be clearly appropriate for the zero transmission or total reflection phenomenon. Since the short circuits shown in the equivalent structure in Fig. 5.33 are not actually present in the phased array (Fig. 5.32), there is some preference for referring to the phenomenon as a *forced* surface wave (resonance), the term forced being interjected to take note of the fact that a surface wave type field is *not* a solution to the homogeneous (no sources) Maxwell's equations for the phased array with no short circuits present (Fig. 5.32). For example, it is clear that excitation of a single waveguide with a propagating waveguide mode will *not* couple or excite the

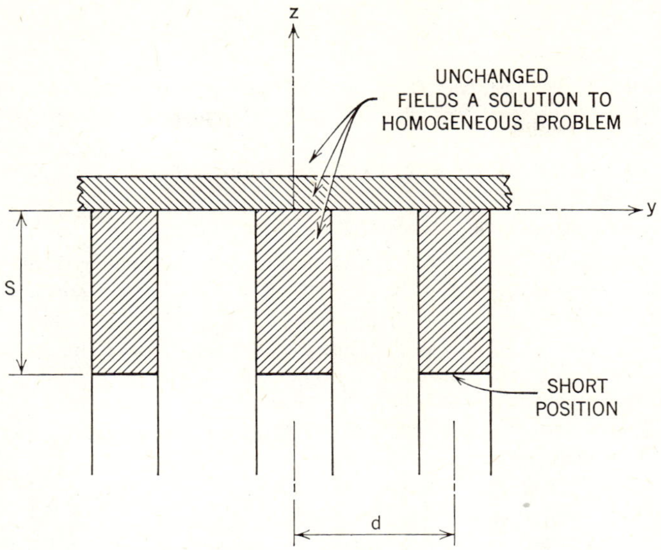

Fig. 5.33. Equivalent surface wave structure.

surface wave, since this wave cannot satisfy reciprocity and couple or excite the waveguide mode. If the surface wave did couple into a waveguide mode, it would decay rather than propagate unattenuated in the y direction.

Further aspects of surface wave excitation will be discussed in Chapters 6–8. In particular, the equivalent structure interpretation and some very significant consequences of this interpretation will be discussed in more detail in Chapter 8.

REFERENCES

1. N. Marcuvitz (ed.), M.I.T. *Waveguide Handbook*, Radiation Laboratory Series, Vol. 10, McGraw-Hill Book Company, New York, 1951.

2. R. I. Primich, "A Semi-infinite Array of Parallel Metallic Plates of Finite Thickness for Microwave Systems," *IRE Trans. Microwave Theory and Techniques*, Vol. 11, pp. 191–193, May 1963.

3. D. S. Jones, "A Critique of the Variational Method in Scattering Problems," *IRE Trans. Antennas and Propagation*, Vol. AP-4, pp. 297–301, May 1956.

4. V. Galindo and C. P. Wu, "Numerical Solutions for an Infinite Phased Array of Rectangular Waveguides with Thick Walls," *IEEE Trans. Antennas and Propagation*, Vol. AP-14, pp. 149–158, March 1966.

5. S. W. Lee, "Radiation from an Infinite Array of Parallel-Plate Waveguides with Thick Walls," *IEEE Trans. Microwave Theory and Techniques*, Vol. 15, No. 6, pp. 364–371, June 1967.
6. R. Mittra, S. W. Lee, and G. F. Van Blaricum, Jr., "A New Residue-Calculus Technique," *Proc. Fall 1967 URSI Meeting*, University of Michigan, Ann Arbor, Mich., p. 83.
7. E. C. Titchmarsh, *The Theory of Functions*, Oxford University Press, London, 1939, pp. 425–427.
8. A. J. Sommerfeld, *Optics*, Academic Press, New York, 1954.
9. V. Galindo and C. P. Wu, "Asymptotic Behavior of the Coupling Coefficients for an Infinite Array of Thin-Walled Rectangular Waveguides," *IEEE Trans. Antennas and Propagation*, Vol. AP-14, pp. 248–249, March 1966.
10. A. R. Neureuther and K. Zaki, "Numerical Solution of Electromagnetic Boundary Value Problems by Means of the Asymptotic Anticipation Method," *Proc. Spring 1967 URSI Meeting*, Ottawa, Ontario, Canada, p. 156.
11. B. Friedman, *Principles and Techniques of Applied Mathematics*, John Wiley and Sons, New York, 1960.
12. R. E. Collin, *Field Theory of Guided Waves*, McGraw-Hill Book Company, New York, 1960, Appendix A.6.
13. G. F. Van Blaricum, Jr., and R. Mittra, "A Modified Residue-Calculus Technique for Solving a Class of Boundary Value Problems," Part I: "Waveguide Discontinuities," and Part II: "Waveguide Phased Arrays, Modulated Surfaces, and Diffraction Gratings," *IEEE Trans. Microwave Theory and Techniques*, Vol. 17, No. 6, pp. 302–318, June 1969.
14. J. R. Pace and R. Mittra, "Generalized Scattering Matrix Analysis of Waveguide Discontinuity Problems," *Proc. 1964 Symposium on Quasi-Optics*, Vol. 14, pp. 177–197.
15. D. D. McCracken, *A Guide to Fortran IV Programming*, John Wiley and Sons, New York, 1965.
16. A. F. Kay and J. F. Nihen, "Scattering and Currents Induced on Sharp and Rounded Corners," *IEEE Trans. Antennas and Propagation*, Vol. AP-14, No. 1, pp. 112–114, January 1966.
17. N. Amitay and V. Galindo, "Application of a New Method for Approximate Solutions and Error Estimates to Waveguide Discontinuity and Phased Array Problems," *Radio Science*, Vol. 3, No. 8, pp. 830–843, August 1968.
18. N. Amitay, C. P. Wu, R. G. Pecina, and J. S. Cook, "On Mutual Coupling and Matching Conditions in Large Planar Phased Arrays," *Proc. 1964 PTGAP International Symposium*, Long Island, N.Y., September 1964, pp. 150–156.
19. V. Galindo and C. P. Wu, "The Relation Between the Far-Zone Pattern of the Singly Excited Element and the Transmission Coefficient of the Principal Lobe in an Infinite Array," *IEEE Trans. Antennas and Propagation*, Vol. AP-14, pp. 397–398, May 1966.
20. B. L. Diamond, "Resonance Phenomena in Waveguide Arrays," *Proc. 1967 PTGAP International Symposium*, Ann Arbor, Mich., October 1967, pp. 110–115. Note: Errata to the symposium record figures are properly included in the figures presented in this book.

21. G. F. Farrell, Jr., and D. H. Kuhn, "Mutual Coupling Effects of Triangular-Grid Arrays by Modal Analysis," *IEEE Trans. Antennas and Propagation*, Vol. AP-14, No. 5, pp. 652–654, September 1966.

22. R. E. Collin, *Field Theory of Guided Waves*, McGraw-Hill Book Company, New York, 1960, pp. 465–469.

23. E. D. Sharp, "A Triangular Arrangement of Planar-Array Elements That Reduces the Number Needed," *IRE Trans. Antennas and Propagation*, Vol. AP-9, No. 2, pp. 126–129, March 1961.

6. The Effects of Dielectrics

1. INTRODUCTION

Antennas are often enclosed in radomes to protect them from the environment. Ordinarily, radomes are designed to have a minimum effect on the antennas they cover. When it is convenient, the radome is placed in the far-field region of the antenna so that its effect may be evaluated by using transmission line approximations. In the case of waveguide type planar phased arrays, because of the geometry, the radomes can conveniently take the form of dielectric sheaths covering the arrays or dielectric loading or plugging inside the waveguides. Since in such configurations the dielectric material is placed in the near field of the array, its presence plays a major role in the aperture field distribution. Therefore it is necessary to consider the radome as an integral part of the array and to reformulate the integral equations of the phased array.

In this chapter we shall discuss the effects of dielectric materials on the radiation properties of arrays. The discussion will be focused primarily on arrays of parallel-plate waveguides because of their structural as well as analytical simplicity, although occasional reference to rectangular waveguide arrays will be made also. A great many of the properties uncovered here may be extrapolated, at least qualitatively, to arrays of more complex elements. Several different geometries will be considered, including covering the array with dielectric slabs, loading the waveguides completely with a dielectric or partially with dielectric plugs, and various combinations thereof.

Two major aspects of the problem will be examined in detail. The first one deals with the impedance matching of the array. Since the use of a dielectric material introduces additional parameters to the problem, as might be anticipated, it is possible to improve the wide scan angle match considerably by judicious choices of the parameters. The second aspect is concerned with the phenomenon of forced aperture resonances or forced surface wave resonances. The appearance of such forced resonances is related to the presence of surface-wave-like modes inside the dielectric slab. Aperture resonances can also occur in planar arrays without dielectric slabs, as discussed in Chapter 5 in connection with the analysis of rectangular waveguide arrays. More detail on this subject is given in Chapter 7 when circular waveguide arrays are considered and in Chapter 8, where there is a rather general treatment of the topic.

The forced aperture (or forced surface wave) resonances are manifested by a total reflection of the incident energy, so that no energy can be transmitted under such conditions. The resonance conditions, in general, cannot be accurately determined without solving the integral equations in one form or another. There are certain situations, however, in which a good estimate of these conditions is possible from a knowledge of the geometry of the problem, and the solution to a related boundary value problem.[14,16]

It is well known by now that the phased array boundary value problem may be regarded as a generalized waveguide discontinuity problem. Viewed in this context, the array aperture can be described as a multiport network with each port representing a mode which may be excited. Frequently, it is sufficient to consider only the propagating modes in such representations. The forced surface wave (or aperture resonance) phenomenon usually occurs when the waveguide supports only one propagating mode and there is only one radiated beam. (This being the usual array operating condition, the occurrence of a total reflection in the normally visible region is of particular significance.) Moreover, in the case of parallel-plate arrays, the phenomenon has been observed only when either a dielectric slab or a dielectric plug is used at the array aperture. The occurrence of resonances is contingent upon the presence of at least one mode which propagates inside the dielectric but is evanescent outside of it. This kind of wave mode may be represented by a shorted port with the short position dependent on the thickness of the dielectric material. The condition of resonance may thus be pictured as a three (or more) port network with one (or more) ports shorted. It is well known from network theory that by adjusting the short positions a condition can be created such that there is no energy transmission between the two remaining ports. The occurrence of forced surface waves may be explained by such a consideration. Another explanation in terms of leaky waves has also been proposed.[15]

Since the reflection coefficient at forced surface wave resonance has a magnitude of 1, the solution to the phased array problem can be related to that of a surface wave propagating above a corrugated structure.[2,17] Some discussion of wave propagation above corrugated structures, both with and without the presence of dielectric materials, was given in Chapter 5, and the topic will be considered further in Chapter 8. Surface wave propagation has certain new, salient features. For example, the wave does not require all the space harmonics for its description, that is, certain space harmonics have vanishing coefficients. They may be of either the forward or the backward wave type.

2. INTEGRAL EQUATION FOR AN ARRAY WITH DIELECTRIC PLUGS

When the array is covered by dielectric slabs or is loaded with dielectric plugs, it is obvious that the integral equations will be different from those for the case with no dielectric material. We shall proceed to derive in some detail integral equations for an array loaded with a single plug. The equations for an array with a single sheath and with multiple sheaths will also be given. The details of the formulation for the latter situations, however, will be dealt with in Appendix I.

Consider the parallel-plate array shown in Fig. 6.1. The array may be scanned in either the E or the H plane, depending on the mode of excitation.

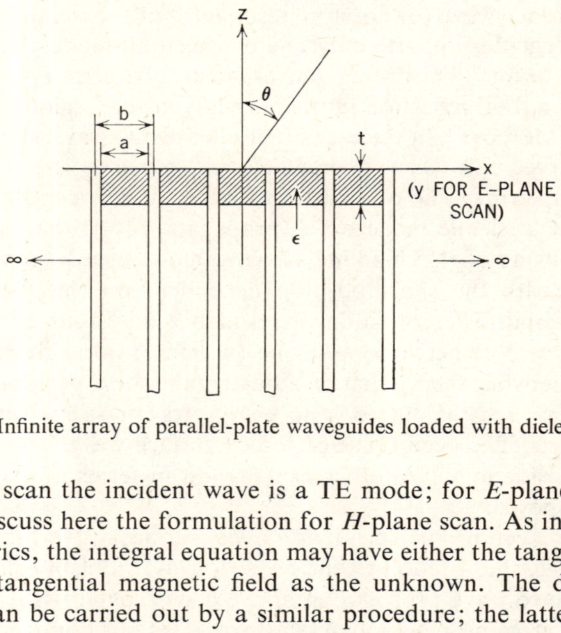

Fig. 6.1. Infinite array of parallel-plate waveguides loaded with dielectric plugs.

For H-plane scan the incident wave is a TE mode; for E-plane scan, a TM mode. We discuss here the formulation for H-plane scan. As in the situation of no dielectrics, the integral equation may have either the tangential electric field or the tangential magnetic field as the unknown. The derivation for either case can be carried out by a similar procedure; the latter situation is treated here. To facilitate comparisons, the integral equation for the unloaded array is repeated:

$$2z_1\Phi_1(x) = \int_{-b/2}^{b/2} \left\{ \sum_{q=1}^{\infty} z_q \Phi_q(x)\Phi_q(x') + \sum_{m=-\infty}^{\infty} Z_m \Psi_m(x)\Psi_m^*(x') \right\} H_x(x')\, dx' \quad (1)$$

Here, $\{\Phi_q(x)\}$ denote the orthonormal modes of the waveguide, and $\{\Psi_n(x)\}$ are the orthonormal Floquet modes for the region exterior to the waveguides; $\{z_q\}$ and $\{Z_m\}$ are the modal impedances for the respective regions. Equation (1) was obtained under the assumption that the incident wave has a unit modal current originated from the waveguide side. In deriving this equation, the waveguide was not required to be free of dielectrics. In fact, when the waveguides are completely filled with dielectrics, the equation is still valid. It is only necessary to use appropriate interior modal impedances to account for the presence of the dielectric. Specifically, the equation for an array with waveguides completely filled with a dielectric of permittivity ε is obtained by replacing z_q by z_q^{ε}. In the case of H-plane scan,

$$z_q^{\varepsilon} = \frac{\omega\mu_0}{\gamma_q^{\varepsilon}} \qquad \text{with} \quad \gamma_q^{\varepsilon} = \sqrt{k^2\varepsilon - (q\pi/a)^2}$$

Numerical results for such a completely loaded array will be presented and discussed later.

We now consider the dielectric plug case.[1,2] There are three distinct regions in the array geometry: the empty waveguide region, the region of waveguide with a dielectric plug, and the free space region. To formulate the problem, it seems natural to divide the space accordingly. It turns out, however, that it suffices to partition the space into two regions, one inside the waveguides and the other outside of them, because a waveguide loaded with dielectric which uniformly fills the waveguide cross section has modal functions identical to those of a waveguide without a dielectric. Consequently the air-dielectric interface does not cause coupling between the waveguide modes. This will become evident as we progress.

Let the incident wave consist of the lowest-order mode, and the air-dielectric interface be used as its phase reference. The extension to a more general excitation is straightforward, as discussed in Chapters 2 and 4. The tangential fields in each region are first described by appropriate normal mode expansions.

For $z \le -t$:

$$\begin{aligned}\mathscr{H}_x(x, z) &= [e^{-j\gamma_1(z+t)} - Re^{j\gamma_1(z+t)}]\Phi_1(x) + \sum_{q=2}^{\infty} i_q e^{j\gamma_q(z+t)}\Phi_q(x) \\ \mathscr{E}_y(x, z) &= -z_1[e^{-j\gamma_1(z+t)} + Re^{j\gamma_1(z+t)}]\Phi_1(x) + \sum_{q=2}^{\infty} z_q i_q e^{j\gamma_q(z+t)}\Phi_q(x)\end{aligned} \tag{2}$$

For $-t \le z \le 0$:

$$\begin{aligned}\mathscr{H}_x(x, z) &= \sum_{q=1}^{\infty}(i_q^{+} e^{-j\gamma_q^{\varepsilon} z} + i_q^{-} e^{j\gamma_q^{\varepsilon} z})\Phi_q(x) \\ \mathscr{E}_y(x, z) &= -\sum_{q=1}^{\infty} z_q^{\varepsilon}(i_q^{+} e^{-j\gamma_q^{\varepsilon} z} - i_q^{-} e^{j\gamma_q^{\varepsilon} z})\Phi_q(x)\end{aligned} \tag{3}$$

For $0 \le z$:

$$\mathscr{H}_x(x, z) = \sum_{m=-\infty}^{\infty} I_m \Psi_m(x) e^{-j\Gamma_m z}$$
$$\mathscr{E}_y(x, z) = -\sum_{m=-\infty}^{\infty} Z_m I_m \Psi_m(x) e^{-j\Gamma_m z} \tag{4}$$

These field components must satisfy the boundary conditions at both $z = 0$ and $z = -t$. Application of the conditions at $z = 0$, using (3) and (4), yields

$$H_x(x) = \mathscr{H}_x(x, 0) = \sum_{q=1}^{\infty}(i_q^+ + i_q^-)\Phi_q(x) = \sum_{m=-\infty}^{\infty} I_m \Psi_m(x)$$
$$E_y(x) = \mathscr{E}_y(x, 0) = -\sum_{q=1}^{\infty} z_q^{\varepsilon}(i_q^+ - i_q^-)\Phi_q(x) = -\sum_{m=-\infty}^{\infty} Z_m I_m \Psi_m(x) \tag{5}$$

By use of the orthonormality relation between the modal functions, it follows that

$$\bar{i}_q = i_q^+ + i_q^- = \int_{-b/2}^{b/2} \Phi_q(x) H_x(x)\, dx$$
$$I_m = \int_{-b/2}^{b/2} \Psi_m^*(x) H_x(x)\, dx \tag{6}$$

Similarly, from application of the continuity conditions at the interface $z = -t$,

$$\mathscr{H}_x(x, -t) = (1 - R)\Phi_1(x) + \sum_{q=2}^{\infty} i_q \Phi_q(x)$$
$$= \sum_{q=1}^{\infty}(i_q^+ e^{j\gamma_q^{\varepsilon} t} + i_q^- e^{-j\gamma_q^{\varepsilon} t})\Phi_q(x)$$
$$\mathscr{E}_y(x, -t) = -z_1(1 + R)\Phi_1(x) + \sum_{q=2}^{\infty} z_q i_q \Phi_q(x) \tag{7}$$
$$= -\sum_{q=1}^{\infty} z_q^{\varepsilon}(i_q^+ e^{j\gamma_q^{\varepsilon} t} - i_q^- e^{-j\gamma_q^{\varepsilon} t})\Phi_q(x)$$

By observing that the right- and left-hand sides of (7) are the field expansions with respect to the same set of modal functions, we may immediately write

$$1 - R = i_1^+ e^{j\gamma_1^{\varepsilon} t} + i_1^- e^{-j\gamma_1^{\varepsilon} t}$$
$$z_1(1 + R) = z_1^{\varepsilon}(i_1^+ e^{j\gamma_1^{\varepsilon} t} - i_1^- e^{-j\gamma_1^{\varepsilon} t}) \tag{8a}$$

and

$$\left.\begin{aligned} i_q &= i_q^+ e^{j\gamma_q^{\varepsilon} t} + i_q^- e^{-j\gamma_q^{\varepsilon} t} \\ z_q i_q &= -z_q(i_q^+ e^{j\gamma_q^{\varepsilon} t} - i_q^- e^{-j\gamma_q^{\varepsilon} t}) \end{aligned}\right\}, \quad q \neq 1 \tag{8b}$$

Equations (8a) and (8b) show that the air-dielectric interface at $z = -t$ is a *simple* one in that the qth mode in the dielectric region couples only into the mode of the same order in the unloaded waveguide region, and vice versa.

The implication of this observation is that the fields at the air-dielectric interface are completely determined when the fields at the array aperture are obtained. Therefore, it is necessary to solve for the fields only at the array aperture. With this in mind, we may eliminate i_q from (8a) and (8b) and express i_q^- in terms of i_q^+ as

$$i_q^- = -\frac{z_q + z_q^\varepsilon}{z_q - z_q^\varepsilon} i_q^+ e^{j2\gamma_q^\varepsilon t} + \delta_{1q} \frac{2z_q}{z_q - z_q^\varepsilon} e^{j\gamma_q^\varepsilon t} \tag{9}$$

and then obtain

$$z_q^\varepsilon(i_q^+ - i_q^-) = \tilde{z}_q i_q - \delta_{1q}\bar{z}_1 \tag{10}$$

where

$$\begin{aligned} \bar{z}_1 &= \frac{2z_1 z_1^\varepsilon}{z_1^\varepsilon \cos \gamma_1^\varepsilon t + jz_1 \sin \gamma_1^\varepsilon t} \\ \tilde{z}_q &= z_q^\varepsilon \frac{z_q + jz_q^\varepsilon \tan \gamma_q^\varepsilon t}{z_q^\varepsilon + jz_q \tan \gamma_q^\varepsilon t} \end{aligned} \tag{11}$$

When (6) and (9) are substituted into (5), there results

$$2\bar{z}_1\Phi_1(x) = \int_{-b/2}^{b/2} \left\{ \sum_{q=1}^{\infty} \tilde{z}_q \Phi_q(x)\Phi_q(x') + \sum_{m=-\infty}^{\infty} Z_m \Psi_m(x)\Psi_m^*(x') \right\} H_x(x')\, dx' \tag{12}$$

This is the desired integral equation for an array loaded with dielectric plugs. A comparison between (1) and (12) shows that these equations have similar expressions, the only difference being for the modal impedances associated with the interior modes, including the one for the incident wave. The exterior modal impedances remain unaltered, since no perturbations are introduced in that region. The equivalent impedance, $\tilde{z}_q$, as given by (11), can be recognized as the input impedance of a transmission line which has a characteristic impedance z_q^ε, propagation constant γ_q^ε, and length t and is terminated in a load impedance z_q. Notice that only quantities pertaining to the qth waveguide mode appear in the expression for $\tilde{z}_q$ in (11).

The free term on the left-hand side of (12) is still proportional to the incident electric field which exists at the array aperture when a wave of unit modal current is incident at the air-dielectric interface. All these facts suggest that the space within the waveguides may be treated as a single region from the onset, as observed earlier, provided that the effects of dielectric plugs are taken into account through the use of appropriate modal impedances, and the incident wave is properly weighted. It may be noted that by setting $t = 0$, corresponding to the case with no dielectric plug, the equation properly reduces to (1). Alternatively, the same result is reached by choosing $\varepsilon = 1$ for the dielectric plug, except for the factor $e^{-j\gamma_1^\varepsilon t}$ on the left side of (12), which accounts for the phase reference of the incident wave at $z = -t$ instead of at $z = 0$.

It should be apparent from this derivation that the results apply, with only minor changes in the integral equations, to a general planar array of any type of waveguide elements.

3. MULTILAYERED PLUGS AND SHEATHS

From the derivation discussed above, it is clear that the integral equation can be generalized by a rather straightforward extension to cases of loading the waveguides with multilayered dielectric plugs or sheaths, or a combination of the two. Frequently, the extension may be carried out by deduction. Consider first the case of a single sheath covering the array as depicted in Fig. 6.2. Since only the exterior region is perturbed by the introduction of the

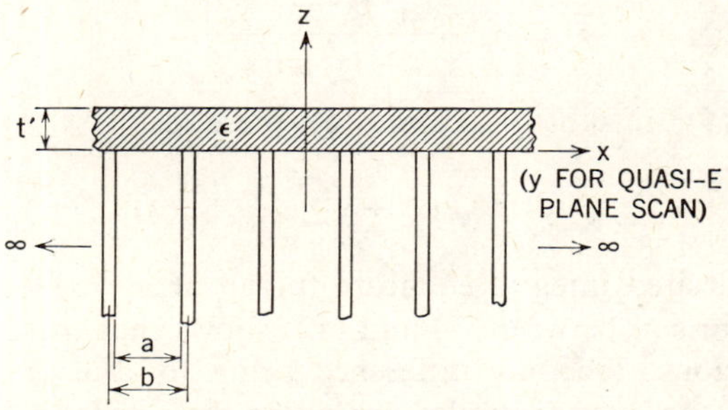

Fig. 6.2. A parallel-plate array covered by a dielectric sheath.

dielectric sheath, it is clear that only the part of the equation pertaining to that region needs to be modified. By inference from the result for the case of plug loading, the equation for the slab case may be written readily as

$$2z_1\Phi_1(x) = \int_{-b/2}^{b/2} \left\{ \sum_{q=1}^{\infty} z_q\Phi_q(x)\Phi_q(x') + \sum_{m=-\infty}^{\infty} \tilde{Z}_m\Psi_m(x)\Psi_m^*(x') \right\} H_x(x')\,dx' \quad (13)$$

where

$$\tilde{Z}_m = Z_m{}^\varepsilon \frac{Z_m + jZ_m{}^\varepsilon \tan \Gamma_m{}^\varepsilon t'}{Z_m{}^\varepsilon + jZ_m \tan \Gamma_m{}^\varepsilon t'} \quad (13a)$$

This again is to be recognized as the input impedance of a transmission line which has a characteristic impedance $Z_m{}^\varepsilon$, propagation constant $\Gamma_m{}^\varepsilon$, and length t', and is terminated in a load impedance Z_m.

The extension to a more general situation of stratified dielectric media follows a similar procedure. For example, if the array is to be covered by a stratified medium of N layers as shown in Fig. 6.3, the integral equation, (13), is still a valid one to use. Only the exterior modal impedances need

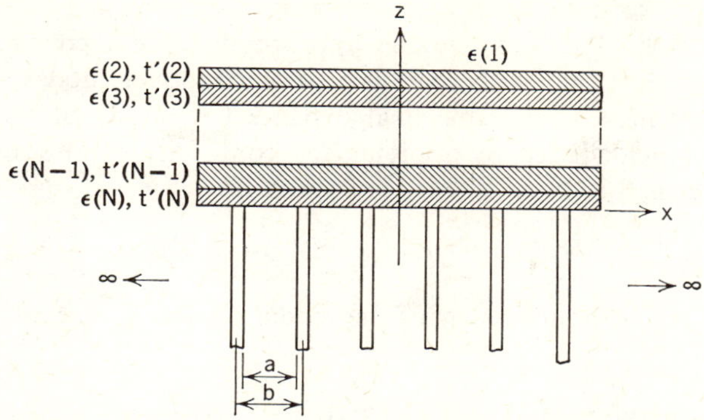

Fig. 6.3. Radiation of an array through N dielectric sheaths.

to be replaced by the appropriate input impedances for the N-layer stratified dielectric as seen at the array aperture. Such modal impedances can be calculated by standard techniques.[4]

The treatment of a multilayer plug loading is equally straightforward. In this case, in addition to the modal impedances in the kernel requiring modification, those in the free term must be changed also. It is recalled that in the formulation of the integral equations the aperture fields were used as the unknown functions. The solutions for the fields at the various dielectric interfaces may be determined from the aperture fields by straightforward modal expansion of the fields. A derivation of the relation between the modal coefficients for a multilayer plug case is given in Appendix I. The simpler case of a single layer is discussed in the next section.

4. REFLECTION AND TRANSMISSION COEFFICIENTS

The integral equations derived in Sections 2 and 3 for arrays that include various configurations of dielectric materials may be solved by using the method of moments discussed in Chapters 3 and 5. The solutions thus obtained are for the unknown fields at the aperture. We now obtain the reflection and transmission coefficients from the approximated aperture fields in the presence of the dielectric materials. Suppose that the following approximation is introduced into (12):

$$H_x(x) \simeq \sum_{m=-M}^{M} \tilde{I}_m \Psi_m(x) \tag{14}$$

and the results are weighted with the set of functions $\Psi_n^*(x)$, $n = -M$, $-(M-1), \ldots, -1, 0, 1, \ldots, M$. This process yields a linear system of equations for the unknowns $\tilde{I}_m$, for which a solution may be effected by matrix inversion. The $\tilde{I}_m$'s are the appropriate modal coefficients of the aperture magnetic field with reference to the Floquet modes, $\Psi_m(x)$. The modal coefficients of the same field expanded in terms of the waveguide modes may be calculated by applying the continuity conditions across the aperture, that is,

$$H_x(x) \simeq \sum_{q=1}^{N} \tilde{\imath}_q \Phi_q(x) \simeq \sum_{m=-M}^{M} \tilde{I}_m \Psi_m(x) \tag{15}$$

By virtue of the orthonormality of the modal functions we find

$$\tilde{\imath}_q \simeq \sum_{m=-M}^{M} C_{qm} \tilde{I}_m \tag{16}$$

where

$$C_{qm} = \int_{-b/2}^{b/2} \Phi_q(x) \Psi_m(x)\, dx$$

The modal coefficients in the unloaded waveguide region may be related to $\tilde{\imath}_q$ by using (8b) and (9):

$$i_q = \frac{z_q^{\varepsilon}}{z_q^{\varepsilon} \cos \gamma_q^{\varepsilon} t + j z_q \sin \gamma_q^{\varepsilon} t}\, \tilde{\imath}_q \qquad \text{for} \quad q \neq 1 \tag{17a}$$

$$R = \frac{z_1^{\varepsilon}}{z_1^{\varepsilon} \cos \gamma_1^{\varepsilon} t + j z_1 \sin \gamma_1^{\varepsilon} t}\, \tilde{\imath}_1 - \frac{z_1^{\varepsilon} \cos \gamma_1^{\varepsilon} t - j z_1 \sin \gamma_1^{\varepsilon} t}{z_1^{\varepsilon} \cos \gamma_1^{\varepsilon} t + j z_1 \sin \gamma_1^{\varepsilon} t} \tag{17b}$$

The transmission coefficients of the array can be expressed directly in terms of the modal coefficients, $\tilde{I}_m$, as in the case when no dielectric is present. These coefficients are usually normalized so that the sum of squares of the reflection coefficient and transmission coefficient(s) is equal to the incident power. With this normalization, the resulting finite scattering matrix for the array aperture junction is unitary. Thus, for $\frac{1}{2} < b/\lambda < 1$, the pertinent transmission coefficients are given by

$$\begin{aligned} T_0 &= \sqrt{Z_0/z_1}\, \tilde{I}_0 && \text{for} \quad 0 \leq \psi \leq \pi \\ T_{-1} &= \sqrt{Z_1/z_1}\, \tilde{I}_1 && \text{for} \quad 2\pi(1 - b/\lambda) \leq \psi \leq \pi \end{aligned} \tag{18}$$

The conservation of power requires

$$|R|^2 + |T_0|^2 = 1 \qquad \text{for} \quad 0 \leq \psi \leq 2\pi(1 - b/\lambda)$$

and

$$|R|^2 + |T_0|^2 + |T_{-1}|^2 = 1 \qquad \text{for} \quad 2\pi(1 - b/\lambda) \leq \psi \leq \pi$$

When the array is covered by dielectric sheaths, (13) is the appropriate equation to solve. The solution by a similar moments expansion gives the modal coefficients of the field at the aperture. However, the modal coefficients in the free space region have to be determined before the transmission coefficients can be calculated. A procedure similar to the one used in obtaining (17a) and (17b) may be used for this purpose. When there is a single sheath covering the array, the resultant transmission coefficients are found to be

$$T_m = \frac{\sqrt{Z_m Z_m^{\varepsilon}}}{Z_m^{\varepsilon} \cos \Gamma_m^{\varepsilon} t' + jZ_m \sin \Gamma_m^{\varepsilon} t'} \tilde{I}_m \tag{18a}$$

This result can be extended to the multilayer situation in a way similar to that used for the multiple-plug case discussed in Appendix I.

5. THE USE OF DIELECTRICS FOR THE IMPROVEMENT OF IMPEDANCE MATCH

In what follows, numerical solutions for the various cases of dielectric loading will be discussed. The adequacy of the moments method has been extensively tested for the unloaded case, as discussed in Chapter 5. The results obtained there may be used as standards for checking the results with dielectric material present by choosing $\varepsilon = 1$ in the more general computer program. This has been done, together with making other checks, so as to make sure that accurate results are obtained.

The use of dielectrics introduces additional physical parameters into the problem. Their presence, of course, plays a major role in the field distribution. Consequently, the choice of values for these parameters, such as the dielectric constant and the thickness of the sheath, greatly influences the array performance. It has been proposed, therefore, by several researchers[5,6] to use dielectrics as a tool for improving the array match characteristics. Magill and Wheeler,[5] for example, proposed placing a thin high-ε dielectric sheet far above the array face for this purpose. Their approach treats the array aperture and the dielectric sheet as coupled only by propagating modes, and their analysis is based on the conventional transmission line approximations. Hence this approach is permissible only when the sheet is located far away from the aperture, so that the effects of nonpropagating modes can be neglected. Some improved match was accomplished in their investigation by using dielectric disk-loaded circular waveguides as the array elements. Since their results were obtained for only a few scan angles which can be simulated in a waveguide,[12] the data were not sufficient to indicate that wide scan angle improvement was in fact achieved. Moreover, as will become clear presently, the use of a dielectric sheet may cause forced surface wave resonances in an array. An analysis employing only a conventional transmission line approximation is not capable of uncovering such a phenomenon.

The idea of using a thin dielectric sheet above the array was also applied by Lee[6] to a rectangular waveguide array. With the assumption that the waveguide walls are infinitesimally thin and the scan is restricted in the *H* plane (in effect, the *H*-plane scan of a parallel-plate array), he was able to derive an analytical solution for the reflection coefficient. Unfortunately the expression is very complex and is valid only when the sheet is located at a distance away from the array aperture.

More recently, Lee and Mittra[7] developed a modified residue calculus technique which enables them to solve infinite systems of equations arising from rectangular waveguide arrays with dielectric loading. The method removes the restriction inherent in Lee's original approach in that the case of completely loading the waveguides may also be handled by this new formulation. However, only the case of *H*-plane scan under thin-wall assumption is tractable by this method at present. Computations based on these techniques have shown that substantial improvement of the match characteristics is indeed achievable over a wide scan range.

In contrast to the limitations inherent in these methods, the moments type solution of the integral equation has the advantage of being very flexible in that it can handle the thick-wall waveguide case, as well as loading or covering by dielectric slabs and a combination thereof in the same formulation. Thus the integral equation formalism appears preferable. It should be emphasized, however, that solutions by alternative methods are extremely important in providing validity checks for numerically obtained results.

The choice of a set of parameters for an optimum match is best made by parameter studies, unless a synthesis procedure is available. The parameter study may be carried out by holding fixed all pertinent factors except one, and performing calculations with this parameter varied over a range of values. The process can be carried out quite effectively by man-machine interaction, thereby narrowing down to the set of values which will yield the desired characteristics. In the following treatment we shall not attempt to match an array; rather we shall show some numerical calculations to illustrate the effects of the dielectric and its geometry on the performance of an array.

5.1 Results with the Waveguides Completely Loaded with Dielectric

Figure 6.4 gives some typical results for the *H*-plane scan direction with the waveguides loaded with $\varepsilon = 0.9$ to $\varepsilon = 3.0$.* The dielectric constant is chosen so that only one mode propagates in the waveguide. The change in R with ε between curves is smooth. Between $\varepsilon = 0.9$ and $\varepsilon = 1.1$, however, the change in $|R|$ is great. This may be attributed to the fact that cut-off of the dominant waveguide mode occurs at $\varepsilon = 0.872$.

* A dielectric with $\varepsilon < 1$ may be realized by a plasma.

We notice that for $\varepsilon \sim 1.3$ the angular response of R versus ψ is nearly flat, both in amplitude and in phase. In fact, for all wavelengths examined there appears to be at least one value of ε for which a nearly flat angular response for R is obtained. It should be noted that, even if the magnitude of R is large, the flatness of the response in both amplitude and phase permits matching the array for all scan angles in the region of flat response, at least at one frequency.

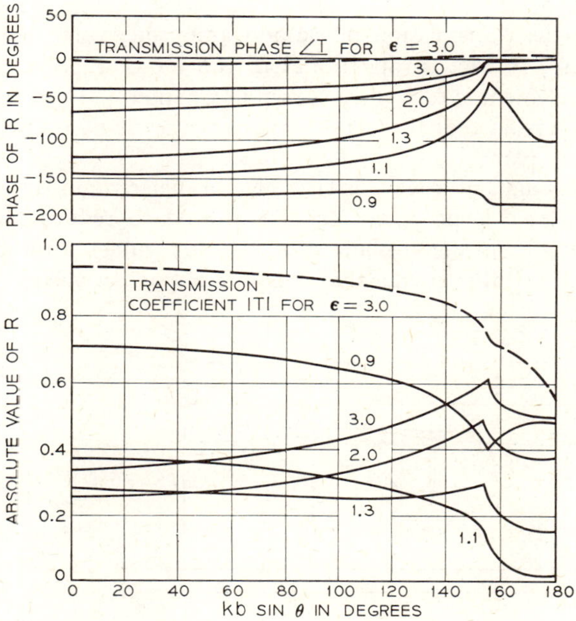

c_n = COUPLING TO nTH REMOVED ELEMENT

ϵ	$\lvert c_0 \rvert$	$\lvert c_1 \rvert$
0.9	0.626	0.015
1.1	0.252	0.080
1.3	0.202	0.094
2.0	0.319	0.106
3.0	0.423	0.081

Fig. 6.4. Dielectric loading inside waveguides for H-plane scanning; $a = 0.937b = 0.5354\lambda$, $b = 0.5714\lambda$.

The discontinuity in slope of these curves at the scan angle θ, determined from $kb \sin \theta = 2\pi(1 - b/\lambda)$, coincides with the onset of a grating lobe at that angle. Notice that the singularity in the derivative of the modulus of R lies on the right side of grating lobe incipience, but on the left side of the phase of R. This is the same result as is found for the thin-wall waveguide case,[8,9] and it may be expected to lead to the same asymptotic behavior of coupling.

The reflection coefficients can be used to calculate the coupling coefficients by a Fourier series transform as described in Chapter 4. A limited computation has been made, and the results are included in Fig. 6.4. The table in the figure shows the self-reflection coefficient, $\mathscr{C}_0$, and the coupling to the adjacent element, $\mathscr{C}_1$, when a single waveguide element is excited and the others are terminated with a perfect match. The adjacent element coupling is found to be smaller by an order of magnitude for the H plane than for the quasi-E plane (cf. Fig. 6.6). For $\varepsilon = 3.0$ the couplings to farther removed elements are also obtained:

$\lvert\mathscr{C}_0\rvert$	$\lvert\mathscr{C}_1\rvert$	$\lvert\mathscr{C}_2\rvert$	$\lvert\mathscr{C}_3\rvert$	$\lvert\mathscr{C}_4\rvert$
0.423	0.081	0.032	0.017	0.013

Figure 6.4 also shows the transmission phase and amplitude curves for $\varepsilon = 3.0$. These curves are, in fact, proportional to the far-field patterns when a single waveguide element is excited as shown in Chapter 4. The very flat phase curve is an indication that the phase center for the singly excited element lies in the aperture plane.

Since the effective wavelength of a wave is shorter in a dielectric medium ($\varepsilon > 1$) than in free space, this property may be utilized to reduce the element spacing. For example, in the H-plane scan, the waveguides without dielectric loading must have widths greater than a half free space wavelength

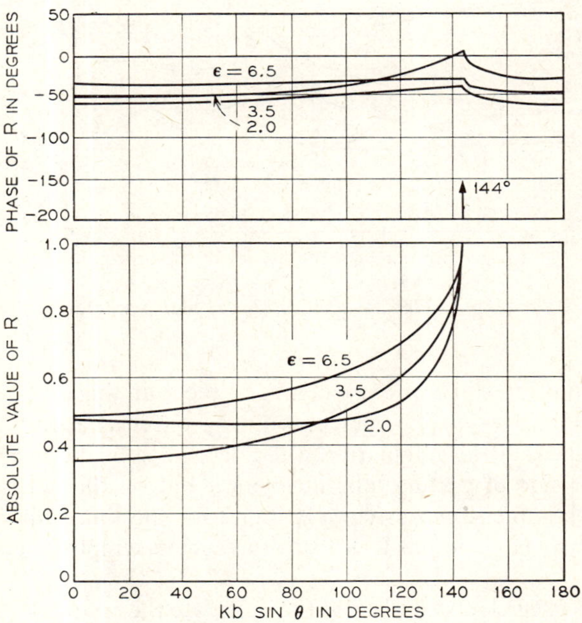

Fig. 6.5. Dielectric loading inside waveguides for H-plane scanning (element spacing $<\lambda/2$); $a = 0.937b = 0.3748\lambda$, $b = 0.400\lambda$.

in order for the lowest-order mode to propagate. This element spacing limits the useful scan angle to less than 90° because of the appearance of grating lobes. The use of a dielectric with proper dielectric constant, on the other hand, can make the width as small as desired. However, more elements will be needed to fill an aperture of given size if smaller elements are used. Figure 6.5 shows the calculations for element spacing $b/\lambda = 0.4$, the waveguide being dielectric loaded. The dielectric constant is chosen so that the wavelength is far removed from the waveguide cut-off value. We see that the variation of R between curves of constant ε is considerably reduced, as compared to the results in Fig. 6.4, when ε is between 0.9 and 1.1.

Some typical results for the quasi-E plane of scan of a rectangular waveguide array are shown next. These are obtained by solving the integral equation for the aperture electric field as

$$2y_0'\Phi_0'(y) = \int_{-c/2}^{c/2} \left\{ \sum_{q=0}^{\infty} y_q'^{\varepsilon}\Phi_q'(y)\Phi_q'(y') + \sum_{m=-\infty}^{\infty} Y_m'\Psi_m'(y)\Psi_m'^{*}(y') \right\} E_y(y')\, dy' \quad (19)$$

The electric field equation is used because of its convenience in considering the effects of irises at the aperture.*

Figure 6.6 gives results for $R(\theta)$, $|\mathscr{C}_0|$, and $|\mathscr{C}_1|$ as a function of ε from $\varepsilon = 0.8$ to $\varepsilon = 1.6$. This range of ε generally showed all the important characteristics observed. Furthermore, a slightly greater value of ε than 1.6 causes the waveguides to multimode (i.e., more than one mode propagates). Notice in the table of coupling coefficients that the adjacent element couplings are greater by an order of magnitude than those of the H-plane scan case, independently of the dielectric constant (see the table in Fig. 6.4). This is true also for the coupling coefficients of farther removed elements, as shown below for $\varepsilon = 1.6$:

$\lvert\mathscr{C}_0\rvert$	$\lvert\mathscr{C}_1\rvert$	$\lvert\mathscr{C}_2\rvert$	$\lvert\mathscr{C}_3\rvert$	$\lvert\mathscr{C}_4\rvert$
0.515	0.245	0.153	0.102	0.075

This behavior is attributable to both element spacing and polarization; that is, by making $b < \lambda/2$, and with an appropriate dielectric loading, H-plane results very similar to these quasi-E-plane results are obtainable. For an example, see Fig. 6.5. (Notice that λ refers to the free space wavelength.)

The curves of $|R|$ in Fig. 6.6 show that total reflection occurs beyond a critical steering phase ($\sim$100°). This is due to the fact that the element spacing, d, is less than $\lambda/[4 - (\lambda/b)^2]^{1/2}$, and no beam is radiated for the range of values of steering phase. Again, the infinite slopes for $|R|$ and for

* The presence of an iris in the array aperture can be easily accounted for in the integral equation, (19), by simply reducing the limits of integration appropriately.

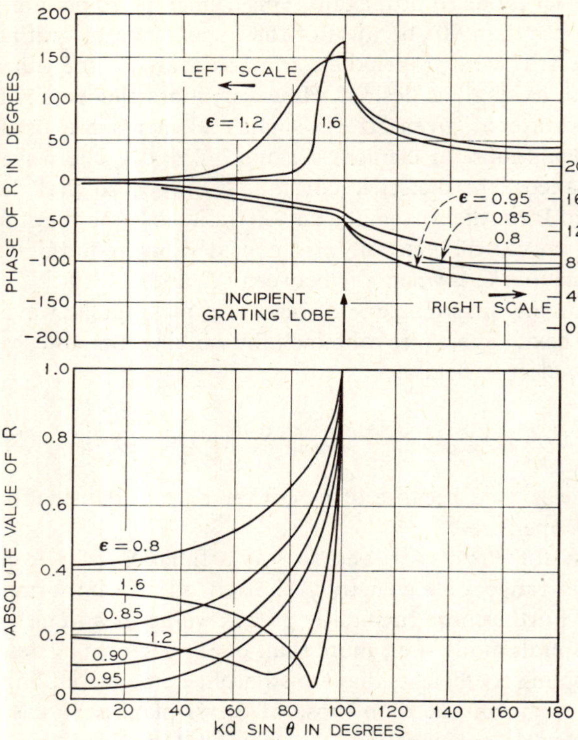

$\mathcal{C}_n$ = COUPLING TO nTH REMOVED ELEMENT

ε	$\lvert\mathcal{C}_0\rvert$	$\lvert\mathcal{C}_1\rvert$
0.8	0.661	0.308
0.85	0.548	0.331
0.90	0.502	0.330
0.95	0.480	0.324
1.2	0.473	0.294
1.6	0.515	0.245

Fig. 6.6. Dielectric loading inside waveguides for quasi-*E*-plane scanning; $a = b = d = 0.5714\lambda$, $h = c = 0.937d = 0.5354\lambda$.

the phase of R, which occur at this critical angle, exhibit the same behavior as is found in the thin-wall analysis[8,9] when $\varepsilon = 1$. Hence the same asymptotic behavior of coupling, $\exp(-jkr)/r^{3/2}$, can be expected for thick-walled dielectric-loaded arrays. (Here r is the distance between the excited and the coupled elements.)

A point of special interest in connection with these curves is the appearance of a resonance near the critical angle. A sharp dip in $|R|$ occurs precisely at the same angle for which the slope of the R phase curves has a maximum. Although the sharpness of the resonance increases gradually with ε, it is interesting that there is no resonance for $\varepsilon < 1.0$.

In Fig. 6.7 we illustrate the effect of a capacitive iris loading (which results when $h < c$), together with dielectric loading. In this case we let $a = b \to \infty$ so that a true E-plane scan in a parallel-plate array is considered. The effect of the iris tends to flatten the phase and amplitude responses, particularly the phase response, at the expense of a somewhat

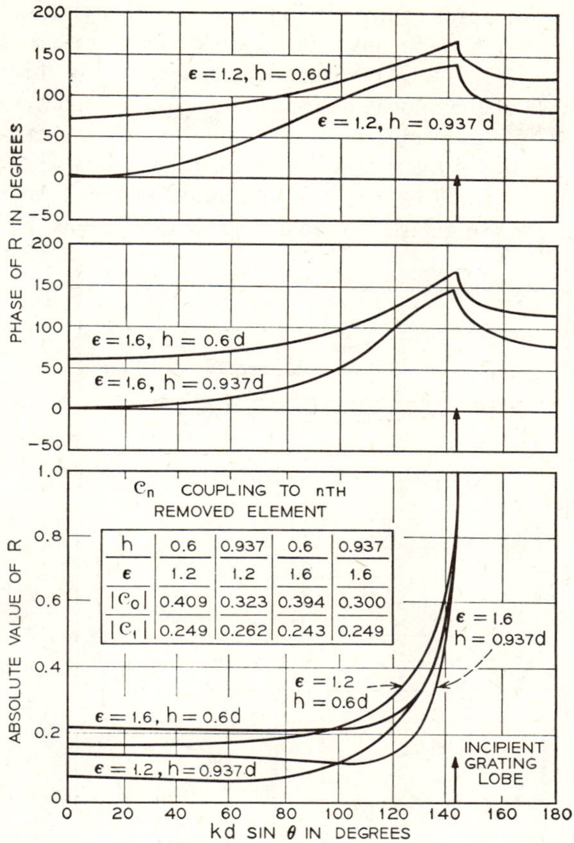

Fig. 6.7. Dielectric-loaded waveguides and iris loaded apertures for E-plane scanning; $a = b = \infty$, $d = 0.400\lambda$, $h = 0.6d = 0.240\lambda$ and $h = 0.937d = 0.3748\lambda$, $c = 0.937d$ ($h = 0.937d \rightarrow$ fully open aperture; $h < 0.937d \rightarrow$ iris at aperture).

greater value of $|R|$. This larger value of $|R|$ can be uniformly reduced, however, by impedance match for the scan region of flat $R(\theta)$.

The solutions of the integral equations (1) and (19) are actually complete solutions of the boundary value problem. The aperture fields, as well as the scattering properties of the array, are determined. It is of interest to examine the behavior of the aperture field distributions. The variation of the aperture electric field with the steering phase, $\psi_y = kd \sin \theta$, is sharper when the waveguides are loaded. Of particular interest is the field change near the critical angle, $\psi_y = kd \sin \theta = (\pi d/\lambda)[4 - (\lambda/b)^2]^{1/2}$, in the quasi-$E$ case, which corresponds to the angle of grazing incidence. This critical angle has the value 99.57° for the results shown in Fig. 6.8. Since the relevant eigenvalue equation[10] for $[E_y(\theta) + E_y(-\theta)]$ has a Hermitian kernel* for $\psi_y = |kd \sin \theta| >$ 99.57°, the phase of $[E_y(\theta) + E_y(-\theta)]$ should be constant in this region. By observing the phase of the approximate field solution at angles greater than the critical angle, we may obtain some evaluation of the errors in this solution. For most angles the errors are small. It is evident, however, that errors in phase occur for θ near the critical angle and also for y near the aperture edge, where a singularity in the true value of E_y exists. Nevertheless,

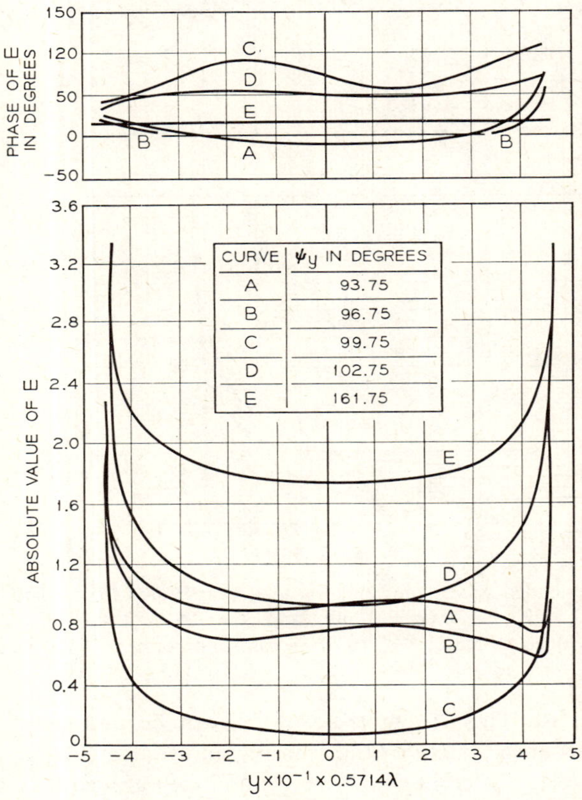

Fig. 6.8. Tangential electric field distribution, $E_y(y)$, in the aperture plane for quasi-E-plane scanning; $\varepsilon = 1.6$, $a = b = d = 0.5714\lambda$, $c = h = 0.937d = 0.5354\lambda$, ψ_y cut-off = 99.57°.

* Since no power is radiated for $|kd \sin \theta| > (\pi d/\lambda)[4 - (\lambda/b)^2]^{1/2}$ when the phasing is directed in the $+\theta$ and $-\theta$ directions simultaneously, the phase array behaves like a closed system, a cavity, and it is easy to show that a Hermitian kernel results. It is well known, then, that the eigenfunctions (the field solution here) of such a kernel have no varying phase.

since R is an averaged quantity over the range of y, these errors do not greatly affect the values of R determined in this solution.

5.2 Thick Dielectric Plug Loading

We next turn our attention to the situation of loading the waveguides with dielectric plugs. In doing so, an additional air-dielectric interface is introduced. At the same time, two new parameters, the dielectric constant and the thickness of the plug, become available for controlling the array properties. The radiation properties of the array may be determined by solving the appropriate integral equations, as discussed in preceding sections. If the plug is sufficiently thick, however, the evanescent modes generated at the array aperture are rapidly attenuated so that they do not interact with the interface located at $z = -t$. As a result, the array properties may be evaluated by accounting for the propagating modes only through the use of a transmission line or scattering matrix method, providing that the properties of an array fully loaded by the same dielectric are already determined. When the dielectric constant is low, there can be only one propagating mode inside the dielectric, and the procedure then is a conventional transmission line calculation. This approach may result in substantial saving of computation with little sacrifice in quality. The following results are obtained on this basis. Later, additional results calculated from the pertinent integral equations will be shown in connection with the discussion of forced surface wave resonances.

In presenting the results, we have kept the dielectric constant fixed and plotted $R(\theta)$ versus $\psi = kb \sin \theta$, with t, the plug thickness, as a parameter. The phase of R is referred to the aperture plane, $z = 0$, although R is the reflection coefficient in the region $z < -t$. In any given figure ε was chosen so that the illustrated results were typical of a wider range of permittivity. With the thick-plug approximation, the results will repeat for every half guide wavelength, so that t is varied over only one-half guide wavelength. The minimum t for which the thick approximation is valid is determined by the relative decay of the first evanescent mode in the distance t. This decay factor, df, is presented with each curve. It is found, generally, that $df \leqslant 0.1$ is sufficient for the thick-plug approximation to be valid. This result is usually satisfied for some $t < \lambda_g/2$, where λ_g is the guide wavelength. Of course, by adding a sufficient number of multiples of a half guide wavelength to t, the results must become valid to any accuracy desired.

Figure 6.9 depicts some typical results with $\varepsilon = 2.0$. Notice that the results are substantially different from those of complete loading owing to the presence of the second dielectric boundary. For any given ε we have found that there exists a thickness, t, for which both the amplitude and the phase of the reflection coefficient are flat over the generally useful region of scan

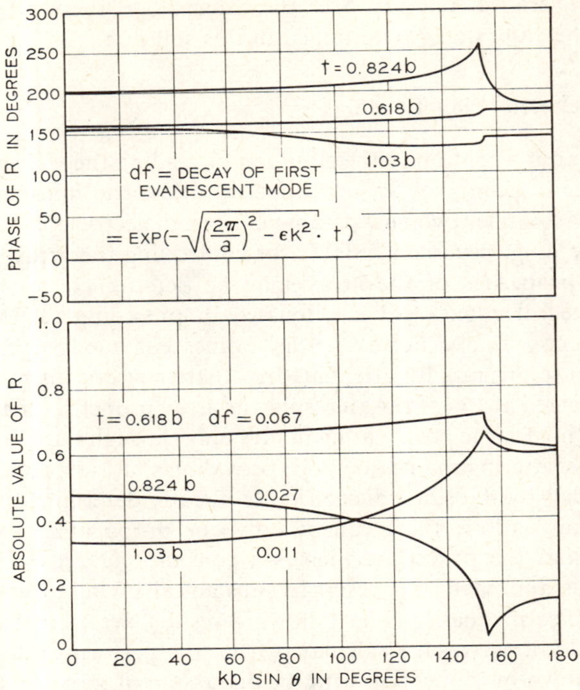

Fig. 6.9. Dielectric plug inside waveguides for H-plane scanning; $\varepsilon = 2.0$, $a = 0.937b = 0.5354\lambda$, $b = 0.5714\lambda$.

angle (the region in which only one lobe radiates). Because of the second interface, however, the improved match tends to become frequency sensitive, that is, the curves of $R(\theta)$ do not remain flat when the frequency is varied. Furthermore, since the nearly flat curve also is found to have high amplitude (see the curve for $t = 0.618b$ in Fig. 6.9), the necessity of matching out a large value of $|R|$ may aggravate the frequency sensitivity problem. The situation seems to become worse with higher dielectric constant.

Qualitatively similar results are obtained in the quasi-E-plane scan case. A set of computations for $\varepsilon = 1.6$ is illustrated in Fig. 6.10. Again, a very flat curve of R is obtained with $t = 1.078d$.

5.3 Thick Dielectric Sheath Covering

The use of a dielectric sheath to cover the array creates a somewhat different situation. The range of scan within which there is only one propagating Floquet mode in the sheath now depends on the dielectric constant.

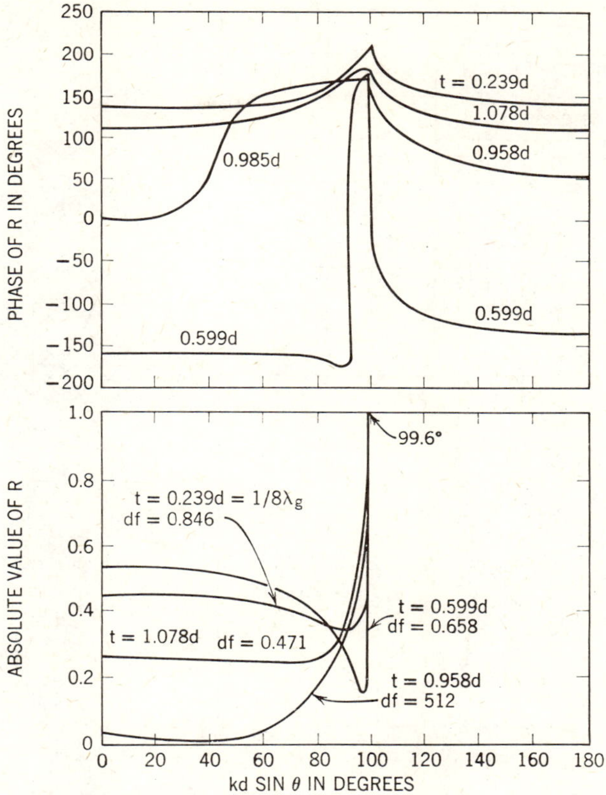

Fig. 6.10. Dielectric plug inside waveguides for quasi-E-plane scanning; $\varepsilon = 1.6$, $\varepsilon_1 = 1.0$, $a = b = d = 0.5714\lambda$, $c = h = 0.937d = 0.5354\lambda$.

(A detailed analysis of this will be made presently.) Also the effective wavelengths of the Floquet modes are functions of the scan angle. Consequently the decay factor, *df*, varies with the scan angle, and the validity of the transmission line approximation, therefore, has to be assessed as the scan angle is varied. Otherwise, the calculation follows a procedure similar to that used for the thick-plug case.

In Fig. 6.11, we show a typical example of the calculation for the quasi-E plane of scan with $\varepsilon = 1.818$ and the thickness of the sheath as a parameter. It is noted that, as in the plug case, a proper choice of the parameters is capable of producing a very flat R over a useful range of scan in both the amplitude and the phase. This is illustrated by the curve for $t' = 0.224d$.

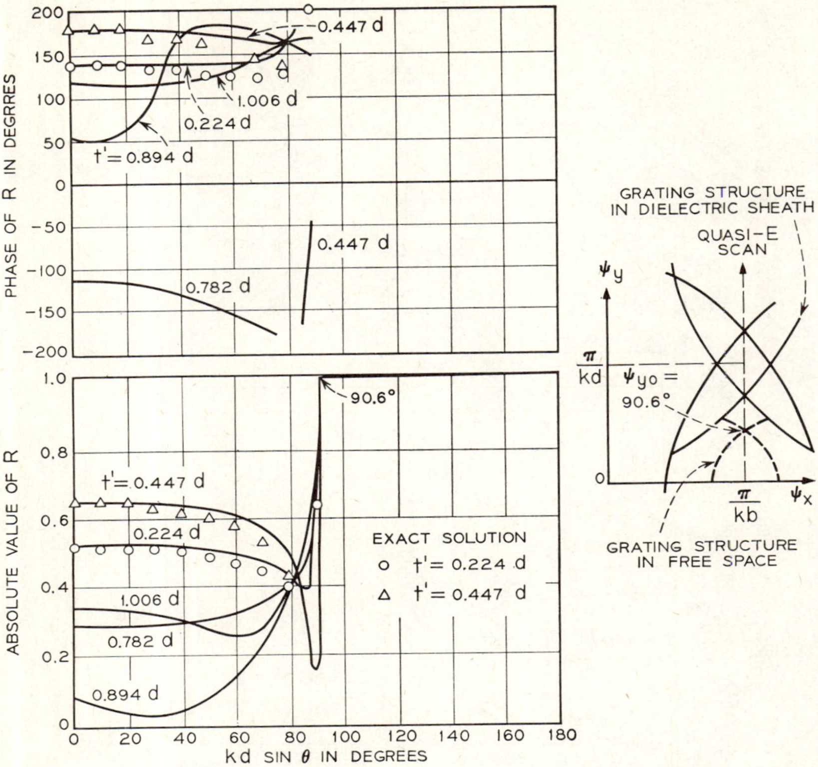

Fig. 6.11. Dielectric sheath outside waveguides for quasi-E-plane scanning; $\varepsilon = 1.818$, $a = b = d = 0.5596\lambda$, $c = h = 0.937d = 0.5243\lambda$.

Solutions of the appropriate integral equation which account for the presence of the dielectric sheath are depicted in Fig. 6.11 by circles and triangles. The results agree very well with solutions in which evanescent mode interaction with the second dielectric boundary at $z = t'$ is neglected. The criterion for the latter approximate calculation is the same as that used for the plug case.

To summarize, we find that dielectric materials offer definite possibilities for improving the match characteristics of an array. Plug or sheath configurations provide additional parameters to adjust and optimize so that improved array performance can be obtained. On the other hand, matching by plugs or sheaths may increase the frequency sensitivity of the array because of the presence of an additional interface. So far, it has been shown that match improvement over a wide scan range is clearly possible at a single frequency. Although it seems plausible that broad-band compensation can be achieved by using multiple sheaths or plugs, this possibility remains to be explored.

When the dielectric constants of a plug or sheath are sufficiently large that more than one mode can propagate within the dielectric, the situation is conducive to the occurrence of extremely interesting forced surface wave phenomena. This is the subject we shall examine in detail in the next section.

6. FORCED SURFACE WAVE EFFECTS

Forced surface wave effects are observed in an array either loaded with dielectric plugs or covered by a dielectric sheath. The conditions for their occurrence in these two situations are somewhat different, but their basic cause seems to be the same. We begin the discussion by considering the sheath case.

The effect of an ordinary dielectric ($\varepsilon > 1$) on wave propagation is to slow down the phase velocity, or equivalently to shorten the wavelength. Hence, when a dielectric sheath is placed over a phased array, two apparent element spacings (in terms of the free space wavelength) have to be considered: one inside the dielectric medium and the other in free space, with the former always greater than the latter. A difference in the element spacings as seen in the two regions results in different scan angles for the appearance or disappearance of grating lobes in the respective regions. Consequently, when the fields are expanded into normal modes according to the Floquet theorem, there will be a range of scan angles over which the number of propagating modes inside the dielectric is larger than that in free space. A mode is said to be propagating when the corresponding z-directed propagation constant is real. The propagation constants of the mth Floquet mode for H-plane scan in the dielectric and in free space are given by

$$\begin{aligned} \Gamma_m &= \sqrt{k^2 - [(2m\pi + \psi)/b]^2} && \text{for free space} \\ \Gamma_m^{\ \varepsilon} &= \sqrt{k^2\varepsilon - [(2m\pi + \psi)/b]^2} && \text{inside the dielectric} \end{aligned} \tag{20}$$

The cut-off frequencies of these modes are determined by setting $\Gamma_m = 0$ and $\Gamma_m^{\ \varepsilon} = 0$, or

$$k = \pm\left(\frac{2m\pi + \psi}{b}\right) \quad \text{and} \quad k\sqrt{\varepsilon} = \pm\left(\frac{2m\pi + \psi}{b}\right) \qquad \text{for} \quad m \begin{matrix} \geq 0 \\ \leq 0 \end{matrix}$$

When $\frac{1}{2} < b/\lambda_\varepsilon < 1$, the zeroth mode is always propagating within the complete scan range $0 \leq \psi \leq \pi$. Only the $m = -1$ mode has a cut-off steering phase, occurring at $\psi = 2\pi(1 - b/\lambda_\varepsilon)$. The remaining Floquet modes are all below cut-off. (Only the range $0 \leq \psi \leq \pi$ need be considered

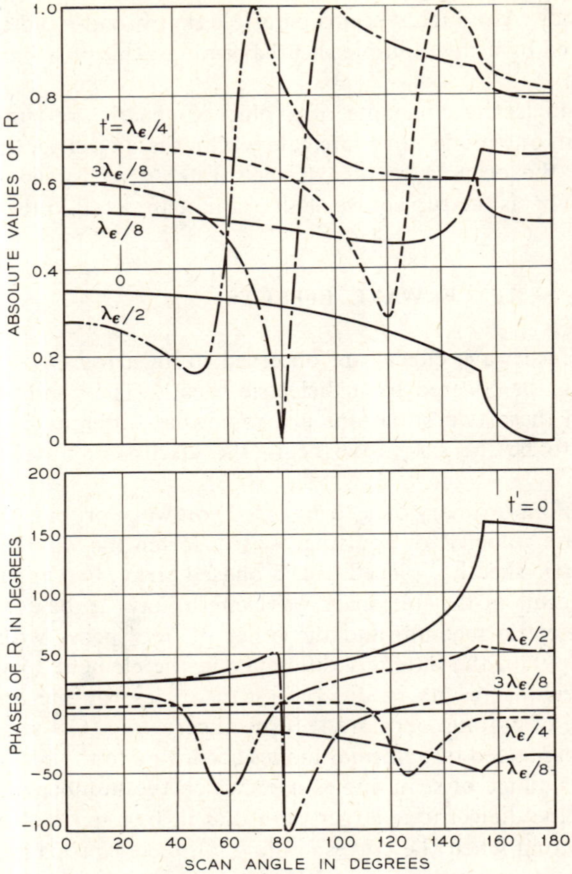

Fig. 6.12. Reflection coefficient R vs. scan angle ψ for H-plane scan with $a/\lambda = b/\lambda = 0.5714$, and $\varepsilon = 3.0625$.

for the radiating elements which possess reflection symmetry with respect to their center. Otherwise, the entire range, $-\pi \leq \psi \leq \pi$, must be considered.) Here, $\lambda_\varepsilon = \lambda/\sqrt{\varepsilon}$ denotes the wavelength of a plane wave in a medium with dielectric constant ε. Corresponding results for free space may be obtained simply by setting $\varepsilon = 1$.

In general, for $m < b/\lambda_\varepsilon < (m + \frac{1}{2})$, where m is an integer, the number of propagating modes will change from $2m + 1$ for $0 \leq \psi < 2\pi(b/\lambda_\varepsilon - m)$ to $2m$ for $2\pi(b/\lambda_\varepsilon - m) \leq \psi \leq \pi$. On the other hand, when $(m + \frac{1}{2}) < b/\lambda_\varepsilon < (m + 1)$, the number of propagating modes will increase by 1, from $2m + 1$ to $2m + 2$, when the scan angle passes from $0 \leq \psi < 2\pi(m + 1 - b/\lambda_\varepsilon)$ to $2\pi(m + 1 - b/\lambda_\varepsilon) \leq \psi \leq \pi$.

The wave modes which are propagating inside the dielectric and are evanescent in free space have a field distribution similar to that of a surface wave. That is, they decay exponentially with increasing z, but propagate energy along the array surface. Such wave modes have profound effects on the radiation characteristics of a phased array, as we shall see in the examples below. It is important to emphasize that a single wave mode of this type by itself is not sufficient to satisfy the boundary conditions. In other words, all the modes are required to constitute a correct solution.

6.1 Sheath-Covered Array Scanned in *H* Plane

Figures 6.12 and 6.13 give the reflection coefficients as a function of scan of a parallel-plate array covered with a single dielectric sheath and scanned

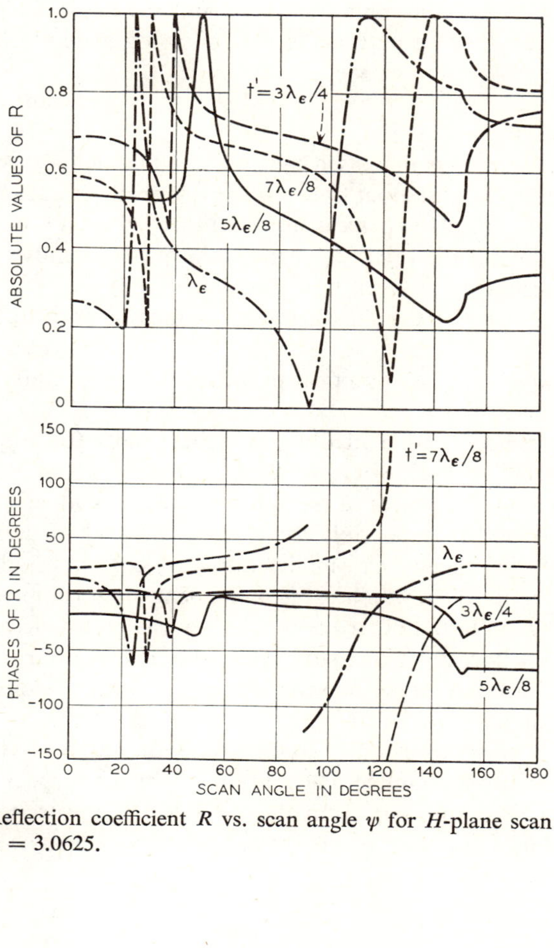

Fig. 6.13. Reflection coefficient R vs. scan angle ψ for H-plane scan with $a/\lambda = b/\lambda = 0.5714$, and $\varepsilon = 3.0625$.

in the H plane. The results are obtained for the following parameters: $b/\lambda = a/\lambda = 0.5714$, $\varepsilon = 3.0625$, with the thickness of the dielectric slab, t', varied over one λ_ε at increments of $\lambda_\varepsilon/8$. Notice that the element spacing is measured in terms of the free space wavelength, λ, for b/λ is the quantity which determines the number of radiated beams at a given scan angle. The element spacing is such that the array radiates one beam for $0 \leq \psi \leq 2\pi(1 - b/\lambda) = 154°$, and two beams when $2\pi(1 - b/\lambda) \leq \psi \leq 180°$. The apparent element spacing in the dielectric medium is $b/\lambda_\varepsilon = 1$. Thus, there are two propagating modes inside the dielectric slab for all scan angles. As a result, we have in $0 \leq \psi < 2\pi(1 - b/\lambda)$ a mode which exhibits a surface-wave-like behavior in that it propagates inside the dielectric and is evanescent in free space. The effect of such a mode is to cause the appearance of sharp resonant peaks in the reflection coefficient at certain scan angles, as is evident from the graphs.

On the basis of the results presented here and some further calculations at different wavelengths, we may make the following general observations:

1. When the thickness of the dielectric sheath is sufficiently small, no resonant peak occurs.
2. When the thickness is increased beyond a critical value, usually in the neighborhood of $3\lambda_\varepsilon/16$, a resonant peak starts to appear at a scan angle close to the value $2\pi(1 - b/\lambda)$ (the peak is usually preceded by a dip).
3. Increasing the sheath thickness causes the peak to move toward the broadside direction, and the peak becomes sharper.
4. A further increase in the sheath thickness makes more than one peak appear.
5. The peaks attain, for all practical purposes, values of unity.

The dielectric constant used in obtaining the results for Figs. 6.12 and 6.13 was chosen to have a value of $\varepsilon = (\lambda/b)^2$. This creates a situation wherein one space harmonic possesses a surface-wave-like field distribution in $0 \leq \psi \leq 2\pi(1 - b/\lambda)$. It is possible for a resonant peak to appear at any scan angle within this range. If a smaller dielectric constant had heen used, the range of scanning over which a resonant peak might appear would be reduced accordingly. On the other hand, a larger value of dielectric constant would give rise to more than one surface-wave-like space harmonic, and it is possible for surface wave resonances to appear with a dielectric sheath thickness even smaller than $3\lambda_\varepsilon/16$.

The reflection coefficients actually attain exactly the value 1 at the resonant peaks. This is shown by the numerical solutions, which approach 1 to within 10^{-6}. It may also draw support from the behavior of the transmission coefficients. Figure 6.14 is typical of the transmission coefficients calculated for

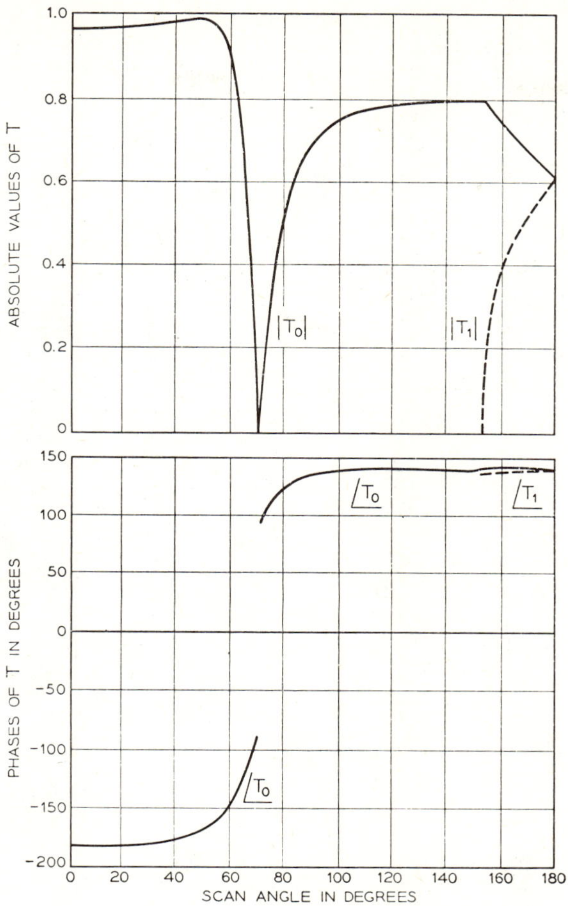

Fig. 6.14. Transmission coefficients T vs. scan angle ψ_x for H-plane scan with $a/\lambda = b/\lambda = 0.5714$, $\varepsilon = 3.0625$, $t' = 0.5\lambda_\varepsilon$.

$b/\lambda = a/\lambda = 0.5714$, $\varepsilon = 3.0625$, and $t' = \lambda_\varepsilon/2$. These transmission coefficients are referred to the air-dielectric interface, $z = t'$. The figure shows the transmission coefficients for both the zeroth-order mode (corresponding to the main beam) and the $m = -1$ space harmonic (corresponding to a grating lobe). It is recalled that T_{-1} has significance as a radiated transmission coefficient only in the scan range $2\pi(1 - b/\lambda) \leq \psi \leq \pi$.

The graph of $|T_0|$ shows a sharp dip at the scan angle of $\psi = 70°$, at which the reflection coefficient attains its peak value of 1. Notice that the phase of T_0 exhibits a discontinuity of 180° at this scan angle because both the real

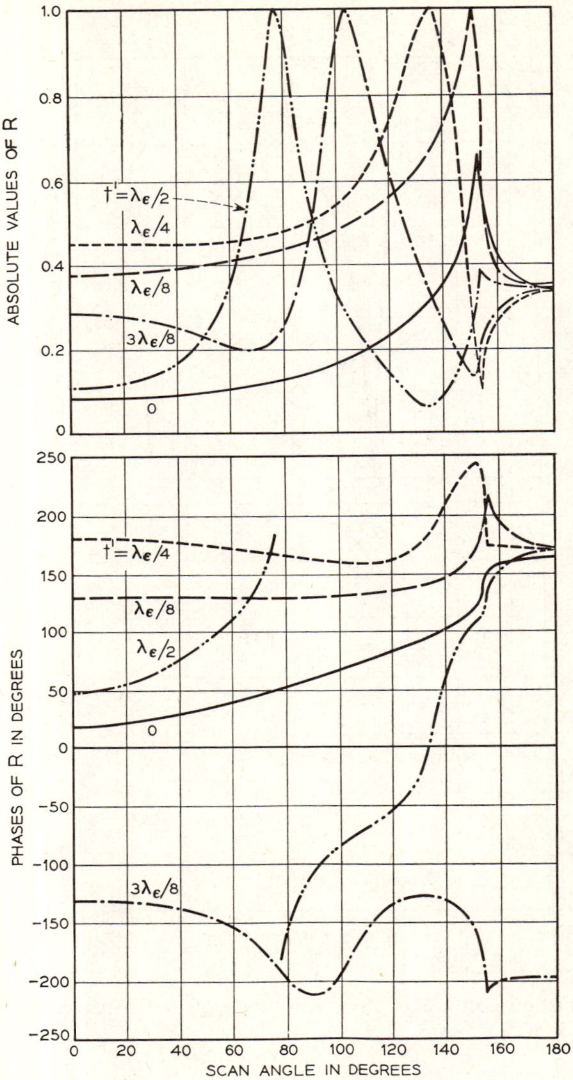

Fig. 6.15. Reflection coefficient R vs. scan angle ψ_y for E-plane scan with $d/\lambda = 0.5714$, $c/d = 0.85$, and $\varepsilon = 3.0625$.

and the imaginary parts of T_0 go through zero and then change sign as the beam is scanned past 70°. This is an indication that $|T_0|$ *does go to zero* rather than approach zero, or, equivalently, that $|R|$ actually attains the value 1. The significance of the difference between $|R|$ approaching 1 and $|R|$ actually attaining 1 lies in whether the match of an array can be improved by network compensation. Moreover, there is an intimate connection between the facts that $|R| = 1$ and that a surface wave can propagate along a related corrugated structure; we discussed this in detail in Chapter 5. Because the solution to the array problem under resonance conditions may be translated into that of surface wave propagation along corrugated surfaces, this phenomenon is sometimes called the forced surface wave effect. On the other hand the resonance is definitely associated with the array aperture; hence it is also termed the forced aperture resonance.

It has already been noted that T_0 is proportional to the radiation pattern in the angular range $0 \leq \theta \leq \sin^{-1}(\lambda/2b)$, when a single element is excited with the rest of the elements terminated in their characteristic impedances. The radiation pattern for the remaining angular range, $\sin^{-1}(\lambda/2b) \leq \theta \leq \pi/2$, can be obtained from the curve of T_{-1} in the scan angle $2\pi(1 - b/\lambda) \leq \psi \leq \pi$, reflected with respect to the $\psi = \pi$ axis. The existence of sharp peaks in the reflection coefficient then is indicative of deep notches in the radiation patterns, and vice versa.

6.2 Sheath-Covered Array Scanned in *E* Plane

The incident wave in the *E*-plane scan is a TEM wave with the electric field polarized in the direction normal to the waveguide walls. Figures 6.15 and 6.16 give results for both the amplitudes and the phases of the reflection coefficients as a function of scan. The set of parameters used for obtaining these results is as follows: $d/\lambda = 0.5714$, $c/d = 0.85$. The array is covered with a single layer of dielectric material, flush with the array aperture. A dielectric constant of $\varepsilon = 3.0625$ is chosen so as to have one surface-wave-like space harmonic present over as wide a scan angle as possible. Relatively thick waveguide walls are used in this case in order that the waveguides will support only the dominant TEM mode. This is necessary, since the element spacing is larger than $\lambda/2$.

From the numerical data we may observe that, in the absence of a dielectric material, the reflection coefficient is relatively flat over a large range of scan angles, except in the vicinity of the grating lobe formation angle (in this case $\psi = 154°$), at which we find a sharp peak. This is in marked contrast with the *H*-plane scan case (see Fig. 6.12). The peak value, however, does not reach 1. When a dielectric sheath of thickness $\lambda_\varepsilon/8$ is used to cover the array, the effect is to raise the level of reflection almost uniformly over all

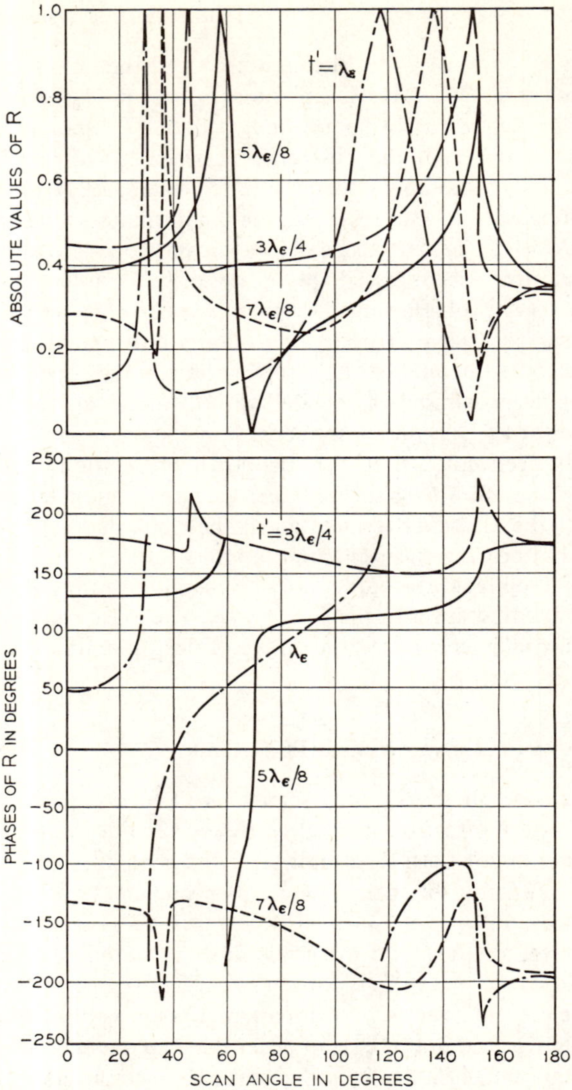

Fig. 6.16. Reflection coefficient R vs. scan angle ψ_y for E-plane scan with $d/\lambda = 0.5714$, $c/d = 0.85$, and $\varepsilon = 3.0625$.

the scan angles. In particular, the peak which occurs at the grating lobe formation angle is seen to reach a value of unity.

Although not presented here, the results for thinner dielectric sheaths show similar tendencies with the exception that the peak values are not necessarily unity. Although the angular position for the appearance of the resonant peaks remains practically unchanged when the dielectric sheath is relatively thin, it starts to shift toward the broadside direction as the thickness is increased beyond a critical value of about $\lambda_\varepsilon/4$. The peaks become sharper as they move toward the broadside direction. A further thickening of the dielectric sheath eventually leads to the emergence of multiple resonant peaks. Investigation of the quasi-E plane of scan leads to a similar conclusion, namely, resonant peaks are present only when the thickness of the dielectric sheath exceeds a certain critical value.

6.3 Plug-Loaded Array Scanned in E or H Plane

We saw in the preceding discussion that the occurrence of a resonant peak in the reflection coefficient is contingent upon the presence of space harmonics which have a surface-wave-like field distribution. If, instead of dielectric sheath covering, the array is loaded with dielectric plugs, a similar situation can be created. The presence of dielectric plugs makes it possible for some of the higher-order waveguide modes generated by the aperture discontinuity to become propagating inside the dielectric, while remaining evanescent in the unloaded waveguide region. This type of mode has a field distribution (in the z direction) analogous to that of the space harmonics, which are responsible for the occurrence of the forced surface wave effects. It may be anticipated, then, that similar phenomena will appear in the plug situation.

Some representative numerical results for the reflection coefficient as a function of the scan angle are presented in Figs. 6.17 and 6.18. These results clearly reveal that dielectric plugs placed inside the waveguides give rise to resonant peaks in the reflection coefficients. There are, however, some distinctive differences between the effects of the dielectric plugs and those of a continuous sheath covering the array. Whereas a continuous sheath invariably causes resonant peaks to appear if its thickness exceeds a certain critical value, such is not the case when plugs are used inside the waveguides. In fact, resonances are found to occur only for certain ranges of the plug thickness, t. In other words, in the plug situation, the occurrence of resonance exhibits a stop-band pass-band type of pattern as t is varied.

The reflection coefficients (as a function of the scan angle) obtained when the waveguides are loaded with dielectric plugs seem to be flatter in general

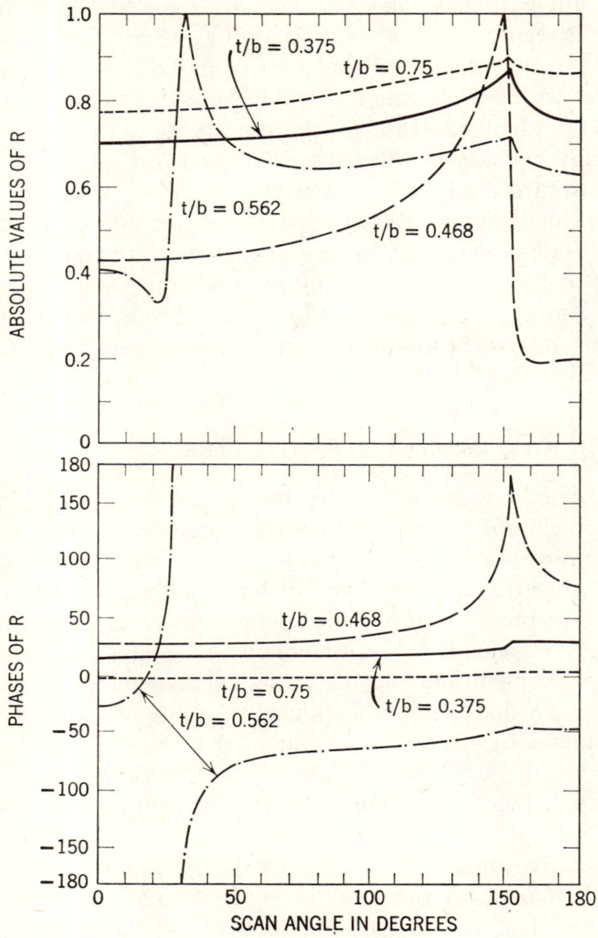

Fig. 6.17. Reflection coefficient R vs. scan angle ψ_x for an array loaded with dielectric plugs scanned in H plane with $a/\lambda = b/\lambda = 0.5714$ and $\varepsilon = 4$.

than those obtained for an array with an exterior sheath. This is of considerable importance from the practical viewpoint of array match improvement.

6.4 Experimental Observation of Forced Surface Wave Resonances

Forced surface wave resonances have been observed experimentally for an array covered with a dielectric sheath.[11] An infinite planar array of thin slots

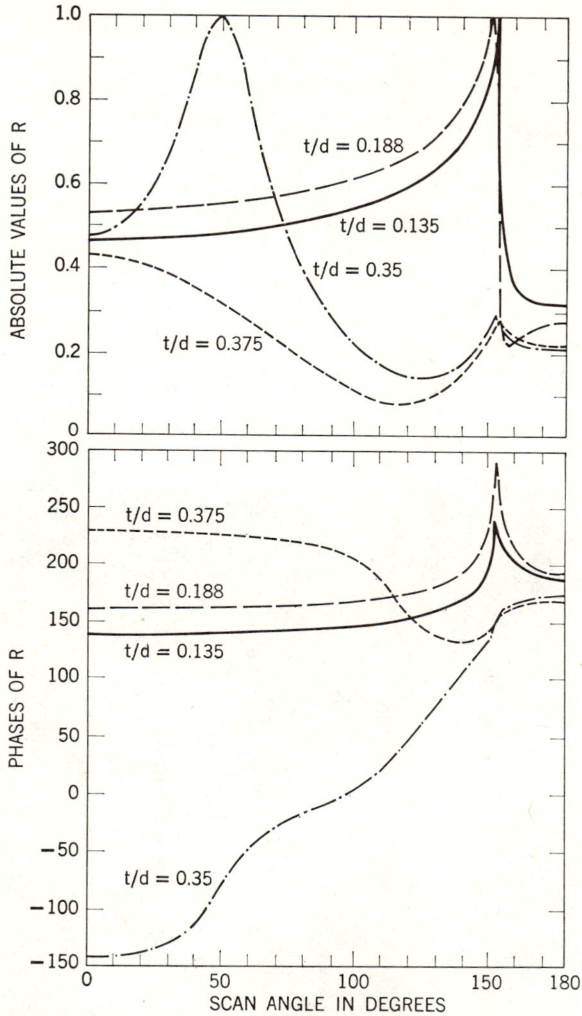

Fig. 6.18. Reflection coefficient R vs. scan angle ψ_y for an array loaded with dielectric plugs scanned in E plane with $d/\lambda = 0.5714$, $c/d = 0.85$, and $\varepsilon = 3$.

fed by rectangular waveguides, as shown in Fig. 6.19*a*, was employed in the study. Rexolite was the dielectric material used for the sheath; the waveguides were completely loaded with Rexolite also. The dielectric constant of Rexolite is approximately 2.55 in the frequency range of the experiment, the L-band. The elements were arranged in a square grid with an element spacing of 4.46 inches. A scan angle in the H plane of the array was simulated in

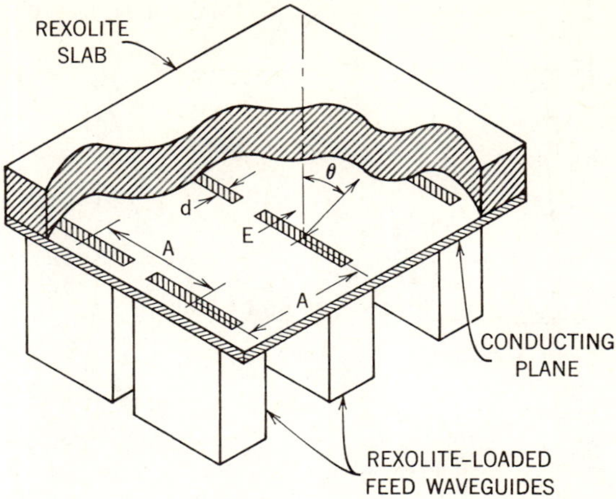

Fig. 6.19*a*. A dielectric-slab-covered slot array.

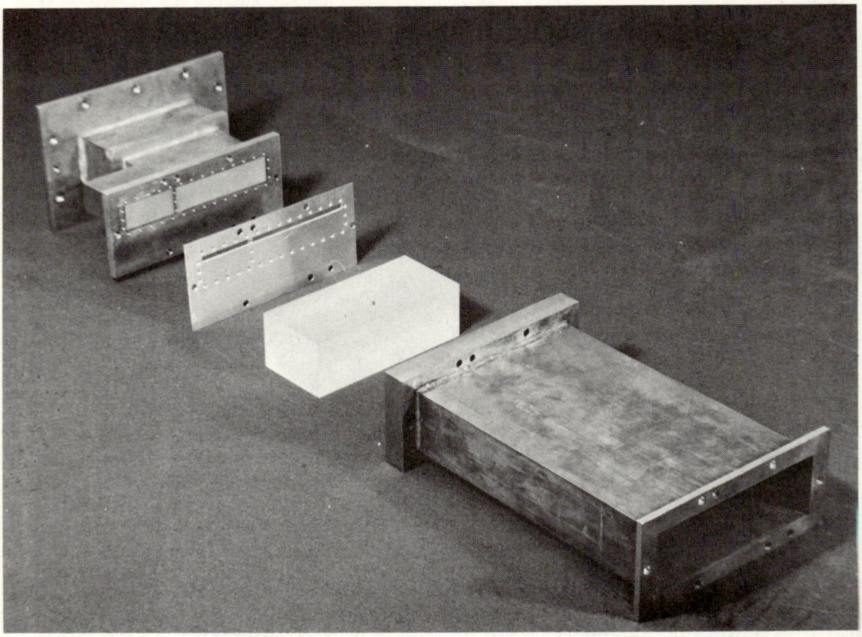

Fig. 6.19*b*. An exploded view of waveguide simulator of slot array.

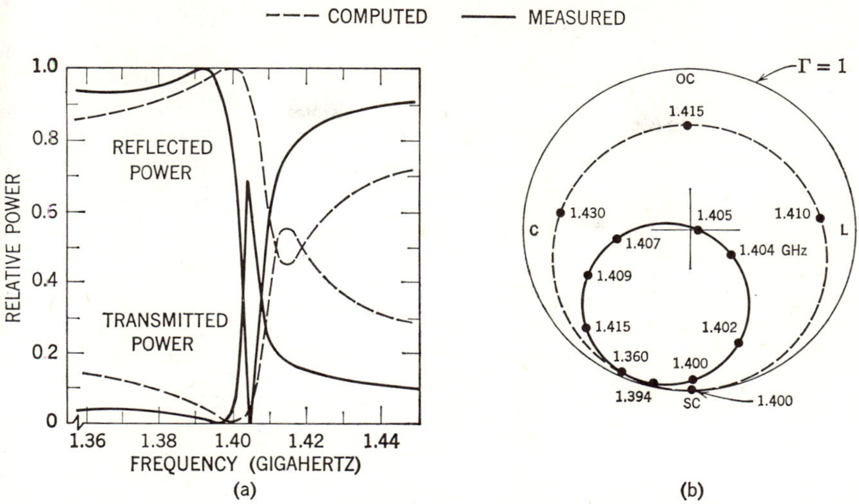

Fig. 6.20. Measured and computed performance for narrow slots, $d/A = 0.028$. (*a*) Reflected and transmitted power. (*b*) Reflection coefficient, Γ.

waveguide simulators,[12] and the reflection coefficient was measured from the feed waveguide as the frequency was varied. A photograph of the simulator is shown in Fig. 6.19*b*.

Figure 6.20 gives the experimental and computational results. The computation was performed by using a variational expression for the input admittance in terms of the aperture electric field, which was assumed to consist of only the incident field. This is equivalent to using one interior mode for expanding the aperture electric field in the moments method applied to the integral equation (see Chapter 3). Aside from the slight deviation between the experimental and the computed data, the presence of a forced surface wave resonance is clearly exhibited by the peak of the reflection coefficient, which has a value of 1.

7. DETERMINATION OF FORCED SURFACE WAVE RESONANCE CONDITIONS

7.1 Small Aperture Case

The scan angles at which resonant peaks appear may be determined accurately by the transverse resonance method only when the radiating aperture is small. This situation applies to the case of *E*-plane scan. In this method, the sum of impedances, looking toward the positive and negative z directions at some reference plane, is set equal to zero, thus yielding a characteristic equation for the resonances.[2,13] It is convenient to use the array aperture as the reference plane. When the radiation aperture is small in comparison to the size of a periodic cell, the "average" impedance looking toward the array side is almost zero. The impedance looking away from the aperture may be approximated by that of the lowest-order mode, in this case

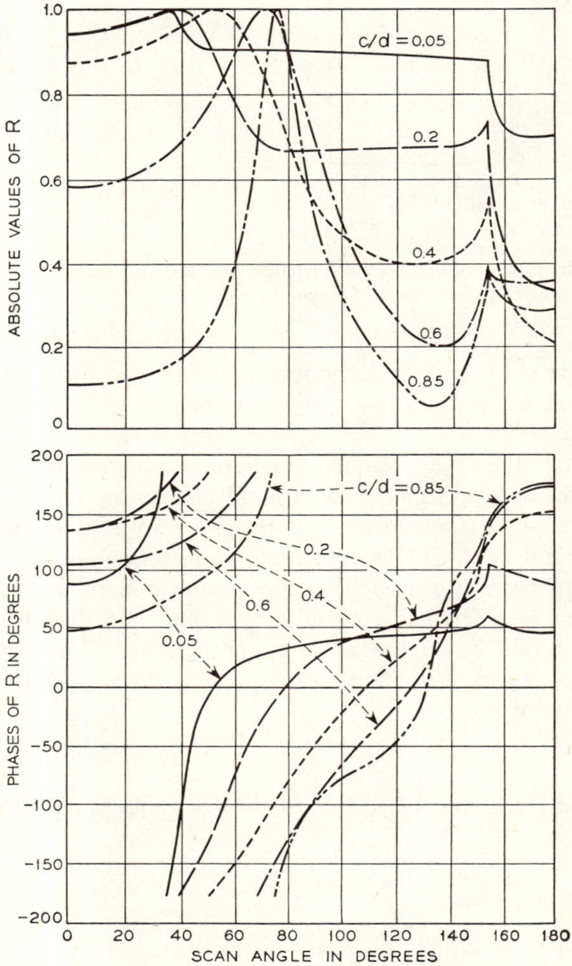

Fig. 6.21. Effects of radiating aperture size (E-plane scan, $d/\lambda = 0.5714$, $\varepsilon = 3.0625$ $t = 0.5\lambda_\varepsilon$).

the $m = -1$ space harmonic. (The zeroth space harmonic is absent at resonance.) The impedance of this mode, from (13a), is given by

$$\tilde{Z}_{-1} = Z_{-1}^{\varepsilon} \frac{Z_{-1} + jZ_{-1}^{\varepsilon} \tan \Gamma_{-1}^{\varepsilon} t'}{Z_{-1}^{\varepsilon} + jZ_{-1} \tan \Gamma_{-1}^{\varepsilon} t'} \tag{21}$$

By setting this expression to zero and using the appropriate modal impedances, we find that the forced surface wave resonance condition is

$$\varepsilon \Gamma_{-1} = -j\Gamma_{-1}^{\varepsilon} \tan \Gamma_{-1}^{\varepsilon} t' \tag{22}$$

Notice that (22) is identical to the characteristic equation for surface waves which propagate along a dielectric sheath on a ground plane.

When this equation is used to determine the scan angle for the appearance of the resonant peak, it is found that excellent agreement with solutions to the pertinent integral equation are obtainable for aperture sizes up to 10% of the size of the periodic cell. For larger apertures the agreement gradually becomes poorer because the impedance looking toward the waveguide side no longer remains negligibly small. The effects of the aperture size on the angular location of resonant peaks are illustrated in Fig. 6.21 for the E plane of scan. The scan angle for the appearance of the resonant peaks is shifted away from the broadside direction as the aperture size is increased. A similar shift in the scan angle for the resonant peak has also been observed in the case of H-plane scan. Moreover, loading the waveguides with a dielectric material also causes similar effects. It is apparent that the resonance phenomenon is strongly dependent on the aperture impedance, which is a function of the array parameters.

7.2 Large Aperture Case

When the waveguide aperture is large in terms of the periodic cell, the transverse resonance technique is useful only as a rough estimate of the conditions of surface wave resonance. An alternative procedure is to apply the scattering matrix formalism, taking into account the scattering property of the array aperture.[14] This approach requires the solution of an auxiliary boundary value problem. However, the resonance conditions are readily determined for *all* scan ranges and plug and sheath thicknesses once the array parameters, such as the element spacing, waveguide width, and dielectric constant, are specified. The procedure therefore offers considerable convenience and economy in computation, particularly when the waveguides are loaded with dielectric plugs, for in such situations, as we have shown, forced surface wave resonances occur considerably less frequently than when the array is covered by a continuous sheath and are harder to detect.

In the scattering matrix formalism, the two interfaces, one at the array aperture and the other at the air-dielectric boundary, are characterized independently first. The interactions among the various modes, including both propagating and nonpropagating modes, are then related through appropriate scattering matrices to arrive at expressions which describe the composite structure.

Consider, for example, the array covered by a dielectric sheath as shown in Fig. 6.2. From the periodicity in the array geometry, the junction between the array elements and the space exterior to them may be regarded as a generalized waveguide discontinuity as shown in Fig. 6.22*a*. The word

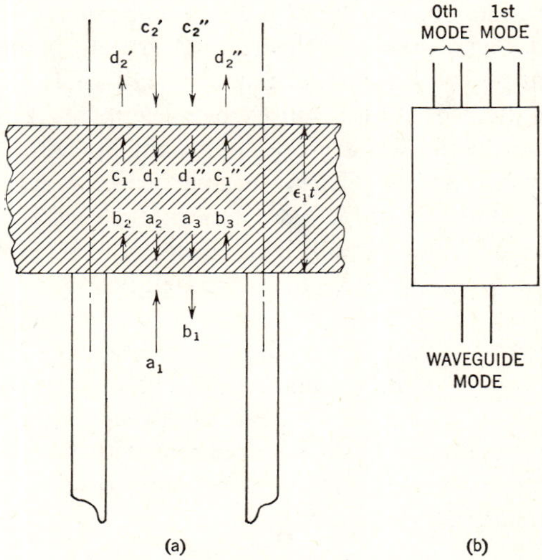

Fig. 6.22. Equivalent networks of an infinite array with dielectric sheath. (*a*) Generalized waveguide discontinuity. (*b*) Equivalent three-port network.

generalized is used here to emphasize the fact that one of the waveguides has unconventional walls specified mathematically by periodic boundary conditions.

This generalized waveguide junction has two consecutive discontinuities. The characterization of this junction, as remarked above, generally requires the consideration of the interaction due to the higher-order modes, particularly if the two discontinuities are close together. When the separation of the discontinuities is sufficiently large, however, since the evanescent modes are usually highly attenuated away from the planes of their excitation, it often suffices to consider just the interaction of the propagating modes. Actually, it will be seen that even quite thin plugs and sheaths can be analyzed accurately by this approach.

An array with element spacing $\frac{1}{2} < b/\lambda < 1$ radiates a single beam when $0 \leq \psi \leq 2\pi(1 - b/\lambda)$ and two beams when $2\pi(1 - b/\lambda) < \psi \leq \pi$. For $1 < \varepsilon < (\lambda/b)^2$, there are two propagating modes inside the dielectric if $2\pi(1 - \sqrt{\varepsilon}\, b/\lambda) \leq \psi \leq \pi$. Therefore, in $2\pi(1 - b/\lambda) \leq \psi \leq \pi$, we have two propagating modes both inside the dielectric and in free space. Under this condition, the array discontinuity may be represented by an equivalent three-port network, as shown in Fig. 6.22*b*. This representation is good when t' is large in terms of the wavelength. On the other hand, when $2\pi(1 - \sqrt{\varepsilon}\, b/\lambda) < \psi < 2\pi(1 - b/\lambda)$, one of these two modes, the $m = -1$ space harmonic, becomes evanescent in free space but remains propagating inside the dielectric sheath. The port corresponding to the $m = -1$ space harmonic does not transmit power, and hence may be regarded as effectively short circuited. This type of mode was shown earlier to be responsible for the occurrence of the forced surface wave resonance.

To derive the scattering matrix for the equivalent network, the discontinuity at the array aperture is described by a 3×3 scattering matrix relating the three propagating waves, one inside the waveguide and two inside the dielectric region. The elements of this scattering matrix are to be determined (e.g., by the integral equation method) with the whole *exterior region* filled with a dielectric of the same dielectric constant as the sheath. If a_i, $i = 1, 2, 3$, denote the incident waves and b_j, $j = 1, 2, 3$, denote the scattered waves, then this relation is given by

$$\begin{bmatrix} b_1 \\ b_2 \\ b_3 \end{bmatrix} = \begin{bmatrix} S_{11} & S_{12} & S_{13} \\ S_{21} & S_{22} & S_{23} \\ S_{31} & S_{32} & S_{33} \end{bmatrix} \begin{bmatrix} a_1 \\ a_2 \\ a_3 \end{bmatrix} \tag{23}$$

The two propagating waves in the dielectric sheath interact with the second discontinuity, located at $z = t'$. Since the second discontinuity, an air-dielectric interface, does not cause intermodal coupling, it is convenient to describe the scattering property of this junction for the individual modes separately by 2×2 matrices. The scattering matrices for the two modes under consideration may be expressed as

$$\begin{bmatrix} d_1' \\ d_2' \end{bmatrix} = \begin{bmatrix} S_{11}' & S_{12}' \\ S_{21}' & S_{22}' \end{bmatrix} \begin{bmatrix} c_1' \\ c_2' \end{bmatrix} \quad \text{for the zeroth Floquet mode} \tag{24a}$$

and

$$\begin{bmatrix} d_1'' \\ d_2'' \end{bmatrix} = \begin{bmatrix} S_{11}'' & S_{12}'' \\ S_{21}'' & S_{22}'' \end{bmatrix} \begin{bmatrix} c_1'' \\ c_2'' \end{bmatrix} \quad \text{for the } m = -1 \text{ Floquet mode} \tag{24b}$$

where the elements of $\{s'_{ij}\}$ and $\{s''_{ij}\}$ are well known and are given by

$$s'_{11} = \frac{Y_0^{\varepsilon} - Y_0}{Y_0^{\varepsilon} + Y_0} = -s'_{22}, \qquad s'_{12} = s'_{21} = \frac{2\sqrt{Y_0^{\varepsilon} Y_0}}{Y_0^{\varepsilon} + Y_0}$$
$$s''_{11} = \frac{Y_{-1}^{\varepsilon} - Y_1}{Y_{-1}^{\varepsilon} + Y_1} = -s''_{22}, \quad \text{and} \quad s''_{12} = \frac{2\sqrt{Y_{-1}^{\varepsilon} Y_{-1}}}{Y_{-1}^{\varepsilon} + Y_{-1}} \tag{24c}$$

Here, c'_i and c''_i, $i = 1, 2$, stand for incident waves, while d'_i and d''_i represent scattered waves. The waves which exist inside the dielectric sheath are related further through the propagation factors by

$$c'_1 = b_2 e^{-j\Gamma_0^{\varepsilon} t'}$$
$$a_2 = d'_1 e^{-j\Gamma_0^{\varepsilon} t'} \tag{25a}$$

and

$$c''_1 = b_3 e^{-j\Gamma_{-1}^{\varepsilon} t'}$$
$$a_3 = d''_1 e^{-j\Gamma_{-1}^{\varepsilon} t'} \tag{25b}$$

where Γ_0^{ε} and $\Gamma_{-1}^{\varepsilon}$ are the z-directed propagation constants of the two waves.

Equations (23)–(25) may now be combined to derive the scattering matrix for the *composite* network:

$$\begin{bmatrix} b_1 \\ d'_2 \\ d''_2 \end{bmatrix} = \begin{bmatrix} S_{11} & S_{12} & S_{13} \\ S_{21} & S_{22} & S_{23} \\ S_{31} & S_{32} & S_{33} \end{bmatrix} \begin{bmatrix} a_1 \\ c'_2 \\ c''_2 \end{bmatrix} \tag{26}$$

With the exception of S_{12}, the explicit expression of the matrix elements S_{ij} need not concern us here and therefore are given in Appendix II.

Now, for the range of the steering phase, $2\pi(1 - \sqrt{\varepsilon}\, b/\lambda) \le \psi \le 2\pi(1 - b/\lambda)$, the $m = -1$ Floquet mode becomes evanescent in free space, and we need only a 2×2 scattering matrix to describe the composite network:

$$[S] = \begin{bmatrix} S_{11} & S_{12} \\ S_{21} & S_{22} \end{bmatrix} \tag{27}$$

The matrix elements remain the same as those in (26).

When the array is scanned to an angle such that a surface wave resonance occurs, the array reflection coefficient attains a magnitude of unity, or, correspondingly, the transmission coefficient goes to zero. This situation implies that

$$S_{21} = 0$$

or

$$s_{21} + s''_{11}(s_{23}s_{31} - s_{21}s_{33}) e^{-j2\Gamma_{-1}^{\varepsilon} t'} = 0$$

Here, s''_{11} is the reflection coefficient of the $m = -1$ Floquet mode at the air-dielectric interface, $z = t'$, and is given in (24c). Under the condition that this mode propagates inside the dielectric but is attenuated in free space, it takes the forms

$$s''_{11} = \frac{\Gamma_{-1}{}^{\varepsilon} - \Gamma_{-1}}{\Gamma_{-1}{}^{\varepsilon} + \Gamma_{-1}} = e^{j2\tan^{-1}(|\Gamma_{-1}|/\Gamma_{-1}{}^{\varepsilon})} \qquad \text{for } H\text{-plane scan} \qquad (28a)$$

and

$$s''_{11} = \frac{\varepsilon\Gamma_{-1} - \Gamma_{-1}{}^{\varepsilon}}{\varepsilon\Gamma_{-1} + \Gamma_{-1}{}^{\varepsilon}} = e^{j2[\pi/2 - \tan^{-1}(\varepsilon|\Gamma_{-1}|/\Gamma_{-1}{}^{\varepsilon})]} \qquad \text{for } E\text{-plane scan} \qquad (28b)$$

Since the scattering matrix of a lossless junction is unitary, we have $(s_{23}s_{31} - s_{21}s_{33}) = -s_{12}^{*}\,|s|$, where $|s|$ is the determinant of the scattering matrix, (23). Taking the above considerations together, we arrive at the conditions of resonance as

$$\tan^{-1}\frac{|\Gamma_{-1}|}{\Gamma_{-1}{}^{\varepsilon}} - \Gamma_{-1}{}^{\varepsilon}t' - \frac{\pi}{2} = \frac{1}{2}\,\underline{/\frac{s_{21}}{s_{23}s_{31} - s_{21}s_{33}}} \qquad \text{for } H\text{-plane scan} \qquad (29a)$$

and

$$\tan^{-1}\frac{\varepsilon\,|\Gamma_{-1}|}{\Gamma_{-1}{}^{\varepsilon}} - \Gamma_{-1}{}^{\varepsilon}t' = \frac{1}{2}\,\underline{/\frac{s_{21}}{s_{23}s_{31} - s_{21}s_{33}}} \qquad \text{for } E\text{-plane scan} \qquad (29b)$$

A similar procedure may be applied for the dielectric plug case. When the dielectric constant of the plug is such that it supports two propagating modes inside the plug while only one mode propagates in the unloaded waveguide region, the corresponding results are

$$\tan^{-1}\frac{|\gamma_{2}|}{\gamma_{2}{}^{\varepsilon}} - \gamma_{2}{}^{\varepsilon}t - \frac{\pi}{2} = \frac{1}{2}\,\underline{/\frac{\bar{s}_{21}}{\bar{s}_{23}\bar{s}_{31} - \bar{s}_{21}\bar{s}_{33}}} \qquad \text{for } H\text{-plane scan} \qquad (30a)$$

and

$$\tan^{-1}\frac{\varepsilon\,|\gamma_{1}|}{\gamma_{1}{}^{\varepsilon}} - \gamma_{1}{}^{\varepsilon}t = \frac{1}{2}\,\underline{/\frac{\bar{s}_{21}}{\bar{s}_{23}\bar{s}_{31} - \bar{s}_{21}\bar{s}_{33}}} \qquad \text{for } E\text{-plane scan} \qquad (30b)$$

where $\{\bar{s}_{ij}\}$ are the elements of the scattering matrix for the array when the waveguides are *completely loaded* with a dielectric of dielectric constant ε. Subscripts $i = 1$ and 3 refer to the first and second propagating modes, respectively, inside dielectric-filled waveguides, while $i = 2$ denotes the ($m = 0$) Floquet mode in free space.

Equations (29a) and (29b), and (30a) and (30b), show that aperture resonance conditions are dependent on the properties of the array elements as displayed by the presence of the matrix elements, s_{ij} and $\bar{s}_{ij}$. When the right-hand sides of (29a) and (29b) are set equal to zero, these equations reduce to those obtainable by the transverse resonance methods as given in (22). The results obtained here apply to the situation in which two modes propagate inside the dielectric. When there are more than two propagating modes, the procedure is still applicable, although the resulting equations become more complex and are less convenient to use when the number of propagating modes is large. In Chapter 7 the case of three propagating modes in the dielectric plugs of a planar array of circular waveguides is discussed, together with the presentation of numerical results.

To determine the resonance conditions in practice, it is convenient to find the thickness of the dielectric sheath or plug for an array by specifying its element spacing and dielectric constant. This can be done by calculating the scattering matrix of the array radiating into a dielectric half space for the sheath case, or of the array completely loaded with a dielectric for the plug case. The integral equation approach can be used for this purpose. It is necessary to perform such a calculation only once at each scan angle of interest for a given element spacing and dielectric constant. The phase angle corresponding to the right-hand side of (29a) and (29b) or (30a) and (30b) is then computed. The determination of the resonant thickness then follows readily.

Some representative results obtained for the E-plane scan are illustrated in Figs. 6.23 and 6.24. Figure 6.23 shows the effects of the waveguide wall thickness, as well as the dielectric constant, on the resonant peak location,

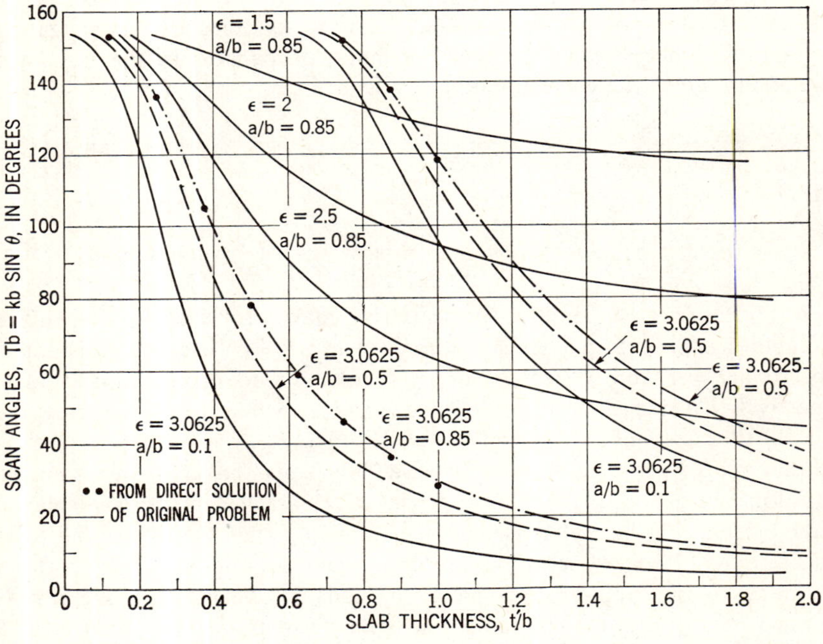

Fig. 6.23. Surface wave resonance conditions for a dielectric-sheath-covered array scanned in the E plane, $b/\lambda = 0.5714$.

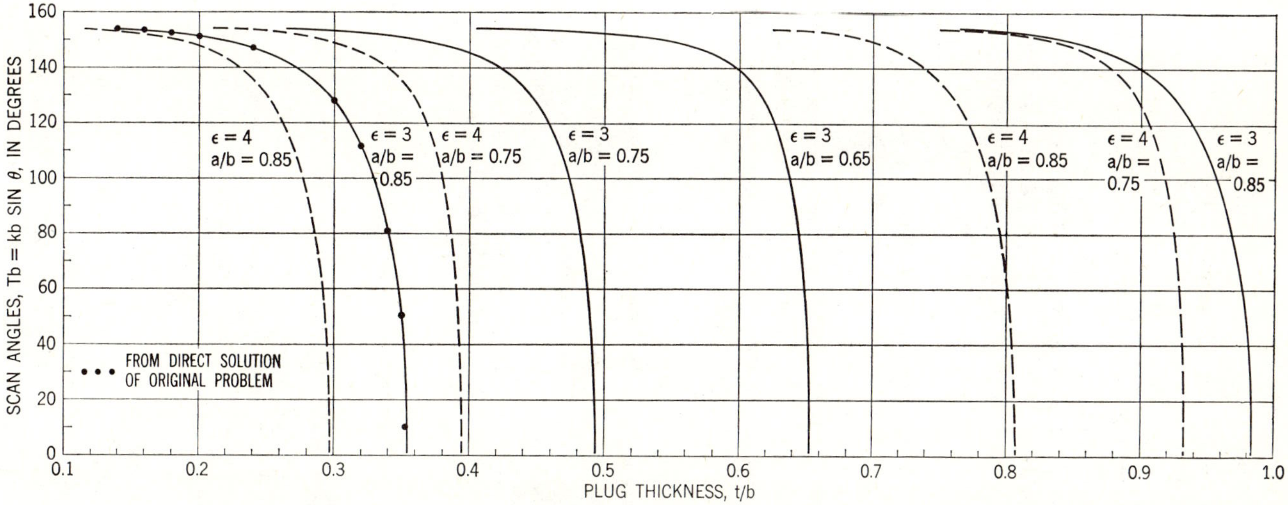

Fig. 6.24. Surface wave resonance conditions for an array loaded with dielectric plugs and scanned in the E plane, $b/\lambda = 0.5714$.

ψ_{sw}, as a function of the sheath thickness. Two sets of curves are presented in this figure. One set of illustrations represents the calculation for a dielectric constant $\varepsilon = (\lambda/b)^2 = 3.0625$, with the waveguide wall thickness varied as a parameter. The graphs clearly show that decreasing the waveguide aperture shifts the resonances toward lower scan angles. Moreover, when thick dielectric slabs are used, there can be two or more scan angles at which resonances may be observed. The other set of curves shows the effects of varying the dielectric constants on the resonance conditions. Notice that the scan range for the resonances depends directly on the dielectric constants.

In Fig. 6.24 we present results for loading the waveguides with dielectric plugs. The effects of varying both the radiation aperture and the dielectric constant are illustrated. A salient feature of the plug case is the appearance of pass-band- and stop-band-like properties as the thickness of the plug is varied. A similar phenomenon has also been observed in the case of *H*-plane scan. Moreover, the resonances occur periodically with increasing plug thickness. The period is π/γ_2^ε for the *H* plane and $\pi/\gamma_1'^\varepsilon$ for the *E* plane. This can be seen from (30a) and (30b), for the arc tangent has an indeterminancy by an integral multiple of π, so that adding $n\pi$ to the arc tangent function and increasing t by $n\pi/\gamma_2^\varepsilon$ or $n\pi/\gamma_1'^\varepsilon$ at the same time leaves the equations unchanged. This is not true, however, for the case in which three or more modes propagate in the dielectric (see Chapter 7).

The data obtained by the scattering matrix approach have been compared with those calculated by a direct moments solution of the original boundary value problems, as illustrated by the dots in Figs. 6.23 and 6.24. The agreement between the two methods is very good. Although the scattering approach is based on the thick-sheath (or plug) assumption, the agreement remains good even when very thin sheaths or plugs are used.

8. THE CAUSE OF FORCED SURFACE WAVE RESONANCES

The existence of forced surface wave resonances has been demonstrated both analytically and experimentally. How the phenomenon may arise has been explained from the viewpoint of microwave circuits. Resonances can occur with or without the presence of dielectrics in planar arrays. In linear arrays, however, resonances have been observed *only* when dielectric sheaths (or plugs) have been used. It is well known that a dielectric sheath can support waves which propagate unattenuated along the sheath in the absence of loss in the dielectric. This suggests that the resonance phenomenon is intimately related to surface wave propagation. Notice that the name forced surface wave resonance is adopted because a surface wave can be propagated along a corrugated surface derived from the array structure.[2,17] This does not imply, however, that a surface wave exists in the original array geometry.

We will show by using reciprocity as an argument that surface waves, if they exist, cannot be excited by the array when only a finite number of elements are excited.

The outstanding feature of the forced surface wave resonances is the *sharp peaks with magnitude* 1 in the reflection coefficient. This, by conservation of energy, is equivalent to the appearance of zeros in the transmission coefficient. Since the element pattern is proportional to the transmission coefficient, the zero transmission is manifested as deep notches or blind spots in the element pattern.

Assume now that surface waves can be propagated along the array aperture. These surface waves cannot be excited by the array under single-element excitation, for, if the surface waves were excited, it would imply that there is coupling between the surface wave modes and the excited element. By reciprocity, then, when a surface wave propagates across the array elements, a portion of its power must be leaked into the elements through coupling. Consequently the surface wave cannot propagate along the array surface without attenuation. This contradicts the original meaning in which the term surface wave is used.

It has been suggested that the occurrence of the forced surface wave effects may be explained in terms of leaky waves which exist in the array structure. Knittel, Hessel, and Oliner,[15] using a dielectric sheath-covered parallel-plate array for *E*-plane scan, have shown that complex roots exist in the characteristic equation derived by the transverse resonance technique. These roots have real parts which are approximately equal to the transverse wave numbers corresponding to the scan angles for resonance. Knittel et al. explain the appearance of element pattern nulls as due to the destructive interference between the direct radiation from the excited waveguide and the component radiated by the leaky wave.

Our understanding of the array resonance phenomenon is still incomplete. For example, why should the reflection coefficient have magnitude 1 at resonance? The reflection coefficients may also possess peaks resembling resonance behavior, but having magnitude less than 1. Are these caused by the same mechanism? Resonances occur with and without the presence of dielectrics, especially in planar arrays. Can these be explained by the same theory? These are some of the important questions which remain to be answered. It is very important to have a thorough understanding of the phenomena in order to find means of avoiding them.

APPENDIX I INTEGRAL EQUATIONS FOR MULTILAYERED PLUG CASE

We consider below the derivation of the integral equation for an array loaded with multiple layers of dielectric material. The purpose is to show

how the modal coefficients in each layer are related to those belonging to the adjacent layer, and in particular to those at the array aperture. A two-layer case as shown in Fig. A.1 is considered here for illustrative purposes. The extension to the case of N layers may be arrived at by induction from the one- and two-layer results.

The derivation of the integral equation follows a line similar to that used in the one-layer case. First, the tangential fields are expanded into normal

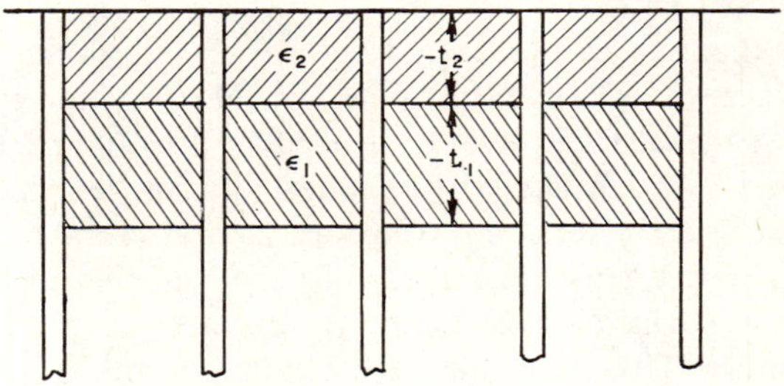

Fig. A.1. An array loaded with two layers of dielectric plug.

modes for the various regions. Thus, under the assumption that the array is excited by a lowest TE mode with the phase referenced to $z = -(t_1 + t_2)$, we have the following:

For $z \leq -(t_1 + t_2)$

$$\mathscr{H}_x(x, z) = [e^{-j\gamma_1(z+t_1+t_2)} - Re^{j\gamma_1(z+t_1+t_2)}]\Phi_1(x)$$

$$+\sum_{q=2}^{\infty} i_q e^{j\gamma_q(z+t_1+t_2)}\Phi_q(x) \quad \text{(A.1a)}$$

$$\mathscr{E}_y(x, z) = -z_1[e^{-j\gamma_1(z+t_1+t_2)} + Re^{j\gamma_1(z+t_1+t_2)}]\Phi_1(x)$$

$$+\sum_{q=2}^{\infty} z_q i_q e^{j\gamma_q(z+t_1+t_2)}\Phi_q(x) \quad \text{(A.1b)}$$

For $-(t_1 + t_2) \leq z \leq -t_2$

$$\mathscr{H}_x(x, z) = \sum_{q=1}^{\infty}[i_q^{1+}e^{-j\gamma_q^{\varepsilon_1}(z+t_2)} + i_q^{1-}e^{j\gamma_q^{\varepsilon_1}(z+t_2)}]\Phi_q(x) \quad \text{(A.2a)}$$

$$\mathscr{E}_y(x, z) = -\sum_{q=1}^{\infty} z_q^{\varepsilon_1}[i_q^{1+}e^{-j\gamma_q^{\varepsilon_1}(z+t_2)} - i_q^{1-}e^{j\gamma_q^{\varepsilon_1}(z+t_2)}]\Phi_q(x) \quad \text{(A.2b)}$$

For $-t_2 \leq z \leq 0$

$$\mathscr{H}_x(x, z) = \sum_{q=1}^{\infty} (i_q^{2+} e^{-j\gamma_q^{\varepsilon_2} z} + i_q^{2-} e^{j\gamma_q^{\varepsilon_2} z}) \Phi_q(x) \tag{A.3a}$$

$$\mathscr{E}_y(x, z) = -\sum_{q=1}^{\infty} z_q^{\varepsilon_2} (i_q^{2+} e^{-j\gamma_q^{\varepsilon_2} z} - i_q^{2-} e^{j\gamma_q^{\varepsilon_2} z}) \Phi_q(x) \tag{A.3b}$$

The field in the exterior $0 \leq z$ region may still be described as in (4). Applying the boundary conditions at the interface $z = -(t_1 + t_2)$ yields

$$1 - R = i_1^{1+} e^{j\gamma_1^{\varepsilon_1} t_1} + i_1^{1-} e^{-j\gamma_1^{\varepsilon_1} t_1} \tag{A.4a}$$

$$i_q = i_q^{1+} e^{j\gamma_q^{\varepsilon_1} t_1} + i_q^{1-} e^{-j\gamma_q^{\varepsilon_1} t_1}, \quad q \geq 2 \tag{A.4b}$$

$$z_1(1 + R) = z_1^{\varepsilon_1}(i_1^{1+} e^{j\gamma_1^{\varepsilon_1} t_1} - i_1^{1-} e^{-j\gamma_1^{\varepsilon_1} t_1}) \tag{A.4c}$$

$$z_q i_q = -z_q^{\varepsilon_1}(i_q^{1+} e^{j\gamma_q^{\varepsilon_1} t_1} - i_q^{1-} e^{-j\gamma_q^{\varepsilon_1} t_1}), \quad q \geq 2 \tag{A.4d}$$

The boundary conditions at $z = -t_2$ require

$$i_q^{1+} + i_q^{1-} = i_q^{2+} e^{j\gamma_q^{\varepsilon_2} t_2} + i_q^{2-} e^{-j\gamma_q^{\varepsilon_2} t_2} \tag{A.5a}$$

$$z_q^{\varepsilon_1}(i_q^{1+} - i_q^{1-}) = z_q^{\varepsilon_2}(i_q^{2+} e^{j\gamma_q^{\varepsilon_2} t_2} - i_q^{2-} e^{-j\gamma_q^{\varepsilon_2} t_2}) \tag{A.5b}$$

The above relations are obtained through mode-by-mode matching, because the modal functions in each region are identical. Finally, from the conditions at the aperture, we have

$$H_x(x) = \sum_{q=1}^{\infty} (i_q^{2+} + i_q^{2-}) \Phi_q(x) = \sum_{m=-\infty}^{\infty} I_m \Psi_m(x) \tag{A.6a}$$

$$E_y(x) = -\sum_{q=1}^{\infty} z_q^{\varepsilon_2} (i_q^{2+} - i_q^{2-}) \Phi_q(x) = -\sum_{m=-\infty}^{\infty} Z_m I_m \Psi_m(x) \tag{A.6b}$$

Equations (A.6a) and (A.6b) indicate that

$$\bar{i}_q^{\,2} = i_q^{2+} + i_q^{2-} = \int_{-b/2}^{b/2} \Phi_q(x) H_x(x)\, dx \tag{A.7a}$$

$$I_m = \int_{-b/2}^{b/2} \Psi_m^*(x) H_x(x)\, dx \tag{A.7b}$$

To find $z_q^{\varepsilon_2}(i_q^{2+} - i_q^{2-})$, to be used in (A.6b), we start from (A.4a)–(A.4d) and eliminate i_q and R to obtain

$$i_q^{1-} = -\frac{z_q + z_q^{\varepsilon_1}}{z_q - z_q^{\varepsilon_1}}\, i_q^{1+} e^{j2\gamma_q^{\varepsilon_1} t_1} + \frac{2z_1}{z_1 - z_1^{\varepsilon_1}}\, e^{j\gamma_1^{\varepsilon_1} t_1} \delta_{1q} \tag{A.8}$$

Substituting this expression into (A.5b) yields

$$-z_q^{\varepsilon_1}(i_q^{1+} - i_q^{1-}) = \tilde{z}_q^1 \bar{i}_q^1 - 2z_1 U_1^{\varepsilon_1}\delta_{1q} \tag{A.9}$$

where

$$\bar{i}_q^1 = i_q^{1+} + i_q^{1-} \tag{A.10a}$$

$$\tilde{z}_q^1 = z_q^{\varepsilon_1} \frac{z_q + jz_q^{\varepsilon_1} \tan \gamma_q^{\varepsilon_1} t_1}{z_q^{\varepsilon_1} + jz_q \tan \gamma_q^{\varepsilon_1} t_1} \tag{A.10b}$$

$$U_q^{\varepsilon_1} = \frac{z_q^{\varepsilon_1}}{z_q^{\varepsilon_1} \cos \gamma_q^{\varepsilon_1} t_1 + jz_q \sin \gamma_q^{\varepsilon_1} t_1} \tag{A.10c}$$

We next use (A.5a), (A.5b), and (A.9) to eliminate $\bar{i}_q^1$. The result is

$$i_q^{2-} = -\frac{\tilde{z}_q^1 + z_q^{\varepsilon_2}}{\tilde{z}_q^1 - z_q^{\varepsilon_2}} i_q^{2+} e^{j2\gamma_q^{\varepsilon_2} t_2} + \frac{2z_1}{\tilde{z}_q^1 - z_q^{\varepsilon_2}} U_1^{\varepsilon_1} e^{j\gamma_1^{\varepsilon_2} t_2} \delta_{1q} \tag{A.11}$$

This relation is finally substituted into (A.6*b*), giving

$$-z_q^{\varepsilon_2}(i_q^{2+} - i_q^{2-}) = \tilde{z}_q^2 \bar{i}_q^2 - 2z_1 U_1^{\varepsilon_1} U_1^{\varepsilon_2} \delta_{1q} \tag{A.12}$$

where

$$\tilde{z}_q^2 = z_q^{\varepsilon_2} \frac{\tilde{z}_q^1 + jz_q^{\varepsilon_2} \tan \gamma_q^{\varepsilon_2} t_2}{z_q^{\varepsilon_2} + j\tilde{z}_q^1 \tan \gamma_q^{\varepsilon_2} t_2} \tag{A.13a}$$

$$U_q^{\varepsilon_2} = \frac{z_q^{\varepsilon_2}}{z_q^{\varepsilon_2} \cos \gamma_q^{\varepsilon_2} t_2 + j\tilde{z}_q^1 \sin \gamma_q^{\varepsilon_2} t_2} \tag{A.13b}$$

and $\bar{i}_q^2$ is as given in (A.7a) in terms of the aperture magnetic field. At this stage, we are ready to derive the integral equation. Using (A.6b) together with (A.7a), (A.7b), and (A.12), we find

$$2z_1 U_1^{\varepsilon_1} U_1^{\varepsilon_2} \Phi_1(x) = \int_{-b/2}^{b/2} \left\{ \sum_{q=1}^{\infty} \tilde{z}_q^2 \Phi_q(x)\Phi_q(x') + \sum_{m=-\infty}^{\infty} Z_m \Psi_m(x) \Psi_m^*(x') \right\} H_x(x')\, dx' \tag{A.14}$$

Notice that $\tilde{z}_q^2$ has the expression of the input impedance of a transmission line terminated in a load impedance $\tilde{z}_q^1$, which is itself the input impedance of the section $-(t_1 + t_2) \leq z \leq -t_2$, with z_q as the load impedance. The expression in (A.14) indicates clearly how the equation may be extended to the general case of N layers.

The solution of (A.14) yields the tangential magnetic field at the aperture. The modal coefficients in terms of the waveguide modes or Floquet modes may be determined by using (A.7a) or (A.7b). From (A.7a) and (A.12) the two components of $\bar{i}_q^2$ are determined as

$$i_q^{2+} = \frac{z_q^{\varepsilon_2} - \tilde{z}_q^2}{2z_q^{\varepsilon_2}} \bar{i}_q^2 + \frac{z_1}{z_q^{\varepsilon_2}} U_1^{\varepsilon_1} U_1^{\varepsilon_2} \delta_{1q} \tag{A.15a}$$

and

$$i_q^{2-} = \frac{z_q^{\varepsilon_2} + \tilde{z}_q^2}{2z_q^{\varepsilon_2}} \bar{i}_q^2 - \frac{z_1}{z_q^{\varepsilon_2}} U_1^{\varepsilon_1} U_1^{\varepsilon_2} \delta_{1q} \tag{A.15b}$$

When these relations are substituted into (A.5a), we find

$$\bar{i}_q^1 = U_q^{\varepsilon_2} \bar{i}_q^2 + j\delta_{1q} 2 \frac{z_1}{z_q^{\varepsilon_2}} U_1^{\varepsilon_1} U_1^{\varepsilon_2} \sin \gamma_q^{\varepsilon_2} t_2 \tag{A.16}$$

This expression gives the modal coefficients at $z = -t_2$ in terms of those at the array aperture. Continuing the process, we obtain from (A.5a) and (A.5b), and (A.15a) and (A.15b),

$$i_q^{1+} = \frac{z_q^{\varepsilon_1} - \tilde{z}_q^1}{2z_q^{\varepsilon_1}} U_q^{\varepsilon_2} \bar{i}_q^2 + \frac{z_1}{z_q^{\varepsilon_1} z_q^{\varepsilon_2}} U_1^{\varepsilon_1} U_1^{\varepsilon_2} \delta_{1q} (z_q^{\varepsilon_2} \cos \gamma_q^{\varepsilon_2} t_2 + j z_q^{\varepsilon_1} \sin \gamma_q^{\varepsilon_2} t_2) \tag{A.17a}$$

and

$$i_q^{1-} = \frac{z_q^{\varepsilon_1} + \tilde{z}_q^1}{2z_q^{\varepsilon_1}} U_q^{\varepsilon_2} \bar{i}_q^2 - \frac{z_1}{z_q^{\varepsilon_1} z_q^{\varepsilon_2}} U^{\varepsilon_1} U^{\varepsilon_2} \delta_{1q} (z_q^{\varepsilon_2} \cos \gamma_q^{\varepsilon_2} t_2 - j z_q^{\varepsilon_1} \sin \gamma_q^{\varepsilon_2} t_2) \tag{A.17b}$$

Finally, we have

$$\begin{aligned} i_q = {} & U_q^{\varepsilon_1} U_q^{\varepsilon_2} \bar{i}_q^2 + j \frac{2z_1}{z_q^{\varepsilon_1} z_q^{\varepsilon_2}} U_1^{\varepsilon_1} U_1^{\varepsilon_2} \delta_{1q} \\ & + j \frac{2z_1}{z_q^{\varepsilon_1} z_q^{\varepsilon_2}} (z_q^{\varepsilon_2} \cos \gamma_q^{\varepsilon_2} t_2 \sin \gamma_q^{\varepsilon_1} t_1 + z_q^{\varepsilon_1} \sin \gamma_q^{\varepsilon_2} t_2 \cos \gamma_1^{\varepsilon_1} t_1) \end{aligned} \tag{A.18}$$

Equation (A.18) is valid also for $q = 1$. The reflection coefficient may be calculated from $1 - R = i_1$.

As can be seen from the above derivation, the procedure is straightforward but tedious. It is easily extended to cases involving more than two layers.

APPENDIX II SCATTERING MATRIX FOR CONSECUTIVE DOUBLE DISCONTINUITIES

The derivation of the scattering matrix (26) may be carried out by substitution and elimination, using (23)–(25). For example, consider the situation wherein $a_1 \neq 0$, $c_2' = 0$, and $c_2'' = 0$. Equations (24a) and (24b) may be rearranged to read

$$\begin{bmatrix} d_1' \\ d_1'' \end{bmatrix} = \begin{bmatrix} s_{11}' & 0 \\ 0 & s_{11}'' \end{bmatrix} \begin{bmatrix} c_1' \\ c_1'' \end{bmatrix} \quad \text{and} \quad \begin{bmatrix} d_2' \\ d_2'' \end{bmatrix} = \begin{bmatrix} s_{21}' & 0 \\ 0 & s_{21}'' \end{bmatrix} \begin{bmatrix} c_1' \\ c_1'' \end{bmatrix} \tag{A.19}$$

Also (25a) and (25b) may be put into matrix form as

$$\begin{bmatrix} c_1' \\ c_1'' \end{bmatrix} = \begin{bmatrix} e^{-j\Gamma_0{}^{\varepsilon}t'} & 0 \\ 0 & e^{-j\Gamma_{-1}{}^{\varepsilon}t'} \end{bmatrix} \begin{bmatrix} b_2 \\ b_3 \end{bmatrix} \tag{A.20a}$$

and

$$\begin{bmatrix} a_2 \\ a_3 \end{bmatrix} = \begin{bmatrix} e^{-j\Gamma_0{}^{\varepsilon}t'} & 0 \\ 0 & e^{-j\Gamma_{-1}{}^{\varepsilon}t'} \end{bmatrix} \begin{bmatrix} d_1' \\ d_2'' \end{bmatrix} \tag{A.20b}$$

Combining the above yields

$$\begin{bmatrix} a_2 \\ a_3 \end{bmatrix} = \begin{bmatrix} (s_{11}'/s_{21}')e^{-j\Gamma_0{}^{\varepsilon}t'} & 0 \\ 0 & (s_{11}''/s_{21}'')e^{-j\Gamma_0{}^{\varepsilon}t'} \end{bmatrix} \begin{bmatrix} d_2' \\ d_2'' \end{bmatrix} \tag{A.21a}$$

and

$$\begin{bmatrix} b_2 \\ b_3 \end{bmatrix} = \begin{bmatrix} e^{j\Gamma_0{}^{\varepsilon}t'}/s_{21}' & 0 \\ 0 & e^{j\Gamma_{-1}{}^{\varepsilon}t'}/s_{21}'' \end{bmatrix} \begin{bmatrix} d_2' \\ d_2'' \end{bmatrix} \tag{A.21b}$$

We next write (23) as

$$\begin{bmatrix} b_1 \\ 0 \\ 0 \end{bmatrix} + \begin{bmatrix} 0 \\ b_2 \\ b_3 \end{bmatrix} = \begin{bmatrix} s_{11} & s_{12} & s_{13} \\ s_{21} & s_{22} & s_{23} \\ s_{31} & s_{23} & s_{33} \end{bmatrix} \left(\begin{bmatrix} a_1 \\ 0 \\ 0 \end{bmatrix} + \begin{bmatrix} 0 \\ a_2 \\ a_3 \end{bmatrix} \right)$$

and then use (A.20b) and (A.21a) to arrive at

$$\begin{bmatrix} 1 & 0 & 0 \\ 0 & e^{j\Gamma_0{}^{\varepsilon}t'}/s_{21}' & 0 \\ 0 & 0 & e^{j\Gamma_{-1}{}^{\varepsilon}t'}/s_{21}'' \end{bmatrix} \begin{bmatrix} b_1 \\ d_2' \\ d_2'' \end{bmatrix}$$
$$= \begin{bmatrix} s_{11} & s_{12} & s_{13} \\ s_{21} & s_{22} & s_{23} \\ s_{31} & s_{32} & s_{33} \end{bmatrix} \left\{ \begin{bmatrix} a_1 \\ 0 \\ 0 \end{bmatrix} + \begin{bmatrix} 0 & 0 & 0 \\ 0 & (s_{11}'/s_{21}')e^{-j\Gamma_0{}^{\varepsilon}t'} & 0 \\ 0 & 0 & (s_{11}''/s_{21}'')e^{-j\Gamma_{-1}{}^{\varepsilon}t} \end{bmatrix} \begin{bmatrix} b_1 \\ d_2' \\ d_2'' \end{bmatrix} \right\} \tag{A.22}$$

Rearranging, we find the elements in the first column of $[S]$. The process may be continued for $a_1 = 0$, $c_1' \neq 0$, and $c_1'' = 0$ to yield the elements in the second column, and for $a_1 = 0$, $c_1' = 0$, and $c_1'' \neq 0$ to yield the elements in the third column. The final results are as follows:

$$S_{11} = \frac{1}{D}\,[s_{11} + s'_{11}(s_{12}s_{21} - s_{11}s_{22})e^{-j2\Gamma_0{}^{\varepsilon}t'} + s''_{11}(s_{13}s_{31} - s_{11}s_{33})e^{-j2\Gamma_{-1}{}^{\varepsilon}t'} + s'_{11}s''_{11}\,|s|\,e^{-j2(\Gamma_0{}^{\varepsilon}+\Gamma_{-1}{}^{\varepsilon})t'}]$$

$$S_{21} = \frac{1}{D}\,(s'_{21}e^{-j\Gamma_0{}^{\varepsilon}t'})[s_{21} + s''_{11}(s_{23}s_{31} - s_{21}s_{33})e^{-j\Gamma_{-1}{}^{\varepsilon}t'}]$$

$$S_{31} = \frac{1}{D}\,(s''_{21}e^{-j\Gamma_{-1}{}^{\varepsilon}t'})[s_{31} + s'_{11}(s_{32}s_{21} - s_{31}s_{22})e^{-j\Gamma_0{}^{\varepsilon}t'}]$$

$$S_{12} = \frac{1}{D}\,(s'_{12}e^{-j\Gamma_0{}^{\varepsilon}t'})[s_{12} + s''_{11}(s_{13}s_{32} - s_{12}s_{33})e^{-j2\Gamma_{-1}{}^{\varepsilon}t'}]$$

$$S_{22} = \frac{1}{D}\,[s'_{22} + s_{22}(s'_{21}s'_{12} - s'_{11}s'_{22})e^{-j2\Gamma_0{}^{\varepsilon}t'} - s''_{11}s'_{22}s_{33}e^{-j2\Gamma_{-1}{}^{\varepsilon}t'} + s''_{11}(s_{23}s_{32} - s_{22}s_{33})(s'_{21}s'_{12} - s'_{11}s'_{22})e^{-j2(\Gamma_0{}^{\varepsilon}+\Gamma_{-1}{}^{\varepsilon})t'}]$$

$$S_{32} = \frac{s'_{12}s''_{21}s_{32}e^{-j(\Gamma_0{}^{\varepsilon}+\Gamma_{-1}{}^{\varepsilon})t'}}{D}$$

$$S_{13} = \frac{1}{D}\,(s''_{12}e^{-j\Gamma_{-1}{}^{\varepsilon}t'})[s_{13} + s'_{11}(s_{12}s_{23} - s_{11}s_{13})e^{-j2\Gamma_0{}^{\varepsilon}t'}]$$

$$S_{23} = \frac{s'_{21}s''_{12}s_{23}e^{-j(\Gamma_0{}^{\varepsilon}+\Gamma_{-1}{}^{\varepsilon})t'}}{D}$$

$$S_{33} = \frac{1}{D}\,[s''_{22} + s_{33}(s''_{21}s''_{12} - s''_{22}s''_{11})e^{-j2\Gamma_{-1}{}^{\varepsilon}t'} - s'_{11}s''_{22}s_{22}e^{-j2\Gamma_0{}^{\varepsilon}t'} + s'_{11}(s''_{21}s''_{12} - s''_{11}s''_{22})(s_{32}s_{23} - s_{33}s_{22})e^{-j2(\Gamma_0{}^{\varepsilon}+\Gamma_{-1}{}^{\varepsilon})t'}]$$

where

$$D = 1 - s'_{11}s_{22}e^{-j2\Gamma_0{}^{\varepsilon}t'} - s''_{11}s_{33}e^{-j2\Gamma_{-1}{}^{\varepsilon}t'} + s'_{11}s''_{11}(s_{22}s_{33} - s_{23}s_{32})e^{-j2(\Gamma_0{}^{\varepsilon}+\Gamma_{-1}{}^{\varepsilon})t'}$$

REFERENCES

1. V. Galindo and C. P. Wu, "Dielectric Loaded and Covered Rectangular Waveguide Phased Arrays," *Bell System Tech. J.*, Vol. 47, pp. 93–116, January 1968.
2. C. P. Wu and V. Galindo, "Surface-Wave Effects on Dielectric Sheathed Phased Arrays of Rectangular Waveguides," *Bell System Tech. J.*, Vol. 47, pp. 117–142, January 1968.

3. C. P. Wu and V. Galindo, "Surface-Wave Effects on Phased Arrays of Rectangular Waveguides Loaded with Dielectric Plugs," *IEEE Trans. Antennas and Propagation*, Vol. AP-16, pp. 358–360, May 1968.

4. J. R. Wait, *Electromagnetic Waves in Stratified Media*, Pergamon Press, New York, 1962, Chapter 2.

5. E. G. Magill and H. A. Wheeler, "Wide-angle Impedance Matching of Planar Array Antenna by a Dielectric Sheet," *IEEE Trans. Antennas and Propagation*, Vol. AP-14, pp. 49–53, January 1966.

6. S. W. Lee, "Impedance Matching of an Infinite Phased Array by Dielectric Sheets," *Electronics Letters*, Vol. 2, pp. 366–368, October 1966.

7. S. W. Lee and R. Mittra, "Radiation from Dielectric-Loaded Arrays of Parallel-Plate Waveguides," *IEEE Trans. Antennas and Propagation*, Vol. AP-16, pp. 513–519, September 1968.

8. C. P. Wu and V. Galindo, "Properties of a Phased Array of Rectangular Waveguides with Thin Walls," *IEEE Trans. Antennas and Propagation*, Vol. AP-14, pp. 163–172, March 1966.

9. V. Galindo and C. P. Wu, "Asymptotic Behavior of the Coupling Coefficients for an Infinite Array of Thin-Walled Rectangular Waveguide," *IEEE Trans. Antennas and Propagation*, Vol. AP-14, pp. 248–249, March 1966.

10. V. Galindo and C. P. Wu, "Numerical Solutions for an Infinite Phased Array of Rectangular Waveguides with Thick Walls," *IEEE Trans. Antennas and Propagation*, Vol. AP-14, pp. 149–158, March 1966.

11. W. S. Gregorwich, A. Hessel, G. H. Knittel, and A. A. Oliner, "A Waveguide Simulator for the Determination of a Phased-Array Resonance," *IEEE G-AP International Symposium Digest*, pp. 134–141, 1968.

12. P. W. Hannan and M. A. Balfour, "Simulation of a Phased Array Antenna in Waveguide," *IEEE Trans. Antennas and Propagation*, Vol. AP-13, pp. 342–353, May 1965.

13. A. A. Oliner and R. G. Maleck, "Speculation on the Role of Surface Waves," in *Microwave Scanning Antennas* (R. C. Hansen, ed.), Vol. 2, Academic Press, New York, 1966, pp. 308–322.

14. C. P. Wu, "Determination of Resonance Conditions in Dielectric-Sheathed or Plug Loaded Phased Arrays," *IEEE Trans. Antennas and Propagation*, Vol. AP-16, pp. 753–755, November 1968.

15. G. H. Knittel, A. Hessel, and A. A. Oliner, "Element Pattern Nulls in Phased Arrays and Their Relation to Guided Waves," *Proc. IEEE*, Vol. 56, pp. 1822–1836, November 1958.

16. N. Amitay and V. Galindo, "Characteristics of Dielectric Covered and Loaded Circular Waveguide Phased Arrays," *IEEE Trans. Antennas and Propagation*, Vol. AP-17, pp. 722–729, November 1969.

17. V. Galindo, "A Generalized Approach to a Solution of Aperiodic Arrays and Modulated Surfaces," *IEEE Trans. Antennas and Propagation*, Vol. AP-16, pp. 424–429, July 1968.

7. Circular Waveguide Planar Phased Arrays

The properties of parallel-plate arrays as well as rectangular waveguide planar arrays were discussed in earlier chapters. Here we shall consider circular waveguides as planar array elements. The practical appeal of circular waveguides, which stems from their symmetry and ease of manufacturing, has led to their frequent deployment in microwave systems.

So far, integral equation formulation has been carried out for rectangular grid arrays. We shall generalize the formulation to include skewed (non-orthogonal) periodic and planar grid arrays of arbitrarily shaped waveguide elements. The numerical results which will be presented here for circular waveguide elements were obtained by a Ritz-Galerkin solution of the infinite array integral equation. Ways of ascertaining the validity of these numerical results will be discussed (see Chapter 3 for a thorough treatment of this subject). The radiation properties of circular waveguide planar arrays, as well as the existence of forced surface wave resonances, will be discussed. These resonances will be shown to exist in the absence as well as in the presence of dielectrics. The dependence of surface wave resonances on the mode of excitation and certain vector symmetries will be discussed. The isolation of these resonance phenomena in the two-dimensional scan plane will also be demonstrated.

A simplified model consisting of a current strip array is used in the analysis of the behavior of the coupling coefficients. It is found that in the array environment the coupling coefficients decay asymptotically as the inverse of the distance squared.

Compound elements (e.g., circular waveguide elements combined with shorted small rectangular elements) will be discussed. Suggestions for the possible elimination of forced surface wave resonances will be outlined as well.

An additional method for the elimination of forced resonances by selectively exciting and shorting the array elements will be discussed in Chapter 8.

1. RITZ-GALERKIN SOLUTION OF THE INTEGRAL EQUATION

Consider the infinite planar array of circular (or any other shape) waveguide elements of Fig. 7.1. The elements are embedded in a conducting ground plane at its interface (plane $z = 0$) with free space and are arranged in a periodic grid along the skewed (nonorthogonal) coordinates η_1 and η_2. Without loss of generality, the η_1 axis is chosen to coincide with the x axis, and the η_2

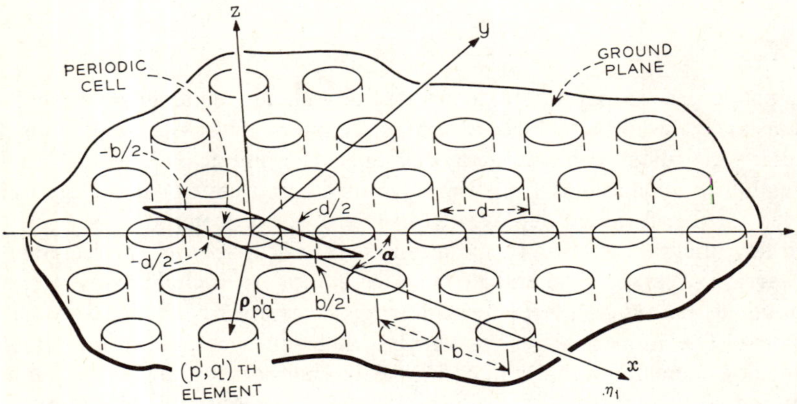

Fig. 7.1. Circular waveguide array geometry.

axis makes an angle, α, with respect to the x (and the η_1) axis. The element location is defined by two indices (p', q') corresponding to a physical location

$$\boldsymbol{\rho}_{p'q'} = p'\, b\hat{\eta}_1 + q'\, d\hat{\eta}_2 \tag{1}$$

where $\hat{\eta}_1$ and $\hat{\eta}_2$ are unit vectors along the η_1 and η_2 axes, while b and d represent the basic periods of the two-dimensional grid along η_1 and η_2. A basic periodic cell,[1] the parallelogram shown in Fig. 7.1, is thereby defined. The array elements are excited uniformly in amplitude with a linear phase taper such that the phase of the (p', q')th element is

$$\psi_{p'q'} = p'\psi_{\sigma_1} + q'\psi_{\sigma_2} \tag{2}$$

where ψ_{σ_1} and ψ_{σ_2} denote the incremental phase shifts (steering phases) between adjacent elements in the η_1 and η_2 directions, respectively. The resulting electromagnetic fields in the (p', q')th and (m', n')th periodic cells satisfy the following periodicity relationship:

$$\mathbf{F}(\boldsymbol{\rho}_{p'q'}) = \mathbf{F}(\boldsymbol{\rho}_{m'n'}) \exp \{j[(m' - p')\psi_{\sigma_1} + (n' - q')\psi_{\sigma_2}]\} \tag{2a}$$

where $\mathbf{F}(\boldsymbol{\rho}_{p'q'})$ may designate the electric or magnetic field at the (p', q')th periodic cell of the grid.* Therefore, except for a phase factor, the fields in all the cells are identical.

In order to formulate this boundary value problem, the exterior (free space) fields are expressed in terms of a complete set of plane wave (of the Floquet type) solutions of Maxwell's equations, $\{\boldsymbol{\Psi}_{pmn} \exp (\pm j\Gamma_{mn}z)\}$. This orthonormal set of vector modes consists of functions which depend on the steering phases, ψ_{σ_1} and ψ_{σ_2}. These functions and the corresponding grating lobe diagram are derived and illustrated in Appendix I. The interior ($z \leq 0$) fields are expressed in terms of the appropriate waveguide complete orthonormal set of vector modes, $\{\boldsymbol{\Phi}_i \exp (\pm j\gamma_i z)\}$.

Following the procedure of Chapter 2 in satisfying the continuity of the transverse electric and magnetic fields within a periodic cell, we obtain an integral equation formulation of the boundary value problem. As shown in Appendix II, the periodic cell consisting of the parallelogram *CDEF* of Fig. 7.2 can be replaced, without loss of generality, by the parallelogram *GHIJ* (or any other periodic contiguous cell containing a *complete* single waveguide aperture).

The resultant integral equation is

$$2 \sum_{i=1}^{2} y_i A_i \boldsymbol{\Phi}_i(\mathbf{r}) = \sum_{i=1}^{\infty} y_i \boldsymbol{\Phi}_i(\mathbf{r}) \iint_A \boldsymbol{\Phi}_i(\mathbf{r}') \cdot \mathbf{E}_t(\mathbf{r}')\, da' + \sum_{p=1}^{2} \sum_{m=-\infty}^{\infty} \sum_{n=-\infty}^{\infty} Y_{pmn} \boldsymbol{\Psi}_{pmn}(\mathbf{r}) \iint_A \boldsymbol{\Psi}^*_{pmn}(\mathbf{r}') \cdot \mathbf{E}_t(\mathbf{r}')\, da' \tag{3}$$

where $\mathbf{E}_t(\mathbf{r}')$ is the unknown tangential electric field at the aperture, A, of the waveguide element. The modal admittances of the circular (or any other) waveguide modes and the periodic free space modes are $\{y_i\}$ and $\{Y_{pmn}\}$, respectively. A more familiar designation for the circular waveguide modes is given by $\{\boldsymbol{\Phi}_{pqs}\}$. These are real functions which consist of both TE ($p = 1$) and TM ($p = 2$) type modes. The subscripts q and s determine the rate of variation of the modal field with respect to the transverse coordinates of the waveguide. For a circular waveguide q and s denote the angular and radial variation, respectively, of the mode. Because of the circular symmetry of this

* The parallelogram in Fig. 7.1 defines the ($p' = 0$, $q' = 0$) cell. The (p', q')th cell is translated by $p'b$ and $q'd$ along the η_1 and η_2 axes, respectively.

element, these modes possess a degeneracy which may be resolved by assigning horizontal and vertical identifiers to the modes that are not circularly symmetric.[2] Despite this multitude of scripts, we can, with complete generality, systematically relabel these modes with a single subscript, $\{\mathbf{\Phi}_i\}$ in (3). The subscript i is chosen so as to increase with the eigenvalues of $\mathbf{\Phi}_i$. In the analysis of the circular waveguide array, we shall allow propagation of only the circular TE_{11} (horizontal and vertical) modes. An appropriate linear combination of these modes will be used for the array excitation, depending on whether a linear or circular polarization is desired.

In the application of the Ritz-Galerkin method we shall employ the set $\{\mathbf{\Phi}_i\}$ for both the expansion of the unknown field, $\mathbf{E}_t$, and for weighting the resulting functional equation. This choice is convenient and preferable when the reflection and radiation characteristics of the array are of prime interest. We thus use the following series expansion for the unknown field, $\mathbf{E}_t$:

$$\mathbf{E}_t = (A_1 + R_1)\mathbf{\Phi}_1 + (A_2 + R_2)\mathbf{\Phi}_2 + \sum_{i=3}^{\infty} D_i \mathbf{\Phi}_i \tag{4}$$

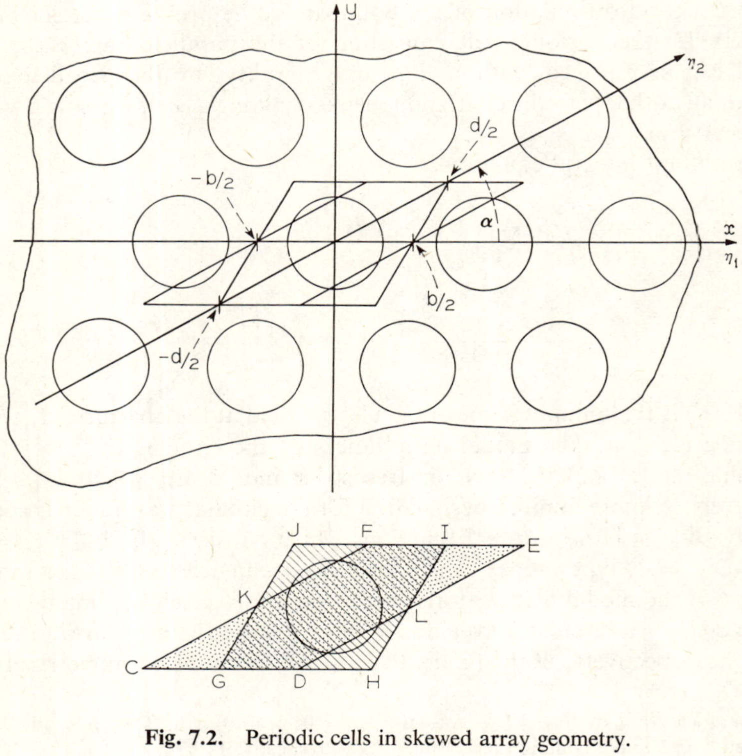

Fig. 7.2. Periodic cells in skewed array geometry.

Equation (3) is reduced (by the Ritz-Galerkin method) to the infinite matrix form:

$$2\begin{bmatrix} y_1A_1 \\ y_2A_2 \\ 0 \\ 0 \\ \cdot \\ \cdot \\ \cdot \end{bmatrix} = [K]\begin{bmatrix} A_1 + R_1 \\ A_2 + R_2 \\ D_3 \\ D_4 \\ \cdot \\ \cdot \\ \cdot \end{bmatrix} \tag{5}$$

where $[K]$ is a square matrix with the (i, q)th element given by

$$k_{iq} = y_i\delta_{iq} + \sum_{p=1}^{2} \sum_{m=-\infty}^{+\infty} \sum_{n=-\infty}^{+\infty} Y_{pmn}C^i_{pmn}C^{q*}_{pmn} \tag{6}$$

In (6) δ_{iq} is the Kronecker delta:

$$\delta_{iq} = \begin{cases} 1 & \text{if} \quad i = q \\ 0 & \text{if} \quad i \neq q \end{cases} \tag{7}$$

and

$$C^i_{pmn} = \iint_{(A)} \mathbf{\Phi}_i(\mathbf{r}) \cdot \mathbf{\Psi}_{pmn}(\mathbf{r})\, da \tag{8}$$

is the coupling coefficient (or scalar product) between the designated waveguide and free space modes. Fortunately, these scalar products, which are derived in Appendix III, can be expressed in closed forms for the circular waveguide modes over a circular aperture. In order to obtain numerical solutions (see Chapter 3) we appropriately truncate the infinite sums in (6), and thus cast (5) into a finite dimensional form. By direct inversion of the finite square matrix, $[K_I]$, we obtain

$$\begin{bmatrix} A_1 + R_1 \\ A_2 + R_2 \\ D_3 \\ \cdot \\ \cdot \\ \cdot \\ D_I \end{bmatrix} = 2[K_I]^{-1}\begin{bmatrix} y_1A_1 \\ y_2A_2 \\ 0 \\ \cdot \\ \cdot \\ \cdot \\ 0 \end{bmatrix} \tag{9}$$

$[K_I]$ is an Ith-order matrix having elements corresponding to those of (6) with the infinite sums appropriately truncated.

In order to use a single mode for the presentation of a linear, elliptical, or circular polarization of excitation of the waveguide, one may redefine the first two modes as $\mathbf{\Phi}_{1N}$ and $\mathbf{\Phi}_{2N}$:

$$\mathbf{\Phi}_{1N} = \frac{A_1}{(|A_1|^2 + |A_2|^2)^{1/2}}\mathbf{\Phi}_1 + \frac{A_2}{(|A_1|^2 + |A_2|^2)^{1/2}}\mathbf{\Phi}_2 \tag{10}$$

$$\mathbf{\Phi}_{2N} = -\frac{|A_2|}{(|A_1|^2 + |A_2|^2)^{1/2}}\mathbf{\Phi}_1 + \frac{A_1\,|A_2|}{A_2(|A_1|^2 + |A_2|^2)^{1/2}}\mathbf{\Phi}_2 \tag{11}$$

with $\mathbf{\Phi}_1$ and $\mathbf{\Phi}_2$ being the horizontal and vertical circular TE_{11} modes, respectively. Thus $A_1 = 0$, $A_2 \neq 0$ or $A_1 \neq 0$, $A_2 = 0$ correspond to linear polarizations of excitation, while $A_1 = \mp jA_2$ corresponds to circular polarizations. This redefinition of the first two modes preserves the orthonormality and completeness of the set of waveguide modes while allowing for flexibility in adjusting the desired array excitation. The reflection coefficients which correspond to these redefined modes are R_{1N} and R_{2N}.

The numerical results will, in general, be presented as a function of scan angle. For convenience, however, the differential phase shifts, which have to be applied between the elements in order to scan the beam, will be used as the independent variables. These quantities are ψ_x in the x direction and ψ_y in the y direction; ψ_x and ψ_y are linearly related to ψ_{σ_1} and ψ_{σ_2} of (2). Furthermore, since we also scan in other radial planes, we introduce the quantity ψ_r. These quantities are related to the directional cosines as follows:

$$\psi_x = \frac{2\pi b}{\lambda}T_x, \qquad \psi_y = \frac{2\pi d \sin\alpha}{\lambda}T_y, \qquad \psi_r = (\psi_x^{\,2} + \psi_y^{\,2})^{1/2} \tag{12}$$

The amount of computation can be reduced if one accounts for the following symmetry of the reflection coefficients as a function of scan:[3]

$$R_i(\psi_x, \psi_y) = R_i(-\psi_x, -\psi_y) \tag{13}$$

Because of the method used in relabeling the waveguide modes, $\{\mathbf{\Phi}_{pqs}\}$, into a single subscript set, $\{\mathbf{\Phi}_i\}$, it turns out that in the newly ordered modes the value of s (the rate of radial variations) stays relatively low and constant with respect to the values of q (the rate of angular variations), which vary over a wider range and attain high values. In observing the convergence of the solutions as a function of the number of waveguide modes, I, used in (9), we note that relatively large changes in the values of R_i occur only when higher values of s are attained with increasing I. These convergence tests as a function of the number of waveguide and free space modes indicate that 18 waveguide modes and 338 free space modes yield several per cent (usually less than 2%) accuracy in the magnitude of the reflection coefficients, R_i, except in the vicinity of ψ_r, where sharp changes of R_i are present.

The values of ψ_r for these sharp changes are accurate to several degrees. As discussed in Chapter 3, the Ritz-Galerkin method of solution has the intrinsic properties of conserving reciprocity and energy. This was tested as a necessary step in ascertaining the validity of the solution.

Among other necessary checks of the validity is the verification of whether the same solution is obtained for certain scan angles and planes of scan that

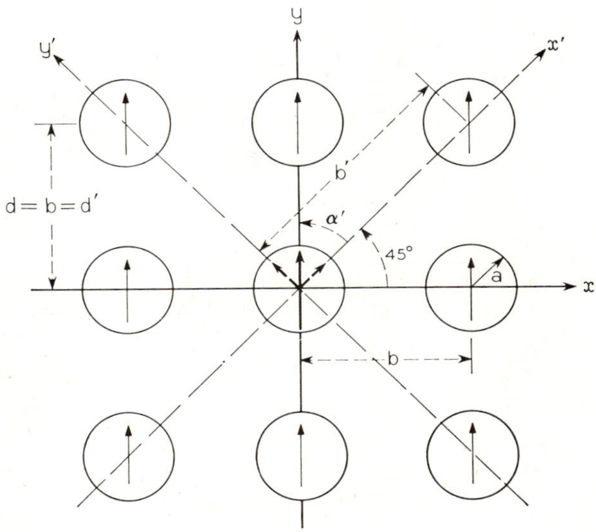

Fig. 7.3. Symmetries in square grid arrays.

are related by symmetry considerations. One such special symmetry check is depicted in Fig. 7.3. A square grid array in the (x, y) coordinates is excited by the vertical TE_{11} $(\mathbf{\Phi}_2)$ mode. In this coordinate system the array parameters are represented in the following way:

$$\left.\begin{aligned} a = \text{waveguide radius}, \quad b = d, \quad \alpha = 90^\circ \\ \mathbf{\Phi}_{1N} = \mathbf{\Phi}_2, \quad \mathbf{\Phi}_{2N} = -\mathbf{\Phi}_1 \end{aligned}\right\} \tag{14}$$

The parameters of the same array, when viewed in the (x', y') coordinate system (which is rotated by 45°), can be alternatively represented as

$$\left.\begin{aligned} b' = \sqrt{2}\, d' = \sqrt{2}\, d, \quad \alpha' = 45^\circ \\ \mathbf{\Phi}'_{1N} = \frac{1}{\sqrt{2}}\{-\mathbf{\Phi}'_1 + \mathbf{\Phi}'_2\}, \quad \mathbf{\Phi}'_{2N} = \frac{1}{\sqrt{2}}\{\mathbf{\Phi}'_1 + \mathbf{\Phi}'_2\} \end{aligned}\right\} \tag{15}$$

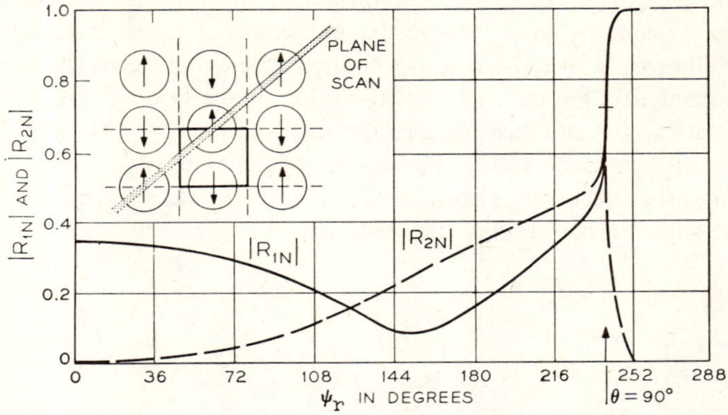

Fig. 7.4. Square grid array: $|R_{1N}|$ and $|R_{2N}|$ vs. ψ_r in the 45° plane of scan ($a = 0.48$, $b = 1$, $d = 1$, $\lambda = 1.5$, and $\alpha = 90°$).

where the reference to the (x', y') system is denoted by primes. The same characteristics should be found for this array at any scan direction regardless of the coordinate system representation. This offers, to a degree, a check on round-off error. Numerically, it was found in this check that the reflection coefficients differed only by a fraction of 1%.

An additional symmetry check is given in Fig. 7.4, where the magnitudes of the reflection coefficients are plotted versus ψ_r for a 45° plane of scan. Here R_{1N} and R_{2N} correspond to the reflection coefficients of $\mathbf{\Phi}_{1N}$ and $\mathbf{\Phi}_{2N}$ as defined by (14). At $\psi_r = 240°$ (shown by a vertical arrow) the main beam is at grazing (i.e., $\theta = 90°$), while for $\psi_r > 240°$ no beam exists in real

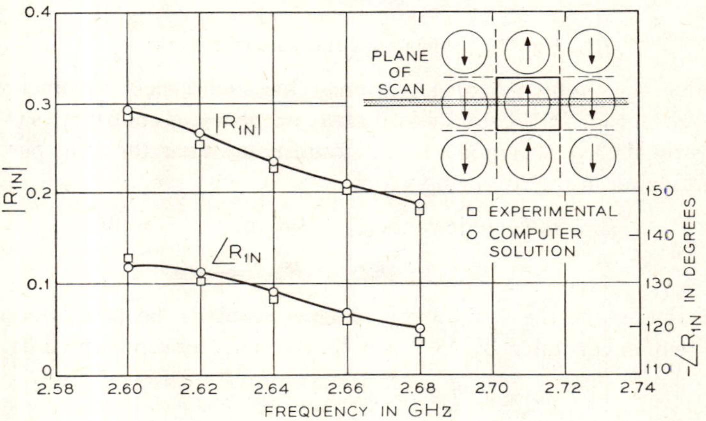

Fig. 7.5. Rectangular grid array: computed and experimental results vs. frequency at H-plane symmetry point ($a = 3.57$ cm, $b = 16.51$ cm, $d = 8.261$ cm, and $\alpha = 90°$).

Fig. 7.6. Rectangular grid array: simulator for H-plane symmetry point.

space and the total incident power is reflected and divided between the two propagating modes, $\mathbf{\Phi}_{1N}$ and $\mathbf{\Phi}_{2N}$. Of special interest is the point $\psi_r = 180° \times \sqrt{2} \simeq 255°$. At this point $\psi_x = \psi_y = 180°$, and the array excitation is as indicated in the inset of Fig. 7.4. This array excitation lends itself to a simple waveguide simulation, using a square waveguide (solid lines) as shown in the inset. It is clear that the horizontal waveguide mode ($\mathbf{\Phi}_{2N}$) cannot be excited. Indeed, the numerical results show that $|R_{2N}| = 0$ at this scan angle.

Figure 7.5 shows a close agreement between experimental and numerical results for a rectangular grid array with vertical polarization excitation scanned in the H plane. The scan angle which corresponds to steering phases $\psi_x = 180°$ and $\psi_y = 0$ can be simulated by the rectangular waveguide (solid line) in the inset. The experimental results were obtained from measurements of an abrupt junction between a circular and a rectangular waveguide, as shown in Fig. 7.6.

2. DIELECTRIC-FREE TRIANGULAR (EQUILATERAL TRIANGLE) GRID ARRAYS

Let us consider the reflection and radiation characteristics of circular waveguide arrays arranged in an equilateral triangle (or hexagonal) grid. This grid is used very widely in practice, since it minimizes the number of elements needed in a given array which is scanned over a circular scan region.[5] Grating lobe incipience or a beam at grazing is designated by a vertical arrow in the illustrations.

Figure 7.7 shows the reflection coefficient of an array in the *E*-plane scan, with vertical polarization excitation. In this plane of scan the horizontal

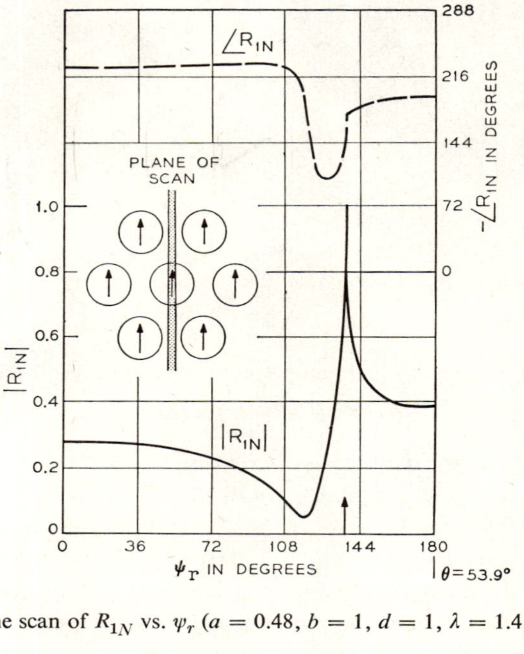

Fig. 7.7. *E*-plane scan of R_{1N} vs. ψ_r ($a = 0.48$, $b = 1$, $d = 1$, $\lambda = 1.4$, and $\alpha = 60°$).

mode is not excited because of symmetry, so that $R_{2N} \equiv 0$. As can be seen, the slope of both the magnitude and the phase of R_{1N} is discontinuous and in fact is singular at a grating lobe incipience. This is consistent with our previous observations in parallel plates and rectangular waveguide arrays. This behavior of R is related to the asymptotic behavior of the mutual coupling coefficients. An analysis of this behavior which extends the linear array case (see Chapter 4) will be discussed later.

A forced surface wave resonance can be seen in the vicinity of grating lobe incipience, where $|R_{1N}| = 1.0$, in Fig. 7.7. Notice that this forced surface wave resonance is extremely sharp and consequently may not be observed in experimental investigations of small finite arrays.

The corresponding transmission coefficient, T_θ,* is shown in Fig. 7.8. The plot of the transmission coefficient is actually the radiation pattern of a

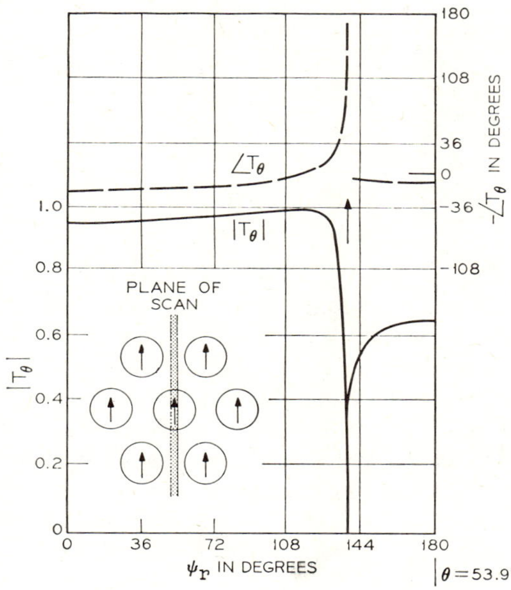

Fig. 7.8. *E*-plane scan of T_θ vs. ψ_r ($a = 0.48$, $b = 1$, $d = 1$, $\lambda = 1.4$, and $\alpha = 60°$).

single element excited in a passive array environment. The radiation pattern exhibits the null which corresponds to a total reflection. Notice that the phase of the transmission coefficient will exhibit a 180° discontinuity when the magnitude has a zero.

The magnitudes of the reflection coefficients of the same array, in a 60° plane of scan, are shown in Fig. 7.9. Again, the singular slope at grating lobe incipience can be observed. As can be seen, the distribution of the reflected power between the two modes is a function of the scan angle.

* The quantities T_θ and T_ϕ are the transmission coefficients of the array scattering matrix and are proportional to the θ and ϕ components of the far field (see Chapter 2).

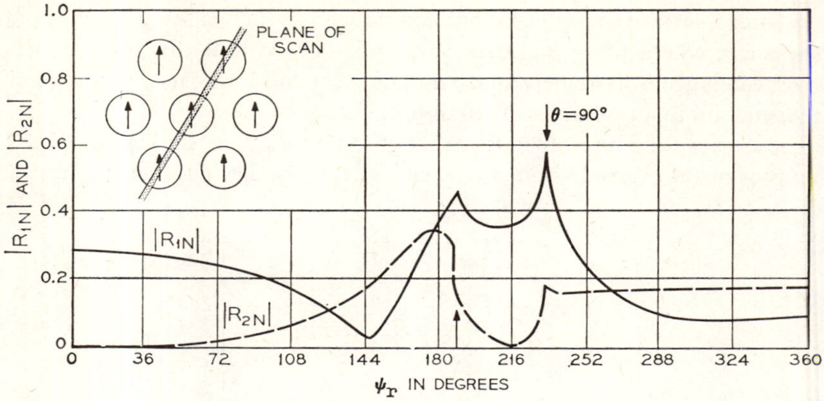

Fig. 7.9. $|R_{1N}|$ and $|R_{2N}|$ vs. ψ_r in the 60° plane of scan ($a = 0.48$, $b = 1$, $d = 1$, $\lambda = 1.4$, and $\alpha = 60°$).

An interesting phenomenon can be observed in the corresponding transmission characteristics, which are shown in Fig. 7.10. The ϕ transmission coefficient, T_ϕ, vanishes before grating lobe incipience (the position of the vertical arrow). The vanishing of T_ϕ at this scan angle can be directly related, when considered with the vector symmetries in the array excitation and geometry, to a forced surface wave resonance in another plane of scan. This can be demonstrated as follows. Let the array excitation consist of the sum of the two TE_{11} modes (equal in phase and magnitude) denoted by the solid arrows in Fig. 7.11. These arrows show the polarization of the electric field. In the 60° scan plane $T_\theta \equiv 0$ from symmetry considerations, and T_ϕ vanishes

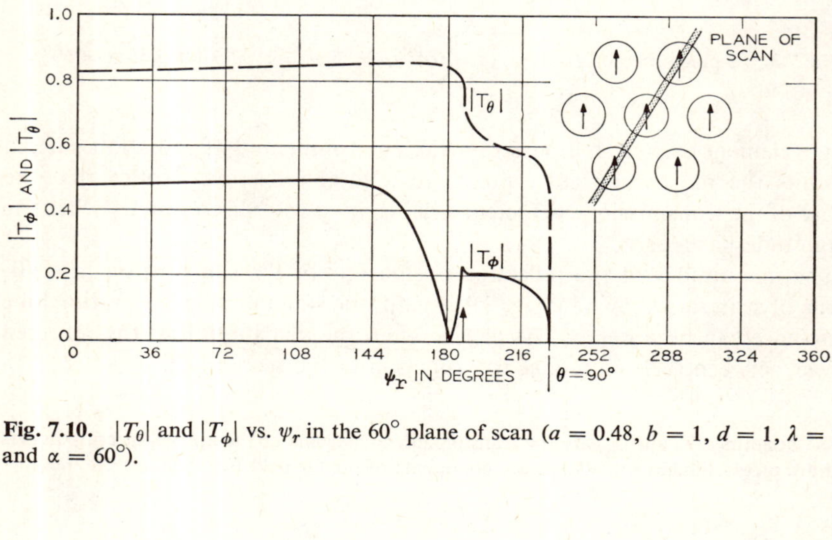

Fig. 7.10. $|T_\theta|$ and $|T_\phi|$ vs. ψ_r in the 60° plane of scan ($a = 0.48$, $b = 1$, $d = 1$, $\lambda = 1.4$, and $\alpha = 60°$).

as shown in Fig. 7.10. Since the vectorial sum of the two TE_{11} modes (solid arrows in Fig. 7.11) is the TE_{11} mode indicated by the dashed arrow, zero transmission or a forced surface wave resonance will occur in the H plane of scan of this polarization. Figures 7.12 and 7.13 indeed show this effect in the H-plane scan, where $|R_{1N}|$ and $|T_\phi|$ attain unity and zero, respectively, before grating lobe incipience.*

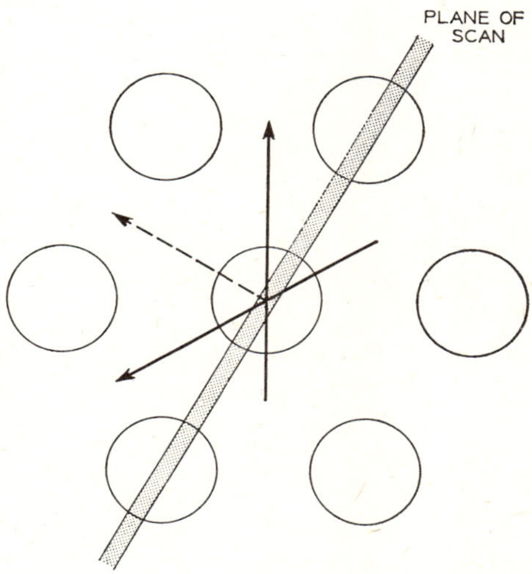

Fig. 7.11. Vector symmetry relationships between transmission zeros and forced surface waves.

Since the forced surface waves are related to the vector symmetries just mentioned, one may anticipate that the scan points at which they occur are isolated. Figure 7.14 demonstrates that indeed the forced resonances in the E and H planes occur at isolated points. The scan around the grating lobe circles in Fig. 7.14 shows that the peak of the total reflection, defined as $|R_T| = (|R_{1N}|^2 + |R_{2N}|^2)^{1/2}$, varies from unity in the H plane ($\delta = 0$), gradually decreases, and then increases again and reaches unity in the E plane ($\delta = 90°$). The anomalous resonance behavior near grating lobe incipience is eliminated when the polarization of the excitation is changed to horizontal, as shown in Fig. 7.15. The implication is that the occurrence of forced surface wave resonances depends strongly on the mode of excitation of the array.

* The difference in the values of ψ_r at which these phenomena occur is inherent in the definition of ψ_r: (12).

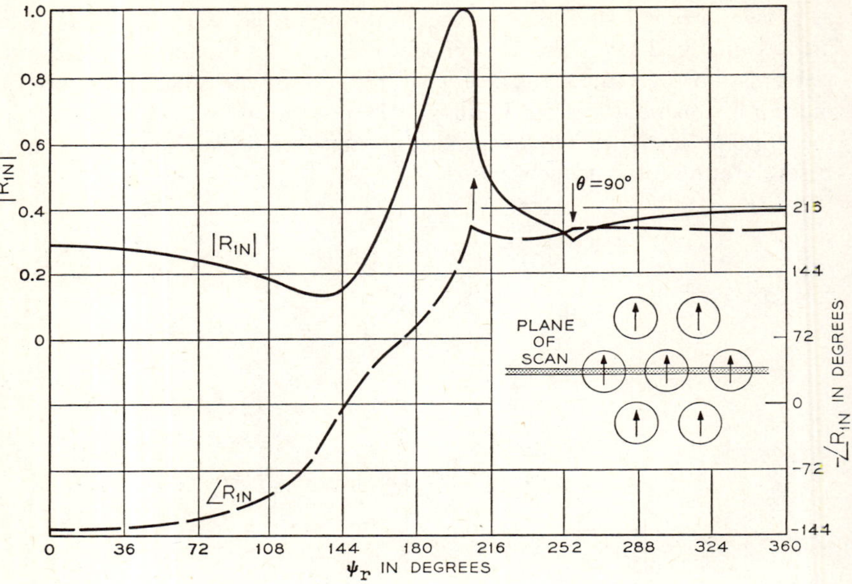

Fig. 7.12. *H*-plane scan of R_{1N} vs. ψ_r ($a = 0.48$, $b = 1$, $d = 1$, $\lambda = 1.4$, and $\alpha = 60°$).

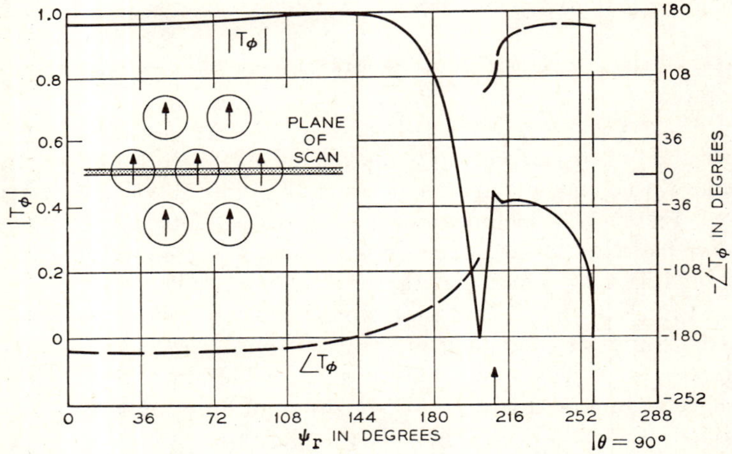

Fig. 7.13. *H*-plane scan of T_ϕ vs. ψ_r ($a = 0.48$, $b = 1$, $d = 1$, $\lambda = 1.4$, and $\alpha = 60°$).

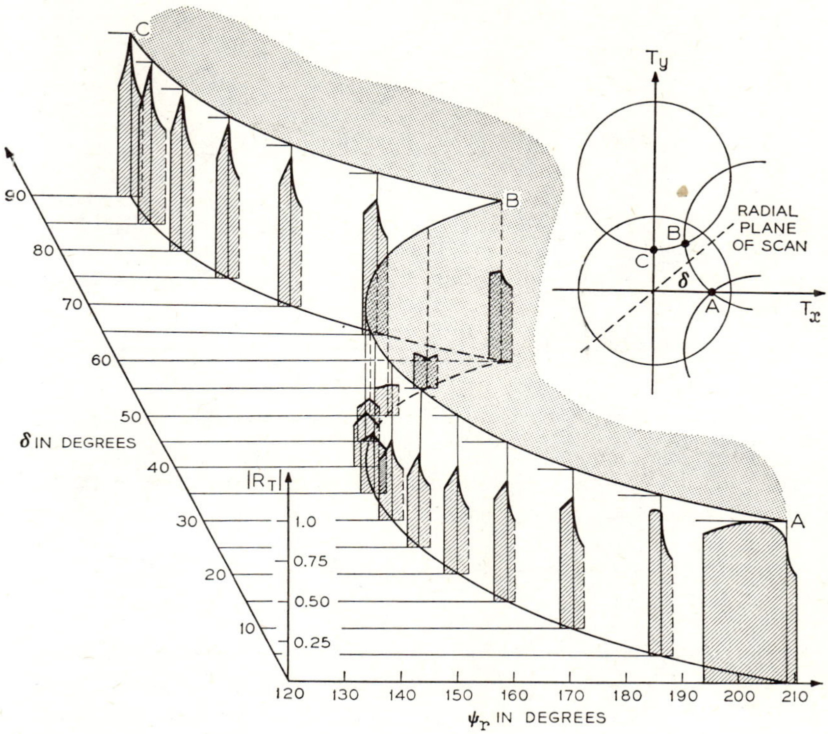

Fig. 7.14. $|R_T|$ vs. ψ_r in the vicinity of grating lobe incipience.

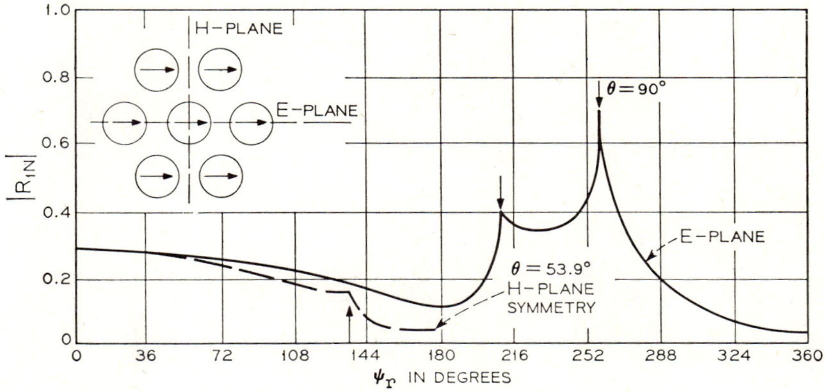

Fig. 7.15. *E*- and *H*-plane scans of $|R_{1N}|$ vs. ψ_r for horizontal polarization ($a = 0.48$, $b = 1$, $d = 1$, $\lambda = 1.4$, and $\alpha = 60°$).

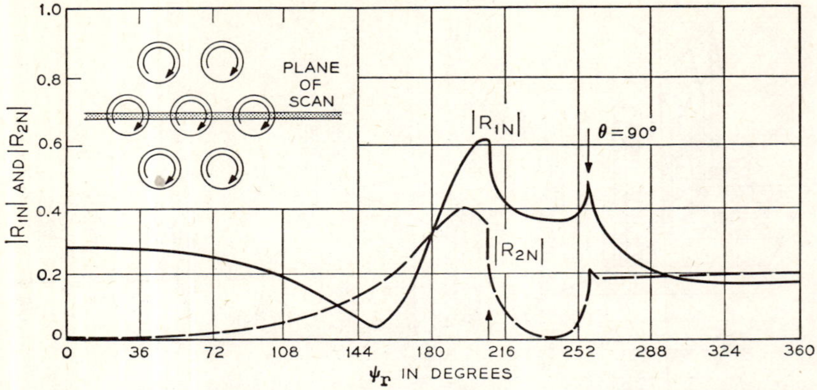

Fig. 7.16. $|R_{1N}|$ and $|R_{2N}|$ vs. ψ_r in the 0° plane of scan. Circularly polarized excitation ($a = 0.48$, $b = 1$, $d = 1$, $\lambda = 1.4$, and $\alpha = 60°$).

Figure 7.16 shows the reflection characteristics of the array with circular polarization excitation. In this case

$$\mathbf{\Phi}_{1N} = -\frac{1}{\sqrt{2}}\{\mathbf{\Phi}_1 - j\mathbf{\Phi}_2\}, \qquad \mathbf{\Phi}_{2N} = -\frac{1}{\sqrt{2}}\{\mathbf{\Phi}_1 + j\mathbf{\Phi}_2\} \tag{16}$$

The incident circularly polarized mode is $\mathbf{\Phi}_{1N}$.

Again the singularity in the slope of these curves at grating lobe incipience can be seen. This behavior does not change with polarization, since the grating lobes are independent of the polarization. The division of the reflected power between the two modes, $\mathbf{\Phi}_{1N}$ and $\mathbf{\Phi}_{2N}$, as a function of scan may be observed as well. It is assumed of course that $\mathbf{\Phi}_{2N}$ "sees" a perfect match,

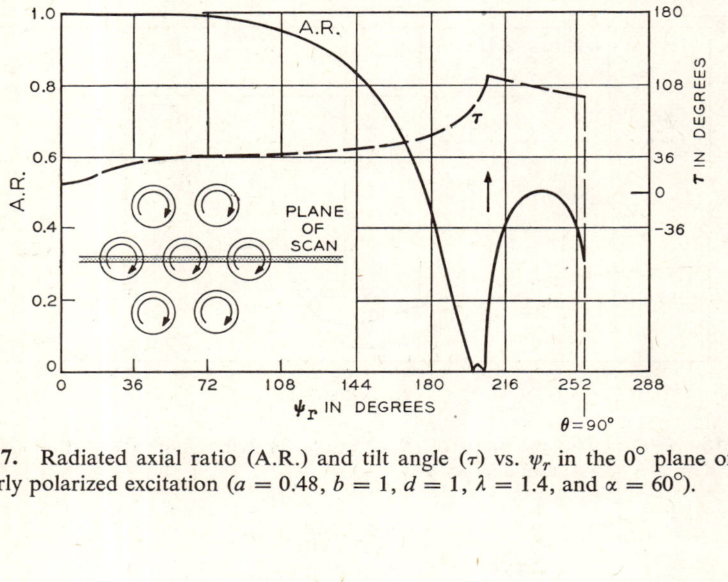

Fig. 7.17. Radiated axial ratio (A.R.) and tilt angle (τ) vs. ψ_r in the 0° plane of scan. Circularly polarized excitation ($a = 0.48$, $b = 1$, $d = 1$, $\lambda = 1.4$, and $\alpha = 60°$).

looking into the waveguides. Figure 7.17 shows the polarization characteristics of the radiation pattern of an array element. The axial ratio, denoted as A.R., is the ratio of the minor to the major axis of the polarization ellipse, while the tilt angle of the major axis, τ, is taken with respect to the ϕ axis. As can be expected from symmetry considerations and as indicated by the plot, the element far-field pattern is circularly polarized around broadside, A.R. = 1.0. It deteriorates to linear polarization, A.R. = 0, at two points before grating lobe incipience. The linear polarization at $\psi_r = 203.5°$ (with $\tau = 90°$) can be related to the H-plane forced surface wave resonance of Figs. 7.12 and 7.13, where T_ϕ vanishes. The null of the axial ratio at $\psi_r = 207°$ (with $\tau = 105°$) is a result of the fact that the difference between the phases of T_θ and T_ϕ is zero (or 180°), so that the ϕ and θ components of the far field are in phase.

A single null of the axial ratio was observed in the 30° plane of scan of this array at grating lobe incipience. It can be related to the E-plane surface wave resonance shown in Figs. 7.7 and 7.8. For radial planes of scan between 0° and 30° the results corresponding to one plane change gradually to those of the other plane. Around grating lobe incipience the axial ratio drops appreciably (to about 0.1) but does not reach zero. Sharp variations of the axial ratio near grating lobe incipience have been observed in other types of planar arrays as well. From the circular symmetry of excitation (circular polarization) and the sixfold symmetry of the array geometry, a 30° sector of scan completely specifies the reflection and radiation characteristics of the array. Over the frequency band given by $1.3 \leq \lambda/b \leq 1.5$ $(b = d)$, qualitatively similar behavior of the radiation and reflection characteristics of the array were observed.

In the brick array of rectangular waveguides described in Chapter 5, surface wave resonances were observed in the H plane. For an identical grid array of circular waveguides excited as shown in Fig. 7.18, we note that a forced resonance is present in the E plane of scan before the first grating lobe. As can be seen, the occurrence of these resonances depends strongly on the shape of the element as well as the array grid. Figure 7.19 demonstrates the point isolation characteristic of the forced surface wave resonances in the two-dimensional scan plane for the 45° grid array. Furthermore, we shall show presently that the point isolation of these forced resonances is preserved even when the TE_{11} waveguide mode, which is orthogonal to the exciting TE_{11} mode, is terminated in a short circuit.* The formulation of the integral equation for the short-circuited orthogonal mode is accomplished by merely changing the effective modal admittances of that mode in (5) and (6). If, for example, $\boldsymbol{\Phi}_2$ (with a propagation constant γ_2) is shorted at a distance l

* In shorting the orthogonal mode we modify the array element from a double- to a single-port device.

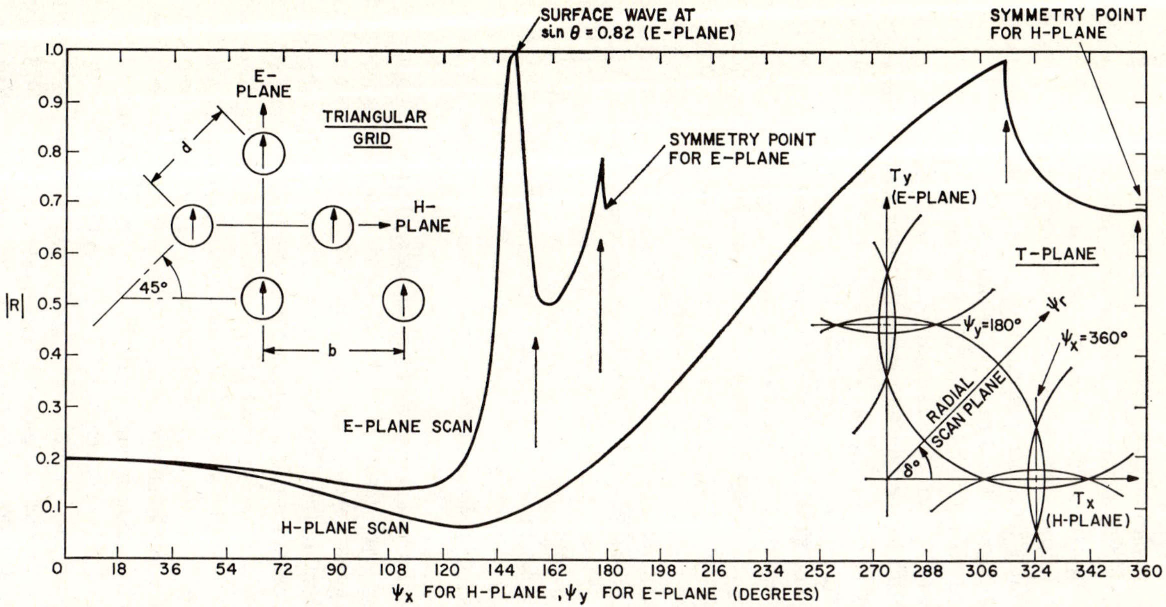

Fig. 7.18. Reflection characteristics in principal planes ($a = 0.356\lambda$, $b = 1.008\lambda$, $d = 0.713\lambda$, and $\alpha = 45°$).

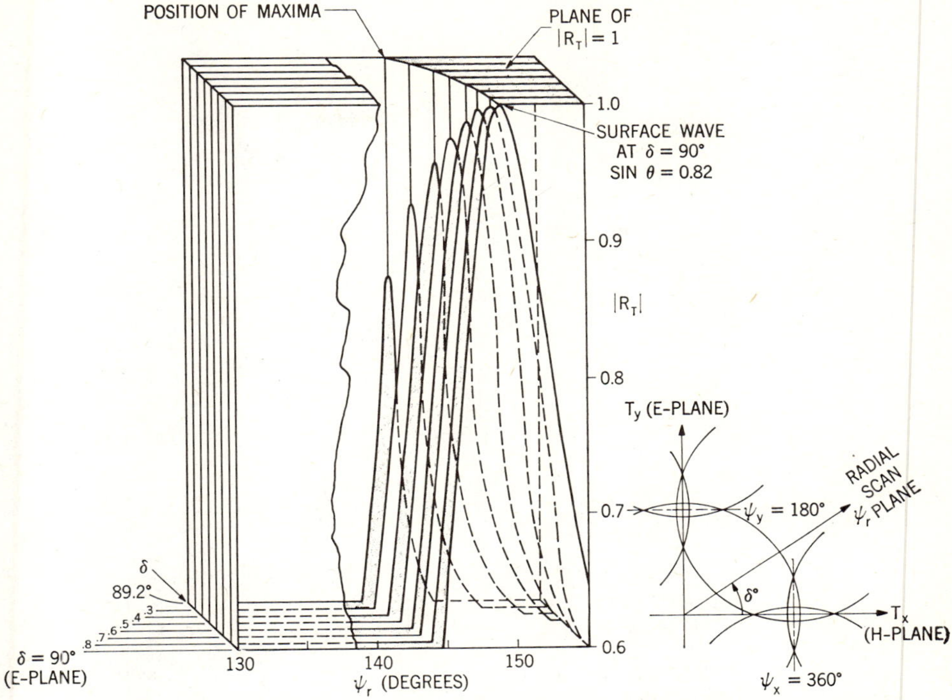

Fig. 7.19. Contour plots of forced surface wave resonances ($a = 0.356\lambda$, $b = 1.008\lambda$, $d = 0.713\lambda$, and $\alpha = 45°$).

away from the aperture, we set

$$A_2 = 0 \tag{17}$$

replace y_2 in (6) by

$$y_2^s = -jy_2 \cot \gamma_2 l \tag{18}$$

and solve the corresponding equations by the procedures described earlier.

Numerical solutions for a single-port circular waveguide array are shown in Fig. 7.20. The point isolation of the forced resonance in the T_x–T_y plane can be clearly seen.

A functional similarity between this single-port circular waveguide array and a single-port rectangular waveguide array (where only the dominant TE_{10} mode can propagate in the rectangular waveguide) can be observed. In both cases y_1 in (3) or (6) is real, whereas y_2 (or y_2^s) is purely imaginary. In (18) we can adjust y_2^s to any value between $-j\infty$ to $j\infty$ by a proper choice of the short-circuit position, l. Therefore we may anticipate that the angle of total reflection (due to the forced resonance) will be isolated in the T_x–T_y plane for the case of the rectangular waveguide phased array as well. Moreover, it may be conjectured that the isolation of the total reflection angle

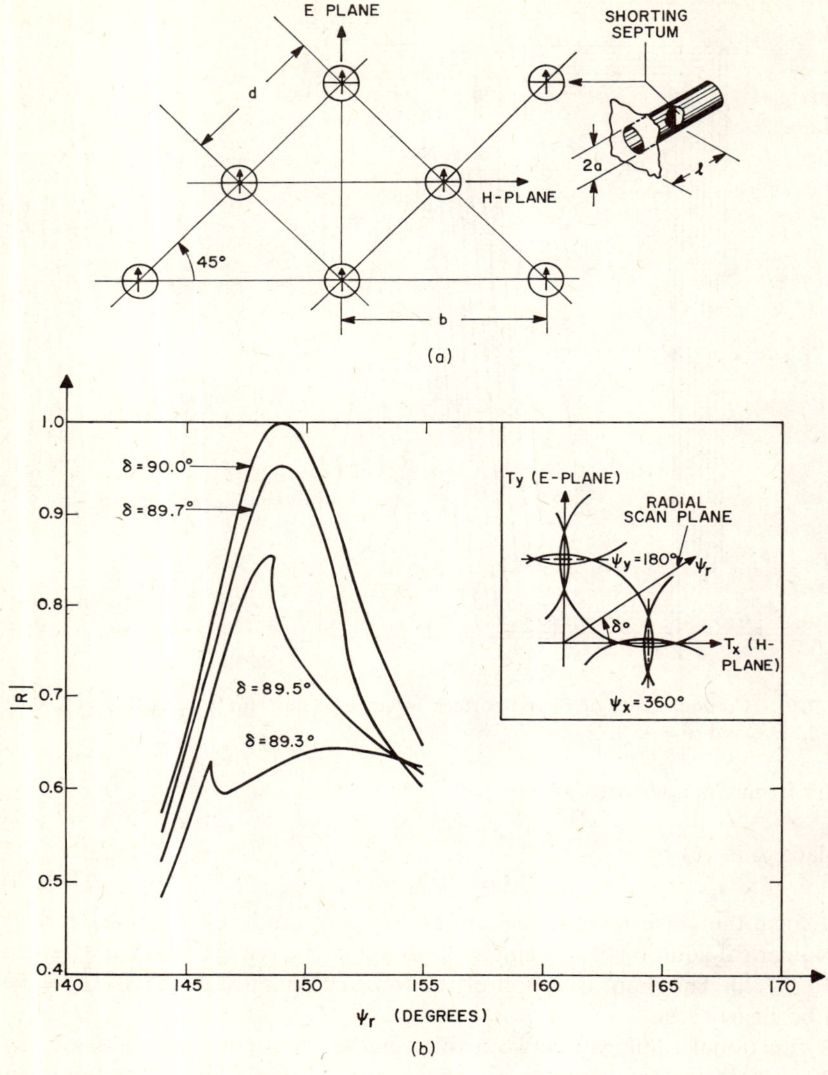

Fig. 7.20. Single-port circular waveguide array. (*a*) Array geometry and excitation. (*b*) Reflection characteristics about the *E*-plane forced resonance [$a = 0.356\lambda$, $b = 1.008\lambda$, $d = 0.713\lambda$, and $l = \frac{5}{8}\lambda_g$ (guide wavelength)].

due to the forced resonance is a general feature of phased arrays, irrespective of the element shape or the number of input ports.

So far, we have observed that total reflections (due to forced surface wave resonances) have occurred when the planes of scan and the symmetry planes of the exciting mode of the element coincided with the symmetry planes of the array grid.* As a result, one may be tempted to suggest that the removal of the above symmetries (e.g., by constructing a periodic grid without plane symmetries) will result in the elimination of the total reflections. A periodic but not plane-symmetric grid of circular waveguides is shown in Fig. 7.21*a*. Both the exciting (horizontal) and the orthogonal (vertical) TE_{11} modes can propagate. The grid was obtained by varying d and α of an equilateral triangular grid while maintaining $d \sin \alpha = 0.5$. As can be seen, the vertical TE_{11} mode can be excited in every plane of scan. In the case of the undeformed equilateral triangular grid array with a similar excitation, isolated forced resonances were observed in both the E and the H planes of scan, while in other planes no such resonances existed (see Figs. 7.7, 7.9, 7.12, and 7.14). The total reflection, $|R_T|$ (of both TE_{11} modes), for the case of the deformed grid is shown in Fig. 7.21*b*. The points of total reflection no longer exist in the E and H planes. Nonetheless, a point of almost total reflection ($|R_T| > 0.99$) appears in the 29.5° plane of scan. The orthogonal (vertical) mode reflection did not identically vanish at this angle of scan. This angle of total reflection was found to be isolated as well.

3. DIELECTRIC-LOADED AND DIELECTRIC-COVERED ARRAYS

The wide use of dielectric materials for the hardening and matching of phased arrays necessitates a more complete understanding of the effects of these materials on the array performance. In Chapter 6, the effects of dielectric materials on the characteristics of rectangular and parallel-plate waveguide arrays were discussed in detail. Many of the results are applicable, at least qualitatively, to other types of arrays.

The results to be discussed in this section were obtained for an equilateral triangular grid array of circular waveguides. Earlier in this book, it was shown that forced surface waves can occur in these and other types of arrays without the presence of dielectric materials loading or covering the array. The effects of dielectric materials on these forced surface waves, as well as

* For example, the planes of symmetry of an equilateral triangular grid array of circular waveguides are the x and y axes, as well as all other planes that are slanted at an integer multiple of 30° with respect to the x axis.

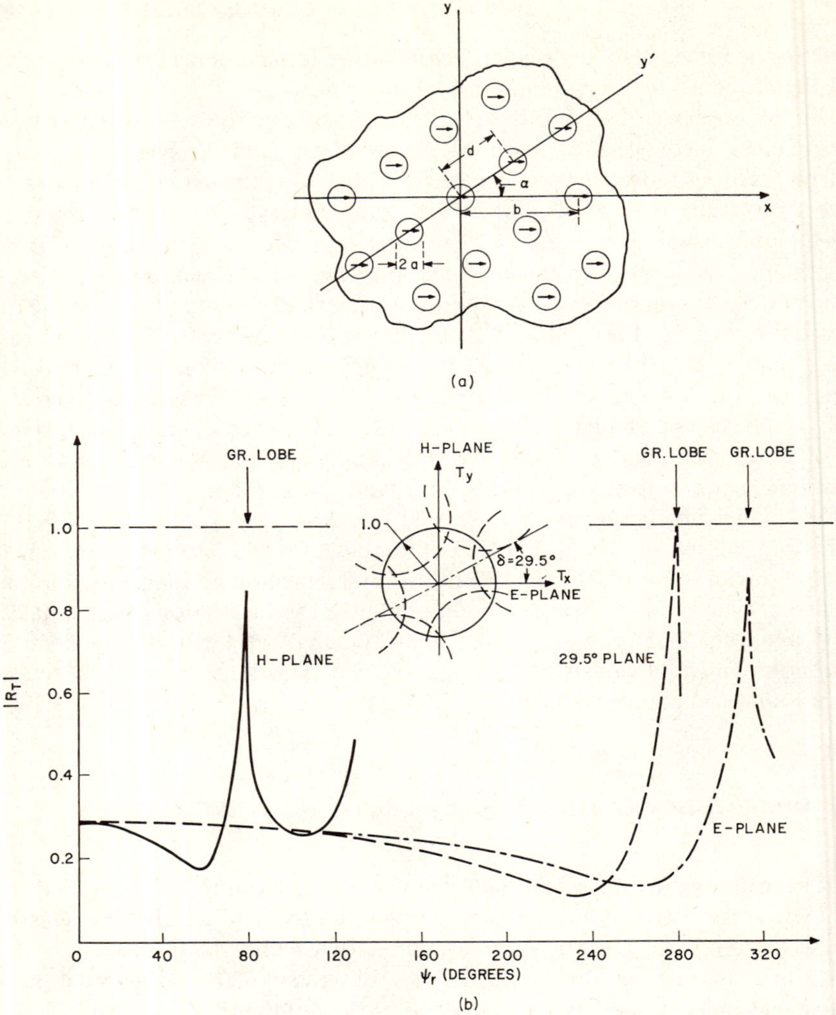

Fig. 7.21. Nonplane symmetric array of circular waveguides. (*a*) Array geometry and excitation. (*b*) Total reflection characteristics in the E, H, and 29.5° planes ($\lambda = 1.4$, $a = 0.43$, $b = 1.732$, $d = 0.894$, and $\alpha = 34°$).

the appearance of forced surface waves that are bound or "trapped"* in the dielectric materials, in either sheath covers or plugs, are discussed here. The occurrence of trapped forced surface waves is contingent upon the presence of dielectric sheath covers or plugs. In contrast, nontrapped forced surface waves are those which occur without the presence of dielectric materials; the introduction of such materials merely shifts the scan angle of their appearance.

The various types of trapped forced surface waves were found to occur at isolated angles in the T_x–T_y scan plane as was true also of dielectric-free arrays.

3.1 Fully Loaded Arrays

In the fully loaded array, the waveguide elements are completely filled with dielectric material up to the aperture at $z = 0$. Let us fully load with dielectric material one of the circular waveguide arrays previously discussed in such a way that only the dominant TE_{11} mode can propagate in the waveguide. The most noticeable change in the array reflection characteristics is that the scan angle of the nontrapped forced surface wave will shift with loading toward broadside. A similar behavior was observed in the case of fully loaded arrays of rectangular waveguides.[6]

In the observation to be described, the waveguide radius, a, was varied simultaneously with the dielectric constant, ε_1, of the loading material so that $\varepsilon_1 a^2 =$ constant. This relationship ensures a fixed ratio between the waveguide operating and cut-off frequencies. The grid dimensions were varied, with the wavelength kept fixed, so that $d/a = b/a =$ constant, in addition to $\varepsilon_1 a^2 =$ constant. Under these conditions, the array is more closely packed with increasing ε_1. The aperture filling factor $(\pi a^2/bd \sin \alpha)$, the ratio between the area of a waveguide aperture and the area of a unit cell, is kept constant. Figure 7.22 shows a typical reflection characteristic of such a fully dielectric-loaded array in the H-plane scan. As a result of the closer element spacing with increasing ε_1, we find that total reflection occurs over continuous ranges of scan angle. The isolated forced surface wave that existed in the absence of dielectric loading ($\varepsilon_1 = 1.0$) merges with the continuous regions of total reflections and does not reappear with increasing values of ε_1.

In the next observation ε_1 and a were varied so that $\varepsilon_1 a^2 =$ constant but the grid dimensions were maintained fixed relative to the free space wavelength, λ ($b/\lambda = d/\lambda =$ constant). In this case the aperture filling factor

* The use of the term trapped forced surface waves is due to the fact that their field distribution resembles that of trapped modes in dielectric slabs.

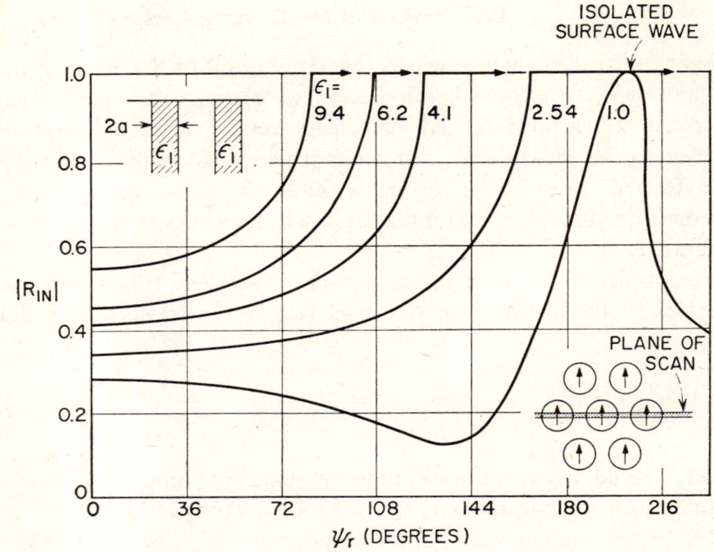

Fig. 7.22. Fully loaded array, H-plane scan ($\varepsilon_1 a^2$ = constant = $0.118\lambda^2$, $b/a = d/a$ = constant = 2.08, and $\alpha = 60°$).

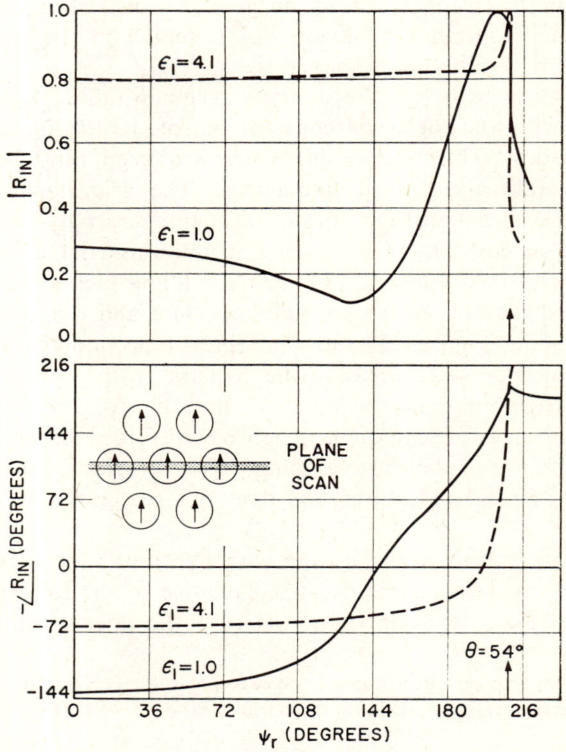

Fig. 7.23. Fully loaded array, H-plane scan ($\varepsilon_1 a^2$ = constant = $0.118\lambda^2$, $b = d = 0.714\lambda$, $a = 0.169\lambda$, $\varepsilon_1 = 4.1$ and $\varepsilon_1 = 1.0$, and $\alpha = 60°$).

decreases with increasing ε_1. Figure 7.23 shows the reflection characteristics in the H-plane scan of such a fully loaded array. As can be seen, the magnitude of R_{1N} increases greatly in real space with increasing dielectric constant, in contrast to the relatively small changes shown in Fig. 7.22. Note also that R_{1N} becomes very flat as a function of ψ_r. The flat reflection characteristics (in both *phase* and amplitude) were observed in other planes of scan as well. The only sharp variations noticed were localized around grating lobe incipience (the vertical arrows in Figs. 7.23 and 7.24). Note that the reflection

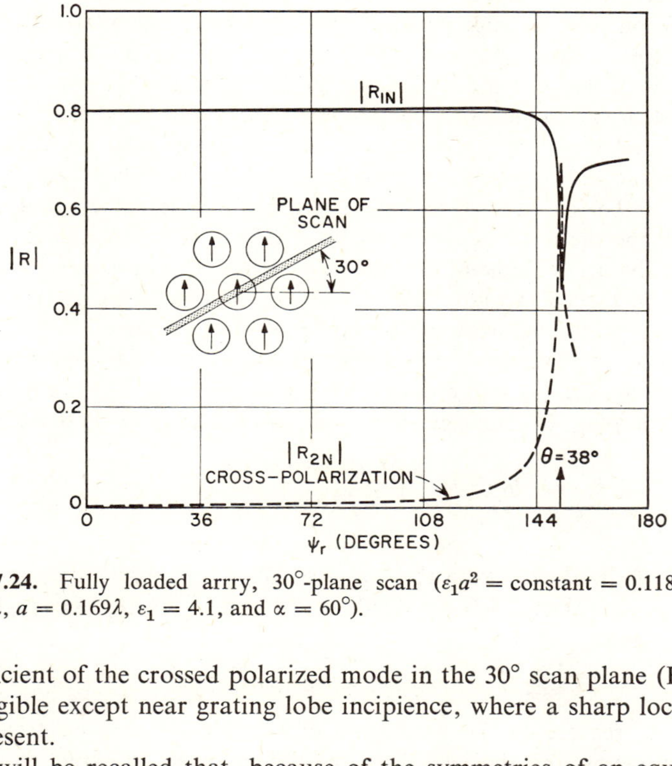

Fig. 7.24. Fully loaded arrry, 30°-plane scan ($\varepsilon_1 a^2$ = constant = $0.118\lambda^2$, $b = d = 0.714\lambda$, $a = 0.169\lambda$, $\varepsilon_1 = 4.1$, and $\alpha = 60°$).

coefficient of the crossed polarized mode in the 30° scan plane (Fig. 7.24) is negligible except near grating lobe incipience, where a sharp localized peak is present.

It will be recalled that, because of the symmetries of an equilateral triangular grid array of circular waveguides, the reflection coefficient at broadside is independent of the polarization of the exciting TE_{11} modes. This fact, together with the flat reflection characteristics of Figs. 7.23 and 7.24, allows us, in principle, to realize a good match over a wide scan region and at least over a narrow frequency band. The transmission coefficients exhibited similarly flat characteristics in both phase and amplitude. This feature, in turn, results in a radiation pattern with relatively constant polarization (e.g., circular) over the same region of scan. The frequency band over which an acceptable performance of an array composed of such relatively small elements can be achieved is narrow. Nevertheless, with a judicious synthesis of a matching network to be inserted behind the aperture one might obtain a good match characteristic as a function of scan over a 10–15% bandwidth. This subject is discussed further in Chapter 9.

3.2 Dielectric-Plugged Phased Arrays

The introduction of dielectric plugs into the waveguide elements of an array for the purpose of improving the match, hardening, or weather protection will greatly affect the array performance.[7,8,9] The plugs are usually placed flush with the aperture. Higher-order modes which are below cut-off in the unloaded waveguides may be able to propagate in the dielectric plug. Since these modes cannot propagate beyond the plug, they appear to be "trapped" inside it. They can strongly influence the *existence* and *position* (with scan) of the forced surface waves. Different higher-order modes will be excited in the dielectric plugs in different planes of scan because of the symmetry of the array element and grid. Therefore, an improvement of the match that is achieved in one plane of scan by the use of a dielectric plug may, in fact, degrade the array performance in other planes.

As was shown in Chapter 6, only minor modifications of the integral and matrix equations, (3)–(6), are required in order to account for the presence of the dielectric plug. On the right-hand side of (3) and (6) y_i is replaced by

$$y_i \to y_i^\varepsilon \frac{y_i + jy_i^\varepsilon \tan \gamma_i^\varepsilon t_1}{y_i^\varepsilon + jy_i \tan \gamma_i^\varepsilon t_1} \tag{19}$$

where y_i^ε and γ_i^ε are the modal admittance and propagation constant, respectively, of the $\mathbf{\Phi}_i$ mode in the dielectric plug (with dielectric constant ε_1 and thickness t_1). In the left-hand side of (3) and (5) we replace $A_i y_i$ by

$$A_i y_i \to A_i \frac{y_i^\varepsilon y_i e^{j\gamma_i t_1}}{y_i^\varepsilon \cos \gamma_i^\varepsilon t_1 + jy_i \sin \gamma_i^\varepsilon t_1} \tag{20}$$

where A_i now represents the amplitude of the ith incident electric field waveguide mode in the dielectric-free region, ε_0, preceding the plug (γ_i being the corresponding propagation constant). The aperture plane is used as the phase reference. The reflection coefficients of the propagating modes in the ε_0 region are related to the aperture field by

$$R_i = \frac{y_i^\varepsilon \iint_A \mathbf{E}_t \cdot \mathbf{\Phi}_i \, da - A_i e^{j\gamma_i t_1}(y_i^\varepsilon \cos \gamma_i^\varepsilon t_1 - jy_i \sin \gamma_i^\varepsilon t_1)}{e^{-j\gamma_i t_1}(y_i^\varepsilon \cos \gamma_i^\varepsilon t_1 + jy_i \sin \gamma_i^\varepsilon t_1)} \tag{21}$$

the phase of R_i being referred to the aperture ($z = 0$) also. Incorporating (19) and (20) into (3) and (6) and then proceeding as before, we obtain a solution for the aperture field, $\mathbf{E}_t$. Equation (21) can then be used to obtain the reflection characteristics.

Figure 7.25 shows $|R_{1N}|$ as a function of scan for various plug thicknesses in the E plane of scan. With the dimensions as indicated and $\varepsilon_1 = 1.8$, the circular TM_{01} mode is "trapped" in the plug. By increasing the plug thickness,

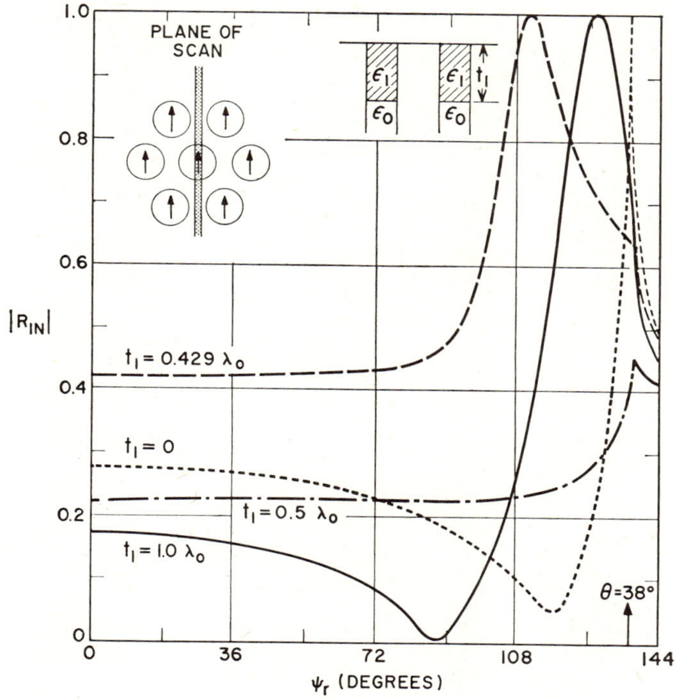

Fig. 7.25. Dielectric-plugged array, E-plane scan; one trapped mode in plug ($b = d = 0.714\lambda$, $a = 0.340\lambda$, $\varepsilon_1 = 1.8$, $\varepsilon = \varepsilon_0$, and $\alpha = 60°$).

t_1, from zero (no plug), we observe that the location of the surface wave resonance shifts toward broadside. For a critical range of plug thickness, we observe that there are no forced surface waves. They reappear, however, as the plug thickness is increased to still higher values. In other words, the dielectric plug exhibits an equivalent "bandpass" characteristic as a function of thickness. This bandpass characteristic can be explained by an equivalent transmission line circuit if we assume that the evanescent modes in the plug (generated at the aperture, $z = 0$) decay sufficiently at $z = -t_1$. We may thus use a transmission line approach to evaluate the forced surface wave angular location, ψ_{sw}, as a function of the thickness.

Figure 7.26*a* shows the equivalent lossless (and consequently unitary, with appropriate normalization) three-port scattering matrix, *S*. Port 3 corresponds to the trapped TM_{01} mode. The elements, $\{s_{ij}\}$, of this matrix can be obtained from the solution of the integral equation, (3), when the dielectric

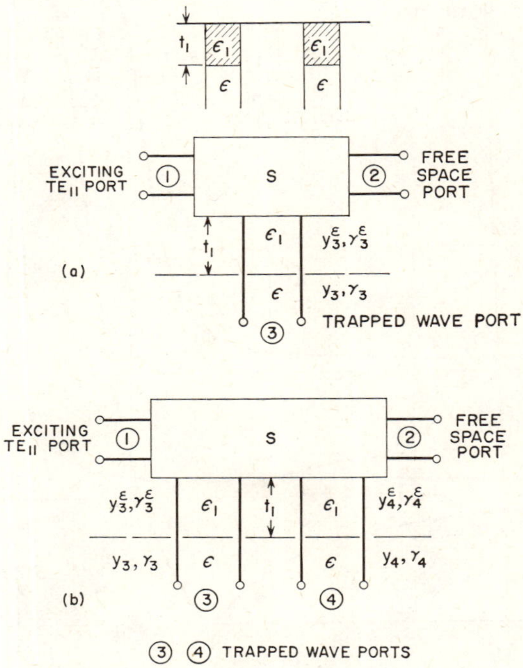

Fig. 7.26. Scattering representation of the array with trapped modes. (*a*) One trapped mode. (*b*) Two trapped modes.

material *totally loads the waveguide* and the various ports are excited individually. The exciting incident field is used as the left-hand side of (3). Since the kernel of the integral equation is independent of the excitation, the matrix inversion needs to be performed only once for *all* incident fields.

Since a forced surface wave is characterized by a zero transmission between ports 1 and 2, this condition is imposed on the scattering matrix equation (see Chapter 4). As a result, we obtain an equation that relates the forced surface wave scan location, ψ_{sw}, to the dielectric thickness, t_1:

$$\frac{y_3^{\varepsilon} - y_3}{y_3^{\varepsilon} + y_3} e^{-2j\gamma_3^{\varepsilon} t_1} = \frac{s_{12}}{s_{12}s_{33} - s_{23}s_{13}} \tag{22}$$

In (22), $\{s_{ij}\}$ are the known elements of the scattering matrix, S (which is a function of the scan angle), y_3^{ε} and γ_3^{ε} are the modal admittance and the propagation constant of the TM_{01} mode in the plug, and y_3 is the admittance of that mode in the air. It can easily be shown that the absolute value of either side of (22) is unity, since y_3^{ε} is real and y_3 is purely imaginary. Once

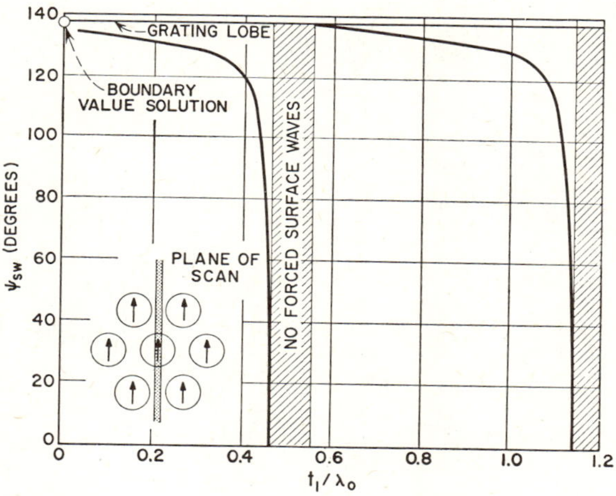

Fig. 7.27. Dielectric-plugged array, E-plane scan; surface wave scan location (ψ_{sw}) versus plug thickness; one trapped mode in plug ($b = d = 0.714\lambda$, $a = 0.340\lambda$, $\varepsilon_1 = 1.8$, $\varepsilon = \varepsilon_0$, and $\alpha = 60°$).

the boundary value problem for the totally loaded waveguide is solved at a given scan angle, we can determine from (22) the thickness, t_1, which will produce a forced surface wave at this angle.

Results for the three-port case, using (22), are shown in Fig. 7.27. Note that there are "bandpass" regions of thickness for which no forced surface wave is excited. These occur because it is possible that, as ψ_r varies from $0 \rightarrow \pi$, the phase of the right-hand side of (22), arg $[s_{12}/(s_{12}s_{33} - s_{23}s_{13})]$, lies within an interval that is smaller than 2π. Since y_3^{ε}, y_3, and γ_3^{ε} are independent of ψ_r, there can be certain ranges of t_1 for which (22) will not be satisfied. Except for very thin plugs, wherein the evanescent modes excited at the aperture cannot be neglected, ψ_{sw} is a periodic function of t_1 with a period of $\pi/\gamma_3^{\varepsilon}$ (half a wavelength of the TM_{01} mode). Figure 7.27 also shows a point obtained directly from the solution of the array boundary value problem. As can be seen, the transmission line approximation holds well even for relatively thin plugs.

The same transmission line approximation can be used when ε_1 is increased to a value such that two modes (the TM_{01} and TE_{21}) are trapped in the plug in the E plane of scan. Figure 7.28 shows $|R_{1N}|$ versus scan for various plug thicknesses in the E plane of scan for a value of $\varepsilon_1 = 2.54$. The element and grid dimensions are identical to the ones discussed previously. Again the

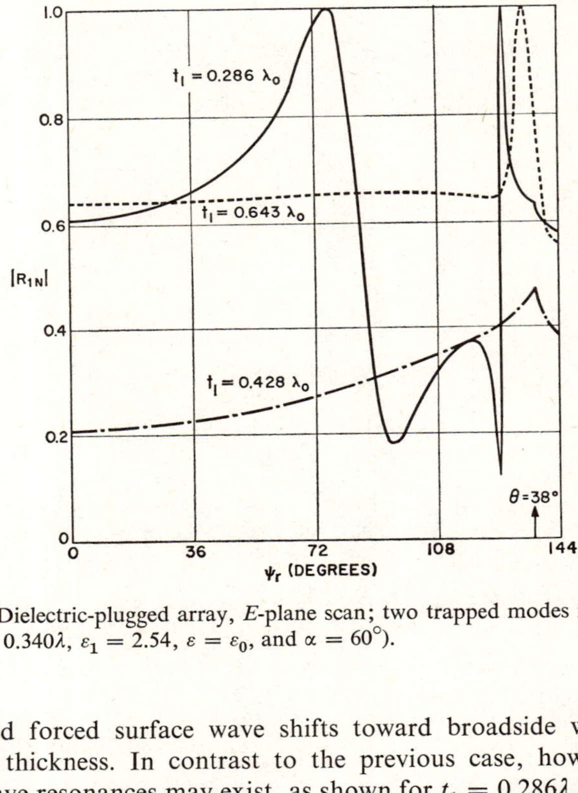

Fig. 7.28. Dielectric-plugged array, E-plane scan; two trapped modes in plug ($b = d = 0.714\lambda$, $a = 0.340\lambda$, $\varepsilon_1 = 2.54$, $\varepsilon = \varepsilon_0$, and $\alpha = 60°$).

nontrapped forced surface wave shifts toward broadside with increasing values of thickness. In contrast to the previous case, however, multiple surface wave resonances may exist, as shown for $t_1 = 0.286\lambda$. As t_1 increases in value, the surface wave resonance disappears and then reappears again for higher values of t_1. We observe, again, that certain "bandpass" regions of t_1 may exist for which no forced surface wave is excited at any scan angle. The equivalent network for a transmission line approximation of this case is shown in Fig. 7.26*b*, where an additional fourth port is added to account for the trapped TE_{21} mode. The equation that relates ψ_{sw} and t_1 is a more complicated transcendental form given by:

$$1 = G_{34}e^{-2j(\gamma_3^\varepsilon+\gamma_4^\varepsilon)t_1} + G_3e^{-2j\gamma_3^\varepsilon t_1} + G_4e^{-2j\gamma_4^\varepsilon t_1} \tag{23}$$

where

$$G_{34} = \frac{s_{12}^*}{s_{12}} \det [S] \cdot \frac{y_3^\varepsilon - y_3}{y_3^\varepsilon + y_3} \cdot \frac{y_4^\varepsilon - y_4}{y_4^\varepsilon + y_4} \tag{24}$$

$$G_3 = \frac{s_{12}s_{33} - s_{13}s_{23}}{s_{12}} \cdot \frac{y_3^\varepsilon - y_3}{y_3^\varepsilon + y_3} \tag{25}$$

and

$$G_4 = \frac{s_{12}s_{44} - s_{14}s_{24}}{s_{12}} \cdot \frac{y_4^\varepsilon - y_4}{y_4^\varepsilon + y_4} \tag{26}$$

Note that y_3^ε, y_3, y_4^ε, y_4, γ_3^ε, and γ_4^ε are the appropriate modal admittances and propagation constants, as indicated in Fig. 7.26*b*. As before, the elements of the scattering matrix are obtained from a solution of (3), with appropriate incident fields chosen. Because the ratio of γ_3^ε to γ_4^ε is, in general, not a rational number, ψ_{sw} is not necessarily periodic with t_1 [see (23)]. In Fig. 7.29 we observe the aperiodicity of ψ_{sw} as a function of t_1. We note, again, the existence of "bandpass" regions, as well as the excellent accuracy of the transmission line approximation for very thin plugs.

3.3 Dielectric-Covered Arrays

The presence of a dielectric sheath or cover (of thickness t_1' and dielectric constant ε_1') over the array, flush with the aperture, can also strongly influence the array performance.[7,8,10] Trapped modes (of the periodic free space type: Ψ_{pmn}) can exist in the dielectric sheath* if ε_1' is sufficiently high. The influence of these trapped modes on the array performance differs markedly from the effects of the trapped waveguide modes in the plugs. The surface wave resonances due to the sheath do *not* exhibit the "bandpass" characteristics that occur in dielectric-plugged arrays. With increasing values of the sheath thickness, t_1', or dielectric constant, ε_1', the angle, ψ_{sw}, at which a dielectric-free surface wave resonance occurs shifts toward broadside but does not vanish. In fact, additional or multiple surface wave resonances appear as either t_1' or ε_1' is increased. As a result, the usable scan region becomes seriously limited for higher values of t_1' or ε_1'.

* The trapped modes in the sheath correspond to grating lobes which are due to an effective element spacing in the dielectric region that is greater than the apparent spacing in free space.

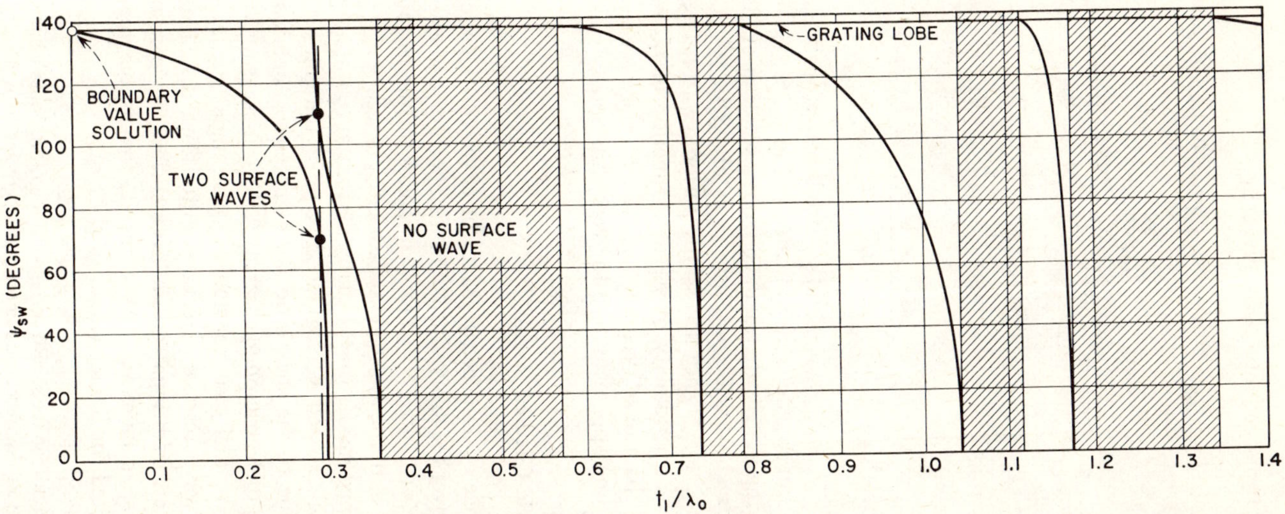

Fig. 7.29. Dielectric-plugged array, *E*-plane scan; surface wave scan location (ψ_{sw}) versus plug thickness; two trapped modes in plug ($b = d = 0.714\lambda$, $a = 0.340\lambda$, $\varepsilon_1 = 2.54$, and $\alpha = 60°$).

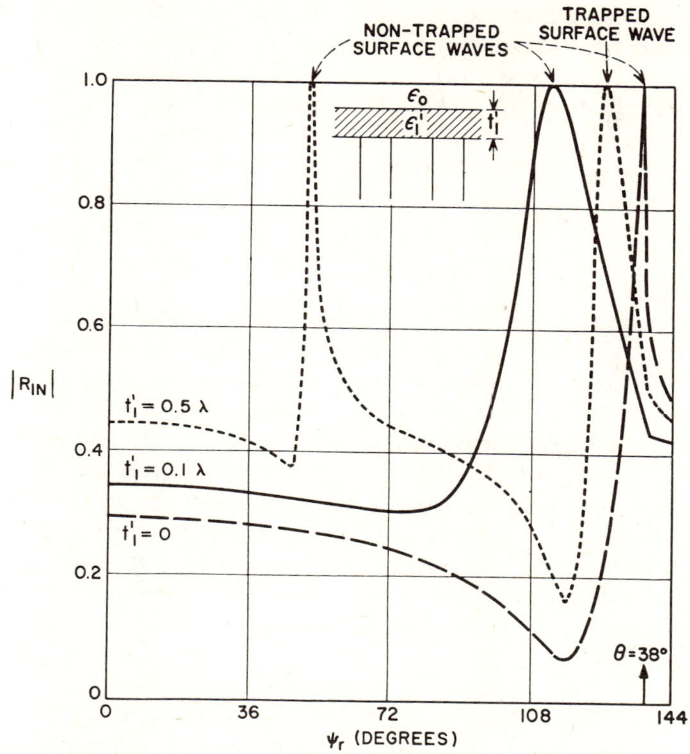

Fig. 7.30. Dielectric-covered array, E-plane scan; one trapped mode in sheath ($b = d = 0.714\lambda$, $a = 0.340\lambda$, $\varepsilon_1' = 2.0$, and $\alpha = 60°$).

The appropriate modification of the integral and matrix equations, (3) and (6), in this case is to replace Y_{pmn} by

$$Y_{pmn} \rightarrow Y_{pmn}^{\varepsilon} \frac{Y_{pmn} + jY_{pmn}^{\varepsilon} \tan \Gamma_{mn}^{\varepsilon} t_1'}{Y_{pmn}^{\varepsilon} + jY_{pmn} \tan \Gamma_{mn}^{\varepsilon} t_1'} \tag{27}$$

where Y_{pmn}^{ε} and $\Gamma_{mn}^{\varepsilon}$ are the modal admittance and the propagation constant, respectively, of the $\mathbf{\Psi}_{pmn}$ mode in the dielectric sheath. The coefficient of the $\mathbf{\Psi}_{pmn}$ mode in the dielectric-free region above the sheath, F_{pmn} (with propagation constant Γ_{mn}), is related to the aperture field, $\mathbf{E}_t$, by

$$F_{pmn} = \frac{Y_{pmn}^{\varepsilon} e^{j\Gamma_{mn} t_1'}}{Y_{pmn}^{\varepsilon} \cos \Gamma_{mn}^{\varepsilon} t_1' + jY_{pmn} \sin \Gamma_{mn}^{\varepsilon} t_1'} \iint_A \mathbf{E}_t \cdot \mathbf{\Psi}_{pmn}^{*} \, da \tag{28}$$

the phase of F_{pmn} being referred to the aperture ($z = 0$).

Figure 7.30 shows the reflection coefficient magnitude as a function of scan, ψ_r, for an E-plane scan of a dielectric-covered array with dielectric constant $\varepsilon_1' = 2.00$ and grid and element dimensions identical to those of the array considered for the plugs in Section 3.2. The effect of increasing the sheath thickness, t_1', is to be noted. As can be seen, for a thin sheath, $t_1' = 0.1\lambda$, the dielectric-free ($t_1' = 0$) surface wave resonance shifts toward broadside, while the magnitude of the R_{1N} curve in the vicinity of the resonance maxima broadens considerably. As the sheath thickness increases to $t_1' = 0.5\lambda$, a further shift toward broadside takes place, together with the appearance of a second surface wave resonance, which occurs as a result of a trapped mode within the sheath. Further increases of t_1' have resulted in three and more surface wave resonances.

An approximate transmission line analysis for the forced resonance scan angle, ψ_{sw}, can also be carried out for this case if we assume that the evanescent modes excited at the aperture decay sufficiently at the other sheath interface ($z = t_1'$). A single trapped mode is assumed to exist in the dielectric; port 3 of Fig. 7.26*a* can represent this trapped mode. One important difference between this case and that for the plugs is that for the sheath the modal admittances and propagation constants associated with the trapped mode, Y_3^ε, Y_3, Γ_3^ε, and Γ_3, are functions of ψ_r. The scattering matrix elements, $\{s_{ij}\}$, are obtained from the appropriate solutions of the integral equation, (3), with the whole half space, $z \geq 0$, loaded with material of dielectric constant ε_1'. Equation (22), appropriately modified for the sheath case, is

$$\frac{Y_3^\varepsilon - Y_3}{Y_3^\varepsilon + Y_3} e^{-2j\Gamma_3^\varepsilon t_1'} = \frac{s_{12}}{s_{12}s_{33} - s_{23}s_{13}} \tag{22a}$$

In (22a), the dominant contribution to the phase variation of the left-hand side comes from the term $e^{-2j\Gamma_3^\varepsilon t_1'}$, where for the E-plane scan we have

$$\Gamma_3^\varepsilon = \sqrt{\varepsilon_1'(2\pi/\lambda)^2 - [(2\pi - \psi_r)/(d \sin \alpha)]^2}, \quad 0 \leq \psi_r \leq \pi \tag{29}$$

Note that Γ_3^ε increases with increasing values of ψ_r. If we assume also that the right-hand side of (22a) varies slowly with ψ_r, by comparison, then a slight increase in the thickness, t_1', will result in a decrease in ψ_{sw}, as shown in Fig. 7.31. We note also that for a sufficiently large value of t_1', the phase, $2\Gamma_3^\varepsilon t_1'$, may go through many multiples of 2π for the indicated range of ψ_r. Consequently, multiple surface wave resonances exist for large thicknesses. For $t_1'/\lambda > 0.9$, as many as three forced surface waves are seen to exist. Note that, with increased values of t_1', ψ_{sw} does not reach broadside but asymptotically approaches a fixed value. The transmission line approximation deviates from the boundary value solutions over a bigger range of thickness values than in the corresponding plug case. This implies that for thin sheaths the boundary value solution has to be used to obtain reliable results. In contrast, in the region where multiple surface waves exist, the approximations seem to improve.

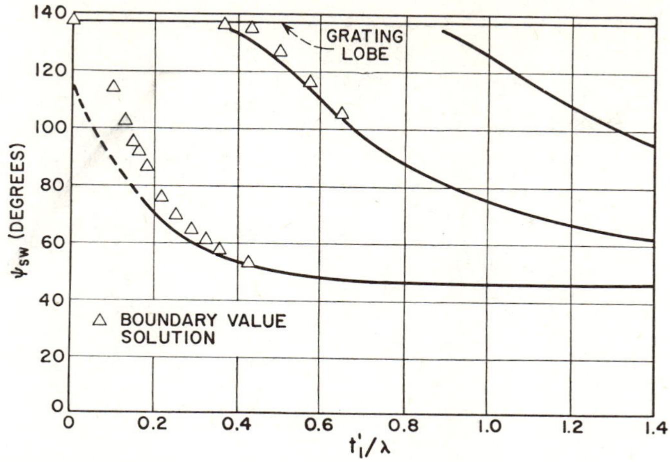

Fig. 7.31. Dielectric-covered array, *E*-plane scan; surface wave location (ψ_{sw}) versus sheath thickness; one trapped mode in sheath ($b = d = 0.714\lambda$, $a = 0.340\lambda$, $\varepsilon_1' = 2.0$, and $\alpha = 60°$).

In a previous section it was found that the forced surface wave resonances occur at isolated points in the scan plane (ψ_x, ψ_y) for dielectric-free arrays. This assertion remains true for both dielectric-covered and dielectric-loaded arrays. Our conclusion is based on the observation of the numerical results in the neighborhood of the surface wave resonance.

4. THE ASYMPTOTIC BEHAVIOR OF THE MUTUAL COUPLING COEFFICIENTS

In this section we shall examine the properties of the mutual coupling coefficients in planar arrays. It was shown in Chapter 4 that the asymptotic behavior of the coupling between a pair of elements which are far removed from each other is governed by the singular nature of the derivative of the active reflection coefficient as a function of scan. In the analysis of planar phased arrays we have also observed the discontinuous nature of the slope of the reflection coefficients when the main beam or a grating lobe is at grazing. This singular behavior of the slope is again expected to play a similar role in the asymptotic behavior of the mutual coupling coefficients. We shall use the model of current elements for simplicity. The array will be arranged in a rectangular grid ($\alpha = 90°$ in Fig. 7.1) for convenience. The results are equally valid, however, for arrays arranged in a skewed doubly periodic grid.[11]

We may recall the following Fourier series relationships (for a single-port input element) between the reflection coefficient, $R(\psi_x, \psi_y)$, and the mutual coupling coefficients, $\{\mathscr{C}_{p'q'}\}$:

$$R(\psi_x, \psi_y) = \sum_{p'=-\infty}^{+\infty} \sum_{q'=-\infty}^{+\infty} \mathscr{C}_{p'q'} e^{j(p'\psi_x + q'\psi_y)} \tag{30}$$

where ψ_x and ψ_y (the differential phase shifts between elements) are related to the scan angle directional cosines, T_x and T_y, by (12). Utilizing (12), we obtain

$$\begin{aligned} \mathscr{C}_{p'q'} &= \frac{1}{4\pi^2} \int_{-\pi}^{\pi} \int_{-\pi}^{\pi} d\psi_x \, d\psi_y \, R(\psi_x, \psi_y) e^{-j(p'\psi_x + q'\psi_y)} \\ &= \frac{k^2 bd}{4\pi^2} \int_{-\pi/kb}^{\pi/kb} dT_x \int_{-\pi/kd}^{\pi/kd} dT_y R(T_x, T_y) e^{-jk(p'bT_x + q'dT_y)} \end{aligned} \tag{31}$$

Our objective is to evaluate this integral in the limit,

$$\lim_{p',q' \to \infty} \mathscr{C}_{p'q'} \equiv \text{asymptotic } \mathscr{C}_{p'q'}$$

We now state (and will show) that, if

(1) the gradient of $R(T_x, T_y)$ is continuous in the (T_x, T_y) plane over the domain of integration, except at the angles where a grating lobe is at grazing [i.e., curves corresponding to $\Gamma_{mn} = 0$ (see e.g. A.8 of Appendix I)], and

(2) the gradient of R normal to the grating lobe grazing curves is singular and proportional to $1/\Gamma_{mn}$,

then the limit of $\mathscr{C}_{p'q'}$ for large values of (p', q') behaves like $e^{-jk\rho}/\rho^2$, where ρ is the radial distance at the aperture plane corresponding to $\rho_{p'q'}$ in Fig. 7.1.

A portion of the grating lobe structure consisting of circular arcs (see Appendix I) in the T_x–T_y plane is shown in Fig. 7.32 for a spacing

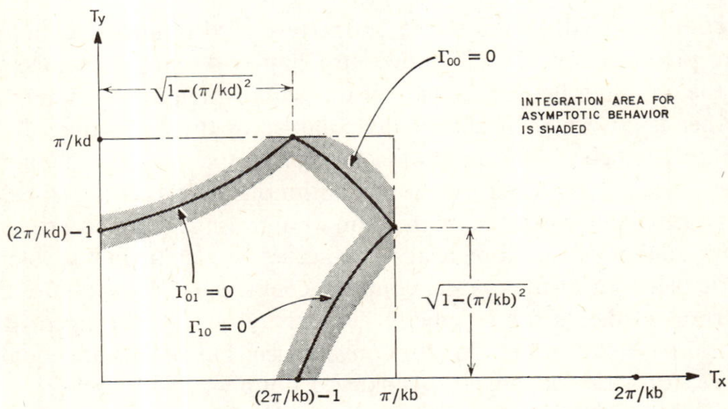

Fig. 7.32. Grating lobe grazing in T_x-T_y plane. Case: $0.5 < (b/\lambda, d/\lambda) < 1$.

$(0.5 < b/\lambda, d/\lambda < 1)$. The actual domain of integration indicated in (31) will reduce, asymptotically for large (p', q'), to the vicinity of the grating lobe curves (the shaded region in Fig. 7.32). This can be observed when integration by parts is used to evaluate (31) and from the fact that the gradient of R is continuous everywhere except on the curves.

Let us now consider a class of planar arrays for which R is known—the periodic current sheet antenna. This array consists of a thin, doubly periodic x-directed current defined on the plane $z = 0$. The coordinates and spacing shown in Fig. 7.32 are assumed. This "nonphysical" array, which has been used by other authors,[12] has three deficiencies as a model. The first is the lack of feeding transmission lines to the array elements. The normalization of input impedance is therefore taken with respect to the free space impedance, Z_0. The second deficiency arises from the usual assumption that the current distribution within a cell (of dimensions $b \times d$) does not vary with the scan angles, (ψ_x, ψ_y) or (T_x, T_y), although in an actual antenna array there would be an aperture field variation with scan. We will assume that, if the current does vary with (T_x, T_y), it varies in such a manner that the previously mentioned conditions on the gradient of R are satisfied. A third deficiency arises because of a singularity in the input impedance in the E-plane scan direction $(T_z = 0)$. This singularity can be removed by placing a ground plane behind the current sheet.[12] Notwithstanding these somewhat objectionable features, we shall take advantage of the closed form expression of R available for this model and proceed to analyze the array.

On the plane $z = 0$, the double Fourier series for an x-directed doubly periodic current sheet with a linear phase taper may be written as

$$J_x(x, y) = 2 \sum_{(r,l)} J_{r,l} e^{-jk(u_r x + v_l y)} \tag{32}$$

where

$$u_r = T_x - \frac{2\pi r}{kb}, \qquad v_l = T_y - \frac{2\pi l}{kd} \tag{33}$$

and the indices (r, l) range from $(-\infty, -\infty)$ to (∞, ∞). We may then readily derive the electric and magnetic fields (see Ref. 12, for example) and determine the complex power, P_c, flowing normal to the array out of a single cell by integrating the Poynting vector:

$$P_c = \int_{-b/2}^{b/2} dx \int_{-d/2}^{d/2} dy\, E_x H_y^* = \sum_{(r,l)} |J_{r,l}|^2 \frac{1 - u_r^2}{W_{r,l}} Z_0 bd \tag{34}$$

where

$$W_{r,l} = \sqrt{1 - u_r^2 - v_l^2} = \frac{\Gamma_{rl}}{k} \tag{35}$$

The result is an equation of the form of a sum of $|I|^2Z$ terms. This can be set equal to an effective current squared, $|J_e|^2$, times an input impedance, Z, to the array element, that is,

$$P_c \equiv |J_e|^2 Zbd \tag{36}$$

Depending on the distribution, $J_x(x, y)$, the definition of J_e may either be quite arbitrary or represent a reasonably natural choice. For example, in the case of a cosine dipole distribution,[12] $I_0 \cos \pi x/L$ ($L =$ dipole length), the natural choice $J_e \equiv I_0$ follows.

Normalizing the currents with respect to J_e so that

$$I_{r,l} \equiv \frac{J_{r,l}}{J_e} \tag{37}$$

we find for the normalized input impedance

$$\frac{Z}{Z_0} = \sum_{(r,l)} |I_{r,l}|^2 \left(\frac{1 - u_r^2}{W_{r,l}}\right) \tag{38}$$

The singularity in the impedance due to the $1/W_{r,l}$ factor is characteristic of the current sheet without a reflector or ground plane present. With such a reflector parallel to the current sheet and positioned at $z = -S$, we may eliminate this singularity. The resulting input impedance with the reflector is

$$\frac{Z}{Z_0} = \sum_{(r,l)} |I_{r,l}|^2\, 2jkS(1 - u_r^2)\left(\frac{\sin kW_{r,l}S}{kW_{r,l}S}\right) e^{-jkW_{r,l}S} \quad (\text{reflector at } z = -S) \tag{39}$$

In either case, (38) or (39), the reflection coefficient, R, is then defined as

$$R = \frac{Z/Z_0 - 1}{Z/Z_0 + 1} \tag{40}$$

It is interesting that the result mentioned by Hannan[13] was for the case in which only the I_{00} term is nonzero.

With R given by (40) we may now proceed to evaluate

$$\lim_{p',q' \to \infty} \mathscr{C}_{p'q'}$$

in (31). First we consider the evaluation of the gradient of R normal to the grating lobe curves. For the subarea of the total rectangular area of integration of (31) which encloses an arc of the (s, t)th grating lobe circle with a center at $(2\pi s/kb, 2\pi t/kd)$, we first translate coordinates to the center of that circle and then convert to cylindrical coordinates (r_{st}, δ_{st}). The gradient normal to the circle is then given simply by

$$\hat{r}_{st}(\hat{r}_{st} \cdot \nabla R) = \hat{r}_{st} \frac{\partial R}{\partial r_{st}} \tag{41}$$

We will evaluate $\partial R/\partial r_{st}$ subject to the assumption that $\partial\, |I_{st}|^2/\partial r_{st}$ does not contribute to the asymptotic behavior of $\mathscr{C}_{p'q'}$. This assumption is reasonably valid, for example, for the constant current sheet,[13] an array of dipoles,[14] and a large class of other current distributions as indicated below:

$$\frac{\partial\,|I_{st}|^2}{\partial r_{st}} \begin{cases} = 0 \text{ (for constant current sheet)} & \\ \sim \text{Nonsingular (array of dipoles with reflector)} & \\ \sim \dfrac{1}{W_{st}{}^p}, \quad p < 1 \text{ for reflector}, \; p < 2 \text{ for no reflector} & \text{No contribution to asymptotic behavior.} \\ \sim \dfrac{1}{W_{st}{}^p}, \quad p = 1 \text{ for reflector}, \; p < 2 \text{ for no reflector} & \text{Does not alter asymptotic behavior.} \end{cases}$$

Under these conditions we find that the gradient of R normal to the grating lobe circles, in the vicinity of these circles ($W_{st} \sim 0$), is proportional to $1/W_{st}$ (both with and without a reflector). Specifically, we find

$$\lim_{W_{st}\to 0} \frac{\partial R}{\partial r_{st}} = \left(\frac{2}{|I_{st}|^2 \sin^2 \delta_{st}}\right)\frac{1}{W_{st}} + O\left(\frac{1}{W_{st}{}^p}\right), \quad p < 1 \quad \text{without a reflector} \tag{42}$$

and

$$\lim_{W_{st}\to 0} \frac{\partial R}{\partial r_{st}} = \left[\frac{-|I_{st}|^2\, 2k^2 S^2 \sin^2 \delta_{st}}{(Z/Z_0 + 1)^2|_{W_{st}=0}}\right]\frac{1}{W_{st}} + O\left(\frac{1}{W_{st}{}^p}\right), \quad p < 1 \quad \text{with a reflector} \tag{43}$$

Without a reflector, the coefficient multiplying $1/W_{st}$ is singular in the E plane of scan, as may be expected. With a reflector, the coefficient is zero in that direction. Hence the E-plane asymptotic behavior for the planar array is not determined by this analysis. (However, the E-plane asymptotic behavior for the linear array was found to be no different from the H-plane behavior.)

In general, we assume therefore that the normal gradient of R, near the grating lobe curves, is equal to a well-behaved function, $f(\delta)$, times $1/W$, that is,

$$\lim_{W_{st}\to 0} \frac{\partial R}{\partial r_{st}} = f_{st}(\delta_{st})\frac{1}{W_{st}} + O\left(\frac{1}{W_{st}{}^p}\right), \quad p < 1 \tag{44}$$

It is physically reasonable, furthermore, to assume that ∇R is continuous everywhere else in the T_x–T_y plane.

With the normal component of the gradient of R given by (44) we proceed to integrate (31) by parts. Thus

$$\mathscr{C}_{p'q'} = \frac{k^2bd}{4\pi^2}\frac{1}{jkq'd}\int_{-\pi/kb}^{\pi/kb} dT_x \int_{-\pi/kd}^{\pi/kd} dT_y \frac{\partial R}{\partial T_y} e^{-jk(p'bT_x+q'dT_y)} \tag{45}$$

since the integrated part vanishes by reason of $R(T_x, \pi/kd) = R(T_x, -\pi/kd)$ (periodicity of R).

For each grating lobe circular arc in the integration area, we define an area, A_{st}, enclosing that arc. The sum of the A_{st}'s is then equal to the total integration area. For each A_{st} we translate to the center of the associated circle and convert to cylindrical coordinates, $(T_x, T_y) \rightarrow (r_{st}, \delta_{st})$, so that (44) becomes

$$\mathscr{C}_{p'q'} = \frac{k^2bd}{4\pi^2}\sum_{(s,t)}\iint_{A_{st}} \frac{r_{st}\,dr_{st}\,d\delta_{st}}{jk\rho_{p'q'}\sin\theta}\left(\frac{\partial R}{\partial r_{st}}\sin\delta_{st} + \frac{\cos\delta_{st}}{r_{st}}\frac{\partial R}{\partial \delta_{st}}\right) \cdot \exp\left[-jk\rho_{p'q'}\cos(\theta-\delta_{st})r_{st}\right] \tag{46}$$

where we define the quantized distances:

$$p'b = \rho_{p'q'}\cos\theta, \qquad q'd = \rho_{p'q'}\sin\theta \tag{47}$$

Note that the integration by parts has resulted in a $1/\rho_{p'q'}$ dependence so far.

Since $\nabla R \cdot \hat{\delta}_{st} = (1/r_{st})(\partial R/\partial\delta_{st})$ is continuous and nonsingular, we find (as for the linear arrays) that

$$\lim_{\rho_{p'q'}\to\infty} \mathscr{C}_{p'q'} = \lim_{\substack{\rho_{p'q'}\to\infty \\ \varepsilon\to 0}} \frac{k^2bd}{4\pi^2}\sum_{(s,t)}\int_{\delta'_{st}}^{\delta''_{st}} \frac{\sin\delta_{st}\,d\delta_{st}}{jk\rho_{p'q'}\sin\theta} \cdot \int_{1-\varepsilon}^{1+\varepsilon} r_{st}\frac{\partial R}{\partial r_{st}}\exp\left[-jk\rho_{p'q'}\cos(\theta-\delta_{st})r_{st}\right]dr_{st} + O\left(\frac{1}{\rho_{p'q'}^{5/2}}\right) \tag{48}$$

where the integration area, A_{st}, has been reduced to the immediate vicinity of the grating lobe curve (e.g., see Fig. 7.32). The reduced area, A_{st}, associated with each grating lobe consists of a circular annulus defined by

$$\delta'_{st} \le \delta_{st} \le \delta''_{st} \quad \text{and} \quad 1-\varepsilon \le r_{st} \le 1+\varepsilon \tag{49}$$

Employing (44) in (48) leads to

$$\lim_{\rho_{p'q'}\to\infty} \mathscr{C}_{p'q'} = \lim_{\substack{\rho_{p'q'}\to\infty \\ \varepsilon\to 0}} \frac{k^2bd}{4\pi^2}\sum_{(s,t)}\int_{\delta'_{st}}^{\delta''_{st}} d\delta_{st}\left\{\frac{f_{st}(\delta_{st})\sin\delta_{st}}{jk\rho_{p'q'}\sin\theta} \cdot \int_{1-\varepsilon}^{1+\varepsilon} \frac{\exp\left[-jk\rho_{p'q'}\cos(\theta-\delta_{st})r_{st}\right]}{\sqrt{1-r_{st}^2}}\,dr_{st}\right\} + O\left(\frac{1}{\rho_{p'q'}^{5/2}}\right) \tag{50}$$

where W_{st} was replaced by $\sqrt{1 - r_{st}^2}$. Upon utilizing the following relation:

$$\lim_{\substack{y\to\infty \\ \varepsilon\to 0}} \int_{1-\varepsilon}^{1+\varepsilon} \frac{e^{-jyu}}{\sqrt{1-u^2}}\, du \sim \sqrt{\pi/2}\, e^{j\pi/4} \frac{e^{-jy}}{y^{1/2}}$$

in (50), we obtain

$$\lim_{\rho_{p'q'}\to\infty} \mathscr{C}_{p'q'} = \lim_{\rho_{p'q'}\to\infty} \left(\sqrt{\pi/2}\, \frac{k^2 bd}{4\pi^2}\, e^{-j\pi/4} \right)$$

$$\cdot \sum_{(s,t)} \int_{\delta'_{st}}^{\delta''_{st}} \frac{f_{st}(\delta_{st}) \sin \delta_{st} \exp\left[-jk\rho_{p'q'} \cos(\theta - \delta_{st})\right]}{(k\rho_{p'q'})^{3/2} \sin\theta \cos(\theta - \delta_{st})}\, d\delta_{st} + O\left(\frac{1}{\rho_{p'q'}^{5/2}}\right) \tag{51}$$

Note that a $1/\rho_{p'q'}^{3/2}$ dependence has been obtained in (51) before the δ integration.

Now, for each value of θ ($\theta \neq 0$ or π, the E-plane scan), there exists at least one grating lobe curve, $(s, t) \equiv (s', t')$, such that

$$\delta'_{st} \leq \theta \leq \delta''_{st} \tag{52}$$

Hence a saddle-point evaluation of (51) gives

$$\lim_{\rho_{p'q'}\to\infty} \mathscr{C}_{p'q'} = \left(\frac{jk^2bd}{4\pi}\right) \left[\sum_{(s',t')} f_{s't'}(\delta_{s't'} = \theta)\right] \frac{e^{-jk\rho_{p'q'}}}{(k\rho_{p'q'})^2} + O\left(\frac{1}{\rho_{p'q'}^{5/2}}\right) \tag{53}$$

Equation (53) describes the asymptotic behavior of the coupling coefficients for this class of planar arrays.

Although the asymptotic decay obtained in (53) does not apply to the E plane, this may be explained in part by the fact that an isolated current dipole has a far-field decay that is faster than $1/\rho$ parallel to the dipole.

These results, as well as those of Chapter 4, are consistent with the physical similarity of the terminated array to the known radiation of a dipole over a lossy earth, wherein the same asymptotic behavior is observed. This leads to the conjecture that all arrays have similar asymptotic mutual coupling behavior.

5. CONCLUDING REMARKS

In designing a practical phased array antenna, one of the challenges is the minimization of the reflected power over a prescribed scan region and frequency band. The reflected power can be reduced by the insertion of matching networks within the elements or their feed lines except when total reflections due to forced surface wave resonances take place. In planar arrays, we have observed that at isolated scan angles total reflections occur for certain polarizations of excitation. In order to eliminate the total reflections in the case of dielectric-free arrays we may utilize the polarization dependence of these total reflection angles. Thus we may choose a polarization of excitation that does not conform with the planes of symmetry in the array geometry. For example, in the equilaterial triangular grid array of circular waveguides we observe that total reflection occurs in the *H* plane of scan with vertical polarization of excitation (Fig. 7.12). When the same array is excited by circular polarization (Fig. 7.16), we notice that at the previous angle of total reflection we can now radiate only about 50% of the power. Total reflection due to forced surface wave resonances has not been observed to date for circular polarization of excitation. It should be noted, however, that even though a point of total reflection can be eliminated by a polarization change in one plane of scan, the possibility remains that points of total reflection may exist in other planes of scan.

Another possible method for the elimination of total reflections due to forced surface wave resonances is to tamper with the aperture field and thereby alter the mutual coupling effects. Thin metallic fences, baffles, and irises (see Chapter 4) have been used[14] to reduce the variations of the active impedance of certain arrays in the *E* plane of scan.

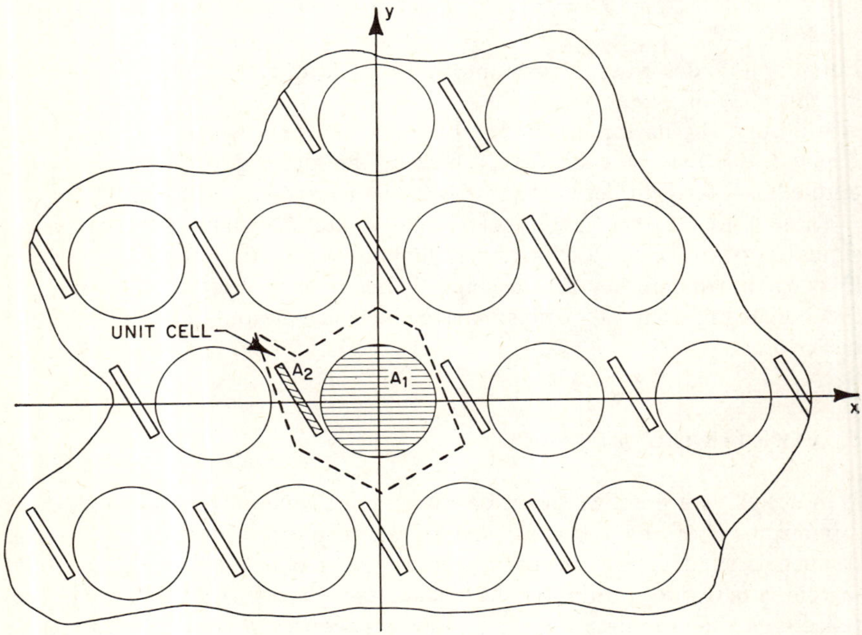

Fig. 7.33. Planar array of compound elements.

Still another way of modifying the mutual coupling effects is to use a compound type of array element, as shown in Fig. 7.33. Here, we add small rectangular waveguide(s) to the unit cell, which contains the circular waveguide. The small rectangular waveguides may operate below cut-off or be shorted at certain distances away from the aperture. In breaking the paths of the ground plane currents, these small rectangular waveguides act mainly as chokes.

The extension of the integral equation formulation to this case is straightforward. As shown in Appendix II, we can deform the shape of the unit cell so that it includes a complete circular and a complete rectangular waveguide aperture, A_1 and A_2 respectively (Fig. 7.33). Let us define the waveguide modes as $\{\mathbf{\Phi}_n^{\bigcirc}\}$ with modal admittances $\{y_n^{\bigcirc}\}$ over A_1, and $\{\mathbf{\Phi}_s^{\square}\}$ with corresponding pertinent admittances (short-circuited or evanescent as the situation may be) $\{y_s^{\square}\}$ over A_2. When only the circular waveguides are excited, the modified version of the integral equation, (3), for the tangential electric field is

$$
\begin{aligned}
2\sum_{i=1}^{2} y_i^{\bigcirc} A_i' \mathbf{\Phi}_i^{\bigcirc}(\mathbf{r}) &= \sum_{i=1}^{\infty} y_i^{\bigcirc} \mathbf{\Phi}_i^{\bigcirc}(\mathbf{r}) \iint_{A_1} \mathbf{\Phi}_i^{\bigcirc}(\mathbf{r}') \cdot \mathbf{E}_t(\mathbf{r}')\, da' \\
&+ \sum_{s=1}^{\infty} y_s^{\square} \mathbf{\Phi}_s^{\square}(\mathbf{r}) \iint_{A_2} \mathbf{\Phi}_s^{\square}(\mathbf{r}') \cdot \mathbf{E}_t(\mathbf{r}')\, da' \\
&+ \sum_{p=1}^{2} \sum_{m=-\infty}^{+\infty} \sum_{n=-\infty}^{+\infty} Y_{pmn} \mathbf{\Psi}_{pmn}(\mathbf{r}) \iint_{A_1 \text{ and } A} \mathbf{\Psi}_{pmn}^{*}(\mathbf{r}') \cdot \mathbf{E}_t(\mathbf{r}')\, da'
\end{aligned} \tag{54}
$$

A corresponding integral equation for the tangential magnetic field (being defined over the entire unit cell) can be easily derived (see Chapter 2). In solving (54) by the Ritz-Galerkin method one should pay special attention to the partitioning of the resultant matrix equation between the circular and rectangular waveguide modes. More explicitly, when the finite sets of functions, $\{\mathbf{\Phi}_n^{\bigcirc}, n = 1, \ldots, N\}$ and $\{\mathbf{\Phi}_s^{\square}, s = 1, \ldots, S\}$, are used for expansion and weighting, the choice of the ratio N/S may strongly influence the accuracy of the solution. The partitioning problem of the compound element array is quite similar to the case of the bifurcated waveguide, for which it was shown[15] that a wrong partitioning may converge to incorrect results.

An additional method for the elimination of forced resonances by selectively exciting and shorting the array elements will be discussed in Chapter 8.

APPENDIX I FLOQUET TYPE WAVE FUNCTIONS IN SKEWED COORDINATES

Consider the periodic array, Figs. 7.1 and 7.2, excited with incremental phase shifts (steering phases) between adjacent cells along the η_1 and η_2 coordinates. A complete set of solutions to the scalar wave equation, each of which varies periodically, according to Floquet's theorem, along the η_1 and η_2 coordinates is

$$S_{mn} = \exp(-j\Gamma_{mn}z) \exp\left[-j\left(\frac{\psi_{\sigma 1}}{b} - \frac{2\pi m}{b}\right)\eta_1\right] \exp\left[-j\left(\frac{\psi_{\sigma 2}}{d} - \frac{2\pi n}{d}\right)\eta_2\right] \tag{A.1}$$

with the integers $m, n = -\infty, \ldots, -1, 0, 1, 2, 3, \ldots, +\infty$. Equation A.1 describes a wave traveling (or decaying) in the $+z$ direction with propagation constant Γ_{mn} [$\exp(j\omega t)$ time convention]. The steering phases, ψ_{σ_1} and ψ_{σ_2}, are directly related to the vector propagation constant, $\mathbf{k} = k\hat{r}$, so that (A.1) can be rewritten as

$$S_{mn} = \exp(-j\Gamma_{mn}z) \exp\left[-j\left(\mathbf{k}\cdot\hat{\eta}_1 - \frac{2\pi m}{b}\right)\eta_1\right] \exp\left[-j\left(\mathbf{k}\cdot\hat{\eta}_2 - \frac{2\pi n}{d}\right)\eta_2\right] \tag{A.2}$$

The free space propagation vector, $\mathbf{k}$, can be expressed in the Cartesian coordinate system as

$$\mathbf{k} = k[T_x\hat{x} + T_y\hat{y} + T_z\hat{z}] \tag{A.3}$$

where T_x, T_y, and T_z are the directional cosines of $\mathbf{k}$ with respect to that system, and $\hat{x}$, $\hat{y}$, and $\hat{z}$ are the unit vectors. The quantities

$$\mathbf{k}\cdot\hat{\eta}_1 = \frac{\psi_{\sigma 1}}{b} \quad \text{and} \quad \mathbf{k}\cdot\hat{\eta}_2 = \frac{\psi_{\sigma 2}}{d} \tag{A.4}$$

are the projections of $\mathbf{k}$ on the reciprocal grid (lattice) coordinates, σ_1 and σ_2, respectively. The unit vectors in the σ_1 and σ_2 directions form a biorthogonal set with $\hat{\eta}_1$ and $\hat{\eta}_2$ (Fig. A.1). To express (A.1) in Cartesian coordinates it can easily be shown that

$$\eta_1 = x - y\cot\alpha, \qquad \eta_2 = \frac{y}{\sin\alpha} \tag{A.5}$$

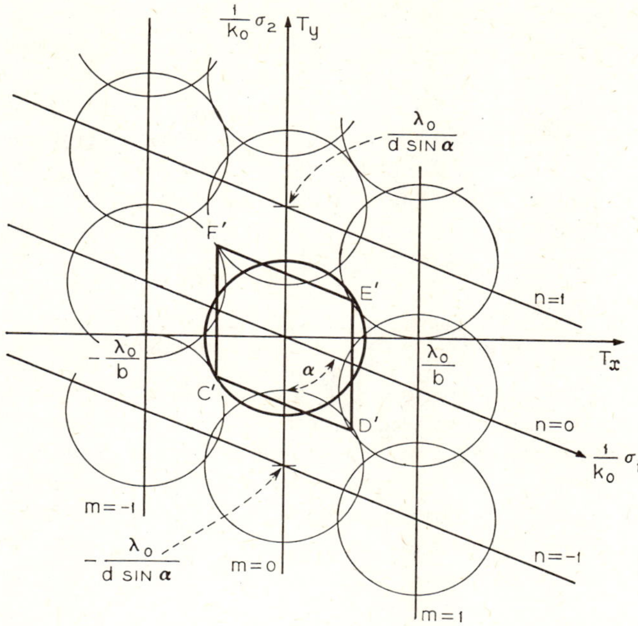

Fig. A.1. Grating lobe diagram in reciprocal lattice coordinates.

so that the substitution of (A.5) into (A.2) yields

$$S_{mn} = \exp\,[-(j\Gamma_{mn}z)]\exp\left[-j\left(kT_x - \frac{2\pi m}{b}\right)x\right] \cdot \exp\left\{-j\left[kT_y - \left(\frac{2\pi n}{d\sin\alpha} - \frac{2\pi m}{b\tan\alpha}\right)\right]y\right\} \tag{A.6}$$

The projections of the propagation constant of the (m, n)th Floquet mode, $\boldsymbol{\kappa}$, along the x and y directions, κ_x and κ_y, respectively, are

$$\begin{aligned}\kappa_x &\equiv \boldsymbol{\kappa}\cdot\hat{x} = kT_x - \frac{2\pi m}{b}\\ \kappa_y &\equiv \boldsymbol{\kappa}\cdot\hat{y} = kT_y - \left(\frac{2\pi n}{d\sin\alpha} - \frac{2\pi m}{b\tan\alpha}\right)\end{aligned} \tag{A.7}$$

Since S_{mn} is a solution to the scalar wave equation, it can be shown that

$$\begin{aligned}\Gamma_{mn} &= (k^2 - \kappa_x^{\,2} - \kappa_y^{\,2})^{1/2}\\ &= \left\{k^2 - \left(kT_x - \frac{2\pi m}{b}\right)^2 - \left[kT_y - \left(\frac{2\pi n}{d\sin\alpha} - \frac{2\pi m}{b\tan\alpha}\right)\right]^2\right\}^{1/2}\end{aligned} \tag{A.8}$$

where the negative imaginary root holds for $\kappa_x^{\,2} + \kappa_y^{\,2} > k^2$.

Each mode, S_{mn}, for which Γ_{mn} is real corresponds to a radiated plane wave of the phased array. The plane wave with the indices $m = 0$ and $n = 0$ is identified with the main beam, while m or $n \neq 0$ corresponds to a radiating grating lobe. As a function of T_x and T_y (or ψ_{σ_1} and ψ_{σ_2}), a given Γ_{mn} may become purely imaginary as it goes through a zero, as in (A.8). In such cases, the related Floquet mode, S_{mn}, becomes evanescent or non-radiating. By plotting the curves obtained by setting the $\Gamma_{mn} = 0$ as a function of T_x and T_y, one obtains a convenient diagram which illustrates these effects. Setting $\Gamma_{mn} = 0$ yields

$$\left(T_x - \frac{m\lambda}{b}\right)^2 + \left[T_y - \left(\frac{n\lambda}{d \sin \alpha} - \frac{m\lambda}{b \tan \alpha}\right)\right]^2 = 1 \tag{A.9}$$

where $\lambda = 2\pi/k$. As a function of T_x and T_y, (A.9) represents a family of circles with unit radius displaced from the origin. This diagram of displaced circles constitutes the well-known grating lobe diagram, Fig. A.1.

Notice that the steering phases, ψ_{σ_1} and ψ_{σ_2}, are related to T_x and T_y through (A.3) and (A.4). The parallelogram $C'D'E'F'$ of Fig. A.1 corresponds to the range of steering phases

$$-\pi \leq \psi_{\sigma_1} \leq \pi, \qquad -\pi \leq \psi_{\sigma_2} \leq \pi \tag{A.10}$$

and is a periodic cell along the σ_1 and σ_2 coordinates.

As mentioned in Section 1, it is possible to define a complete orthonormal set of vector modes, $\{\mathbf{\Psi}_{pmn}\}$, over the parallelogram $CDEF$ (Fig. 7.2). The tangential electromagnetic field at the plane $z = 0^+$ can be expressed by a Fourier series of this set of modes, which consists of both TE ($p = 1$) and TM ($p = 2$) modes (transverse to z). These modes, $\{\mathbf{\Psi}_{1mn}\}$ and $\{\mathbf{\Psi}_{2mn}\}$, are given by

$$\mathbf{\Psi}_{1mn} = \frac{\exp\left[-j(x\kappa_x + y\kappa_y)\right]}{(bd \sin \alpha)^{1/2}} \left\{\frac{\kappa_y}{\kappa_r}\hat{x} - \frac{\kappa_x}{\kappa_r}\hat{y}\right\}\Bigg|_{m,n} \tag{A.11}$$

$$\mathbf{\Psi}_{2mn} = \frac{\exp\left[-j(x\kappa_x + y\kappa_y)\right]}{(bd \sin \alpha)^{1/2}} \left\{\frac{\kappa_x}{\kappa_r}\hat{x} + \frac{\kappa_y}{\kappa_r}\hat{y}\right\}\Bigg|_{m,n} \tag{A.12}$$

with $\kappa_r = (\kappa_x^2 + \kappa_y^2)^{1/2}$. The quantities κ_x and κ_y are functions of (m, n) as given by (A.7). The orthonormality of this set of vector modes is defined by the following scalar products:

$$\langle \mathbf{\Psi}_{1mn}, \mathbf{\Psi}^*_{1pq} \rangle = \iint_{\substack{\text{parallelogram} \\ CDEF}} \mathbf{\Psi}_{1mn} \cdot \mathbf{\Psi}^*_{1pq}\, dx\, dy = \delta_{mn,pq} \tag{A.13}$$

where

$$\delta_{mn,pq} = \begin{cases} 1 & \text{for } m = p \text{ and } n = q \\ 0 & \text{otherwise} \end{cases} \tag{A.14}$$

$$\langle \Psi_{1mn}, \Psi^*_{2pq} \rangle = 0 \tag{A.15}$$

and

$$\langle \Psi_{2mn}, \Psi^*_{2pq} \rangle = \delta_{mn,pq} \tag{A.16}$$

APPENDIX II THE INVARIANCE OF THE SCALAR PRODUCT WITH THE SHAPE OF THE PERIODIC CELL

The orthonormality and completeness of the set $\{\Psi_{pmn}\}$ of Floquet modes, (A.11) and (A.12), need not be defined over a specific periodic cell such as the parallelogram *CDEF* of Fig. 7.2. This fact is especially significant when a periodic cell intercepts parts of more than one circular (or other type of element) aperture. It will be shown that the orthonormality of $\{\Psi_{pmn}\}$ can be preserved over a properly deformed periodic cell which contains only a single waveguide aperture.

Using (A.4), we can rewrite (A.11) as

$$\Psi_{1mn} = \mathbf{F}(m, n) \exp\left[-j\left(\frac{\psi_{\sigma_1} - 2\pi m}{b}\eta_1 + \frac{\psi_{\sigma_2} - 2\pi n}{d}\eta_2\right)\right] \tag{A.17}$$

The scalar product between two TE modes is

$$\begin{aligned}
\langle \Psi_{1mn}, \Psi^*_{1pq} \rangle &= \delta_{mn,pq} \\
&= \iint_{CDEF} \Psi_{1mn} \cdot \Psi^*_{1pq}\, dx\, dy \\
&= \iint_{CDEF} \Psi_{1mn} \cdot \Psi^*_{1pq} \sin\alpha\, d\eta_1\, d\eta_2 \\
&= \mathbf{F}(m, n) \cdot \mathbf{F}^*(pq) \sin\alpha \\
&\quad \times \iint_{CDEF} \exp\left\{j\left[\frac{2\pi(m-p)}{b}\eta_1 + \frac{2\pi(n-q)}{d}\eta_2\right]\right\} d\eta_1\, d\eta_2
\end{aligned} \tag{A.18}$$

The integral over the parallelogram $CDEF$ in (A.18) can be divided into three (or more) integrals over the triangles CGK and LEI and the polygon $GDLIFK$ (Fig. 7.2):

$$\iint_{CDEF} = \iint_{CGK} + \iint_{LEI} + \iint_{GDLIFK} \tag{A.19}$$

Because of the array periodicity, the triangles DHL and LEI are displaced by b with respect to the triangles CGK and KFJ, respectively, along the η_1 direction. Thus, for example, if the η_1 coordinate of the points within the triangle DHL is η_1', given by

$$\eta_1' = \eta_1 + b \qquad \text{with} \quad d\eta_1' = d\eta_1 \tag{A.20}$$

then

$$\begin{aligned}
&\iint_{DHL} \exp\left\{j\left[\frac{2\pi(m-p)}{b}\eta_1' + \frac{2\pi(n-q)}{d}\eta_2\right]\right\} d\eta_1\, d\eta_2 \\
&= \exp\left[j2\pi(m-p)\right] \iint_{CGK} \exp\left\{j\left[\frac{2\pi(m-p)}{b}\eta_1 + \frac{2\pi(n-q)}{d}\eta_2\right]\right\} d\eta_1\, d\eta_2 \\
&= \iint_{CGK} \exp\left\{j\left[\frac{2\pi(m-p)}{b}\eta_1 + \frac{2\pi(n-q)}{d}\eta_2\right]\right\} d\eta_1\, d\eta_2
\end{aligned} \tag{A.21}$$

Similarly

$$\begin{aligned}
&\iint_{LEI} \exp\left\{j\left[\frac{2\pi(m-p)}{b}\eta_1 + \frac{2\pi(n-q)}{d}\eta_2\right]\right\} d\eta_1\, d\eta_2 \\
&\qquad = \iint_{KFJ} \exp\left\{j\left[\frac{2\pi(m-p)}{b}\eta_1 + \frac{2\pi(n-q)}{d}\eta_2\right]\right\} d\eta_1\, d\eta_2
\end{aligned} \tag{A.22}$$

Thus

$$\begin{aligned}
&\iint_{CDEF} \exp\left\{j\left[\frac{2\pi(m-p)}{b}\eta_1 + \frac{2\pi(n-q)}{d}\eta_2\right]\right\} d\eta_1\, d\eta_2 \\
&\qquad = \iint_{GHIJ} \exp\left\{j\left[\frac{2\pi(m-p)}{b}\eta_1 + \frac{2\pi(n-q)}{d}\eta_2\right]\right\} d\eta_1\, d\eta_2
\end{aligned} \tag{A.23}$$

and the orthonormality and completeness of the set $\{\Psi_{pmn}\}$ is preserved over the new periodic cell, *GHIJ*. This is the two-dimensional analog of the single-dimensional Fourier series whereby the functions and coefficients are independent of the initial value of the period. In fact, the two-dimensional periodic cell can be deformed to any shape which contains a single waveguide, provided that the area of the cell stays the same and the parts which cause the deformation are translated by integer multiples of b or d along the η_1 or η_2 coordinate, respectively.

APPENDIX III THE SCALAR PRODUCT OF CIRCULAR AND CARTESIAN WAVE FUNCTIONS

Here, we shall derive the scalar product of Cartesian and more generalized circularly symmetric wave functions.[16] Consider the array of coaxial waveguides, Fig. A.2*a*, arranged in a skew-symmetric grid. A magnified view of each element is shown in Fig. A.2*b*. Thin circular irises are present at the aperture.

The orthonormalized free space wave functions[2] are

$$\Psi_{1mn} = \frac{l}{\sqrt{S_\alpha}}\left(\hat{x}\frac{\kappa_y}{\kappa_r} - \hat{y}\frac{\kappa_x}{\kappa_r}\right) \exp\left[-j(x\kappa_x + y\kappa_y)\right] \tag{A.24}$$

and

$$\Psi_{2mn} = \frac{1}{\sqrt{S_\alpha}}\left(\hat{x}\frac{\kappa_x}{\kappa_r} + \hat{y}\frac{\kappa_y}{\kappa_r}\right) \exp\left[-j(x\kappa_x + y\kappa_y)\right] \tag{A.25}$$

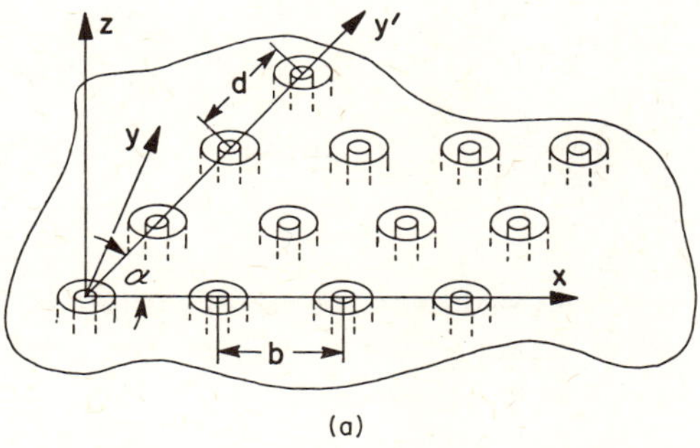

Fig. A.2*a*. Phased array of coaxial waveguide elements arranged in a skew-symmetric grid.

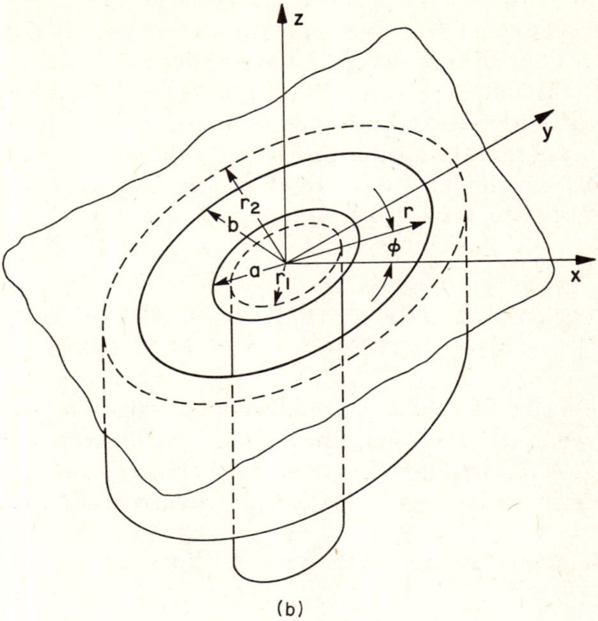

Fig. A.2*b*. Coaxial waveguide element with thin circular irises.

where $S_\alpha = bd \sin \alpha$ is the area of a periodic cell,

$$\kappa_r^2 = \kappa_x^2 + \kappa_y^2$$

$$\kappa_x = \frac{\psi_x}{b} - \frac{2\pi m}{b}$$

$$\kappa_y = \frac{\psi_y}{d \sin \alpha} - \left(\frac{2\pi n}{d \sin \alpha} - \frac{2\pi m}{b} \cot \alpha\right), \qquad -\infty \leq m, \quad n \leq +\infty$$

and ψ_x/b and $\psi_y/(d \sin \alpha)$ are the phase shifts per unit length in the x and y directions. The complete orthonormal set of vector modes of the coaxial waveguide[2] consists of TE and TM type modes. The TE vector modes of the transverse electric field are

$$ {}_H^V\mathbf{\Phi}_{1qs} = \hat{r}\frac{q}{r} Z_q(\beta_s r)\begin{cases} \sin q\phi \\ -\cos q\phi \end{cases} + \hat{\phi}\beta_s Z_q'(\beta_s r)\begin{cases} \cos q\phi \\ \sin q\phi \end{cases} \tag{A.26}$$

and the TM vector modes are

$$ {}^{V}_{H}\mathbf{\Phi}_{2qs} = -\hat{r}\alpha_s \bar{Z}_q'(\alpha_s r)\begin{Bmatrix} \sin q\phi \\ \cos q\phi \end{Bmatrix} + \hat{\phi}\frac{q}{r}\bar{Z}_q(\alpha_s r)\begin{Bmatrix} -\cos q\phi \\ \sin q\phi \end{Bmatrix} \tag{A.27} $$

with $q, s = 0, 1, 2, \ldots, \infty$. In these expressions $Z_q(\beta_s r)$ and $\bar{Z}_q(\alpha_s r)$ are particular linear combinations of Bessel and Neumann functions that satisfy the boundary conditions in the coaxial waveguide.[2] Primes denote derivatives with respect to the argument. The left-hand super- and subscripts H and V designate the degenerate "horizontal" or "vertical" mode with angular variation $\cos q\phi$ or $\sin q\phi$, respectively, of the radial component of the field. Expressing the free space modes, (A.24) and (A.25), in cylindrical coordinates yields

$$ \mathbf{\Psi}_{1mn} = \frac{1}{\sqrt{S_\alpha}}[\hat{r}\sin(\phi_1 - \phi) - \hat{\phi}\cos(\phi_1 - \phi)]\exp[-j\kappa_r\cos(\phi - \phi_1)] \tag{A.28} $$

and

$$ \mathbf{\Psi}_{2mn} = \frac{1}{\sqrt{S_\alpha}}[\hat{r}\cos(\phi_1 - \phi) + \hat{\phi}\sin(\phi_1 - \phi)]\exp[-j\kappa_r\cos(\phi - \phi_1)] \tag{A.29} $$

where

$$ \phi_1 = \tan^{-1}\frac{\kappa_y}{\kappa_x} $$

A typical scalar product (or coupling coefficient) between the waveguide and free space mode for the element shown in Fig. A.2*b* is

$$ {}^{V}_{H}C^{1qs}_{1mn} = \langle {}^{V}_{H}\mathbf{\Phi}_{1qs}, \mathbf{\Psi}_{1mn}\rangle = \int_a^b \int_0^{2\pi} {}^{V}_{H}\mathbf{\Phi}_{1qs}\cdot\mathbf{\Psi}_{1mn} r\,dr\,d\phi \tag{A.30} $$

Upon substitution of (A.26) and (A.28) into (A.30) and integration with respect to ϕ, one obtains an integral of the form

$$ I = \int_a^b \left[\beta_s\kappa_r Z_q'(\beta_s r)J_q'(\kappa_r r) + \frac{q^2}{r^2}Z_q(\beta_s r)J_q(\kappa_r r)\right] r\,dr \tag{A.31} $$

where $J_q(x)$ is the Bessel function of order q and argument x. Equation (A.31) can be integrated in closed form[17] to

$$ I = \left\{\frac{r[\kappa_r^2\beta_s J_q(\kappa_r r)Z_q'(\beta_s r) - \beta_s^2\kappa_r J_q'(\kappa_r r)Z_q(\beta_s r)]}{\kappa_r^2 - \beta_s^2}\right\}\Bigg|_a^b \tag{A.32} $$

The final expressions for scalar products between the wave functions of the two regions are given by

$$\langle {}^{V}_{H}\mathbf{\Phi}_{1qs}, \mathbf{\Psi}_{1mn}\rangle = \frac{2\pi\beta_s(-j)^{q-1}}{(\kappa_r^2 - \beta_s^2)\sqrt{S_\alpha}}\{b[\beta_s Z_q(\beta_s b)J_q'(\kappa_r b) - \kappa_r Z_q'(\beta_s b)J_q(\kappa_r b)] - a[\beta_s Z_q(\beta_s a)J_q'(\kappa_r a) - \kappa_r Z_q'(\beta_s a)J_q(\kappa_r a)]\}\begin{cases}\cos q\phi_1 & (V)\\ \sin q\phi_1 & (H)\end{cases} \tag{A.33}$$

$$\langle {}^{V}_{H}\mathbf{\Phi}_{1qs}, \mathbf{\Psi}_{2mn}\rangle = \frac{2\pi q(-j)^{q-1}}{\kappa_r\sqrt{S_\alpha}}[Z_q(\beta_s b)J_q(\kappa_r b) - Z_q(\beta_s a)J_q(\kappa_r a)]\begin{cases}\sin q\phi_1 & (V)\\ -\cos q\phi_1 & (H)\end{cases} \tag{A.34}$$

$$\langle {}^{V}_{H}\mathbf{\Phi}_{2qs}, \mathbf{\Psi}_{1mn}\rangle = \frac{2\pi q(-j)^{q-1}}{\kappa_r\sqrt{S_\alpha}}[\bar{Z}_q(\alpha_s b)J_q(\kappa_r b) - \bar{Z}_q(\alpha_s a)J_q(\kappa_r a)]\begin{cases}\cos q\phi_1 & (V)\\ -\sin q\phi_1 & (H)\end{cases} \tag{A.35}$$

$$\langle {}^{V}_{H}\mathbf{\Phi}_{2qs}, \mathbf{\Psi}_{2mn}\rangle = \frac{2\pi\alpha_s(-j)^{q-1}}{(\kappa_r^2 - \alpha_s^2)\sqrt{S_\alpha}}\{b[\alpha_s\bar{Z}_q(\alpha_s b)J_q'(\kappa_r b) - \kappa_r\bar{Z}_q'(\alpha_s b)J_q(\kappa_r b)] - a[\alpha_s\bar{Z}_q(\alpha_s a)J_q'(\kappa_r a) - \kappa_r\bar{Z}_q'(\alpha_s a)J_q(\kappa_r a)]\}\begin{cases}\sin q\phi_1 & (V)\\ \cos q\phi_1 & (H)\end{cases} \tag{A.36}$$

The nondegenerate TEM mode, $\mathbf{\Phi}_{200}$, must be integrated separately with the results

$$\langle \mathbf{\Phi}_{200}, \mathbf{\Psi}_{2mn}\rangle = \sqrt{\frac{2\pi}{S_\alpha \ln(r_2/r_1)}}\,\frac{(-j)}{\kappa_r}[J_0(\kappa_r a) - J_0(\kappa_r b)] \tag{A.37}$$

and

$$\langle \mathbf{\Phi}_{200}, \mathbf{\Psi}_{1mn}\rangle = 0 \tag{A.38}$$

The radial dependence of the waveguide wave functions in (A.26) is as follows:

$$Z_q(\beta_s r) = \frac{\sqrt{\pi\varepsilon_q}}{2}\,\frac{J_q(\beta_s r)N_q'(\beta_s r_1) - N_q(\beta_s r)J_q'(\beta_s r_1)}{\{[J_q'(\beta_s r_1)/J_q'(\beta_s r_2)]^2[1 - (q/\beta_s r_2)^2] - [1 - (q/\beta_s r_1)^2]\}^{1/2}} \tag{A.39}$$

where $N_q(x)$ is the Neumann function of order q and argument x, and

$$\varepsilon_q = \begin{cases}1 & \text{for } q = 0\\ 2 & \text{for } q \neq 0\end{cases}$$

The quantities $\{\beta_s\}$ are the roots of the following characteristic equation:

$$J_q'(\beta_s r_2)N_q'(\beta_s r_1) - N_q'(\beta_s r_2)J_q'(\beta_s r_1) = 0 \tag{A.40}$$

Similarly,

$$Z_q(\alpha_s r) = \frac{\sqrt{\pi\varepsilon_q}}{2}\frac{J_q(\alpha_s r)N_q(\alpha_s r_1) - N_q(\alpha_s r)J_q(\alpha_s r_1)}{\{[J_q(\alpha_s r_1)/J_q(\alpha_s r_2)]^2 - 1\}^{1/2}} \tag{A.41}$$

with $\{\alpha_s\}$ being the roots of

$$J_q(\alpha_s r_2)N_q(\alpha_s r_1) - N_q(\alpha_s r_2)J_q(\alpha_s r_1) = 0 \tag{A.42}$$

For circular waveguides we have $r_1 = 0$, and the above functions reduce to

$$Z_q(\beta_s r) = \sqrt{\varepsilon_q/\pi}\,\frac{1}{\sqrt{(\beta_s r_2)^2 - q^2}}\frac{J_q(\beta_s r)}{J_q(\beta_s r_2)} \tag{A.43}$$

with $\{\beta_s\}$ being the roots of $J_q'(\beta_s r_2) = 0$, and

$$Z_q(\alpha_s r) = \sqrt{\varepsilon_q/\pi}\,\frac{J_q(\alpha_s r)}{\alpha_s r_2 J_{q+1}(\alpha_s r_2)} \tag{A.44}$$

with $\{\alpha_s\}$ being the roots of $J_q(\alpha_s r_2) = 0$.

It should be mentioned that, when no irises are present at the aperture (i.e., $r_1 = a$ and $r_2 = b$), there is no coupling between the interior TM and free space TE modes. It is evident from (A.41) and (A.42) that $Z_q(\alpha_s a) = Z_q(\alpha_s b) = 0$ and therefore $\langle {}_H^V\boldsymbol{\Phi}_{2qs}, \boldsymbol{\Psi}_{1mn}\rangle = 0$ in (A.35). The closed form integration of the mode coupling coefficients has many applications in other radiation and waveguide discontinuity problems also.

REFERENCES

1. L. Brillouin, *Wave Propagation in Periodic Structures*, Dover Publications, New York, 1953, pp. 94–121.
2. N. Marcuvitz (ed.), *Waveguide Handbook*, M.I.T. Radiation Laboratory Series, Vol. 10, McGraw-Hill Book Company, New York, 1951, pp. 66–89.
3. E. C. DuFort, "Finite Scattering Matrix for an Infinite Array Antenna," *Radio Science*, Vol. 2, pp. 19–27, January 1967.
4. N. Amitay and V. Galindo, "On Energy Conservation and the Method of Moments in Scattering Problems," *IEEE Trans. Antennas and Propagation*, Vol. AP-17, pp. 747–751, November 1969.
5. E. D. Sharp, "A Triangular Arrangement of Planar Array Elements that Reduces the Number Needed," *IEEE Trans. Antennas and Propagation*, Vol. AP-9, pp. 126–129, March 1961.
6. B. L. Diamond, "Resonance Phenomena in Waveguide Arrays," *IEEE G-AP International Symposium Digest*, pp. 110–115, 1967.

7. V. Galindo and C. P. Wu, "Dielectric Loaded and Covered Rectangular Waveguide Phased Array," *Bell System Tech. J.*, Vol. 47, pp. 93–116, January 1968.

8. N. Amitay and V. Galindo, "Characteristics of Dielectric Loaded and Covered Circular Waveguide Phased Arrays," *IEEE Trans. Antennas and Propagation*, Vol. AP-17, pp. 722–729, November 1969.

9. C. P. Wu and V. Galindo, "Surface Wave Effects on Phased Arrays of Rectangular Waveguides Loaded with Dielectric Plugs," *IEEE Trans. Antennas and Propagation*, Vol. AP-16, pp. 358–360, May 1968.

10. C. P. Wu and V. Galindo, "Surface Wave Effects on Dielectric Sheathed Phased Arrays of Rectangular Waveguides," *Bell System Tech. J.*, Vol. 47, pp. 117–142, January 1968.

11. V. Galindo and C. P. Wu, "On the Asymptotic Decay of Coupling for Infinite Phased Arrays," *Proc. IEEE*, Vol. 56, pp. 1872–1880, November 1968.

12. L. Stark, "Radiation Impedance of a Dipole in an Infinite Planar Phased Array," *Radio Science*, Vol. 1, pp. 361–378, March 1966.

13. P. W. Hannan, "The Ultimate Decay of Mutual Coupling in a Planar Array Antenna," *IEEE Trans. Antennas and Propagation*, Vol. AP-14, pp. 246–248, March 1966.

14. A. A. Oliner and R. G. Malech, in *Microwave Scanning Antennas* (R. C. Hansen, ed.), Vol. II, Academic Press, New York, 1964, Chapter 4.

15. R. Mittra, "Relative Convergence of the Solution of a Doubly Infinite Set of Equations," *J. Research* (*Radio Science*), Series D, Vol. 67D, pp. 245–254, March–April 1963.

16. N. Amitay and V. Galindo, "On the Scalar Product of Certain Circular and Cartesian Wave Functions," *IEEE Trans. Microwave Theory and Techniques*, Vol. MTT-16, pp. 265–266, April 1968.

17. W. R. Smythe, *Static and Dynamic Electricity*, McGraw-Hill Book Company, New York, 1950, pp. 173–177.

8. Finite Arrays, Edge Effects, and Aperiodic Arrays

In the preceding chapters, we discussed, for the most part, the analyses and solutions of infinite phased arrays. An interesting exception was the discussion of a finite group of parallel-plate elements excited within an infinite array environment (Chapter 4). A rather generalized extension of those results is possible and will be made in this chapter. This extension is desirable, since the *edge* effects in finite arrays, whether terminated in a ground plane or placed in an infinite array environment, become extremely significant under at least two specific conditions. One condition is, clearly, that in which very small arrays are under consideration. A second condition is associated with the situation of forced *surface waves*, which are found to occur in certain arrays. In this chapter we will treat the theory and analysis of arrays under both of these conditions.

The case of a small finite parallel-plate array (with and without dielectric plugs or a dielectric sheath cover) in an infinite ground plane will be discussed. Integral equations for such arrays will be derived and solved by the Ritz-Galerkin method. The effects due to the finite size of these arrays will be compared to the results found for infinite arrays.

The case of a finite array embedded in a *modulated* or *corrugated* infinite surface can be treated in accordance with a general theory of aperiodic arrays.[1] The general aperiodic array is assumed to consist of a finite group of excited elements surrounded by *inactive* waveguides with any termination impedance (including short circuits), which varies from element to element in a periodic manner.

A problem that is related to this modulated surface is the infinite phased array wherein *all* the elements are excited and all are terminated in their characteristic impedances. It is shown that the boundary value problem solution to any aperiodic array problem can be found from the related phased array boundary value solution, *uniquely* and *explicitly*. This is true regardless of whether the solutions are obtained by *exact*, *numerical*, or *experimental* analysis.

Of particular interest is the "dispersion" relation, for any given periodically modulated surface, which can be found from the solution to the related phased array. The connection between this dispersion relation and the surface wave effect observed in the operation of phased arrays will be made clear.

1. APERIODIC ARRAYS AND MODULATED SURFACES

Although considerable effort has been expended in obtaining solutions to *periodic* phased arrays, certain types of aperiodic arrays are also of great interest. For example, there is the case wherein a *finite* group of elements in an infinite periodic array becomes short circuited or develops some terminal impedance other than the characteristic impedance. This condition may develop accidentally by failure of the diode or other components in the feed network of the array. In some circumstances an aperiodic arrangement of elements may be used *purposefully*[2] in order to suppress the occurrence of grating lobes and possibly surface waves. This will be discussed further.

Another case of interest is that in which only a finite group of elements in an infinite periodic array is excited. The remaining infinite number of elements may be terminated in any impedance, including short circuits. (The case in which the infinite remaining elements are terminated in their characteristic impedance was discussed in Chapter 4.) In fact, the infinite remaining elements can be terminated in impedances which vary from element to element in any manner so long as the elements that are asymptotically distant from the finite group of excited elements have *periodic* terminal impedance variations from element to element. We call this a modulated surface. This case is of interest for two reasons. The first reason is that we have here a type of *edge effect* wherein a possibly small finite array is immersed in an infinite modulated surface impedance. Second, we shall see that a solution to this problem will lead to a solution to the homogeneous problem of surface wave propagation on a modulated corrugated surface.

1.1 Solution to the General Aperiodic Array

An arbitrary aperiodic array is shown in Fig. 8.1. It is derived from an infinite array of identical elements by aperiodically shorting* the elements.

* Actually, an aperiodic *arbitrary termination* of the waveguides could be allowed. However, to simplify the subsequent analysis, we will assume a perfect short-circuited termination. We will show later how the results may be simply generalized to include the arbitrary termination.

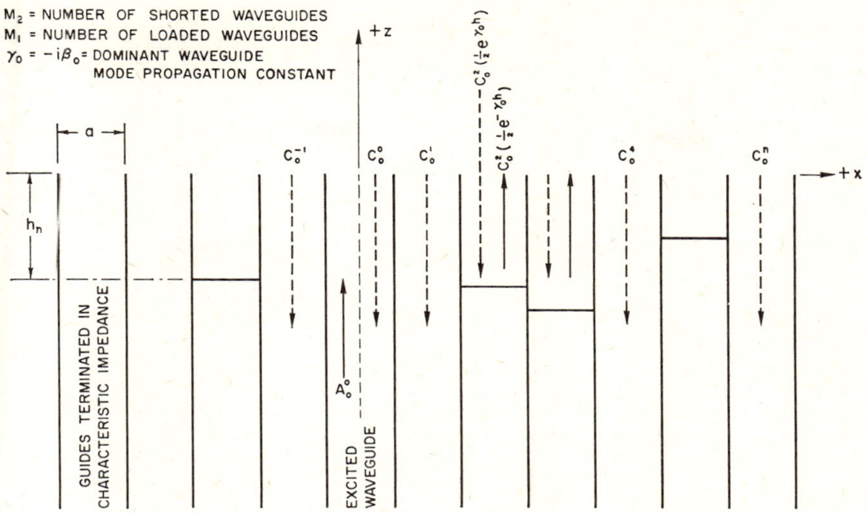

Fig. 8.1. Aperiodic array.

The shorting depth, h_n, need not be fixed but may vary with n from element to element. At large distances from the single excited element in Fig. 8.1, $|n|$ very large, we require that the array be periodic. This may take a form wherein either the elements are shorted at a periodically modulated depth ($M_2 = \infty$) or they are terminated in their characteristic impedance ($M_1 = \infty$).

For a particular array of thin-walled parallel plates, the above two related problems were solved first by Lee,[3] using a Weiner-Hopf method. Lee's results suggested that the solution may be obtained from a superposition of the solutions to the periodic phased array problem (see Fig. 8.2). The latter approach is more general in that it enables us to extend the results to any array-element type. Moreover, it permits a derivation of the dispersion relation mentioned earlier. The superposition method of solution will be treated in the subsequent discussion.

We will consider the case for $M_1 = \infty$ first, since this solution is more straightforward. The case for $M_2 = \infty$ (i.e., the case in which for large $|n|$ (Fig. 8.2) the elements are short circuited at a depth in the waveguides which varies periodically with n) will lead naturally, as will be seen, to a solution for the dispersion relation for the periodically modulated surface.

1.1.1 The Case with a Finite Number of Elements Short Circuited. With reference to Fig. 8.1, we allow a single element in the array to be excited with complex amplitude A_0^0. The subscript denotes the dominant mode; the superscript, the location of the waveguide, in this case at the center. The

ψ = kb sin θ, where θ = scan angle away from broadside
$R(\psi)$ = reflection coefficient (all guides excited as a phased array)

$$C_{p0} = \frac{1}{2W} \int_{-\pi}^{\pi} R(\psi) e^{-jp\psi} \, d\psi$$

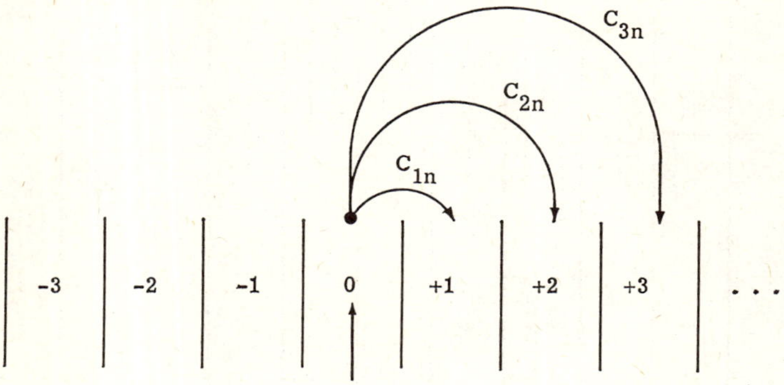

$$C_{pn} = \frac{1}{2\pi} \int_{-\pi}^{\pi} I_n(\psi) e^{-jp\psi} \, d\psi$$

I_n = modal coefficients of interior field (all guides excited as a phased array)

Fig. 8.2. Periodic array.

coefficients of the dominant propagating modes in the waveguides are as designated in Fig. 8.1. The distance to a short, h_n, is assumed sufficiently large so that all evanescent waveguide modes excited at the aperture ($z = 0$) are fully decayed at the short. Actually, the reflection of a finite number of evanescent modes from the short circuit (or any waveguide terminal impedance) can be taken into account.[1,3] This subject is discussed in Appendix I. The short position, h_n, may vary from element to element. The coefficients C_0^n are the coupling coefficients (for the zeroth, or dominant, waveguide mode and the nth element) with the short circuits present.

In Fig. 8.2 the *periodic* phased array is depicted. The coupling coefficients designated here are to be distinguished from those designated earlier. These are defined for the case in which no elements are short circuited; for example, $\mathscr{C}_{mn}$ designates the coupling to the mth removed waveguide (from the excited waveguide) in the nth waveguide mode.

The notation used and the schematic diagrams of the figures imply that the analysis is being presented for a *linear* array. The results are equally valid, however, for *planar* (two-dimensional) arrays, and the necessary simple extension for the planar array results is shown in Appendix II.

The basic problem, then, is to determine the C_0^n when M_2 is finite and greater than zero, assuming that the solution for $R(\psi)$, the reflection coefficient of the periodic array excited as a phased array, is known. Here ψ is the differential phase shift between elements. As shown in several studies[4,5] and in earlier chapters, we can obtain the $\mathscr{C}_{n0}$ from

$$\mathscr{C}_{n0} = \frac{1}{2\pi}\int_{-\pi}^{\pi} R(\psi)e^{-jn\psi}\,d\psi \tag{1}$$

Referring again to the aperiodic array in Fig. 8.1 (the case $M_1 = \infty$), we may use the periodic phased array coupling coefficients, $\mathscr{C}_{n0}$, and simply sum all the contributions that all waves incident upon the apertures ($z = 0$) have in coupling to a particular waveguide, say the nth, with a wave traveling into that waveguide. The unshorted waveguides have inward-traveling waves with coefficients given by $C_0{}^n$, and the shorted waveguides have inward-traveling waves with coefficients defined as

$$C_0{}^n \tfrac{1}{2} e^{\gamma_0 h_n}$$

(The constant γ_0 is taken as purely imaginary for the propagating mode.) This superposition leads to

$$\begin{aligned} C_0{}^n &= A_0{}^0\mathscr{C}_{n0} + \sum_{m\in M_2} (C_0{}^m \tfrac{1}{2} e^{-\gamma_0 h_m})\mathscr{C}_{n-m,0} \qquad \text{for} \quad n \in M_1 \\ C_0{}^n \tfrac{1}{2} e^{\gamma_0 h_n} &= A_0{}^0\mathscr{C}_{n0} + \sum_{m\in M_2} (C_0{}^m \tfrac{1}{2} e^{-\gamma_0 h_m})\mathscr{C}_{n-m,0} \qquad \text{for} \quad n \in M_2 \end{aligned} \tag{2}$$

where the symbol $n \in M_1$ denotes those values of n belonging to the set of M_1 waveguides.

Because the second equation of (2) contains a finite number of unknowns, $C_0{}^n$, assuming that the $\mathscr{C}_{n0}$ are known, it may be solved as an M_2th-order matrix for the $C_0{}^n$ in terms of the $\mathscr{C}_{n-m,0}$. The first equation is used to determine the remaining $C_0{}^n$ for $n \in M_1$.

The higher (evanescent) mode coefficients, $C_m{}^n$, are similarly obtained from the solution of the periodic array.[5,6] If the higher-order mode coefficients for the periodic array excited as a phased array are I_m, then (see Fig. 8.2) the higher-order mode coupling coefficients $\mathscr{C}_{nm}$ are found, as in (1), by

$$\mathscr{C}_{nm} = \frac{1}{2\pi}\int_{-\pi}^{\pi} I_m(\psi)e^{-jn\psi}\,d\psi \tag{3}$$

By the same principle of superposition as was used in (2), it follows that a set of linear equations for $C_m{}^n$ may be written (M_2 finite). Hence, a *complete* solution for the periodic phased array, in the sense that all the $I_m(\psi)$ and $R(\psi)$ are known, can be used for determining a *complete* boundary value solution for the aperiodic array problem. This will be seen to apply to the case when $M_2 = \infty$ also.

It should be mentioned here that, since $M_1 = \infty$, or for large $|n|$ the waveguides are terminated in their characteristic impedance, the *behavior* of the $C_0{}^n$ and $\mathscr{C}_{n0}$ must be identical in the asymptotic region for $|n| \to \infty$. It was pointed out in earlier chapters that for a linear array[6] the coefficients $\mathscr{C}_{n0}$, and now also $C_0{}^n$ $(n \in M_1)$, decay as $1/r^{3/2}$, where r is the distance from the excited element, and similarly for the planar array[7] as $1/r^2$. This should be apparent from physical considerations and can also be shown using (2).

1.1.2 The Case with an Infinite Number of Elements Short Circuited. We now will consider the case in which M_2, the number of shorted waveguides, is infinite. With reference to Fig. 8.1, it is necessary to assume that the shorting position h_n, at least for large $|n|$, varies in depth *periodically* with position n. For simplicity, we will assume in the following discussion that this periodicity in shorting positions is valid for all waveguide locations, n, although this is not a requirement.

The solution for the case in which M_1 is finite and M_2 is infinite cannot be obtained directly from (2), for then an infinite set of linear equations for the $C_0{}^n$ would result and this cannot be regarded as a solution. Instead, we shall use a somewhat different approach based on the periodic properties of the array reflection coefficients. If we now multiply each equation of (2) by $e^{jn\psi}$ and make use of the inverse Fourier series transform of (1)

$$\sum_{n=-\infty}^{\infty} \mathscr{C}_{n0} e^{jn\psi} = \sum_{n=-\infty}^{\infty} \mathscr{C}_{n-m,0} e^{j(n-m)\psi} = R(\psi) \tag{4}$$

and the fact that

$$\sum_{n \in M_1} + \sum_{n \in M_2} = \sum_{n=-\infty}^{\infty} \tag{5}$$

(i.e., the sum of the waveguides $M_1 + M_2$ = the total infinite number of waveguides in the array), we can directly, and uniquely, derive from (2) the following result:

$$\sum_{m \in M_1} C_0{}^m e^{jm\psi} + \sum_{m \in M_2} \{C_0{}^m \tfrac{1}{2}[e^{\gamma_0 h_m} - R(\psi) e^{-\gamma_0 h_m}]\} \cdot e^{jm\psi} = A_0{}^0 R(\psi) \tag{6}$$

A similar result was first obtained by Lee for parallel-plate arrays with thin walls. The derivation here shows, however, that the result is applicable to any type of waveguide element. Furthermore, it is more general in that the shorting position (or, more generally yet, the *terminal impedance*) may be an arbitrary function of m (waveguide position). In addition, the solution for the case in which M_2 is infinite can be derived from this equation.

As an example, consider the array in which $h_m = h$ independently of m, that is, h_m is kept constant with waveguide position. Then the factor $[e^{\gamma_0 h} - R(\psi) e^{-\gamma_0 h}]$ becomes independent of position m and may be factored out from the sums over M_2 in (6) to yield

$$\sum_{m\in M_1} C_0{}^m e^{jm\psi} + [e^{\gamma_0 h} - R(\psi)e^{-\gamma_0 h}]\cdot \sum_{m\in M_2} (C_0{}^m \tfrac{1}{2} e^{-\gamma_0 h})e^{jm\psi} = A_0{}^0 R(\psi) \quad (7)$$

We next divide (7) by $[e^{\gamma_0 h} - R(\psi)e^{-\gamma_0 h}]$, multiply (7) by $e^{-jn\psi}$, and then integrate from $\psi = -\pi$ to $\psi = +\pi$. By virtue of the following orthogonality relation:

$$\int_{-\pi}^{\pi} e^{j(m-n)\psi}\, d\psi = 2\pi\delta_{mn}$$

we can eliminate the term containing the infinite sum, that is, $\sum_{m\in M_2}$, and obtain

$$\frac{1}{2\pi}\sum_{m\in M_1} C_0{}^m \int_{-\pi}^{\pi} \frac{e^{j(m-n)\psi}}{e^{\gamma_0 h} - R(\psi)e^{-\gamma_0 h}}\, d\psi = A_0{}^0 \frac{1}{2\pi}\int_{-\pi}^{\pi} \frac{R(\psi)e^{-jn\psi}}{e^{\gamma_0 h} - R(\psi)e^{-\gamma_0 h}}\, d\psi \quad (7a)$$

for $n \in M_1$, and

$$\frac{1}{2\pi}\sum_{m\in M_1} C_0{}^m \int_{-\pi}^{\pi} \frac{e^{j(m-n)\psi}}{e^{\gamma_0 h} - R(\psi)e^{-\gamma_0 h}}\, d\psi + C_0{}^n \tfrac{1}{2} e^{-\gamma_0 h}$$
$$= A_0{}^0 \frac{1}{2\pi}\int_{-\pi}^{\pi} \frac{R(\psi)e^{-jn\psi}}{e^{\gamma_0 h} - R(\psi)e^{-\gamma_0 h}}\, d\psi \quad (7b)$$

for $n \in M_2$.

Equation 7a may be solved as an M_1th-order matrix equation for the $C_0{}^n$ when $n \in M_1$. For $n \in M_2$, (7b) may be solved *subsequently*, since the summation in this equation can be carried out directly after finding the $C_0{}^n$ ($n \in M_1$) from (7a).

It should be clear that the division of (7) by the factor $[e^{\gamma_0 h} - R(\psi)e^{-\gamma_0 h}]$ is permissible only when

$$[e^{\gamma_0 h} - R(\psi)e^{-\gamma_0 h}] \neq 0 \quad (8)$$

for any ψ in the range $-\pi \leq \psi \leq \pi$. This singularity will, however, be integrable in ψ and can, in fact, be shown to determine the asymptotic behavior, for large $|n|$, of the $C_0{}^n$ and $C_m{}^n$.[3,6,7] This will be discussed further in a later section of this chapter.

Actually, by setting the left-hand side of (8) equal to zero, we have the characteristic equation solution for the corrugated surface of waveguide elements which are short circuited at depth h. With reference to Fig. 8.1, we do, in fact, couple into a surface wave on the periodic short-circuited structure. This corresponds to the anomalous result of total reflection, which is frequently reported for periodic phased arrays.[8,9] Note that $R(\psi_0) = 1e^{2\gamma_0 h}$ or $|R(\psi_0)| = 1$ (total reflection) satisfies (8) with the right-hand side equal to zero. Hence, if total reflection of a phased array occurs, so that $R(\psi_0) = e^{j\phi}$, it is possible to place a true short circuit at $h = j(\phi + 2\pi l)/2\gamma_0$, $l =$ integer, and to find a surface wave solution to the homogeneous problem of the corrugated structure.[8–10]

When h_m is an arbitrary *periodic* function of m and not a constant, it is still possible to "factor" an appropriate function of $R(\psi)$ from the *infinite* sum, $\sum_{m \in M_2}$, as we have done in (7), and thus obtain a general solution for the case $M_2 = \infty$. The process of "factoring" $R(\psi)$ from the infinite sum [so as to obtain an M_1th-order matrix to solve as in (7)] is straightforward in principle but is not literally a simple factorization. The algebraic complexity of this "factorization" procedure increases with the number of waveguide elements in a period of the periodic function h_m.

Consider, for example, the case for which

$$\begin{aligned} h_m &= h = \text{constant} \qquad \text{for} \quad m \text{ even} \\ h_m &= c = \text{constant} \qquad \text{for} \quad m \text{ odd} \end{aligned} \tag{9}$$

This choice corresponds to the modulated surface of Karjala and Mittra[11] and of Hessel and Hochstadt,[12] when the array is a set of parallel plates. Let the group $M_2 = M_2{}^e + M_2{}^0$, where $m \in M_2{}^e$ if m is even, and $m \in M_2{}^0$ if m is odd. Furthermore, define $R'(\psi) \equiv R(\psi + \pi)$. It is then possible to write (6) *twice*, once for ψ and once for $\psi + \pi$ so as to form the linearly independent set of equations

$$(e^{\gamma_0 h} - Re^{-\gamma_0 h}) \sum_{m \in M_2{}^e} (C_0{}^m \tfrac{1}{2} e^{jm\psi}) + (e^{\gamma_0 C} - Re^{-\gamma_0 C}) \sum_{m \in M_2{}^0} (C_0{}^m \tfrac{1}{2} e^{jm\psi})$$
$$= A_0{}^0 R - \sum_{m \in M_1} C_0{}^m e^{jm\psi} \tag{10a}$$

$$(e^{\gamma_0 h} - R'e^{-\gamma_0 h}) \sum_{m \in M_2{}^e} (C_0{}^m \tfrac{1}{2} e^{jm\psi}) - (e^{\gamma_0 C} - R'e^{-\gamma_0 C}) \sum_{m \in M_2{}^0} (C_0{}^m \tfrac{1}{2} e^{jm\psi})$$
$$= A_0{}^0 R' - \sum_{m \in M_1} C_0{}^m (-1)^m e^{jm\psi} \tag{10b}$$

Using (10a) and (10b), we may solve for

$$\sum_{m \in M_2{}^e} C_0{}^m \tfrac{1}{2} e^{jm\psi} \quad \text{and} \quad \sum_{m \in M_2{}^0} C_0{}^m \tfrac{1}{2} e^{jm\psi}$$

and then proceed as before to find $C_0{}^m$. In the process, we find a determinantal equation

$$\mathscr{D}_2 = (e^{\gamma_0 h} - Re^{-\gamma_0 h})(e^{\gamma_0 C} - R'e^{-\gamma_0 C}) + (e^{\gamma_0 C} - Re^{-\gamma_0 C})(e^{\gamma_0 h} - R'e^{-\gamma_0 h}) \tag{11}$$

which is identical to the characteristic equation for a modulated corrugated surface.

As we will discuss further, if the above determinant vanishes for some ψ, then we have coupled into a surface wave along the corrugated surface. Relation (11) is in fact the same as that found by Karjala and Mittra and by

Hessel and Hochstadt, except that they derived the relation for a thin-walled parallel-plate array. In (11), we have generalized the characteristic determinantal relation, $\mathscr{D}_2 = 0$, to include any array, independently of its geometry. The functional dependence of $R(\psi)$ which is needed may be determined analytically or experimentally.

1.1.3 The Generalized Modulated Surface Dispersion Relation and Arbitrary Terminal Impedances. It is possible by induction (see Appendix III) to obtain a general dispersion relation for the arbitrary periodically modulated surface. When the shorting position, h_m, is asymptotically periodic (i.e., far from the exciting element or elements) over P elements, then we find, by a similar procedure, that the determinant of the appropriate linear set of P equations is

$$\mathscr{D}_P = \det \{(e^{\gamma_0 h_m} - R_n e^{-\gamma_0 h_m})e^{jm(n-1)(2\pi/P)}\} = \det \{\mathscr{D}_{P_{nm}}\}$$

where n = row position of the $\mathscr{D}_{P_{nm}}$ matrix element,

m = column position of the $\mathscr{D}_{P_{nm}}$ matrix element,

P = order of determinant, periodicity, and number of linear equations, and

$$R_n = R[\psi + (n-1)(2\pi/P)].$$

If we set this determinant equal to zero, we have the dispersion relation for a *periodic arbitrarily modulated surface.*

If

$$\mathscr{D}_P = \det \{[e^{\gamma_0 h_m} - R_n(\psi)e^{-\gamma_0 h_m}]e^{jm(n-1)(2\pi/P)}\} = 0$$

for some real ψ in the interval $-\pi \leq \psi \leq \pi$, this means that a surface wave will propagate on the modulated surface.

If the short circuits are replaced by *arbitrary terminal impedances* characterized by a complex reflection coefficient $= \Gamma_m$, referred to a depth h_m, the generalized determinant is*

$$\mathscr{D}_P^{\Gamma} = \det \{(e^{\gamma_0 h_m} - \Gamma_m R_n e^{-\gamma_0 h_m})e^{jm(n-1)(2\pi/P)}\}$$

The assumption of an arbitrary terminal impedance, or reflection coefficient, could have been made at the start of the analysis with very little additional complication.

The inclusion of Γ_m, with the possibility that $|\Gamma_m| < 1$, has some interesting implications. We would expect that if $|\Gamma_m| < 1$ (for any m in the P periodic waveguides in the asymptotic region), it would not be possible for a surface

* We will use the notation $\mathscr{D}_P^{\Gamma}$ for a determinant in which Γ is not necessarily equal to 1. The notation $\mathscr{D}_P$ will refer exclusively to *short-circuited* waveguide terminations in all asymptotically located waveguides.

wave to propagate unattenuated when a *finite* number of waveguides are excited (this case is to be distinguished from the *phased array* case in which an infinite number of waveguides are excited). This is easily demonstrated for the simple case $P = 1$, wherein

$$\mathscr{D}_1^{\Gamma} = [e^{\gamma_0 h} - \Gamma R(\psi) e^{-\gamma_0 h}]$$

The product $\Gamma R(\psi)$ can never have a magnitude equal to unity if $|\Gamma| < 1$, even if for some ψ_0 we get total reflection, $|R(\psi_0)| = 1$. Hence in this case $\mathscr{D}_1^{\Gamma} \neq 0$ *for any* ψ, and no surface waves are excited.

It should be added that, even if

$$\mathscr{D}_P = \det\{(e^{\gamma_0 h_m} - R_n e^{-\gamma_0 h_m}) e^{jm(n-1)2(\pi/P)}\} = 0$$

we may still solve for the C_0^m and C_n^m. For example, when $h_m = h =$ constant, we can still integrate:

$$\int_{-\pi}^{\pi} \frac{e^{j(m-n)\psi}}{e^{\gamma_0 h} - R(\psi) e^{-\gamma_0 h}}\, d\psi \quad \text{and} \quad \int_{-\pi}^{\pi} \frac{R(\psi) e^{-jn\psi}}{e^{\gamma_0 h} - R(\psi) e^{-\gamma_0 h}}\, d\psi$$

even with a zero of $[e^{\gamma_0 h} - R(\psi_0) e^{-\gamma_0 h}]$. We will find, if such a zero exists, that we have *coupled* to a surface wave and the C_0^m (*and* C_n^m) *will not decay asymptotically* with m. This will be discussed further in a later section.

1.2 The Fields in the Open Region—the Radiation Fields

The determined coefficients, C_0^n for the dominant waveguide propagating mode and C_m^n for the mth waveguide mode, determine the fields for the waveguide region ($z \leq 0$) in either the $M_1 = \infty$ or the $M_2 = \infty$ case.

Whether $M_1 = \infty$ or $M_2 = \infty$, the total field at the aperture plane ($z = 0$) is known from a knowledge of the coefficients C_m^n. The total fields in the near and far zones for the open region ($z > 0$) can be determined by an appropriate Fourier transform of the field in the aperture plane.[13] In the cases in which a surface wave solution to

$$\det\{(e^{\gamma_0 h_m} - R_n e^{-\gamma_0 h_m}) e^{jm(n-1)(2\pi/p)}\} = 0 \tag{12}$$

exists for one or more discrete values of ψ (see Section 1.1), the transformed aperture field will contain poles, the residues of which will yield the coupling to the surface waves. An alternative approach for determining the total field by simple superposition is essentially identical to that described next for determining the far-zone radiated fields.

If only the far-zone or radiation fields of the aperiodic array are desired, a much simpler and more straightforward approach exists which requires a knowledge of only the dominant mode coefficients, C_0^n. A solution for the periodic phased array problem yields,[4,8] in addition to the reflection coefficient $R(\psi)$, a transmission coefficient, $T(\psi) = T(kb \sin \theta)$ (see Fig. 8.2), which can be shown[14] to be proportional to the far-zone pattern of a single element excited in the infinite array when $M_2 = 0$ (no waveguides short circuited). In the aperiodic case with $M_2 > 0$, it is then possible to add the contributions from all elements which have waves traveling toward the array aperture. These include the short-circuited elements and the centrally excited element. The appropriate coefficient $(C_0{}^n \tfrac{1}{2} e^{-\gamma_0 h_n})$ and displacement phase factor $(e^{+jn\psi})$ must be multiplied by $T(\psi)$ for each element. The complex sum (infinite for $M_2 = \infty$) determines the far-zone radiation field for the single $(n = 0)$ element excited in the presence of the infinite aperiodic structure:

$$f(\theta) = \left[A_0{}^0 + \sum_{n \in M_2} (C_0{}^n \tfrac{1}{2} e^{-\gamma_0 h_n}) e^{jn\psi} \right] T(\psi) \tag{13}$$

where $\psi = ka \sin \theta$.

As mentioned earlier, a similar simple superposition of the near-zone solutions of the periodic phased array will give the total field solution to the aperiodic array in the $(z > 0)$ region. Far-zone radiation patterns, coupling coefficients, reflection coefficients, and other results for a finite array in an infinite corrugated surface will be illustrated in this chapter when *edge effects* are discussed in more detail.

1.3 The Suppression of Surface Waves and Improving Match

The occurrence of surface waves in an infinite phased array has been shown to be intimately related to a homogeneous boundary value solution of a corresponding corrugated structure. In fact, once the phased array solution for the reflection coefficient, $R(\psi)$, is found for all ψ in the range $-\pi \leq \psi \leq \pi$, the characteristic determinant, $\mathscr{D}_P$ or $\mathscr{D}_P{}^\Gamma$, is found. The determinant $\mathscr{D}_P$ will indicate precisely whether a surface wave will propagate on any given corrugated structure. The total field solution, interior and exterior to the waveguides, can also be found for the corrugated structure.

This information also provides us with a design approach which can lead to a phased array that does not produce surface waves in the visible region and that is reasonably well matched. In fact, the dispersion relationship $\mathscr{D}_P$ or $\mathscr{D}_P{}^\Gamma$ provides a means for designing and testing the use of *periodically* placed chokes or other impedances in a large array for the purpose of suppressing forced surface waves. The input impedance to such a periodically choked array can then be found from a knowledge of the input impedance to the related unchoked array. After the presence of excited surface waves is removed from the large array (by chokes or other means), we can consider systematic methods[15,16] for improving the match versus scan characteristics of the phased array (see Chapter 9 also). The chokes or periodically placed terminal impedances can also be used for matching purposes.

1.3.1 Periodically Choked Arrays. If we *periodically* short-circuit an element of the infinite phased array, we will of course alter the impedance match of the remaining elements. The periodically placed short circuits can also be used to suppress the surface waves that may be excited when the array is fully excited (all elements active in the infinite array).

The solution for the periodically terminated array shown schematically in Fig. 8.3 was first presented by Dufort.[17a,b] This array differs from those we

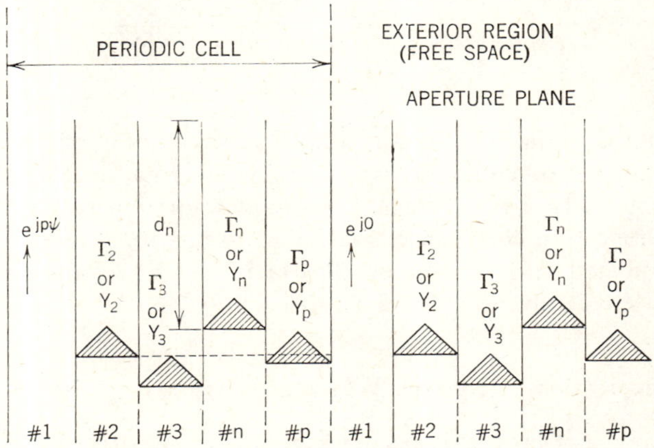

Fig. 8.3. Periodically terminated array—period of P elements.

have been considering in that here an *infinite* number of elements are excited. The situation treated earlier, in which a period of P elements were short circuited (or otherwise terminated) in a *modulated* fashion, is actually related to this case, as may be expected, but we will analyze this array with a somewhat different approach.

The solution for the periodically terminated array of Fig. 8.3 can be obtained in terms of the solution for the (fully excited) phased array by superposition of the latter solution. Instead of terminating the elements in Fig. 8.3 with the terminal admittances* Y_n (referred to a depth d_n), we will temporarily terminate all the P elements in their characteristic impedance. As each element, separately, is excited by an input wave of unit amplitude, the waves reflected into each of the other elements and transmitted to free

* An equivalent terminal *reflection*, Γ_n, can of course be used. If a short circuit is desired, we simply have $Y_n = -j \cot(j\gamma_0 d_n)$ or $\Gamma_n = -e^{2\gamma_0 d_n}$, $j\gamma_0$ being the waveguide propagation constant (purely imaginary) for the propagating mode.

space can be computed. These *scattering* coefficients form the elements of an S matrix[17c] of $(P + 1)$th order for the periodically excited array. Once the S matrix is determined, we can find the solution for any set of terminations or excitations.

We will now superimpose the solutions for the fully excited array as indicated in Fig. 8.4. The object is to superimpose a set of solutions such that only the *qth element* is excited after superpositioning the excitations. The linear phase taper in Fig. 8.4 is the mth such excitation ($m = 0, 1, 2, \ldots,$

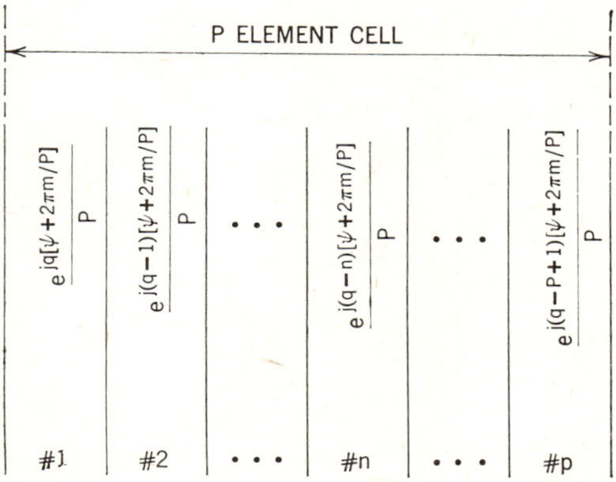

Fig. 8.4. Superimposed linear excitations; n = running index of element, m = fixed for any one linear phase excitation, q = fixed for all elements and excitations (determines the resultant excited element).

$P - 1$) to be superimposed. The total excitation on any one element is clearly

$$\delta_{qn} = \sum_{m=0}^{P-1} \frac{1}{P} e^{j(q-n)[\psi+(2\pi m/P)]} \tag{14}$$

where δ_{qn} is the Kronecker delta. If $R(\psi)$ is the reflection coefficient for the fully excited array element with e^{j0} excitation, it is clear, by superposition, that

$$\mathscr{R}_{n,q}(\psi) = \sum_{m=0}^{P-1} R\left(\psi + \frac{2\pi m}{P}\right) \frac{e^{j(q-n)[\psi+(2\pi m/P)]}}{P} \tag{15}$$

is the reflection into the nth element when the qth element is excited in the periodically excited array.

If $T_q(\psi)$ is the transmission coefficient relative to the qth element of a *fully* excited array,* then

$$\tau_q(\psi) = \sum_{m=0}^{P-1} \frac{T_q[\psi + (2\pi m/P)]}{P} \tag{16}$$

* If $T(\psi)$ is the transmission coefficient relative to the element excited by a wave with e^{j0} coefficient in the fully excited array, then $T(\psi)e^{jq\psi} \equiv T_q(\psi)$.

where τ_q is the transmission coefficient of the qth element of the P element cell of Fig. 8.5.

Since the excitation of the waveguides is unchanged for a wave incident from the exterior region, the scattering or S matrix for the $P + 1$ ports is known. With the S matrix determined, it is then possible to terminate the array as shown in Fig. 8.3 and to solve for the complex reflection coefficient of the excited element and other characteristics of the array. If the $P - 1$ terminations are chosen as short circuits

$$\Gamma_n = -e^{2\gamma_0 d_n}$$

then the short-circuit positions, d_n, are chosen so as to eliminate any surface

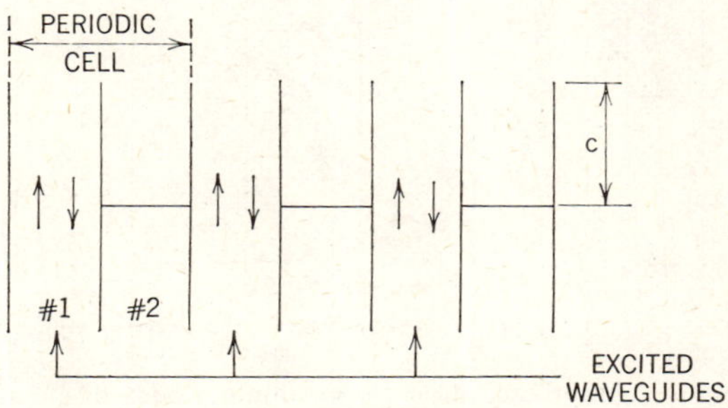

Fig. 8.5. Periodically shorted array, $P = 2$.

wave associated with the fully excited array and to improve the match[17(b),18] as a function of differential phase shift or scan angle ψ.

1.3.2 Periodically Choked Arrays and the Characteristic Determinant. Actually, once the surface wave effect has been eliminated, it is possible to employ other very useful *synthesis* methods to improve the match of the array. Some such synthesis techniques are discussed in Chapter 9. The most significant objective is, therefore, to eliminate a surface wave that may occur in the fully excited array. It is generally possible to do this by employing a *periodically choked* array as shown in Fig. 8.3 ($\Gamma_n = -e^{2\gamma_0 d_n}$) and utilizing the S matrix derived in the preceding section.

An alternative analytical approach is to use the characteristic determinant*

$$\mathscr{D}_P = \det\{(e^{\gamma_0 d_m} - R_n e^{-\gamma_0 d_m})e^{jm(n-1)(2\pi/P)}\}$$

(see Section 1.1.3) to determine the short positions, d_m, so that no surface wave will be excited on the periodically choked array.

* If $|\Gamma_n| \neq 1$, the determinant $\mathscr{D}_P{}^\Gamma$ (see Section 1.1.3) can be used. However, as was shown earlier, it is unlikely that a surface wave will be excited on such an array. Of course, an array of this type is lossy and therefore generally undesirable.

How and why the determinant, $\mathscr{D}_P$, is used to determine an appropriate short position will be illustrated for the case $P = 2$, in which only one choke and one active element are used. We consider an array such as that shown schematically in Fig. 8.5. In this array, one short position, d_2, is fixed at $d_2 \equiv C$, while the other waveguide is excited. It is conceivable that there exists a short position for guide 1,

$$d_1 \equiv h$$

such that the resulting structure, a modulated short-circuited surface wave structure, will support a surface wave. For example, Karjala and Mittra[11] provide solutions for h and C such that a surface wave will exist for the thin-walled parallel-plate array. Of course setting $\mathscr{D}_2 = 0$ [see (11)] provides just that solution when $R(\psi)$ for the thin-walled parallel-plate array[4] is used—in full agreement with Karjala and Mittra.

The design objective is, then, clearly to choose a value of C for d_2 such that

$$\mathscr{D}_2 = (e^{\gamma_0 h} - Re^{-\gamma_0 h})(e^{\gamma_0 C} - R'e^{-\gamma_0 C}) + (e^{\gamma_0 C} - Re^{-\gamma_0 C})(e^{\gamma_0 h} - R'e^{-\gamma_0 h}) \neq 0$$

for all possible values of h and of ψ in the visible region. If it is assumed that the shorting depth, h, is limited to a range of one guide wavelength, $\lambda_g = 2\pi/j\gamma_0$, the problem is three dimensional in the range $(-\pi \leq \psi \leq \pi)$ and

$$\left(-\frac{\lambda_g}{2} + n\lambda_g\right) \leq h \leq \left(\frac{\lambda_g}{2} + n\lambda_g\right), \quad n \text{ any integer}$$

We develop a *surface* in the (ψ, h, C) space by use of

$$\mathscr{D}_2 = 0$$

If this surface projects over a range greater than λ_g on the C axis, then no solution for C exists such that $\mathscr{D}_2 \neq 0$, and we must set $P = 3$ and look for a solution for *two* shorting positions which give $\mathscr{D}_3 \neq 0$. If the surface generated by setting $\mathscr{D}_2 = 0$ does not project in this way, we may choose any C *outside* the projected range. With this value of C, $\mathscr{D}_2$ *cannot* be zero, and therefore no surface wave can be generated.

1.3.3 Other Methods for Suppressing Surface Waves. There are, of course, methods for eliminating the surface waves which occur in infinite phased arrays other than those discussed here. The most obvious approach is simply to redesign the basic waveguide element. This approach is quite practical when the "cause" of the surface wave is understood to some extent.

For example, the presence of dielectric plugs and sheaths in the waveguides is known to cause surface wave resonances in the array in some circumstances (see Chapters 6 and 7 for an extended discussion on these effects). If a change in the dielectric constant or thickness of a plug or sheath can eliminate the surface wave, this is generally much simpler to implement than the methods discussed in this chapter. In the case of a dielectric sheath, in theory at least, it is possible to raise the sheath away from the aperture plane[19] so that no evanescent free space modes couple to the sheath. If this is accomplished, no surface wave will propagate on the sheath. A similar procedure could of course be followed with a waveguide plug by moving it away from the aperture plane into the waveguide. Of course these adjustments are not always practical; furthermore, surface waves are often generated independently of the presence of dielectric materials.[20,21]

Another approach, often used to reduce grating lobes in a fixed (or nonscanning) array, is that of using randomly spaced elements.[22,23] This method usually involves thinning or removing elements from the central portion of the array. The reduction in the number of elements thus allows for substantial cost reduction. However, for a scanning array, the randomness of spacing makes it extremely difficult to program and implement scanning or phase shifting for each element. An innovation[2] which helps to overcome many of these difficulties is accomplished by choosing the element locations of a random array in such a manner that the array is a selection of elements of a periodic array. An example of such an array is a periodic array in which randomly selected elements are short circuited (possibly at random depths). Such an array would not support propagation of a surface wave.

The analysis of such an array could proceed along the lines discussed in Sections 1.1.1–1.1.3 if we consider the random or aperiodically shorted elements to be finite and surrounded by an infinite modulated corrugated structure. Alternatively, the *finite* array can be surrounded by an infinite ground plane. The analysis of finite arrays of this type and the *edge* effects associated with these arrays, together with some results, will be discussed in Section 2.

2. EDGES EFFECTS AND FINITE ARRAYS

The discussion and analysis of infinite phased arrays throughout this book is based on the idea that for very large arrays, at least, most of the "centrally" located elements behave as if they were placed in an infinite array environment. The validity of this concept has already received some quantitative support for the case of parallel-plate phased arrays, as discussed in Chapter 4. Strictly and realistically speaking, all arrays may be viewed as a *finite* group of elements that are placed in some sort of infinite environment. The nature of the infinite environment can vary and will determine what we call the edge effects of the array.

The *edge effects* are, by definition, the changes in the input impedance or reflection coefficient, the transmission coefficient or radiation pattern, and the aperture fields of a given element arising because of its proximity to the edge of the array. Since, in fact, all elements of any array are affected by the finiteness of the array, this terminology has significance only in the practical sense. Some numerical data will therefore be presented to illustrate these effects.

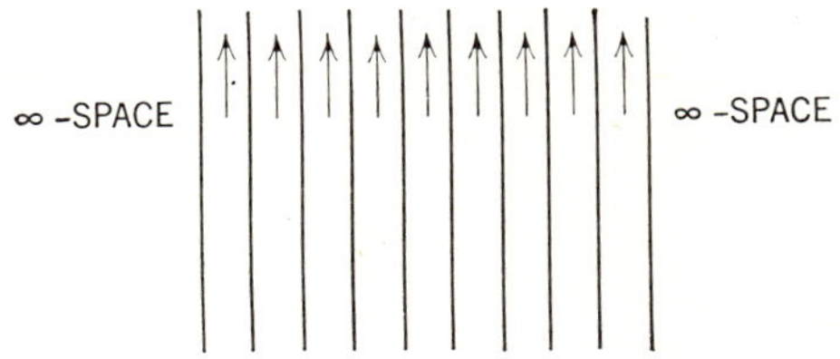

Fig. 8.6*a*. Finite array in infinite space.

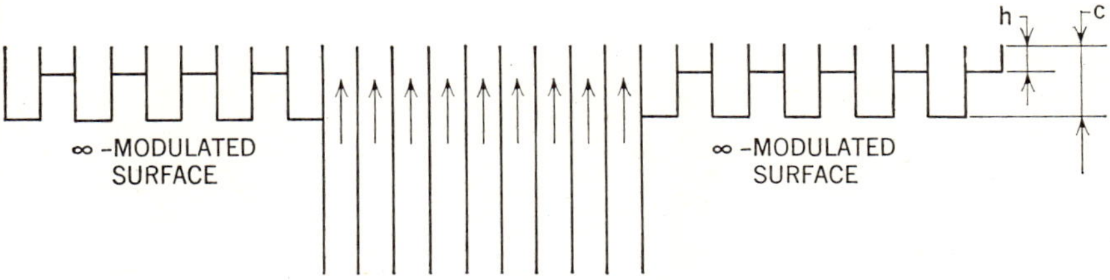

Fig. 8.6*b*. Finite array in modulated surface.

Depending on the environment and size of the array, the *edge* effects can significantly influence its performance. In some circumstances, these effects are significantly independent of the array size.

Several types of array environments of importance are illustrated schematically in Figs. 8.6–8.7. In Fig. 8.6*a*, the finite group of waveguides is immersed completely in free space. The excited waveguides themselves form the boundary geometry of the array. Numerical techniques similar to those employed throughout much of this book could be used effectively to solve array problems with this geometry, although with considerably more effort.

Other array edges such as those shown in Figs. 8.6*b*–8.7*b* are more readily analyzed by the techniques described in this book. For example, the finite array immersed in an infinite, passive, and terminated set of elements such as that illustrated in Fig. 8.7*b* was analyzed[4] in Chapter 4. The mathematical techniques for analyzing the finite array that is embedded in a short-circuited modulated surface such as that shown in Fig. 8.6*b* were largely developed[1] in the first part of this chapter. Finally, the array that is terminated in a ground plane, as shown in Fig. 8.7*a*, will be treated by very similar mathematical methods[5,24] later in this chapter.

Although the terminating surfaces which surround the finite arrays in Figs. 8.6*b*–8.7*b* are actually infinite in dimension, they often can be *cut to a finite size without affecting the predicted performance* of any excited element in the array, central or edge in position. That is, after performing the mathematical analysis of the arrays in Figs. 8.6*b*–8.7*b*, we can determine the

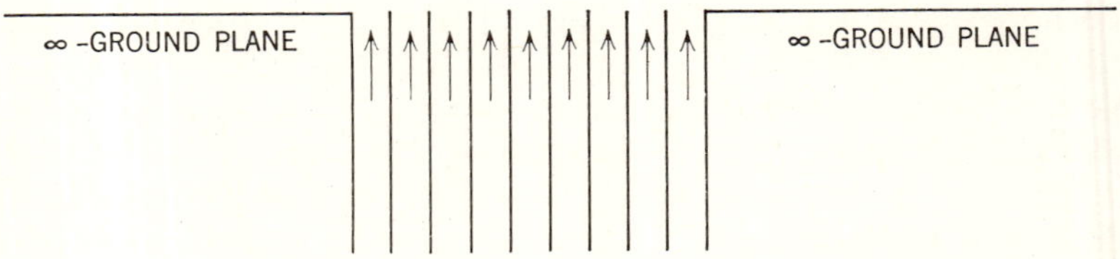

Fig. 8.7*a*. Finite array in ground plane.

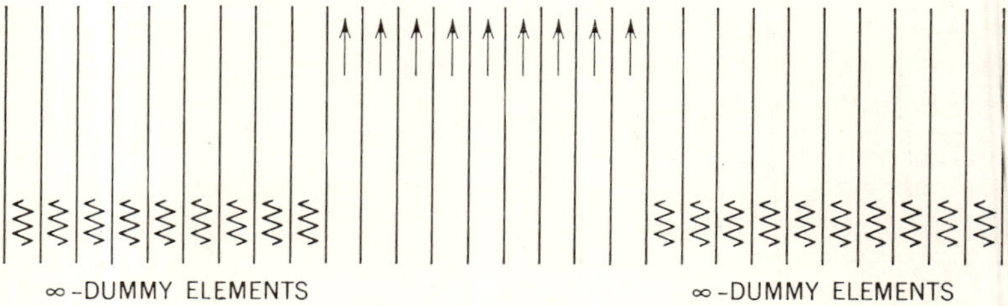

Fig. 8.7*b*. Finite array in passive elements.

amplitude of the fields coupled along the edges of the modulated surface, the passively terminated waveguides, and the ground plane. At some distance from the finite array, the fields along the edge (modulated surface, ground plane, etc.) may decay to a sufficiently small value so that, by abruptly removing the remaining portion of the edge (modulated surface, ground plane, etc.), we will have a sufficiently small effect on the finite array performance. The net result is a finite array with a *finite edge* and a prescribed or at least *predictable* edge effect. Under some circumstances, we shall see that we cannot remove any portion of the modulated surface edge without affecting the finite array performance.

The edge effects for a rectangular waveguide array in the H and quasi-E planes of scan have been discussed for the case of Fig. 8.7*b*. The results for finite arrays of five and of nine elements were presented in Chapter 4. The results for a *semi-infinite* array of excited elements are also presented. It was shown that, if $\mathscr{C}_m$ is the coupling coefficient from the zeroth element to the

mth element of an array, the reflection back into the zeroth element of an array of $M + N + 1$ elements is given by

$$R(M, N) = \sum_{m=-M}^{N} \mathscr{C}_m e^{-jm\psi} \tag{17}$$

at the phase scan angle ψ, when there are M excited elements to the left and N excited elements to the right of the zeroth element. If we let $M = 0$, we obtain the result for an edge element. It was concluded from the computations in Chapter 4 that an element removed by at least five elements from the edge will behave as if it were infinitely removed from the edge, for most practical applications.

We can conclude, therefore, that a finite array of rectangular waveguides can be surrounded by five dummy or passive elements (in the H or quasi-E plane of scan) in order to *simulate the effect of an infinite array.* Since the fields and coupling coefficients away from the excited elements decay asymptotically as $1/r^{3/2}$ for the linear array[6] and $1/r^2$ for the planar array,[7] it is *always* possible to terminate this type of array at some point with a finite number of dummy elements without significantly affecting the array performance.

One difficulty with using a passively terminated array of dummy elements at the edge is that these elements absorb power and thus introduce loss. Also, although the edge effects caused by these dummy elements are predictable, these effects are not controllable. In other words, unless the basic element design is changed, no control parameters are available if we terminate the array as shown in Fig. 8.7*b*.

The ground plane termination shown in Fig. 8.7*a* does not have easily controlled parameters which will allow for a *designed* edge effect. Although the ground plane edge termination will not have the losses that the dummy element termination will have, the results are somewhat more difficult or tedious to predict or to compute numerically.

The "corrugated" or modulated edge termination shown schematically in Fig. 8.6*b* is essentially a lossless termination and also has many parameters that can be controlled to design a "controlled" edge effect. An important difficulty here (aside from greater physical complexity and cost over the ground plane) is that it is sometimes possible to excite surface waves along the edges.

2.1 Finite Array in a Modulated Surface Environment

If we desire to immerse a finite array in a modulated surface environment such as that shown schematically in Fig. 8.6*b* and then terminate the modulated surface so that it is a *finite* modulated surface edge, we must use some of the mathematical techniques discussed in the earlier sections of this chapter to determine the edge effects and to avoid propagation of surface waves along the modulated surface.

Finding a solution to the array shown in Fig. 8.6*b* (when the modulated surface is infinite in extent) means that we must determine the coefficients $C_0{}^m$ (see Fig. 8.1). If $m \in M_1$, the excited waveguides, then $C_0{}^m$ denotes the reflection into the *m*th waveguide when only the zeroth guide is excited. The results for all M_1 waveguides excited may be determined by simple superposition. If $m \in M_2$, one of the short-circuited waveguides, then $(C_0{}^m \frac{1}{2} e^{\gamma_0 h_m})$ determines the reflection into the *m*th waveguide (with short circuit at depth h_m). We desire, therefore, when considering a truncation of the infinite modulated surface, to determine the behavior of the $C_0{}^m$ when $m \in M_2$ and, in particular, to determine the *decay* of the $C_0{}^m$ with increasing $|m|$. Thus we can determine a sufficiently large $|m|$ in which to truncate the modulated surface without affecting the finite array performance significantly. Furthermore, we can adjust the short circuit positions, h_m, and their period P (see Section 1.3.2) so as to control the $C_0{}^m$ near the edge of the array.

To determine the $C_0{}^m$ we make use of (6), which is repeated here for convenience:

$$\sum_{m \in M_1} C_0{}^m e^{jm\psi} + \sum_{m \in M_2} \{C_0{}^m \tfrac{1}{2}[e^{\gamma_0 h_m} - R(\psi) e^{-\gamma_0 h_m}]\} e^{jm\psi} = A_0{}^0 R(\psi) \tag{6}$$

where $M_2 = \infty$ and M_1 is finite. If the period of the modulated surface includes P short-circuited elements, we develop a set of P linear equations from (6) by considering the equation for P different values of ψ:

$$\psi_n = \psi + (n-1)\frac{2\pi}{P}, \quad n = 1, 2, \ldots, P$$

In Section 1.3.2, we discussed the determinant, $\mathscr{D}_P$, of this linear set of equations and some of the implications concerning the propagation of surface waves along the modulated surface. A solution for the case $h_m =$ constant $= h$ was carried through in (7). It may prove useful now to follow the solution through for the case in which $P = 2$, $h_m = h$, C (see Fig. 8.6*b*), so that the extension to arbitrary P will become clearer.

In (10a) and (10b), we have the pair of linear equations for the case $P = 2$. Using Cramer's rule, we solve for the

$$\sum_{m \in M_2{}^e} \quad \text{and} \quad \sum_{m \in M_2{}^0} \text{terms}$$

expressed in matrix notation as

$$\begin{bmatrix} \sum_{m\in M_2{}^e} (C_0{}^m \tfrac{1}{2} e^{jm\psi}) \\ \sum_{m\in M_2{}^0} (C_0{}^m \tfrac{1}{2} e^{jm\psi}) \end{bmatrix} = \begin{bmatrix} \dfrac{e^{\gamma_0 C} - R' e^{-\gamma_0 C}}{\mathscr{D}_2} & \dfrac{-(e^{\gamma_0 C} - R e^{-\gamma_0 C})}{\mathscr{D}_2} \\ \dfrac{-(e^{\gamma_0 h} - R' e^{-\gamma_0 h})}{\mathscr{D}_2} & \dfrac{e^{\gamma_0 h} - R e^{-\gamma_0 h}}{\mathscr{D}_2} \end{bmatrix} \times \begin{bmatrix} A_0{}^0 R - \sum_{m\in M_1} C_0{}^m e^{jm\psi} \\ A_0{}^0 R' - \sum_{m\in M_1} C_0{}^m (-1)^m e^{jm\psi} \end{bmatrix} \tag{18}$$

where $\mathscr{D}_2$ is as given in (11). The *infinite* sums $\sum_{m\in M_2{}^e}$ and $\sum_{m\in M_2{}^0}$ are now eliminated, and the $C_0{}^m$ $(m \in M_1)$ solved for by multiplying (18) through by $e^{-in\psi}$ $(n \in M_1)$ and integrating over $-\pi \leq \psi \leq \pi$. A set of M_1 linear equations is developed and solved by familiar methods.

Once the $C_0{}^m$ $(m \in M_1)$ are determined, the $C_0{}^m$ $(m \in M_2)$ can be determined *directly* by multiplying (18) through by $e^{-in\psi}$ $(n \in M_2{}^e$ or $M_2{}^0)$ and integrating as before. We note that the solutions for $C_0{}^m$ $(m \in M_2)$ or $C_0{}^m$ $(m \in M_1)$ are proportional to an integral of the form given by

$$\int_{-\pi}^{\pi} \frac{F(\psi) e^{-jn\psi}}{\mathscr{D}_2(\psi)}\, d\psi \tag{19}$$

Here $F(\psi)$ denotes an analytic function. In fact, it is easy to see that this is true for any P. Hence, regardless of the period of the modulated surface, we can always set

$$C_0{}^n = \int_{-\pi}^{\pi} \frac{F(\psi) e^{-jn\psi}}{\mathscr{D}_P(\psi)}\, d\psi \quad (n \in M_2) \tag{20}$$

The asymptotic behavior of the $C_0{}^n$ for very large n will thus be determined by the behavior[6,25] of the quotient, $F(\psi)/\mathscr{D}_P(\psi)$—in particular, by the zeros, if there are any, of the determinant, $\mathscr{D}_P(\psi)$.

If $1/\mathscr{D}_P(\psi)$ is not singular, the asymptotic behavior of the $C_0{}^n$ is determined by the discontinuous functions:[6,7]

$$R_q(\psi) = R\left[\psi + (q-1)\frac{2\pi}{P}\right], \quad q = 1, \ldots, P$$

and we have the behavior $1/q^{3/2}$ or $1/q^2$ (or $1/r^{3/2}$ or $1/r^2$ where r = spacing × q) for linear[6] or planar[7] arrays, respectively.

If the determinant $\mathscr{D}_P(\psi)$ has a zero, then $F(\psi)/\mathscr{D}_P(\psi)$ in (20) is more sharply discontinuous than the $R_q(\psi)$, and the $C_0{}^n$ will consequently decay more slowly than $1/r^{3/2}$ or $1/r^2$. In a linear array the $C_0{}^n$ will *not decay*, and in a planar array the coefficients will decay as $1/r^{1/2}$ away from the excited portion of the array. In either case, the "decay" is the result of a surface wave excited on the array. In these cases it is either impossible or very difficult to truncate the modulated surface edge without affecting the finite array performance. The singularities in $1/\mathscr{D}_P(\psi)$ can be avoided, however, by some of the methods discussed in the earlier part of this chapter.

We have not actually proved that the singularity in $1/\mathscr{D}_P(\psi)$ leads to a *surface wave* propagation. All we can say is that the theory is *self-consistent.*[10] The theory could lead to other interesting results. For example, it is interesting to determine the analytic behavior of $R(\psi)$ in the *vicinity* of the angle, ψ, of surface wave excitation. Referring to (7) for $P = 1$, we note that

$$\mathscr{D}_1(\psi) = e^{\gamma_0 h} - R(\psi)e^{-\gamma_0 h}$$

is zero at the singular point along the ψ axis where $|R| = 1$. However, not only must $|R| = 1$ at this point but also $R(\psi)$ must have the appropriate analytic behavior[25] so that the $C_0{}^n$ decay accordingly as a surface wave. It is, of course, plausible to extend these ideas to two-dimensional scanning or the planar array case[26,27] (see Chapter 7), wherein $\mathscr{D}_1 = \mathscr{D}_1(\psi_x, \psi_y)$.

If we consider the $P = 1$ case, a short-circuit depth with $h = 0.5$ waveguide wavelength (see Fig. 8.6*b* with $h = C$) is of particular interest, since it simulates a true ground plane. In a true ground plane *all* waveguide modes are short circuited, whereas in the modulated surface only the propagating mode is short circuited (although a finite number of higher-order modes could also be short circuited, as discussed in Section 1). Some results for this case were computed by Lee[3] for an array of thin-walled parallel-plate waveguides propagating only in the TEM mode (hence scanning in the E plane). The geometry for this array is the same as that shown in Fig. 8.6*b*.

Figure 8.8 shows plots of the radiated magnetic field, H_y (see Fig. 8.1 also), for a small array of five excited elements with element spacing $a =$ 0.45 guide (and free space in this case) wavelength. Each plot shows the radiation pattern with a different element excited. When the center element is excited, pattern (*a*), the radiation pattern as a function of scan angle, θ ($\psi = ka \sin \theta$), is symmetric. The patterns become asymmetric, however, when either the near-edge or the edge element is excited, thus indicating a very strong *edge effect.* (It is conceivable that using other short depths or periods, P, may alter this result so that the edge effects become less pronounced.)

When more elements are used in an array, the edge effects tend to become less pronounced. This assertion is verified in Fig. 8.9, which shows the radiation pattern for the excitation of the center and near-center elements in a nineteen-element parallel-plate array (using the same spacing and shorting depth as for the array of Fig. 8.8). We notice comparatively good

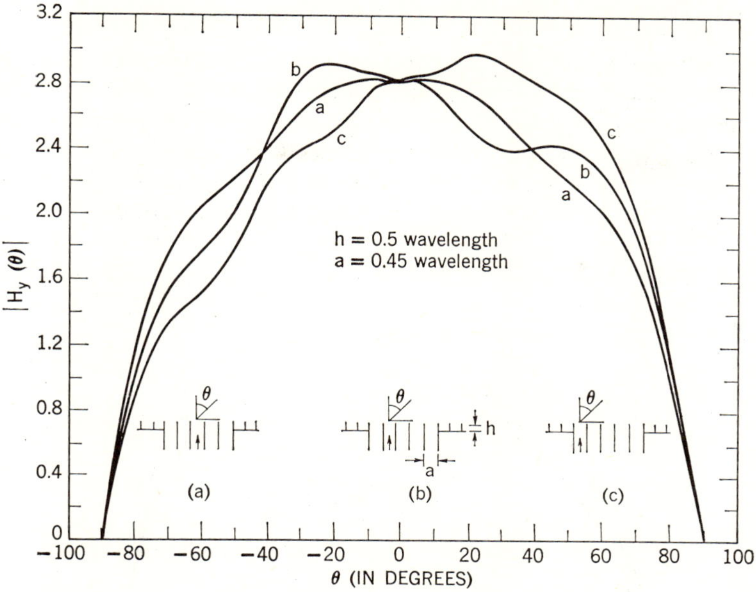

Fig. 8.8. Normalized radiation patterns for the array of five elements with excitation shown in (*a*), (*b*), and (*c*).

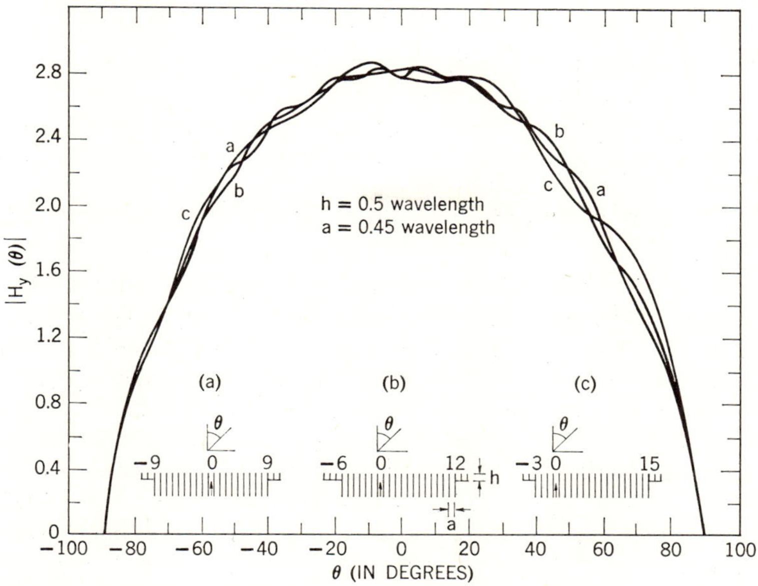

Fig. 8.9. Normalized radiation patterns for the array of nineteen elements with excitation shown in (*a*), (*b*), and (*c*).

symmetry of the radiation pattern for element excitation as close as four elements from the edge. It thus appears that the *fourth* element from the edge is very little affected by the edge and behaves as if it were in an infinite array. One useful implication of this result is that infinite array coupling coefficients, $\mathscr{C}_{pn}$ (see Fig. 8.2), can be measured in an array of nineteen elements up to $\mathscr{C}_{12,n}$ with little error. In other words, the coupling between

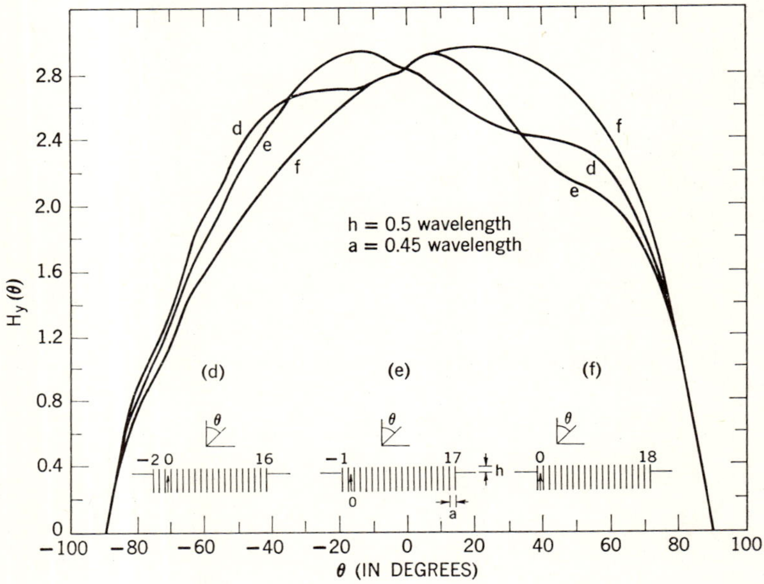

Fig. 8.10. Normalized radiation patterns for the array of nineteen elements with excitation shown in (*d*), (*e*), and (*f*).

elements at opposite ends of the array and each removed from the edge by three elements will be very much the same as that measured for equally spaced elements in an infinite array. This gives a measure of the array size necessary to obtain accurate coupling measurements up to a given order of coupling.

The radiation patterns for elements closer than three elements from the edge of the nineteen-element parallel-plate array begin to be greatly affected by the edge termination. This is illustrated in the radiation patterns plotted in Fig. 8.10. Radiation patterns of $|H_y|$ are plotted for elements excited at the edge or one element or two elements removed from the edge. It is clear from the asymmetry of the patterns that the edge has a considerable effect. These results and the conclusions drawn from them are in good agreement with the results and conclusions found in Chapter 4 for the passively terminated edge (which is the type of edge shown in Fig. 8.7*b*).

The results found for the nineteen-element array and illustrated in Figs. 8.9 and 8.10 with plots of the radiation pattern can also be illustrated by plotting the modulated surface coupling coefficients, $C_0{}^m$ (see Fig. 8.1). The magnitude and the phase of the $C_0{}^m$ are plotted in Figs. 8.11 and 8.12, respectively, for the nineteen-element parallel-plate array. The spacing of

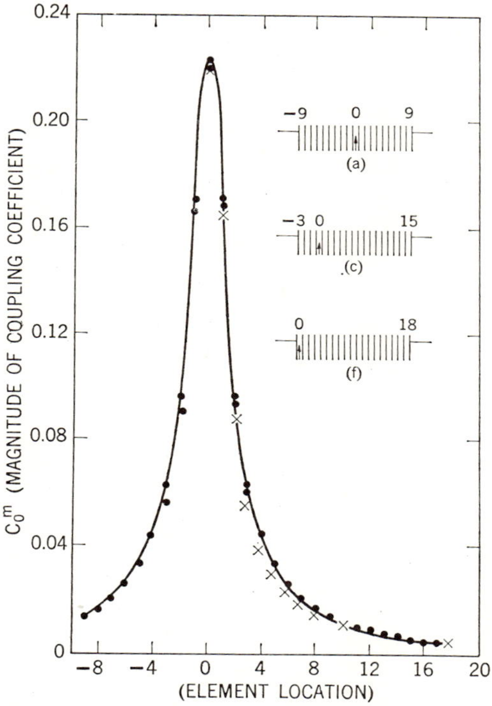

Fig. 8.11. Magnitudes of coupling coefficients for the array of nineteen elements.

elements, $a = 0.45\lambda$, and the short-circuit depth, $h = 0.5\lambda$, are the same as for the array of Figs. 8, 9, and 10.

In Fig. 8.11, we have plotted the $|C_0{}^m|$ determined for three different excitations. In one case, the center element of the array is excited; in the second, an element that is fourth from the edge is excited; and finally the edge element is excited. We should recall that the $C_0{}^m$, as distinguished from the infinite phased array coupling coefficients $\mathscr{C}_{0n}$, are theoretically *not translation invariant*. Hence we would expect the $C_0{}^m$ determined for edge element excitation to differ from those for center element excitation. We note in Fig. 8.11, however, that the magnitudes of $C_0{}^m$ determined for all

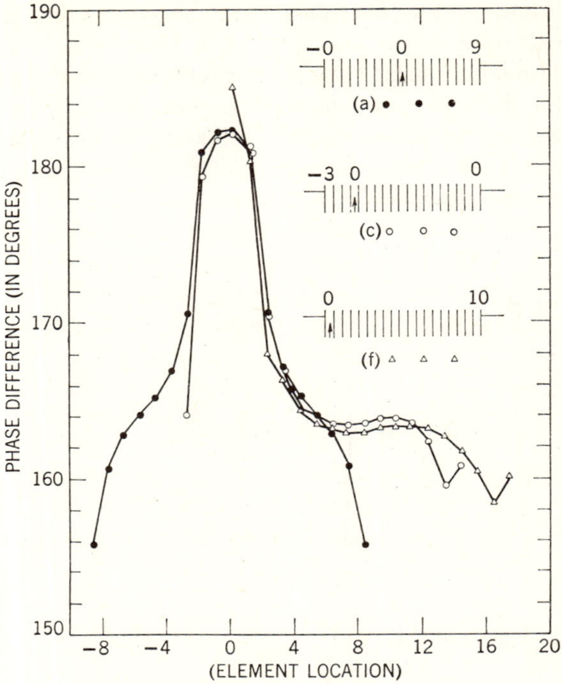

Fig. 8.12. Phase differences of the coupling coefficients between two adjacent elements for the array of nineteen elements.

three excitations fall very closely on the same approximating curve. In fact, this result appears to be valid for all $C_0{}^m$ ($m \in M_1$).

The results in Fig. 8.11 are consistent with those found for the radiation patterns depicted in Figs. 8.9 and 8.10. In fact, from the results for $|C_0{}^m|$ in Fig. 8.11, one might infer that the edge effect in this array is negligibly small. It turns out, however, that the phase differences of the coupling coefficients, plotted in Fig. 8.12, appear to depend more strongly on element excitation.

Figures 8.9–8.12 depict the finite array characteristics with modulated or corrugated surface termination when only a single element in the array is excited. The analysis discussed earlier and in this part of the chapter was presented for single-element excitation but could, of course, very readily be extended to include directly the effects of more than one waveguide excited. However, it is clear that the principle of linear superposition applies in this case and that the results for a finite phased array excitation, as depicted in

Fig. 8.6*b* or the inset to Fig. 8.13, are obtainable by superposition of the results depicted earlier.

The results shown in Fig. 8.13 are for the same thin-walled parallel-plate array discussed in connection with the earlier figures when all *five* elements of this array are excited with *equal amplitude* and a linear phase taper $\psi = (2\pi/\lambda)a \sin \theta$, where a = element spacing and θ = beam pointing direction from the array normal. Two aspects of this figure are significant. The first is that the reflection coefficient for each element varies significantly for the five-element array. As the array is small, these results can be expected.

The second point of interest is the general shape of all the curves—in particular, the $|R_0|$ curve for the center element. These curves approximate the E-plane scan results for the infinite phased array discussed in Chapter 4. In fact, as the number of elements is increased it may be expected that R_0, the center element reflection coefficient, will more nearly approximate the infinite phased array reflection coefficient. We note, interestingly, that for a waveguide spacing of $a = 0.45\lambda$ (less than 0.5λ) there exists a region of scan in the visible region ($\theta \lesssim 90°$) for which *total* reflection occurs in the infinite phased array. This corresponds to the higher reflections observed in Fig. 8.13 for the higher scan angles. It is important to note, in this respect, that the characteristic determinant

$$\mathscr{D}_1 = e^{\gamma_0 h} - R(\psi = ka \sin \theta)e^{-\gamma_0 h}$$

(where $\gamma_0 = jk$ and R is the reflection coefficient for the infinite phased array) does *not have a zero* for $h = 0.5\lambda$. Hence, even though $|R| = 1$ in the total reflection region for the infinite phased array, a surface wave is *not* excited on the corrugated surface with this short-circuit depth. The reason is that $R(\psi)$ has a limited *phase* variation over the scan range and the condition $\mathscr{D}_1(\psi) = 0$ cannot be satisfied.

As was mentioned earlier, the short-circuit depth of 0.5λ only approximates a ground plane surface termination, since only the dominant (TEM in this case) mode is short circuited. The results for a true ground plane termination will be discussed next.

2.2 Finite Array in an Infinite Ground Plane

The ground plane termination of a finite array such as is shown schematically in Fig. 8.7*a* is probably the most commonly used termination for arrays. It is simpler and less expensive to construct than the edge terminations of Figs. 8.6*b* and 8.7*b*, and it affords some "shielding" for the feed structure, as compared to the abrupt termination shown in Fig. 8.6*a*.

For very small arrays, the ground plane makes it possible to obtain an approximate analytical solution that is simpler than usual in some respects.

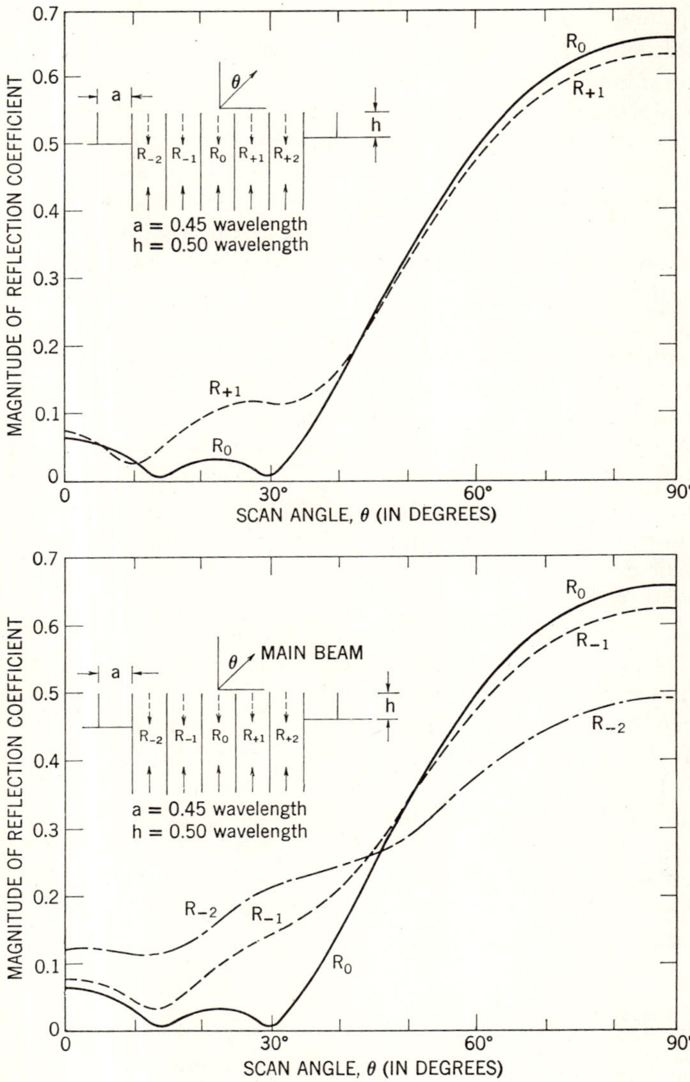

Fig. 8.13. Reflection coefficients of the scanning array with five elements.

The tangential electric field in the small array aperture can be approximated in a variational expression, and good results frequently are obtainable with simple approximations.[28] The Kirchhoff or *incident* field approximation is often used with success.

When it is desirable to compute the coupling effects between several array elements and to include the effects of dielectric plugs and covers over the array, more sophisticated approximations of the Ritz-Galerkin type discussed earlier become necessary. The basic functions for approximation must represent the field over the *entire* array shown in Fig. 8.7*a*, since Floquet's theorem is no longer applicable. If the unknown aperture function is the tangential *electric* field, then, because of the boundary conditions, only the field over the aperture region needs to be considered. On the other hand, since the tangential *magnetic* field is not zero on the ground plane, formulation in terms of the magnetic field requires representing the unknown field over an *infinite* region. For this reason we invariably solve for the electric field as the unknown aperture field in this type of problem.

For a very large array that is terminated by a ground plane we may choose a suitable finite number of elements near the edges of the array and represent these elements separately from the remaining *central* elements. We may presume that Floquet's theorem applies to these central elements and solve this problem, except for a single unknown coefficient, separately from the edge element problem.

A distinguishing feature for *any finite array*, regardless of the nature of its termination, is that the part of the integral equation kernel related to the *exterior* region ($z \geq 0$) field description will be an *integral* of *continuous* eigenfunctions, as opposed to the *summation* of *discrete* eigenfunctions that occurs for both the interior region ($z \leq 0$) description and the exterior region description when Floquet's theorem is applicable. We will examine some of the analytical aspects of this problem in the next section. Numerical results for the particular cases of E-plane and H-plane scan of a parallel-plate array will be given. We will also examine the effects on the array behavior of placing dielectric plugs and covers over these arrays.

2.2.1 Finite Array E-Plane Scan. Many important features of the radiation, impedance, and coupling characteristics of finite arrays with a ground plane edge termination can be illustrated by analyzing the parallel-plate array shown in Fig. 8.14 for an E-plane scan.[29] Although the coupling between two[30] and three[31] such waveguides was computed earlier by means of the geometrical theory of diffraction, we will use the Ritz-Galerkin variational solution of the appropriate integral equation. As we shall see, the integral equation kernel will contain contributions from both discrete and continuous spectra.

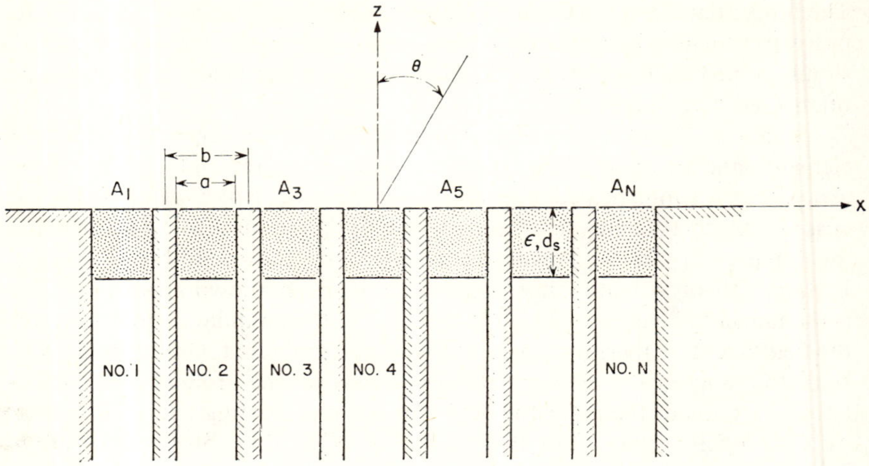

Fig. 8.14. Finite array of parallel-plate waveguides.

One interesting result determined from a solution of the ground plane terminated array, we will find, is that the coupling between two elements decreases as $1/r^{1/2}$ (see Fig. 8.14) for large r, the separation between the waveguides. This is expected, since we know that the currents on the ground plane and fields near the plane decrease asymptotically as $1/r^{1/2}$. The decrease is considerably *slower*, however, than in the case of dummy element termination (see Fig. 8.7*d*), where the fields and coupling decrease as $1/r^{3/2}$ (for linear arrays). In general, therefore, we can expect to need a *larger* finite ground plane edge.

In order to form the appropriate Fredholm first type integral equation for the finite array shown in Fig. 8.14, we will describe the tangential fields in the entire aperture of the array at $z = 0$ first in terms of the *discrete* spectrum of the waveguide region and then in terms of the *continuous* spectrum of the free space region. When we equate these two formulations across the *entire* boundary aperture

$$A = \bigcup_{p=1}^{N} A_p = A_1 \cup A_2 \cup A_3 \cup \ldots \cup A_N \tag{21}$$

we can easily obtain an integral equation with the boundary conditions intrinsically satisfied (see Chapter 2).

Since the transverse magnetic field is in the x direction and there is no variation in this direction for any of the field components, we can describe the entire field structure with TM modes only (see Chapter 4). If we assume that the pth waveguide (see Fig. 8.14) is excited by the dominant waveguide mode in that guide, $\phi_{0p}(y)$, with a given complex coefficient α_p and reflection coefficient* R_p, we may describe the tangential electric field by

$$E_y(y) = \sum_{p=1}^{N} \alpha_p(1 + R_p)\phi_{0p}(y) + \sum_{p=1}^{N} \sum_{n=1}^{\infty} V_{np}\phi_{np}(y) \tag{22}$$

Notice that double subscripts are used in the modal functions and modal voltages. The first one refers to the modal order; the second, to the waveguide position. When the array is composed of identical waveguides with equal element spacing, the modal functions are given by

$$\phi_{np}(y) = \begin{cases} \sqrt{(2 - \delta_{0n})/a} \cos \dfrac{n\pi}{a}(y - pb), & pb \le y \le pb + a \\ 0, & \text{otherwise} \end{cases} \tag{23}$$

Equation 22 is valid across the entire aperture, A; in fact, it is valid over $-\infty \le y \le \infty$ because of the boundary conditions.

The tangential magnetic field is obtained readily (see Chapters 2 and 4) in the form

$$H_x(y) = -\sum_{p=1}^{N} \alpha_p \tilde{Y}_{0p}(1 - R_p)\phi_{0p}(y) + \sum_{p=1}^{N} \sum_{n=0}^{\infty} \tilde{Y}_{np} V_{np}\phi_{np}(y) \tag{24}$$

where

$$\tilde{Y}_{np} = Y_n^{\varepsilon} \frac{Y_n + jY_n^{\varepsilon} \tan \gamma_n^{\varepsilon} d_s}{Y_n^{\varepsilon} + jY_n \tan \gamma_n^{\varepsilon} d_s} \tag{25}$$

with Y_n and Y_n^{ε}, respectively, the modal admittances of an empty waveguide and of a waveguide filled with a dielectric of relative permittivity ε. These admittances are given by

$$Y_n^{\varepsilon} = \frac{\omega\varepsilon_0\varepsilon}{\gamma_n^{\varepsilon}}, \qquad \text{where} \quad \gamma_n^{\varepsilon} = \sqrt{\varepsilon k^2 - (n\pi/a)^2} \tag{26}$$

The γ_n are the waveguide propagation constants.

The voltage coefficients, V_{np}, are readily obtained in terms of the aperture field as

$$\begin{aligned} V_{np} &= \int_A \phi_{np}(y)E_y(y)\,dy, \quad n > 0 \\ V_{0p} &\equiv \alpha_p(1 + R_p) = \int_A \phi_{0p}(y)E_y(y)\,dy \end{aligned} \tag{27}$$

* The definition of "reflection coefficient" is somewhat arbitrary when more than one waveguide is excited. Most often the term is defined as the reflection when a *single* element guide is excited by a wave of unit amplitude, the others being inactively terminated (i.e., $\alpha_p = \delta_{pq}$ for determining R_q). See Ref. 34 and Chapter 7 for further discussion.

With the definition of $\phi_{np} = 0$ outside of A_p, the integration in (27) can be taken over either

$$A_p \quad \text{or} \quad A = \bigcup_{p=1}^{N} A_p$$

The representation of the field of the finite array in the exterior region ($z \geq 0$) will require a *continuous* spectrum of eigenfunctions, since this array is not periodic. Actually, even if the array were periodic, since the excitation is not periodic, a continuous spectral representation would be required. This is an alternative approach to the aperiodic array analysis employed earlier.[5]

Two approaches to the derivation of the exterior representation are possible. One of them starts with the Fourier integral spectral representation and can lead to the appropriate Green's function for the aperture field on a ground plane[5] (we want to represent the fields in the exterior region in terms of the aperture field). The other approach[24,29] utilizes Green's theorem to obtain the representation directly in terms of the Green's function. The Fourier integral representation approach will be used here, since it is very similar to the discrete summation representations employed so far for periodic structures and excitations. When no dielectrics are present in the exterior region, we shall see that the continuous summation (integral) can be placed in closed form (the usual cylindrical Green's function).

A Fourier representation for the tangential electrical field on A is

$$E_y(y) = \frac{1}{2\pi} \int_{-\infty}^{\infty} dk_y \int_{-\infty}^{\infty} dy' e^{-jk_y(y-y')} E_y(y') \tag{28}$$

We have used, in (28), the Fourier transform and its *inverse* transform to form a *unit operator* on $E_y(y')$. This representation can be interpreted as the sum of contributions from a continuous set of modes pertinent to the exterior region. The modal functions are given by

$$\psi_{k_y}(y) = \frac{1}{\sqrt{2\pi}} e^{-jk_y y} \tag{29}$$

where k_y is a continuous summation index.

The field everywhere in the exterior region is then readily obtained[13] as

$$\mathscr{E}_y(y, z) = \frac{1}{2\pi} \int_{-\infty}^{\infty} dk_y \int_{-\infty}^{\infty} dy' \exp\left[-jk_y(y - y')\right] \exp\left(-j\sqrt{k^2 - k_y^2}\, z\right) E_y(y') \tag{30}$$

By Maxwell's equations, one finds the tangential magnetic field as

$$\mathscr{H}_x(y, z) = -\frac{\omega\varepsilon_0}{2\pi} \int_{-\infty}^{\infty} dk_y \int_{-\infty}^{\infty} dy' \times \frac{\exp\left[-jk_y(y - y')\right]}{\sqrt{k^2 - k_y^2}} \exp\left(-j\sqrt{k^2 - k_y^2}\, z\right) E_y(y') \tag{31}$$

In particular, at $z = 0$,

$$H_x(y) = -\frac{\omega\varepsilon_0}{2\pi}\int_{-\infty}^{\infty} dk_y \int_{-\infty}^{\infty} dk' \frac{e^{-jk_y(y-y')}}{\sqrt{k^2 - k_y^{\,2}}} E_y(y') \tag{32}$$

We could place (32) in a form analogous to the expressions used earlier for the discrete spectrum cases:

$$H_x(y) = -\int_{-\infty}^{\infty} dk_y \int_{-\infty}^{\infty} dy' Y_{k_y}\, \psi_{k_y}(y)\psi_{k_y}^{*}(y')E_y(y') \tag{33}$$

where

$$Y_{k_y} = \frac{\omega\varepsilon_0}{\sqrt{k^2 - k_y^{\,2}}} \tag{34}$$

are the modal admittances for the continuum modes given by (29).

For the present dielectric-free exterior region, it is possible to sum or integrate the kernel of (32) in closed form.* That is, we find[32] that

$$H_0^{(2)}(k\,|y - y'|) = \frac{1}{\pi}\int_{-\infty}^{\infty} dk_y \frac{e^{-jk_y(y-y')}}{\sqrt{k^2 - k_y^{\,2}}} \tag{35}$$

where $H_0^{(2)}(k\,|y - y'|)$ is the outgoing (for $e^{j\omega t}$ time convention) Hankel function or the cylindrical Green's function for a current on an infinite ground plane.

Using (25) and the fact that $E_y(y') = 0$ outside of A, we can write (32) as

$$H_x(y) = -\frac{\omega\varepsilon_0}{2}\int_A dy'\, H_0^{(2)}(k\,|y - y'|)E_y(y') \tag{36}$$

If we now equate (24) and (36) so as to ensure continuity of tangential H_x across the boundary aperture, A, we arrive [using (27) also] at the integral

* In order to develop a kernel for the integral of (31) it is necessary to *reverse the order of integrations* for the repeated integrals. Earlier (e.g., see Chapter 2), we reversed the order of an integration and summation in forming an integral equation, the summation forming the kernel of the resultant equation. In earlier formulations the summation could not be performed in closed form and consequently the reversal of order of operations was essentially symbolic. In the present case, as indicated in (34), the summation (integration) can be made in closed form so this reversal becomes important. Sometimes however, the interchange of order of integrations may lead to difficulties, and care should be exercised in doing this. A more complete discussion of this problem from a mathematical point of view is beyond the scope of this presentation but can be found in Ref. 33. We will, however, concern ourselves with some "numerical" or "computational" aspects of this problem in Section 2.2.2.

equation for the unknown E_y:

$$2\sum_{p=1}^{N} \alpha_p \overline{Y}_{0p}\phi_{0p}(y)$$

$$= \int_A \left\{ \sum_{p=1}^{N} \sum_{n=0}^{\infty} \tilde{Y}_{np}\phi_{np}(y)\phi_{np}(y') + \frac{\omega\varepsilon_0}{2} H_0^{(2)}(k\,|y - y'|) \right\} E_y(y')\, dy' \quad (37)$$

where

$$\overline{Y}_{0p} = \frac{Y_0 Y_0^{\varepsilon}}{Y_0^{\varepsilon}\cos \gamma_0^{\varepsilon} d_s + jY_0 \sin \gamma_0^{\varepsilon} d_s} \quad (38)$$

This integral equation is in the same form as those obtained earlier (see Chapters 2 and 4) except that the exterior portion of the kernel has been summed in closed form.

Numerical solutions to (37) have been obtained by use of the Ritz-Galerkin method of approximation. A pulse basis of the type discussed earlier (Chapters 3 and 5) was employed. It was found that about ten sampling pulse functions per aperture or waveguide, corresponding to about twenty samplings per wavelength, were adequate for an accuracy of a few per cent in determining the reflection coefficients, R_p, and other pertinent parameters.

The reflection coefficients, R_p, are of course determined from the second part of (27) once the approximate $E_y(y)$ is obtained. The total y component of the electric field in the exterior region can be found from (30), and $\mathscr{H}_x(y, z)$, given in (31), is the total magnetic field in the exterior region. Since

$$k_y(y - y') + \sqrt{k^2 - k_y^2}\,z = \bar{k}(\bar{\rho} - \bar{y}') \quad (39)$$

where

$$k_z = \sqrt{k^2 - k_y^2}, \quad \bar{\rho} = y\hat{y} + z\hat{z}, \quad \text{and} \quad \bar{y}' = y'\hat{y} \quad (40)$$

we may rewrite (31) as

$$\mathscr{H}_x(y, z) = -\frac{\omega\varepsilon_0}{2}\int_A H_0^{(2)}(k\,|\bar{\rho} - \bar{y}'|)E_y(y')\, dy' \quad (41)$$

If we allow only the qth element to be excited with a wave of unit amplitude by setting $\alpha_p = \delta_{pq}$, we can obtain a normalized radiation pattern relative to the incident power. This facilitates a comparison with the radiation pattern of a single element excited in an infinite array.[14] In this case our primary interest is the *far zone* of the exterior region field, the radiated field. We obtain this far-zone field by using the asymptotic form of $H_0^{(2)}(k\,|\bar{\rho} - \bar{y}'|)$ for large $|\bar{\rho}|$ in (41). Upon normalization, as indicated, we obtain

$$T(\theta) = \sqrt{\gamma_0/2\pi}\int_A e^{-j(\psi/b)y'}E_y(y')\, dy' \quad (42)$$

where $\psi = kb \sin \theta$. The notation $T(\theta)$ is adopted by analogy with the transmission coefficient designation for a single-element excitation in an infinite array. Of course, by superposition of $T(\theta)$, obtained from excitation of the various elements in Fig. 8.14, we can obtain the far-zone pattern for any excitation of the array.

The numerical results for the radiation admittance of a single waveguide ($N = 1$) radiating from the ground plane are presented in Fig. 8.15. They

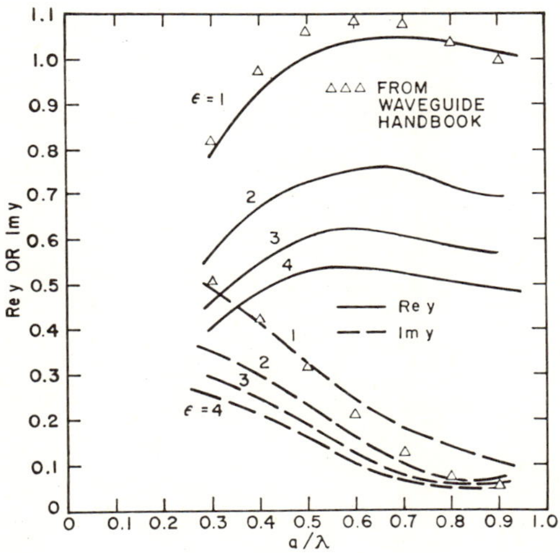

Fig. 8.15. Normalized radiation admittances of a parallel-plate waveguide.

show the effect of waveguide dimension a and of loading the waveguide completely ($d_s = \infty$) with dielectric material. The results, taken from the *Waveguide Handbook*[28] for $\varepsilon = 1$, were computed for an assumed aperture field, E_y, equal to the incident field (the Kirchhoff approximation) in a variational expression.

Although the Kirchhoff assumption for E_y yields satisfactory results for the radiation admittances from the variational expression, the more accurate results for $E_y(y)$ obtained from the Ritz-Galerkin solution of the pertinent integral equation show, as indicated in Fig. 8.16*a*, considerable deviation of the E_y field from the incident field [which is uniform in amplitude and phase; see (23) for $n = 0$]. The deviation becomes greater with higher dielectric constant.

The reflection coefficient of a single waveguide loaded with a dielectric plug is presented in Fig. 8.16*b*. Both the amplitude and the phase are shown

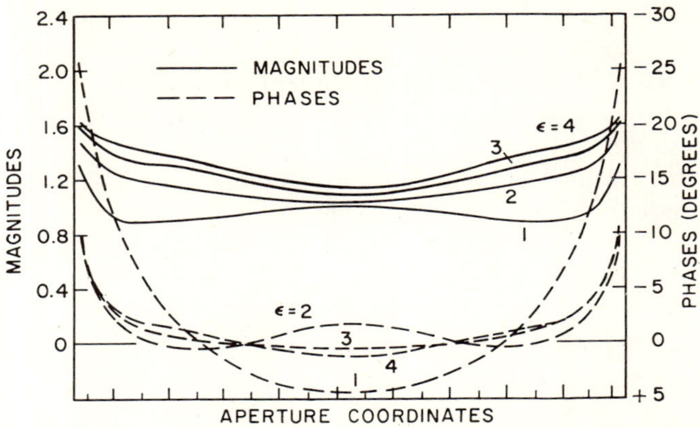

Fig. 8.16*a*. Aperture electric fields of a parallel-plate waveguide.

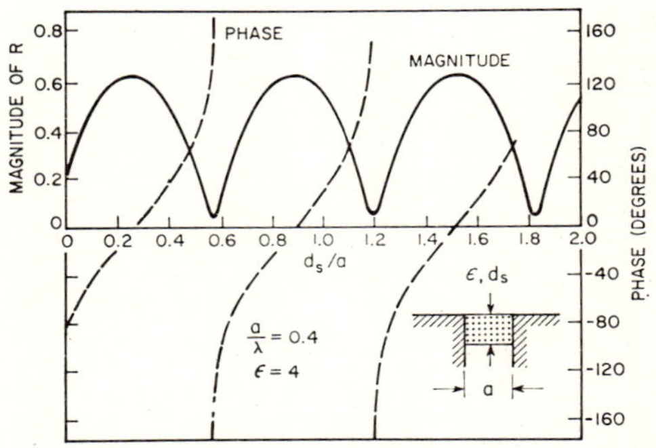

Fig. 8.16*b*. *R* of a parallel-plate waveguide loaded with a dielectric plug.

as functions of thickness, d_s, of the plug for fixed values of a/λ and ε. These parameter values are chosen so that only one mode can propagate in the dielectric plug and in the empty waveguide region. Except for very small thicknesses,[34,35] the reflection is a periodic function of the thickness.

We next present the coupling between two empty guides. The coupling is defined as the modal voltage induced in one element when the other is excited with a unit modal voltage. We show in Fig. 8.17 the coupling calculated as a function of the separation between the two waveguides. We note first that the coupling decreases with increasing values of a/λ. Moreover, the results indicate that the coupling decreases monotonically with the separation and that for large separations, say of the order of $5a$, the decrease is in proportion to $s^{-1/2}$, where s denotes the separation. This rate of decay is to be contrasted with that of infinite array cases, in which the asymptotic

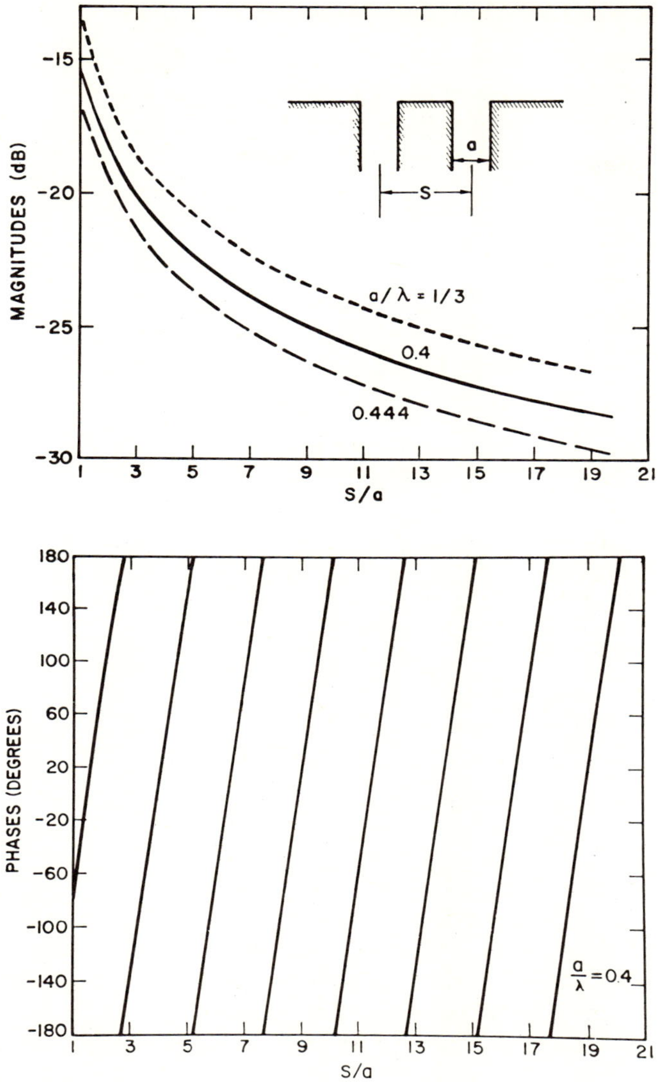

Fig. 8.17. Coupling between two parallel-plate waveguides.

coupling was found to behave as $s^{-3/2}$. The phase variation versus distance, on the other hand, forms a set of parallel lines with the separation between each pair of these lines equal to the free space wavelength. This characteristic behavior remains essentially unchanged down to very small values of s/a.

The mutual coupling in a three-waveguide array is illustrated in Fig. 8.18. The result for the two-element case is also included for comparison. We see that the coupling between the excited edge element and the center one

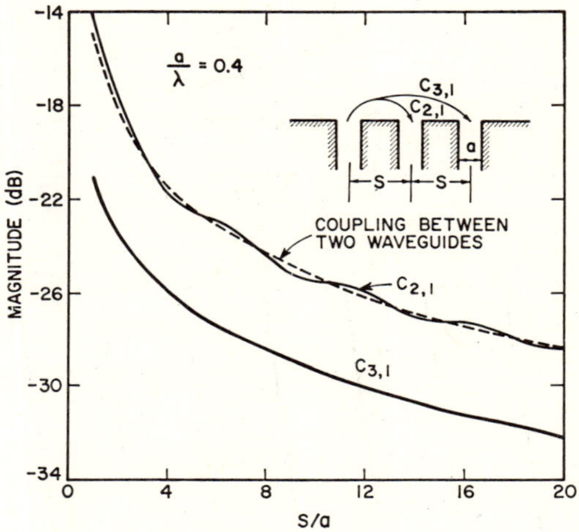

Fig. 8.18. Mutual coupling in a three-waveguide array.

displays an oscillatory behavior due to the presence of the other edge element. The effect, however, seems to be relatively small, so that it may be negligible as a first approximation. On the other hand, the coupling between the two edge waveguides is diminished by about 1–2 dB from the value obtained when the presence of the center element is ignored.

The effect of loading the waveguides with dielectric plugs is considered next. Figure 8.19 shows the results for a representative calculation using $a/\lambda = 0.4$ and $\varepsilon = 4$. The choice of these parameters makes it possible for two propagating modes to exist inside the dielectric plug, whereas only one mode can propagate in the empty waveguide region. This situation causes one of the two modes inside the dielectric to "see" a reactive termination when looking toward the waveguide side, so that, whenever that mode is excited, all the energy will have to be reflected back to the aperture and be reradiated. This process may sometimes introduce strong multiple reflections between the excited and receiving waveguides. As a result, the characteristics of

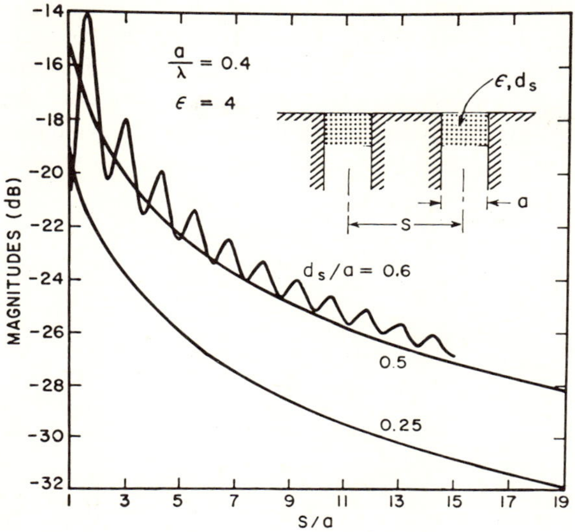

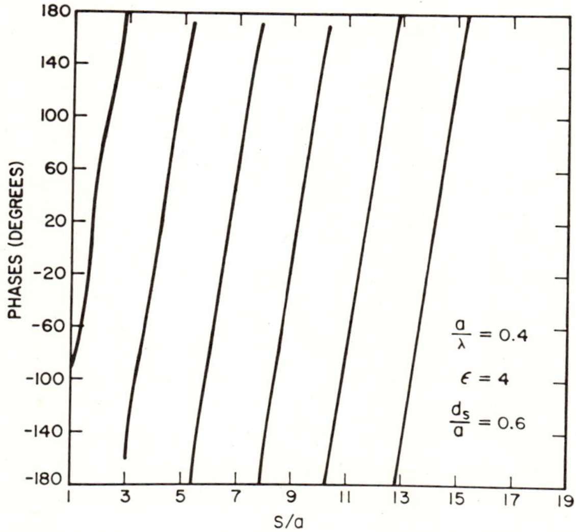

Fig. 8.19. Coupling between two waveguides loaded with dielectric plugs.

coupling are significantly altered. For example, the coupling may be enhanced or suppressed, depending on the choice of parameters, as exemplified by the results for the cases of $d_s/a = 0.5$ and $d_s/a = 0.25$. Moreover, when the parameters fall into a certain "critical region," the coupling may even exhibit an oscillatory behavior, as the curve for $d_s/a = 0.6$ shows. This "critical region" may be more vividly displayed by considering the coupling

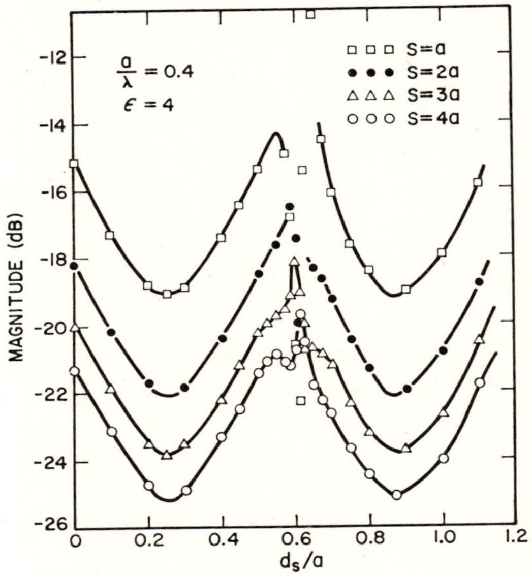

Fig. 8.20. Coupling between two waveguides loaded with dielectric plugs versus plug thickness.

as a function of the plug thickness. Figure 8.20 shows the results for such a computation, using the separation between the waveguides as a parameter.

When the coupling is studied as a function of frequency, a similar oscillatory behavior is observed. This result is illustrated in Fig. 8.21. We note that the resonant behavior is confined in a small frequency band and that, for $d_s/a = 0.625$, a maximum coupling of about -8 dB is reached for $\lambda/a = 2.48$ and a minimum of -22.2 dB at $\lambda/a = 2.52$. Of course, different maximum and minimum couplings will occur at different wavelengths if other values of d_s/a are used. It is important to point out also that the resonant behavior is observed only when "suitable" values of plug thickness are used.

The element patterns, $T(\theta)$ [see (42)], of the *central* element in finite arrays of various sizes are shown in Fig. 8.22. The results show that in a five-element array the pattern compares well with that of an infinite array

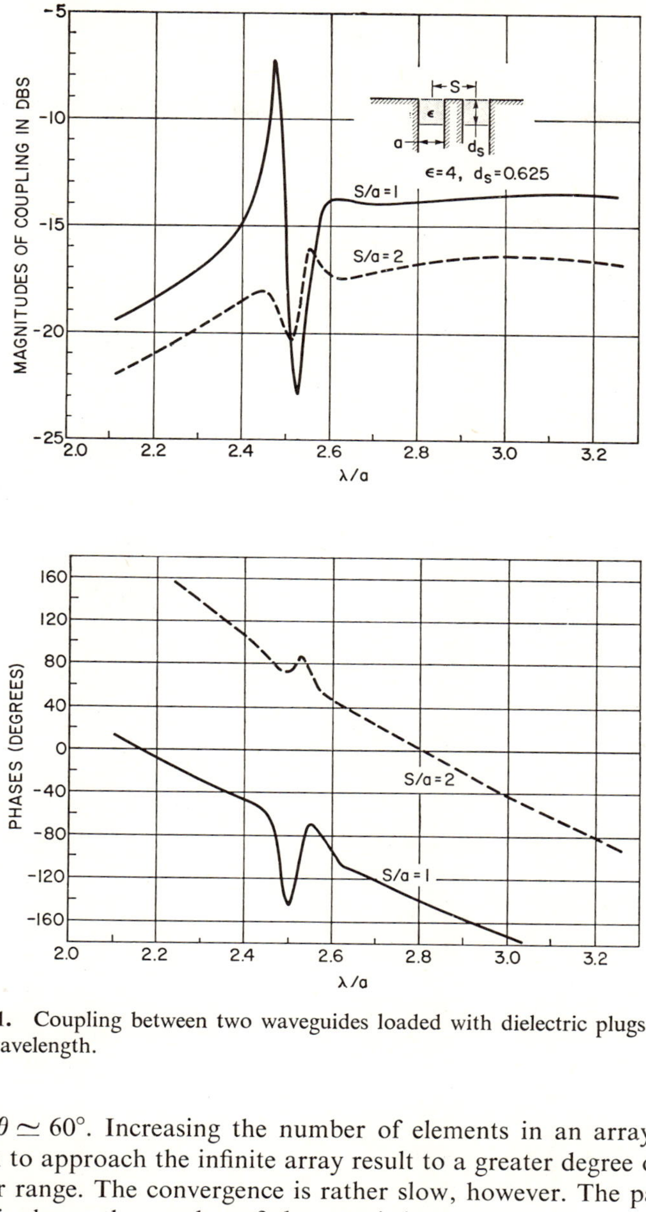

Fig. 8.21. Coupling between two waveguides loaded with dielectric plugs as a function of the wavelength.

up to $\theta \simeq 60°$. Increasing the number of elements in an array causes the pattern to approach the infinite array result to a greater degree over a wider angular range. The convergence is rather slow, however. The patterns have more ripples as the number of elements is increased.

The reflection coefficients for the same array with all the elements excited as a phased array ($\alpha_p = 1e^{jp\psi}$) are shown in Fig. 8.23. For the element

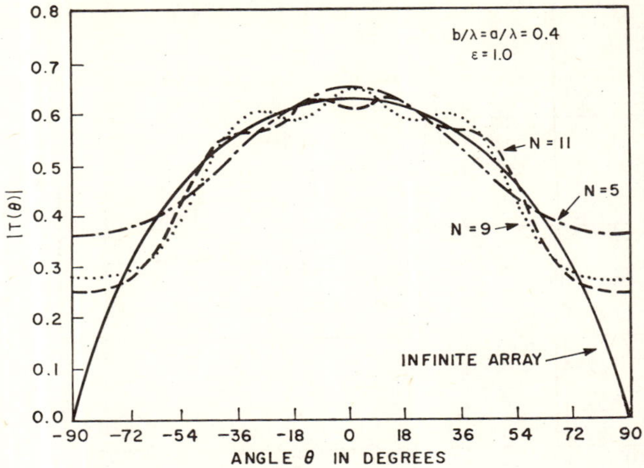

Fig. 8.22. Radiation patterns, $T(\theta)$, of center elements.

spacing of $b/\lambda = 0.4$, the reflection coefficient for the infinite array shows a sharp discontinuity in slope when the beam is at grazing ($\psi = 144°$). This discontinuity is more closely approximated as N is increased, in much the same way as it is more closely approximated when more elements are excited in the finite array embedded in an infinite array of dummy elements (see Chapter 4 and Fig. 8.7*b*).

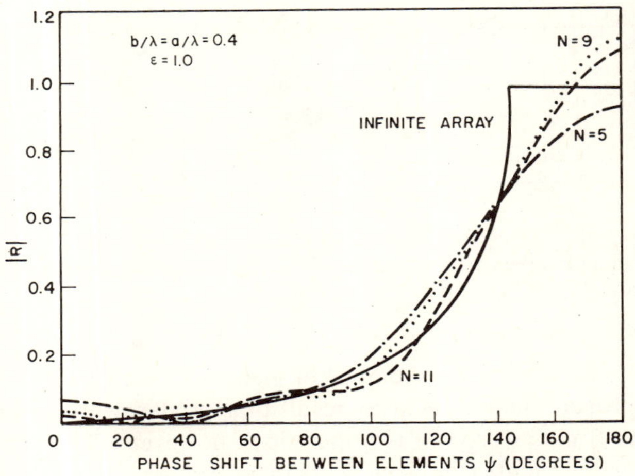

Fig. 8.23. Reflection coefficients, R, of center elements with all elements excited with uniform amplitude and linear phase slope, ψ degree/element; $\psi = kb \sin \theta$.

The element patterns for different element locations in a fifteen-element array are compared in Fig. 8.24. The results show clearly that the pattern of the edge element deviates considerably from that of an infinite array case. As we move toward the center of the array, the pattern becomes more symmetric with respect to broadside in addition to showing a closer match with the infinite array pattern. The fifth element has an acceptable range of more than $\theta \simeq +60°$. Figure 8.25 shows a corresponding comparison for

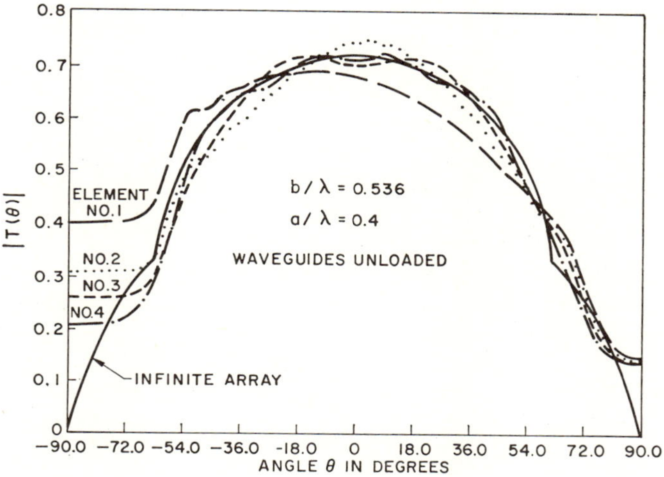

Fig. 8.24. Element patterns, $T(\theta)$, in a fifteen-element array. (Element 1 is furthest to the left ($-y$); see Fig. 8.14.)

the reflection coefficients, and a trend similar to that of the patterns is revealed.

Because the patterns of each element are different owing to the difference in mutual coupling, which depends on the element location, the principle of pattern multiplication becomes applicable only as an approximation. The radiation pattern of the array must then be calculated directly, taking into explicit account each element pattern. This is accomplished by the superposition of each element pattern, $T(\theta)$, from (42) as discussed earlier, taking into account the fact that $T(\theta)$ varies with element position. The resultant superimposed patterns, $T_t(\theta)$, of a nine-element uniformly illuminated array are depicted in Fig. 8.26, showing the effects of scan. Although the results are expected to be different from what would be obtained if the principle of pattern multiplication were applied using, say, the central element pattern

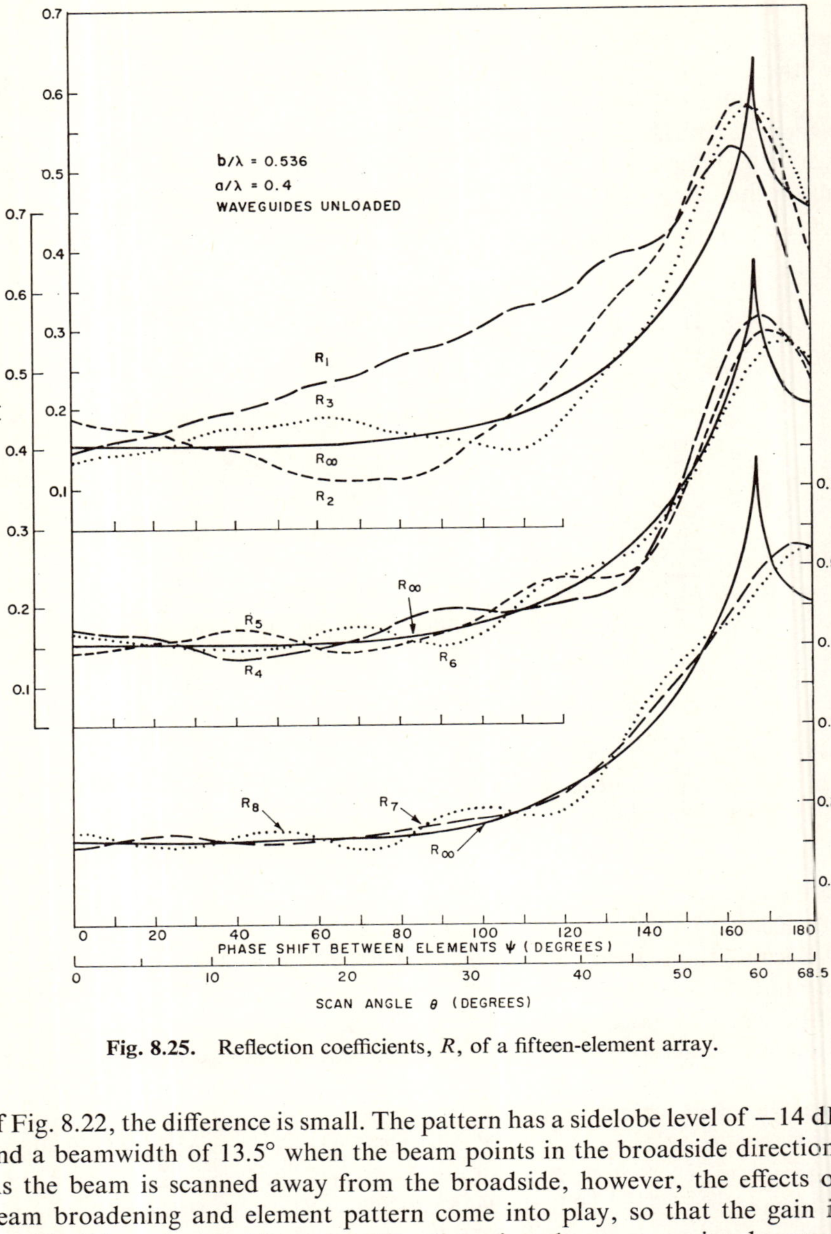

Fig. 8.25. Reflection coefficients, R, of a fifteen-element array.

of Fig. 8.22, the difference is small. The pattern has a sidelobe level of −14 dB and a beamwidth of 13.5° when the beam points in the broadside direction. As the beam is scanned away from the broadside, however, the effects of beam broadening and element pattern come into play, so that the gain is considerably reduced. Moreover, the distortion due to scanning becomes evident when the beam is steered far away from broadside.[36]

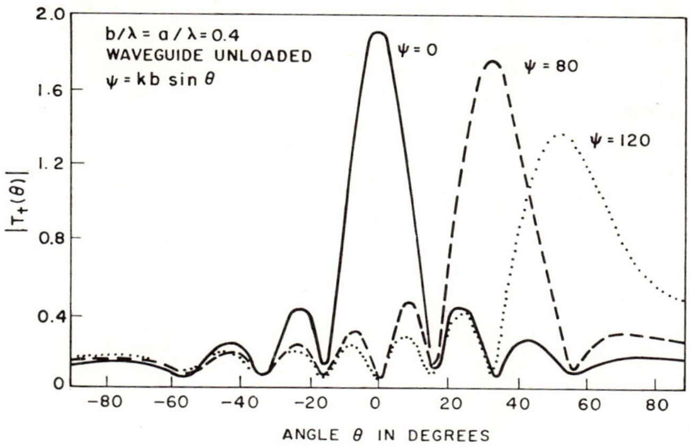

Fig. 8.26. Radiation patterns of a nine-element array.

We next turn to the situation of loading the waveguide with dielectric plugs near the aperture. Suitable values are chosen for the pertinent parameters in the dielectric plugs so that an aperture resonance is present. The condition of resonance may be determined by applying a generalized transmission line approximation[21,37] (see Chapters 6 and 7). The radiation patterns of the center elements in arrays of various sizes are presented in Fig. 8.27

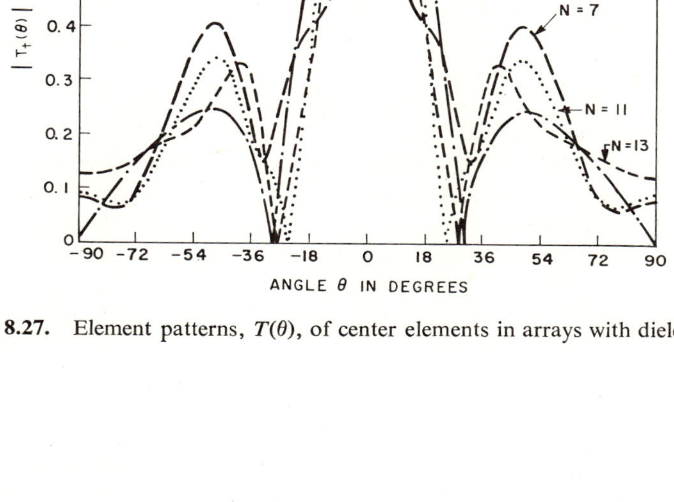

Fig. 8.27. Element patterns, $T(\theta)$, of center elements in arrays with dielectric plugs.

for $\varepsilon = 4$, $d_s/a = 0.6375$, and $b/\lambda = a/\lambda = 0.4$. The infinite array pattern displays a minimum or "resonance" at $\theta = 24°$ for this set of parameters. The pattern for a seven-element array exhibits what appears to be a corresponding minimum at $\theta \approx 30°$. The minimum does not reach zero, however. As more elements are added to the array, we observe that these minima appear to converge toward the zero of the infinite array. It seems that more than

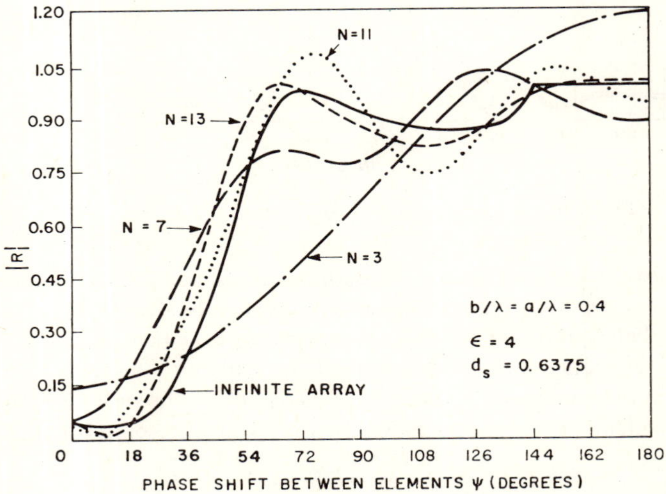

Fig. 8.28. Reflection coefficients of center elements in arrays with dielectric plugs.

thirteen elements are required for the center element to have a pattern which is reasonably close to that of an element in an infinite array. A similar comment holds true for the corresponding reflection coefficients, as illustrated in Fig. 8.28. The minima of the radiation are manifested here as maxima which become greater and sharper with increasing finite array size, N.

In view of the fact that considerably larger arrays are needed to simulate an infinite array environment when the array parameters are such as to permit the occurrence of an aperture resonance, it is not surprising to find that the elements of a ten-element array have substantially different impedance and radiation properties for different positions in the finite array. Figures 8.29 and 8.30 illustrate the element patterns and the reflection coefficients, respectively, for a ten-element array. The dielectric plugs are chosen to have $\varepsilon = 4$ and $d_s/a = 0.6425$, so that a resonance will occur at

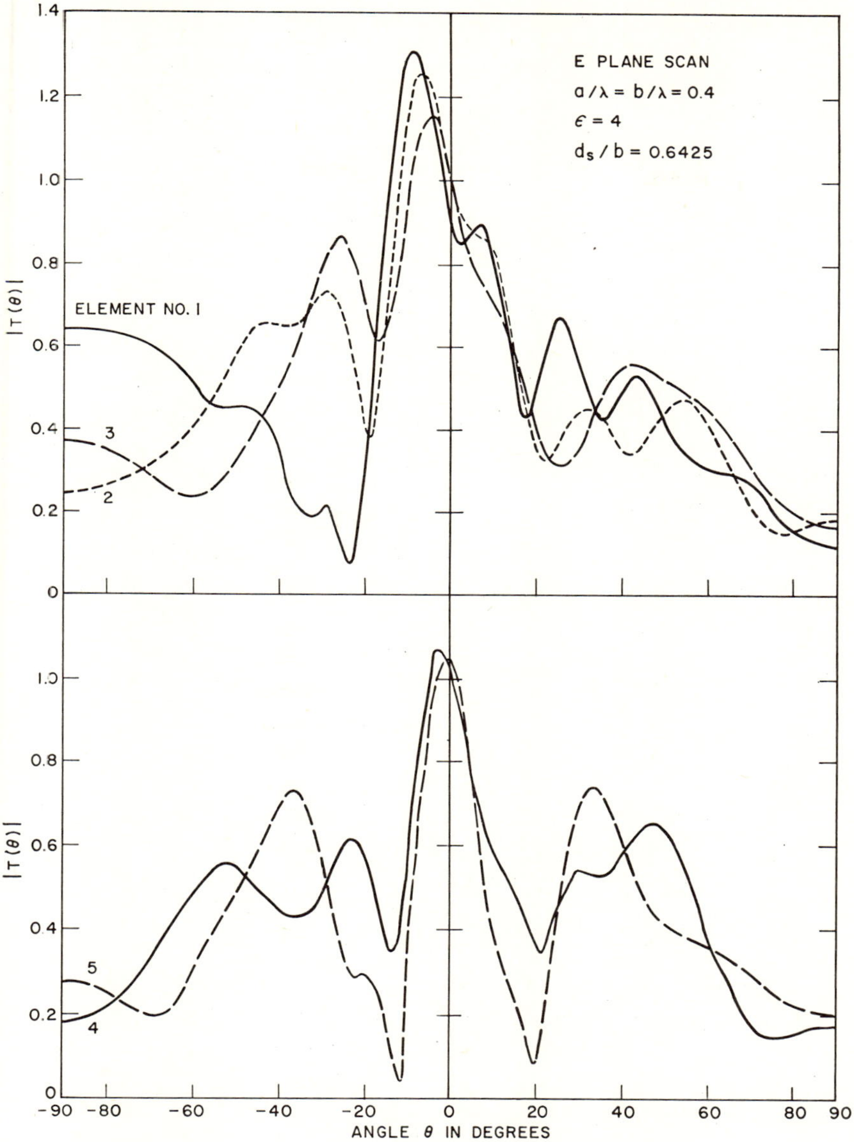

Fig. 8.29. Element patterns, $T(\theta)$, of a ten-element array loaded with dielectric plugs.

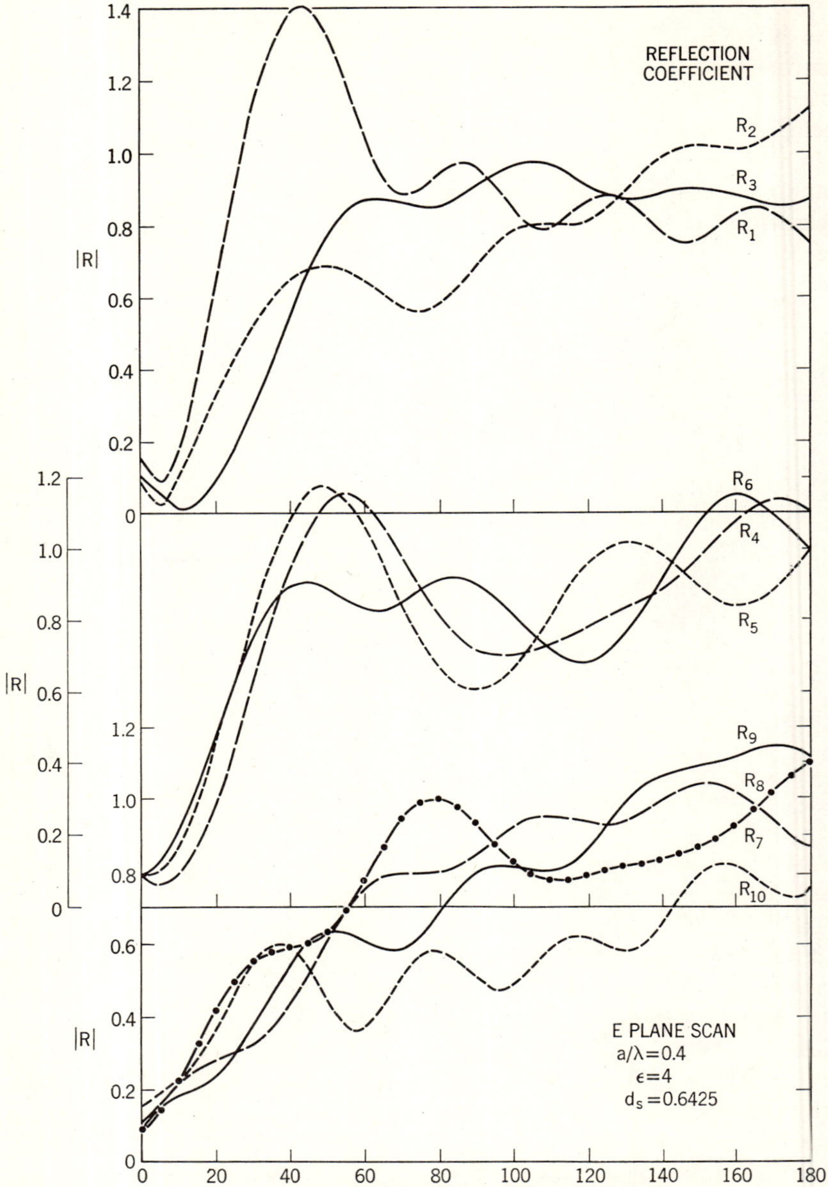

Fig. 8.30. Reflection coefficients of a ten-element array loaded with dielectric plugs.

$\psi = 30°$ $(\theta = 12°)$ in the related infinite array. The wide variation in the results from element to element is quite evident. Moreover, it should be observed that not all the elements display a sharp peak in the reflection coefficient, as might be anticipated in large arrays.

In regard to the radiation patterns for an eleven-element array under scanning conditions, we see that, since the waveguides are loaded with dielectric plugs in order to cause a resonant condition to exist (Fig. 8.31), the beam has a substantially suppressed gain with high sidelobes when it is

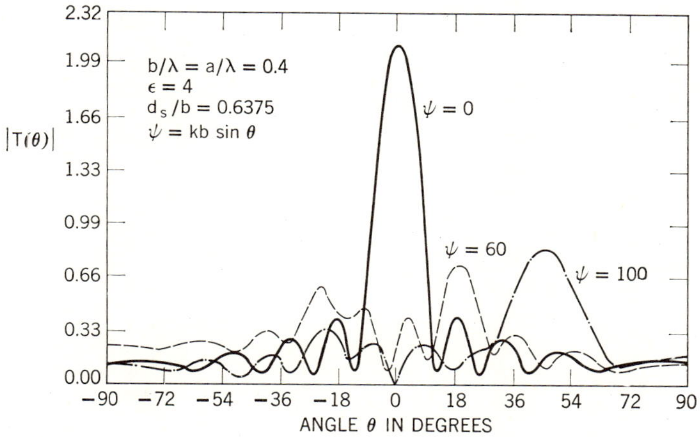

Fig. 8.31. Radiation patterns, $T(\theta)$, of an eleven-element array.

scanned into the vicinity of resonance. The gain in this scan region naturally would be suppressed even more if larger arrays were used.

The appearance of array resonances is a manifestation of the importance of mutual coupling effects. We examine next the coupling coefficients of two arrays, one with unloaded waveguides and the other with dielectric plugs with their thickness and dielectric constant set in the resonant region. The results for the unloaded case are shown in Fig. 8.32*a*. Several observations may be made for this situation. First, the coupling decreases *monotonically* with the element separation. Also, the coupling between any two elements is rather insensitive to their location in an array and the size of the array, provided neither element lies at the edge. These results are in agreement with other calculations[3,4] discussed earlier in this chapter. On the other hand, when array elements contain dielectric plugs with values of thickness and dielectric constant in the critical range required to produce array resonance, the behavior of the coupling is very different. The coupling coefficients do not

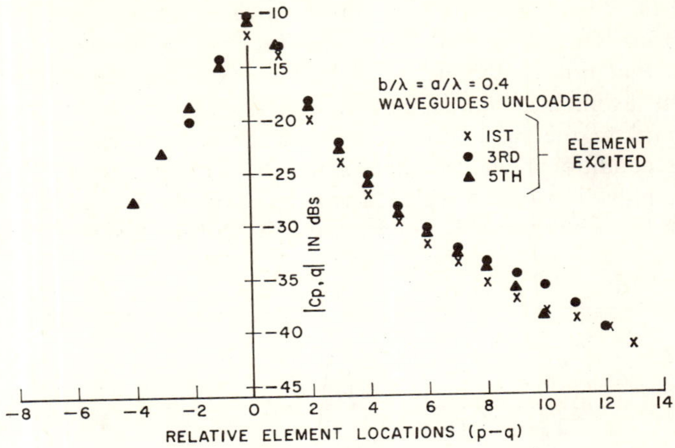

Fig. 8.32*a*. Mutual coupling in a fifteen-element array with $\varepsilon = 1$.

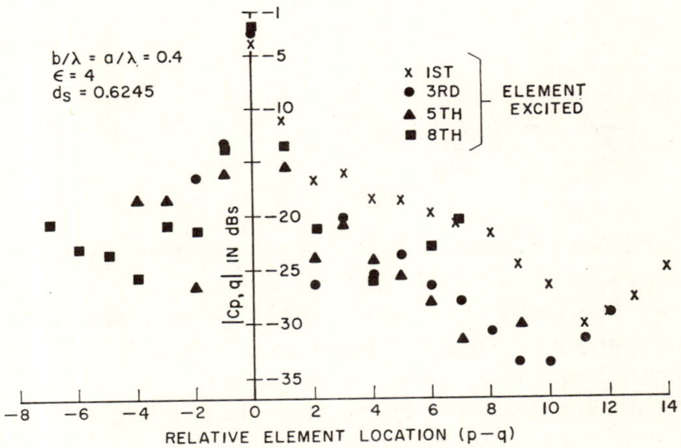

Fig. 8.32*b*. Mutual coupling in a fifteen-element array with dielectric plugs.

necessarily decay smoothly with element separation, and they are considerably dependent on the size of the array and the location of the excited element in the array. The example shown in Fig. 8.32*b* illustrates part of this behavior.

2.2.2 Finite Array H-Plane Scan. A discussion of the finite parallel-plate array in the *H* plane of scan[38,39] is of interest in terms of both the mathematical formulation of the relevant integral equation and some of the new physical results. The geometry for this problem is essentially the same as that for the E-plane scan except that, in consistency with the discussions in Chapters 4 and 5, we will scan the array in the x–z plane and, in addition, allow for a more general geometry, as depicted in Fig. 8.33.

In the present treatment, we will excite the waveguides with the lowest-order TE mode (TE_{10}) so that both the incident and the scattered fields are all TE and invariant with the y coordinate. Thus the only field components that exist for the finite array are the E_y, H_x, and H_z components.

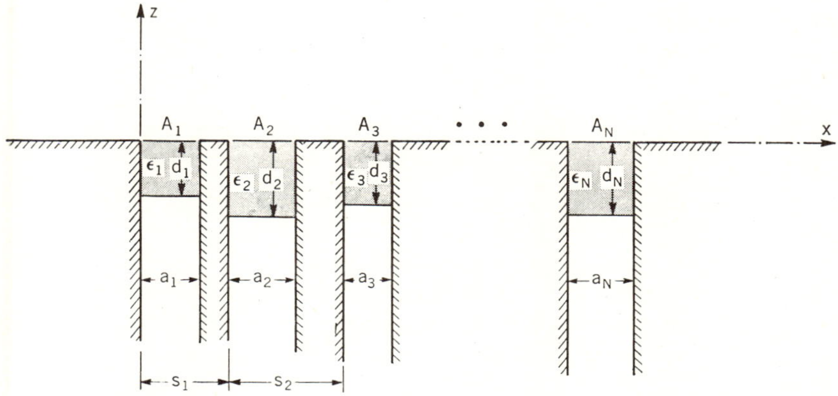

Fig. 8.33. A finite parallel-plate array for H-plane scan.

In deriving an equation for the tangential fields in the x–z plane we may solve directly for either the unknown E_y or H_x component. The essential difference between the procedures for obtaining these fields is that E_y is equal to zero outside of the aperture region, $A = \bigcup_{n=1}^{N} A_n$ (see Fig. 8.33), whereas H_x is not zero outside this interval. In other words, we must solve for H_x over an *infinite interval.*

Solving for H_x over an infinite interval is not as prohibitive as it may appear at first. The key to solving for H_x by the method of moments is to approximate H_x in the asymptotic region, $|x|$ very large, by the *known* asymptotic behavior of H_x in this region (except for unknown complex constants). For example, we may approximate

$$H_x(x) = \begin{cases} \sum_{n=1}^{N} a_n u_n(x), & x_{-1} \le x \le x_{+1} \\ \dfrac{C_{-1}}{x^{3/2}}, & x \le x_{-1} \\ \dfrac{C_{+1}}{x^{3/2}}, & x_{+1} \le x \end{cases} \tag{43}$$

where the finite interval ($x_{-1} \le x \le x_{+1}$) is chosen sufficiently large. The asymptotic $x^{-3/2}$ behavior will be discussed later for this problem. The functions $u_n(x)$ may be pulse or any other type of functions* which are linearly independent and rapidly convergent in the finite interval. A total of $N + 2$ unknown coefficients exists for this representation.

If E_y is solved for directly, we need to represent E_y, in the method of moments, only over the aperture, A. However, when we formulate the integral equation for E_y in the H-plane scan, we shall see that certain convergence problems arise which will require special care when obtaining a numerical solution. These convergence problems are associated with the necessity of including TE modes or fields in the solution for E_y. They occur for *any* problem in which tangential $\mathbf{E}_t$ is solved for and in which an *infinite continuum or a discrete* number of modes are necessary to represent $\mathbf{E}_t$. This problem arises for both the exterior ($z \ge 0$) and interior ($z \le 0$) representations for $\mathbf{E}_t$.

As with the E-plane scan problem of Section 2.2.1, we may represent the fields in the exterior region in terms of the aperture field, $E_y(x)$, by either the spectral or the Green's function approach. The spectral approach will lead to the Green's function [e.g., see (34)] representation, as we observed earlier. We will take the spectral approach in order to draw an analogy with the discrete spectrum representation for the infinite array case.

We showed in Section 2.2.1 that the electric field in $z \ge 0$ may be represented as

$$\mathscr{E}_y(x, z) = \frac{1}{2\pi}\int_{-\infty}^{\infty} dk_x \int_{-\infty}^{\infty} dx' \exp\left[-jk_x(x - x')\right] \exp\left(-j\sqrt{k^2 - k_x^{\,2}}\, z\right) E_y(x') \tag{44}$$

This equation reduces to the following when $z = 0$:

$$E_y(x) = \frac{1}{2\pi}\int_{-\infty}^{\infty} dk_x \int_{-\infty}^{\infty} dx'\, e^{-jk_x(x-x')} E_y(x') \tag{45}$$

Using Maxwell's curl equation, we readily obtain an equation for H_x:

$$\begin{aligned}\mathscr{H}_x(x, z) &= +\frac{1}{j\omega\mu_0}\frac{\partial \mathscr{E}_y(x, z)}{\partial z} \\ &= -\frac{1}{2\pi}\int_{-\infty}^{\infty} dk_x \int_{-\infty}^{\infty} dx' \frac{\sqrt{k^2 - k_x^{\,2}}}{\omega\mu_0} \\ &\quad \times \exp\left[-jk_x(x - x')\right] \exp\left(-j\sqrt{k^2 - k_x^{\,2}}\right) \mathscr{E}_y(x')\end{aligned} \tag{46}$$

* Certain rather broad restrictions on the choice of basis functions, $u_n(x)$, will be discussed later in this section.

In particular, at $z = 0$ we obtain

$$H_x(x) = \frac{-1}{2\pi}\int_{-\infty}^{\infty} dk_x \int_{-\infty}^{\infty} dx' \left(\frac{\sqrt{k^2 - k_x^{\,2}}}{\omega\mu_0}\right) e^{-jk_x(x-x')} E_y(x') \tag{46a}$$

We observe that, in analogy with the discrete spectrum formula for infinite arrays, (46) has contributions from the *continuum* eigenfunctions, $(1/\sqrt{2\pi})e^{-jk_x x}$ [see (29)], and that the continuum TE type modal admittances are represented by

$$Y_{k_x} = \frac{\sqrt{k^2 - k_x^{\,2}}}{\omega\mu_0} \tag{47}$$

Because of the particular form of these TE modal admittances, difficulties in convergence will be encountered *if* we attempt to reverse the order of integrations. [We will see shortly that the same difficulty exists also in the discrete case for the interior representation of $E_y(x)$ $(z \leq 0)$.] The reversal of the order of integrations in (32), leading to (35), we recall, involved special mathematical considerations which are beyond the scope of this book. In the present case, if we attempt to integrate (46a) with respect to k_x first, we find even more serious convergence problems for this integral.

There are, however, two reasons for attempting to reverse the order of integrations in (46a). One is to put the ultimate integral equation in a more standard form. The second and more important reason is to enable us to perform one integration in closed form if this is possible, as it was with (32). If we introduce the differential operation

$$\left(\frac{\partial^2}{\partial x^2} + k^2\right)\frac{e^{-jk_x(x-x')}}{\sqrt{k^2 - k_x^{\,2}}} = \sqrt{k^2 - k_x^{\,2}}\, e^{-jk_x(x-x')} \tag{48}$$

we find that we can utilize (35) and transform (46a) into

$$H_x(x) = -\frac{1}{2\omega\mu_0}\left(\frac{\partial^2}{\partial x^2} + k^2\right)\int_A H_0^{(2)}(k\,|x - x'|)E_y(x')\,dx' \tag{49}$$

or

$$\mathscr{H}_x(x, z) = -\frac{1}{2\omega\mu_0}\left(\frac{\partial^2}{\partial x^2} + k^2\right)\int_A H_0^{(2)}(k\,|\bar{\rho} - \bar{x}'|)E_y(x')\,dx' \tag{50}$$

as in (40) and (41). The differentiation in (49) and (50) cannot be brought under the integral sign in the usual sense.[33]

The same problems of convergence arise when we attempt to represent $E_y(x)$ and $H_x(x)$ in the region $z \leq 0$. The fields inside the waveguides are most conveniently described in terms of the normal modes of the waveguides. Let $\phi_{pn}(x)$ and Y_{pn} denote, respectively, the nth orthonormal modal function and the modal admittance of the pth waveguide. The modal functions are defined in such a way that they are equal to the usual modal functions when $x \in A_p$, but zero elsewhere. We may then write the tangential fields for the pth waveguide as

$$E_y(x) = \alpha_p(1 + R_p)\phi_{p1}(x) + \sum_{n=2}^{\infty} V_{pn}\phi_{pn}(x) \tag{51}$$

and

$$H_x(x) = -Y_{p1}\alpha_p(1 - R_p)\phi_{p1}(x) + \sum_{n=2}^{\infty} Y_{pn}V_{pn}\phi_{pn}(x) \tag{52}$$

where we have assumed that the lowest-order mode ($n = 1$ here) having a modal voltage of α_p is incident in the pth waveguide and R_p is the reflection in that mode. Utilizing the orthonormality relation among the modal functions, we find the unknown modal voltages in terms of the aperture field as

$$V_{pn} = \int_A \phi_{pn}(x')E_y(x')\,dx' \tag{53}$$

with $V_{p1} = \alpha_p(1 + R_p)$ for the dominant mode.

Substituting (53) into the expression of H_x in (52) leads, for the pth guide, to

$$\begin{aligned} H_x(x) &= -2Y_{p1}\alpha_p\phi_{p1}(x) + \sum_{n=1}^{\infty} Y_{pn}\phi_{pn}(x)\int_A \phi_{pn}(x')E_y(x')\,dx' \\ &= -2Y_{p1}\alpha_p\phi_{p1}(x) + \left(\frac{\partial^2}{\partial x^2} + k^2\right) \\ &\quad \cdot \int_{A_p} \sum_{n=1}^{\infty} \frac{Y_{pn}}{\gamma_{pn}^{\ 2}}\,\phi_{pn}(x)\phi_{pn}(x')E_y(x')\,dx' \end{aligned} \tag{54}$$

where $\gamma_{pn} = \sqrt{k^2 - (n\pi/a_p)^2}$ are the propagation constants of the nth mode in the pth waveguide, a_p being the width of this waveguide. In arriving at (54), we used the relation

$$\left(\frac{\partial^2}{\partial x^2} + k^2\right)\phi_{pn}(x) = \gamma_{pn}^{\ 2}\phi_{pn}(x) \tag{54a}$$

in order to interchange the order of integration and summation without incurring a divergent integral. This is analogous to the operation performed in (48). We next match the boundary conditions on H_x across *all* the apertures, using (49) and (54), to obtain

$$\begin{aligned} 2\sum_{p=1}^{N} Y_{p1}\alpha_p\phi_{p1}(x) = \left(\frac{\partial^2}{\partial x^2} + k^2\right)\int_A \Bigg\{ &\sum_{p=1}^{N}\sum_{n=1}^{\infty} \frac{Y_{pn}}{\gamma_{pn}^{\ 2}}\,\phi_{pn}(x)\phi_{pn}(x') \\ &+ \frac{1}{2\omega\mu_0}\,H_0^{(2)}(k\,|x - x'|)\Bigg\} E_y(x')\,dx' \end{aligned} \tag{55}$$

This is an *integrodifferential* equation for an array of N empty waveguides with the pth one excited by modal voltage α_p.

The derivation of this equation for the situation of loading the waveguides with dielectric plugs follows a similar procedure. As in earlier problems in which dielectric materials were added, we merely change the admittances in (55) so that the free term on the left-hand side of the equation has the admittances changed to

$$Y_{p1} \to \bar{Y}_{p1} = \frac{Y_{p1} Y_{p1}^{\varepsilon}}{Y_{p1}^{\varepsilon} \cos \gamma_{p1}^{\varepsilon} d_p + j Y_{p1} \sin \gamma_{p1}^{\varepsilon} d_p}$$

and the interior or waveguide admittances on the right-hand side become

$$Y_{pn} \to \tilde{Y}_{pn} = Y_{pn}^{\varepsilon} \frac{Y_{pn} + j Y_{pn}^{\varepsilon} \tan \gamma_{pn}^{\varepsilon} d_p}{Y_{pn}^{\varepsilon} + j Y_{pn} \tan \gamma_{pn}^{\varepsilon} d_p} \tag{56}$$

where

$$Y_{pn}^{\varepsilon} = \frac{\gamma_{pn}^{\varepsilon}}{\omega \mu_0} \quad \text{and} \quad \gamma_{pn}^{\varepsilon} = \sqrt{k^2 \varepsilon_p - (n\pi/a_p)^2} \tag{56a}$$

The convergence problems associated with (55) are not completely eliminated by preserving the order of integration and differentiation in this equation. When we attempt to solve (55) by the method of moments, this problem arises again and care must be taken in solving this equation. The same difficulties are present for both the internal summation or discrete spectral part of the kernel of (55) and the continuous part. This should be evident from a comparison of the spectral derivation of the continuous part [(44)–(49)] and the derivation of the discrete part [(51–(55)].

We note, for example, that both parts of the kernel of (55) have *logarithmic* singularities at $x = x'$. This is precisely the same type of singularity that both parts of the kernel had in the integral equation for the E-plane scan [(37)], although in that case, unlike (55), there is no differential operation after the integration.

In the E-plane scan case, we note that $E_y(y)$ had a singularity at the aperture edges. This is a consequence of the edge boundary conditions. In the H-plane scan case, however, $E_y(x)$ is not singular except in the second derivative of E_y with respect to x at the aperture edge. That is, as illustrated in Fig. 8.34, $E_y(x) = 0$ at the edge, but $(\partial/\partial x)E_y(x)$ is discontinuous and $(\partial^2/\partial x^2)E_y(x)$ is singular at the edge [as $H_x(x)$ is singular at the edge]. This comparative *smoothness* of $E_y(x)$ in the H-plane scan ensures the existence or convergence of the right-hand side of (55).

For example, it is clear from (54a) that, if an approximation for $E_y(x')$ was excessively singular in x', so that the terms $\int_A \phi_{pn}(x')E_y(x')\,dx'$ behaved as

$$\lim_{n\to\infty} \int_A \phi_{pn}(x') E_y(x')\,dx' \sim \text{constant}$$

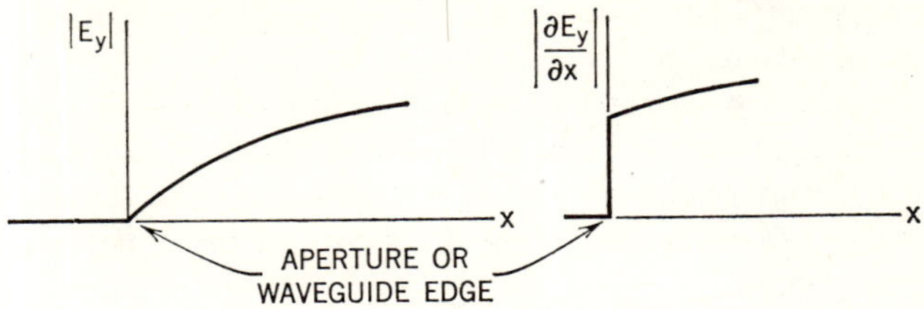

Fig. 8.34. Edge behavior of E_y in H-plane scan.

that is, were *independent of n* for large n, then

$$\left(\frac{\partial^2}{\partial x^2} + k^2\right)\int_A \sum_{p=1}^{N} \sum_{n=1}^{\infty} \frac{Y_{pn}}{\gamma_{pn}^2} \phi_{pn}(x)\phi_{pn}(x')E_y(x')\,dx'$$
$$= \sum_{p=1}^{N} \sum_{n=1}^{\infty} Y_{pn}\phi_{pn}(x)\int_A \phi_{pn}(x')E_y(x')\,dx' \sim \frac{1}{|x - x'|^2} \qquad (57)$$

In other words, the right-hand side of (55) would behave as the *second derivative* of the logarithmic singularity.* Such behavior is inconsistent, however, with the analytic behavior of the incident field, the left-hand side of (55), and will give an incorrect solution.

We must therefore ensure in any approximation for $E_y(x)$, such as in the method of moments, that $E_y(x)$ is a *sufficiently smooth* function so that

$$\sum_{p=1}^{N} \sum_{n=1}^{\infty} Y_{pn}\phi_{pn}(x)\int_A \phi_{pn}(x')E_y(x')\,dx' \qquad (58)$$

will converge uniformly for all $x \in A$. The "smoother" $E_y(x)$ is, the more rapidly the series expansion of $E_y(x)$ in $\phi_{pn}(x)$ will converge.[25] As an example, if we approximate $E_y(x)$ as a series of pulse functions as illustrated in Fig. 8.35*a*, we find that (58) will diverge or be singular for some x. If, on the other hand, we smooth the pulses by using a trapezoidal (or Simpson or higher-order) rule integration as shown in Fig. 8.35*b*, the approximate $E_y(x)$ has the same smoothness (or maximum order of *bounded* derivative) as the true E_y. This is illustrated in Fig. 8.34. In this case, we find that (58) will converge.

Exactly the same situation will prevail for the exterior or continuous part of the kernel of (55). This is readily shown by transforming $H_0^{(2)}(k\,|x - x'|)$ back into its spectral form, (34), and considering the dependence on the

* This is true, of course, since, for large n, $Y_{pn} \sim n$, and

$$\sum_{n=1}^{\infty} n\phi_{pn}(x)\phi_{pn}(x') \sim \frac{1}{|x - x'|^2}$$

as a result.[40]

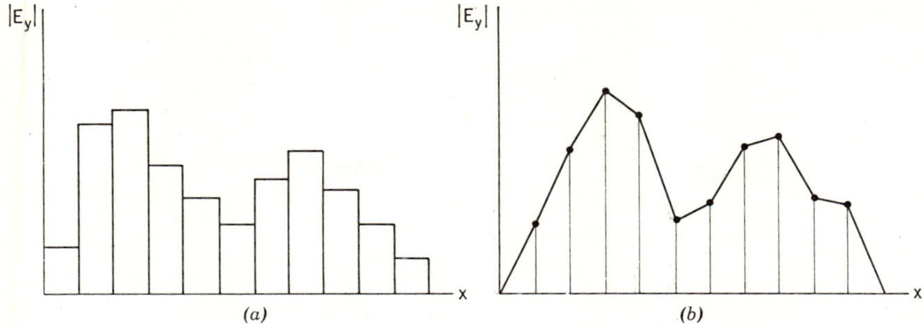

Fig. 8.35. Smoothness approximations for E_y. (*a*) Pulse basis. (*b*) Trapezoid basis.

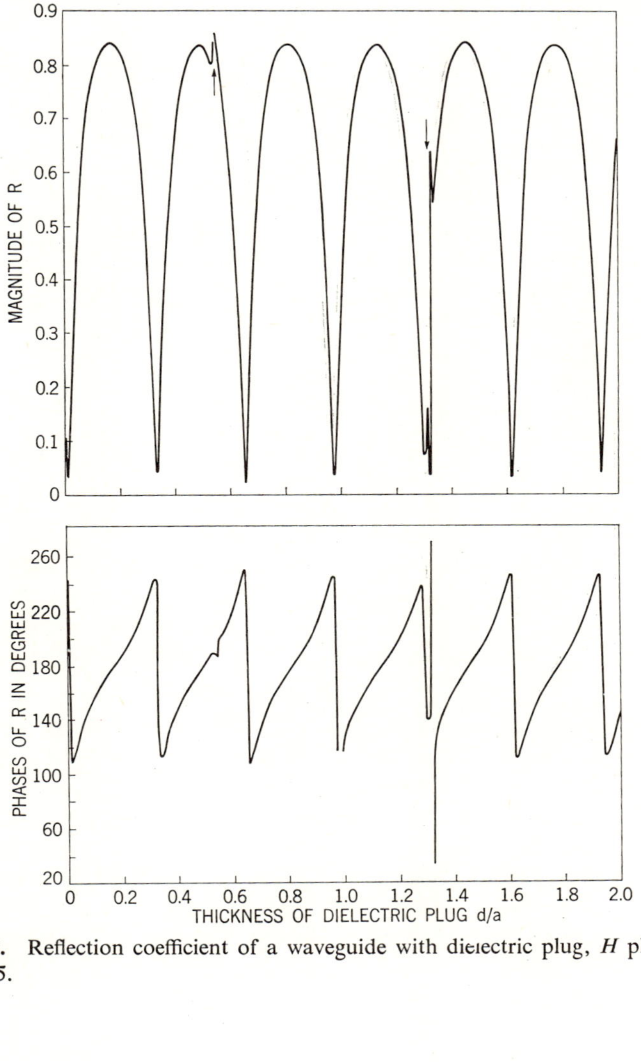

Fig. 8.36. Reflection coefficient of a waveguide with dielectric plug, *H* plane; $\varepsilon = 6.0$, $\lambda/a = 1.5$.

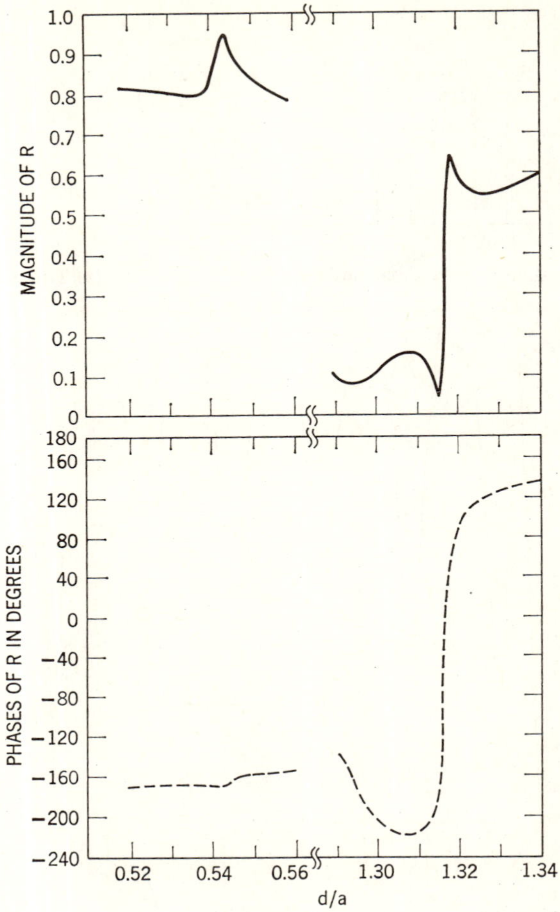

Fig. 8.37. Details of R versus d/a for $\varepsilon = 6$ and $\lambda/a = 1.5$.

continuous modal index, k_x. It is important to observe that this convergence problem and its proper handling by appropriate approximations for the aperture field apply equally well to general planar array problems *independently of the array element type* considered.

The actual numerical solutions to (55) obtained for one or more elements in the ground plane (Fig. 8.33) are of interest in themselves, since the H-plane and E-plane scans exhibit certain different properties. The variation of the reflection coefficient, R_p [(53)], is considered first for a single waveguide in the ground plane, $N = 1$ and $p = 1$, excited in the dominant mode. Figure 8.36 exhibits the results for $a/\lambda = 1.5$ and $\varepsilon = 6.0$, with d/a the independent variable. The dielectric constant is chosen so that the third-order mode, $\phi_3(x)$, is propagating in the dielectric [the second-order mode, $\phi_2(x)$, is suppressed because of symmetry] and is evanescent in the air-filled region.

The reflection coefficient shows a smooth standing-wave-like variation versus d/a over the entire range of d_1 considered, except in the vicinities of $d/a \simeq 0.54$ and $d_1/a_1 \simeq 1.31$, where sharp spikes appear. The details of the reflection coefficient near these spikes are shown in Fig. 8.37.

The maxima (or minima) of the standing-wave-like pattern are equally displaced at a distance given by $\pi/\gamma_1^{\varepsilon}$, where γ_n^{ε} is the propagation constant

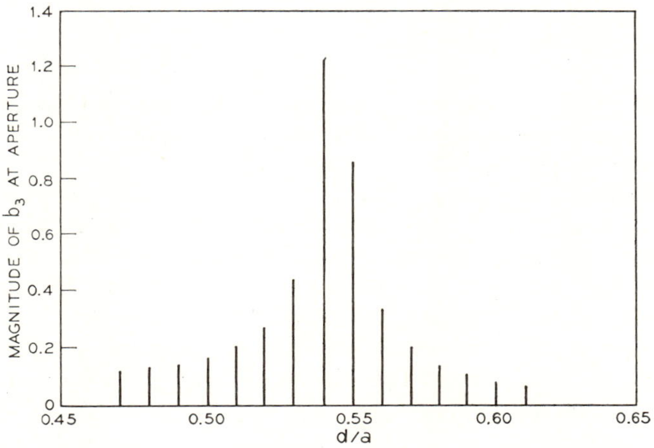

Fig. 8.38. Magnitude of third modal coefficients versus d/a for $\varepsilon = 6$ and $\lambda/a = 1.5$.

of the nth mode of a dielectric-loaded waveguide. The separation between the two spikes, Δd, is found to be obtainable from the relation $\gamma_3^{D}(\Delta d) = \pi$. This observation suggests that, for most dielectric thicknesses, the third-order mode is only weakly excited, so that the radiation impedance of the waveguide is determined primarily by the fundamental mode. Only when the dielectric plug has a certain thickness is the third-order mode excited strongly enough to influence the reflection coefficient of the fundamental mode. In Fig. 8.38 we show the solutions for the third-order modal coefficients versus d to demonstrate that this is indeed the case.

The magnetic field anywhere in the exterior region ($z \geq 0$) may be obtained from (50) once a solution for $E_y(x)$ is obtained. In particular, we are interested in the *far*-zone radiation patterns, which can be obtained from (50) by using the asymptotic expression for the Hankel function, $H_0^{(2)}(k\,|\bar{\rho} - \bar{x}'|)$, when the argument is large. We find

$$E_y(\rho, \theta) \approx \sqrt{k/2\pi\rho}\; e^{-j[k\rho-(3\pi/4)]} \cos\theta \cdot \int_A e^{j(k\sin\theta)x'} E_y(x')\, dx' \qquad (59)$$

and

$$H_\theta(\rho, \theta) = \eta_0 E_y(\rho, \theta)$$

where η_0 is the characteristic admittance of free space.

It is useful *to normalize* the radiation pattern for *unit incident* power. When this is done and when the method of moments approximation

$$E_y(x') \approx \sum_{n=1}^{Q} b_n \phi_{1n}(x') \tag{60}$$

is used, we obtain a useful expression for the normalized far-zone radiation field, $T(\theta)$:

$$T(\theta) = \frac{2k}{\sqrt{2\pi\gamma_1}} \cos\theta \sum_{n=1,3,5,\ldots}^{Q} b_n \frac{\cos[(ka/2)\sin\theta]}{(n\pi/a)\{1 - [(2a/n\lambda_0)\sin\theta]^2\}} \tag{61}$$

We note that (60), using the waveguide modal functions as a basis, satisfies the smoothness criteria established earlier for $E_y(x')$ approximations. This is generally true for *all* waveguide arrays.

The normalized radiation patterns, $T(\theta)$, for several dielectric plug thicknesses, $\lambda/a = 1.5$, and $\varepsilon = 6$ are depicted in Fig. 8.39*a*. It is interesting that no sharp variations in pattern are evident here, even when parameters which allow sharp spikes in the reflection coefficient are used. Nevertheless, dips occur for some parameter values as illustrated in Fig. 8.39*b*. The previously discussed[14] relationship between the radiation pattern of the singly excited element and the reflection coefficient is, however, clearly a property of only large or infinite arrays.

We consider next some numerical results obtained from (55) for two identical waveguides (see Fig. 8.33). We first observed the coupling between two waveguides containing no dielectric (one guide only is excited, i.e., $\alpha_p = \delta_{1p}$). These results are shown in Fig. 8.40 for several values of λ/a as a function of waveguide separation, s/a. Generally speaking, we find that coupling between TE modes (*H* plane) is somewhat *weaker* than coupling between TM modes (*E* plane). For sufficiently large separations, in fact, we find that the coupling is proportional to $S^{-3/2}$ in the *H* plane, whereas, asymptotically, it is proportional to $S^{-1/2}$ in the *E* plane. We find this to be true also when dielectric plugs of various thicknesses are included in the waveguides. After an initial (small separation, S, between the guides) oscillation that occurs at certain critical thicknesses, as discussed earlier for single-waveguide radiation, the asymptotic coupling here decays as $S^{-3/2}$ also.

The $S^{-3/2}$ behavior for TE (*H*-plane) mode coupling can be explained as a consequence of the fact that the tangential aperture electric field can, by Huygens' principle,[41] always be replaced by *magnetic dipoles*:

$$\mathbf{M} = \mathbf{E} \times \hat{z} = \hat{x} E_y \tag{61}$$

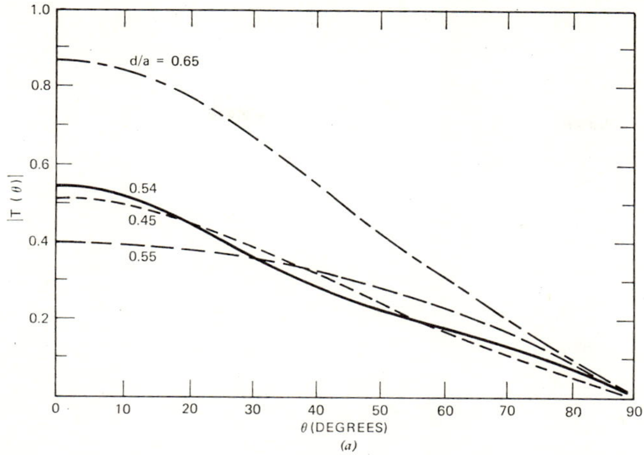

Fig. 8.39*a*. Normalized radiation patterns, $T(\theta)$, of a waveguide with dielectric plugs, H plane. $\lambda/a = 1.5$, $\varepsilon = 6$.

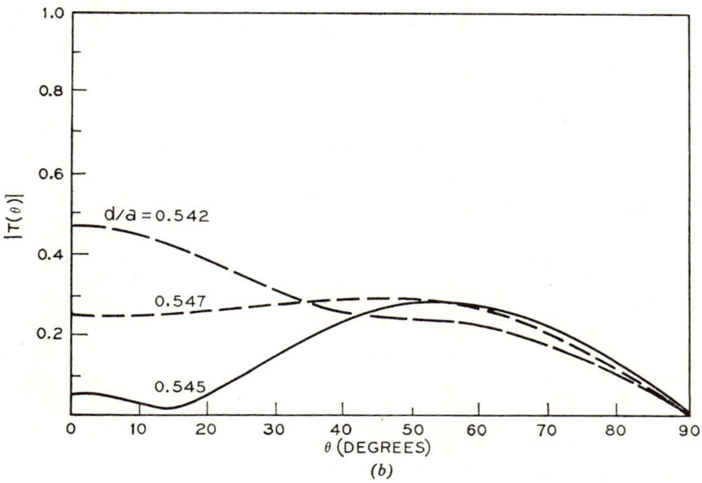

Fig. 8.39*b*. Normalized radiation patterns, $T(\theta)$, of a waveguide with dielectric plugs, H plane. $\lambda/a = 1.5$, $\varepsilon = 6$.

over a metal surface. When this is done with the aperture field in Fig. 8.33, we have a finite region of $\hat{x}$-directed magnetic dipoles radiating on a ground plane. It is well known that such dipoles, like electric dipoles, have a *zero far-zone* radiation field in the direction of their orientation. It thus follows that the dominant far-zone field in this direction will decay as the next term in the appropriate asymptotic expansion, in this case the $S^{-3/2}$ term. The

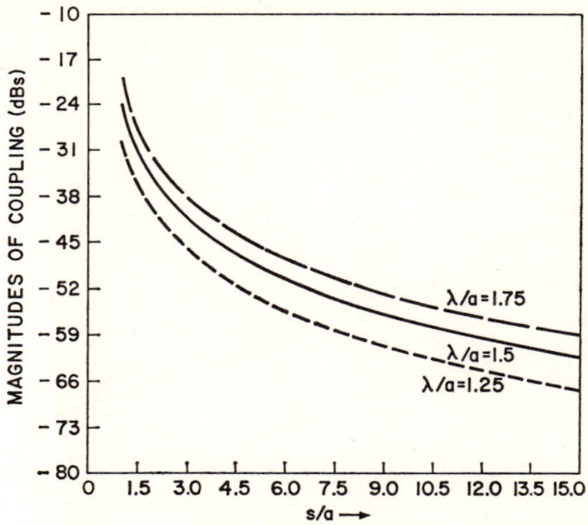

Fig. 8.40. Coupling versus separation between two unloaded waveguides, H plane.

origins of this $S^{-3/2}$ behavior are thus substantially different from those of the $S^{-3/2}$ behavior obtained for the coupling between elements in an array with infinitely many elements.

With $\varepsilon = 6$ and $\lambda/a = 1.5$, we observe, in Fig. 8.42, the coupling between two closely spaced waveguides ($S/a = 1.0$) as a function of plug thickness (these are the same parameters used earlier for Figs. 8.36–8.41). As the thickness of the plug is varied, the coupling reveals sharp resonance characteristics

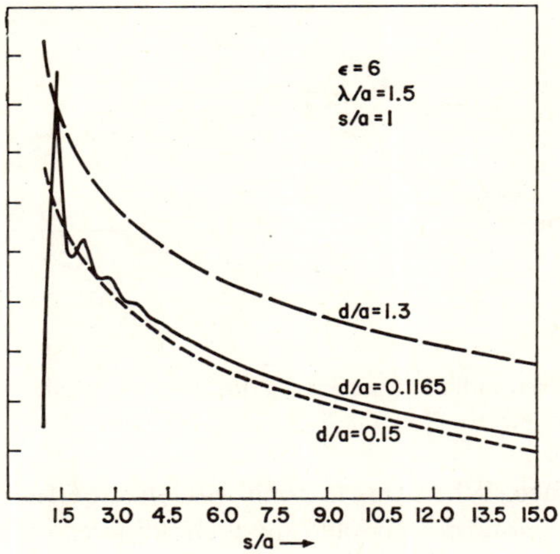

Fig. 8.41. Coupling versus separation between two unloaded waveguides with dielectric plugs, H plane (same scale as in figure 8.40).

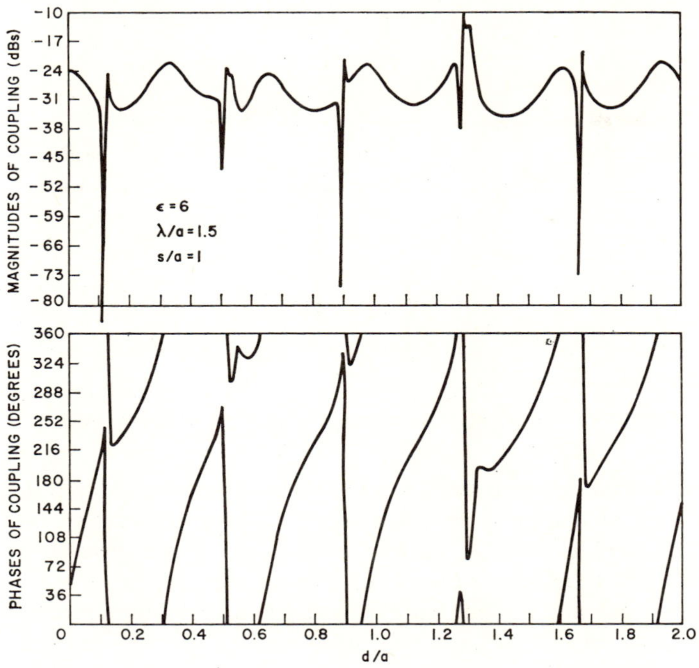

Fig. 8.42. Coupling versus d/a between two waveguides with dielectric plugs, H plane.

over various ranges of d. Examination of the propagation constants for the second and third propagating modes, γ_2^{ε} and γ_3^{ε}, indicates that the resonances are caused by the second- and third-order modes resonating *separately*. Note that in this case the second-order mode, $\phi_2(x)$, is not suppressed because of symmetry. The resonances due to the third-order mode occur near those of the second-order mode at $d/a \simeq 0.52$ and 1.29. The other sharp variations at $d/2 \simeq 0.12$, 0.9, and 1.67 are produced by the second-order mode alone.

2.2.3 Radiation from a Waveguide through a Dielectric Sheath on a Ground Plane. The problem of a waveguide radiating through a dielectric sheath covering the infinite ground plane, as shown in Fig. 8.43, is considerably different from the dielectric plug problem discussed in Sections 2.2.1 and 2.2.2. Two principal differences are associated with this sheath problem.[42]

First of all, the continuous spectral distribution (Fourier integral representation) for the field description in the region $z \geq 0$ is not integrable into a closed form Green's function. This introduces considerable difficulty in the numerical handling of the problem. In fact, the Green's function pertinent

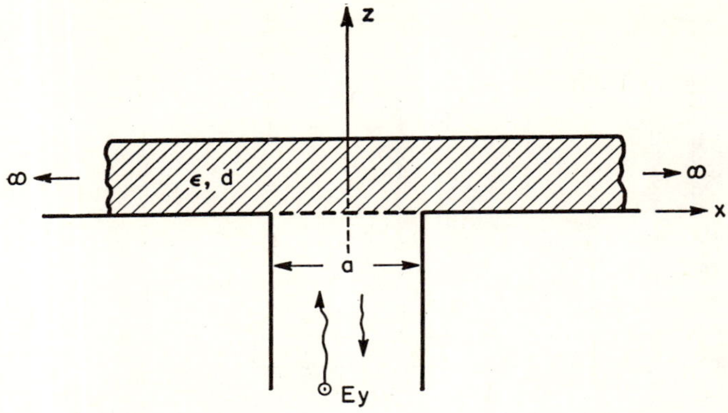

Fig. 8.43. Parallel-plate waveguide covered by a dielectric slab.

to the region of a dielectric slab located above an infinite ground plane has been discussed at length in the literature.[43,44] These discussions have been limited, however, to the asymptotic evaluation of the Fourier integral representation of the Green's function in the far-zone region.

Also significant (and evident from the spectral distribution representation) is the fact that *surface waves* may be excited on the dielectric sheath in much the same manner as for the corrugated surfaces discussed in Section 2 1. In this respect, this geometry (Fig. 8.43) must be designed so that no surface waves are excited when the covered ground plane is used as a finite or truncated terminating structure of a finite phased array.

We shall see, in fact, that the exterior representation for this structure has all the properties of the discrete summation representation for the $z \geq 0$ region of an infinite array. In this case, the *discrete* sum becomes a *continuous* integration in the same manner as discussed for the dielectric plugs case in Sections 2.2.1 and 2.2.2, and the physical interpretations of evanescent modes, trapped modes, radiating modes, etc., all remain, in essence, the same as in earlier chapters.

For the polarization indicated in Fig. 8.43, the incident TE_{10} wave excites only TE scattered fields (see Section 2.2.2 for the H-plane scan case). With the fields invariant in the y direction, Maxwell's equations reduce to

$$j\omega\mu_0 H_x = \frac{\partial E_y}{\partial z}, \qquad -j\omega\mu_0 H_z = \frac{\partial E_y}{\partial x}$$

and (62)

$$\left(\frac{\partial^2}{\partial x^2} + \frac{\partial^2}{\partial z^2} + k^2\varepsilon_i\right)E_y = 0, \qquad \varepsilon_i = \begin{cases} \varepsilon & \text{for } 0 \leq z \leq d \\ 1 & \text{for } z \geq d \end{cases}$$

for this geometry. If we now represent

$$E_y(x) = \frac{1}{2\pi}\int_{-\infty}^{\infty} dk_x \int_A dx'\, e^{-jk_x(x-x')} E_y(x') \tag{63}$$

where the transform of $E_y(x)$ is

$$e_y(k_x) = \frac{1}{2\pi}\int_A dx'\, e^{jk_x x'} E_y(x') \tag{64}$$

we readily find $\mathscr{E}_y(x, z)$ and $\mathscr{H}_x(x, z)$ in terms of $E_y(x)$ by applying the boundary conditions at $z = d$, which require the continuity of the tangential $\mathscr{E}_y(x, d)$ and $\mathscr{H}_x(x, d)$ components and the radiation condition at $z \to \infty$. We thus obtain

$$\begin{aligned}\mathscr{E}_y(x, z) &= \int_A dx'\, G_\varepsilon(x, z; x', 0) E_y(x') \\ \mathscr{H}_x(x, z) &= \frac{-1}{\omega\mu_0}\int_A dx'\, j\frac{\partial}{\partial z} G_\varepsilon(x, z; x') E_y(x')\end{aligned} \tag{65}$$

where the appropriate Green's function is

$$G_\varepsilon(x, z; x') = \frac{1}{2\pi}\int_{-\infty}^{\infty} dk_x\, e^{-jk_x(x-x')} \times \begin{cases} \dfrac{\gamma_\varepsilon \cos\gamma_\varepsilon(d-z) + j\gamma \sin\gamma_\varepsilon(d-z)}{\gamma_\varepsilon\cos\gamma_\varepsilon d + j\gamma\sin\gamma_\varepsilon d}, & 0 \le z \le d \\ \dfrac{\gamma_\varepsilon e^{-j\gamma(z-d)}}{\gamma_\varepsilon\cos\gamma_\varepsilon d + j\gamma\sin\gamma_\varepsilon d}, & d \le z \end{cases} \tag{66}$$

and the z-directed propagation constants inside and outside of the dielectric are, respectively,

$$\begin{aligned}\gamma_\varepsilon &= \sqrt{\varepsilon k^2 - k_x^{\,2}} = -j\sqrt{k_x^{\,2} - \varepsilon k^2} \\ \gamma &= \sqrt{k^2 - k_x^{\,2}} = -j\sqrt{k_x^{\,2} - k^2}\end{aligned} \tag{67}$$

(the appropriate branches for the time convention, $e^{+j\omega t}$, being indicated). At $z = 0$, the tangential aperture fields are therefore given by (63) for $E_y(x)$, and for $H_x(x)$ by

$$H_x(x) = -\int_A dx' \int_{-\infty}^{\infty} dk_x \left(\frac{\gamma_\varepsilon}{2\pi\omega\mu_0}\right) \frac{\gamma + j\gamma_\varepsilon \tan\gamma_\varepsilon d}{\gamma_\varepsilon + j\gamma\tan\gamma_\varepsilon d}\, e^{-jk_x(x-x')} E_y(x') \tag{68}$$

It will be noted that the orders of integrations in the above equations have been changed, so that the very serious questions of convergence discussed in Section 2.2.2 become pertinent here also. The reader may view this order of integration as symbolic and assume that in any actual numerical computation the appropriate basis functions and procedures for computation discussed in Section 2.2.2 will be used. The form of (68) is significant in its similarity to the *infinite* sheath-covered array problems (see Chapters 6 and 7). We note that the infinite *sum*, $\sum_{m=-\infty}^{\infty}$, of the infinite array problem is now a continuous *integration*, $\int_{-\infty}^{\infty} dk_x$, and also that the modal admittance for the continuous spectral distribution appears as

$$\frac{\gamma_\varepsilon}{\omega\mu_0} \frac{\gamma + j\gamma_\varepsilon \tan \gamma_\varepsilon d}{\gamma_\varepsilon + j\gamma \tan \gamma_\varepsilon d} \tag{69}$$

which is identical in form to the modal admittances for the discrete modes in the case of a sheath over an infinite array.

It should be added that the forms of (68) and (69) are essentially the same for the alternative orthogonal polarization excitation (i.e., in the E plane or y-z plane). The admittances of the continuous spectral *modes* will, for this alternative polarization, be proportional to the *inverse* of the γ_ε and γ z-directed propagation constants defined by (67). [See e.g., (25), (26), and (32)–(34).]

With the now familiar interior ($z \leq 0$) representation of $E_y(x)$ and $H_x(x)$ in terms of the normal modes of the single waveguide considered (see Fig. 8.43), we may readily write the tangential electric and magnetic fields in the aperture in terms of the electric field as

$$E_y(x) = (1 + R)\phi_1(x) + \sum_{n=2}^{\infty} \phi_n(x) \int_A \phi_n(x')E_y(x')\, dx'$$

$$1 + R = \int_A \phi_1(x')E_y(x')\, dx' \tag{70}$$

and

$$H_x(x) = -2Y_1\phi_1(x) + \sum_{n=1}^{\infty} Y_n\phi_n(x) \int_A \phi_1(x')E_y(x')\, dx' \tag{71}$$

respectively. We then obtain the desired integral equation for $E_y(x)$ by matching tangential $H_x(x)$ across the aperture

$$2Y_1\phi_1(x) = \int_A dx' \left\{ \sum_{n=1}^{\infty} Y_n\phi_n(x)\phi_n(x') + \frac{1}{\omega\mu_0} \tilde{G}_\varepsilon(x, x') \right\} E_y(x') \tag{72}$$

where

$$\tilde{G}_\varepsilon(x, x') = \frac{1}{2\pi} \int_{-\infty}^{\infty} dk_x \gamma_\varepsilon \frac{\gamma + j\gamma_\varepsilon \tan \gamma_\varepsilon d}{\gamma_\varepsilon + j\gamma \tan \gamma_\varepsilon d}\, e^{-jk_x(x-x')} \tag{73}$$

As already discussed, careful note of the *symbolic* nature of the order of integration and representation in (72) and (73), as regards the convergence of these integrals, should be made.

The contour of integration in both (63) and (73) can be taken along the real axis, as shown in Fig. 8.44, if we take the appropriate branches of γ as

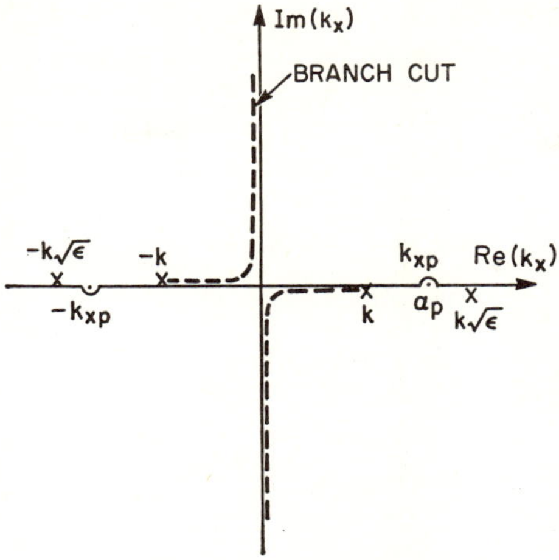

Fig. 8.44. Integration contour.

indicated in (67). We note that the form of the integrand in (73) is such that there is a branch point at the zeros of γ ($k_x = \pm k$), but there is none at the zeros of γ_ε ($k_x = \sqrt{\varepsilon}k$) because the integrand is an even function of γ_ε.

In evaluating (73), we must also take into account any pole singularities, $k_x = k_{xp}$, of the integrand as indicated in Fig. 8.44. The zeros of

$$\gamma_\varepsilon + j\gamma \tan \gamma_\varepsilon d\big|_{k_x = k_{xp}} = 0 \tag{74}$$

it should be noted, are the same as those for the homogeneous or characteristic equation solution of a dielectric slab on a ground plane.[45] The existence of such zeros indicates that surface waves may be excited on the slab for this polarization. The residue of these poles will be discussed in more detail later.

The present form of the exterior portion of the kernel, (73), can be reformulated so that a *singular* part is singled out from the integral. This is analogous to what was done earlier in Chapter 5 for the discrete interior and exterior parts of the kernel in which a quasi-static part was extracted. The same manipulation can also be made for the discrete part of (72), and similarly for both parts of the kernel for the E-plane alternative polarization. In the present case we note that, when $k_x^2 > k^2\varepsilon$, both γ and γ_ε become purely imaginary, and the integrand of $\tilde{G}_\varepsilon(x, x')$ becomes

$$j\,|\gamma_\varepsilon| \frac{|\gamma| + |\gamma_\varepsilon| \tanh |\gamma_\varepsilon|\, d}{|\gamma_\varepsilon| + |\gamma| \tanh |\gamma_\varepsilon|\, d}\, e^{-jk_x(x-x')} \tag{75}$$

By virtue of the relation

$$\tanh u \approx 1 \qquad \text{for} \quad \text{large } u$$

it is possible to find a number, say M, sufficiently large so that, when $|\gamma_\varepsilon|\, d \geq M$,

$$\frac{|\gamma| + |\gamma_\varepsilon| \tanh |\gamma_\varepsilon|\, d}{|\gamma_\varepsilon| + |\gamma| \tanh |\gamma_\varepsilon|\, d} \approx 1 \tag{76}$$

In the limit, as $M \to \infty$, we obtain for $\tilde{G}_\varepsilon(x, x')$

$$\tilde{G}_\varepsilon(x, x') = \frac{1}{2}\left(\frac{\partial^2}{\partial x^2} + \varepsilon k^2\right) H_0^{(2)}(\sqrt{\varepsilon}\, k\, |x - x'|)$$

$$+ \frac{1}{2\pi} \int_{-\infty}^{\infty} dk_x \gamma_\varepsilon \frac{(\gamma - \gamma_\varepsilon)(1 - j \tan \gamma_\varepsilon d)}{\gamma_\varepsilon + j\gamma \tan \gamma_\varepsilon d}\, e^{-jk_x(x-x')} \tag{77}$$

We have in effect separated the *singularity* of $\tilde{G}_\varepsilon(x, x')$ from the "quasi-static" part of $\tilde{G}_\varepsilon(x, x')$—the remaining *rapidly converging* integration.

In the integrals of (63), (73), and (77) we recognize that, as $|k_x|^2$ varies from 0 to ∞, the different ranges of $|k_x|$ contribute different types of waves to the spectrum of the aperture fields. In the range $|k_x|^2 \leq k^2$ we have the portion of the spectrum associated with propagating waves in the air and dielectric regions. In the range $k^2 \leq k_x^{\,2} \leq \varepsilon k^2$ we have waves which are *trapped* in the dielectric, that is, are propagating in the dielectric and evanescent in air. The remaining waves, in the range $|k_x|^2 \geq \varepsilon k^2$, are evanescent everywhere. In the integral of (77), we note that the zeros of γ_ε now contribute a branch point to the integrand. The branch cuts in Fig. 8.44 may now be a line connecting $-k$ with $-\sqrt{\varepsilon}\, k$ and $+k$ with $+\sqrt{\varepsilon}\, k$. The same contour and branch cuts are applicable also to the E-plane alternative polarization. The surface wave poles, $k_x = k_{xp}$, will lie on the branch cuts.

The purpose of introducing the transformation is, as earlier for the quasi-static transformation in the infinite array, to bring the kernel into a suitable form for numerical evaluation. The infinite range or limits of the integrals in (73) must be truncated in a numerical evaluation. The integration in (77), however, converges much more rapidly than that in (73) as the limits grow to $\pm\infty$. Hence, after the transformation to (77), it is seen that the infinite integral may, to a good approximation, be reduced to a finite range, and the calculation is considerably simplified.

Once the aperture field, $E_y(x)$, is found from a solution of (72), the reflection coefficient, R, is obtained, as earlier, from (70). The total fields for all $z \geq 0$ can be found from (65). The evaluation of (65) is of particular interest in the far-zone region, $kr \gg 1$. This far-zone field is found from (65) by a saddle-point integration.

We consider first the field in $z \geq d$. Let

$$k_x = k \sin w, \qquad \gamma = k \cos w \tag{78}$$

This transformation maps the complex k_x plane into the strip $-\pi \leq \text{Re } w \leq \pi$ in the complex w plane, as shown in Fig. 8.45. Substitution of (78) into (65) yields

$$E_y(\rho, \theta) = k \int_C dw \cos w e^{-jk\rho \cos (w-\theta)} e_y(k \sin w)$$
$$\times \frac{\sqrt{\varepsilon - \sin^2 w}}{\sqrt{\varepsilon - \sin^2 w} \cos kd\sqrt{\varepsilon - \sin^2 w} + j \cos w \sin kd\sqrt{\varepsilon - \sin^2 w}} \tag{79}$$

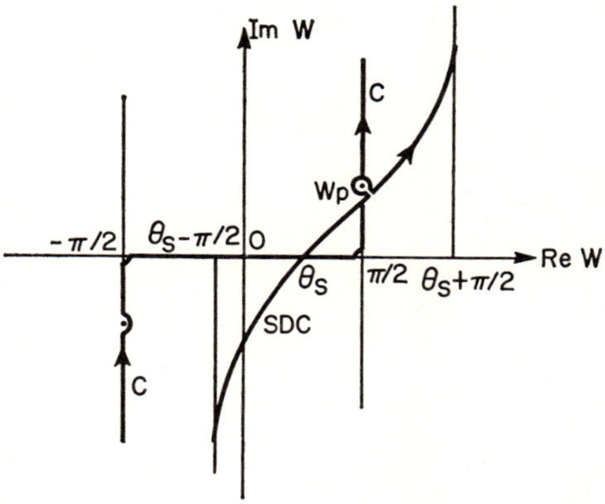

Fig. 8.45. Contour of integration, C, for far-zone fields; $k_x = k \sin w$.

where e_y is given by (64). The contour of integration, C, is as indicated in Fig. 8.45. The coordinates (x, z) are related to (ρ, θ) by

$$x = \rho \sin \theta, \qquad z - d = \rho \cos \theta$$

with ρ and θ measured in relation to the point $(0, d) = (x, z)$.

To apply the saddle-point method of integration, the contour of integration, C, is deformed into the steepest descent path (S.D.P.). If, in the process of contour deformation, any poles of the integrand are crossed, the original integral is equal to the integral along the S.D.P. plus the residue contribution from such poles, according to the Cauchy residue theorem. The possibilities of surface wave and leaky wave poles in the configuration of the problem have been extensively analyzed.[43,44] In particular, the effects of so-called leaky waves (complex poles) on the radiation patterns of line sources have been carefully investigated. The integral along the S.D.P. may be approximated by the dominant contribution occurring at the saddle points, which are located at the zeros of

$$\frac{d}{dw}\cos(w-\theta_s)=0 \quad \text{or} \quad w=\theta_s$$

The procedure of applying the saddle-point method is relatively standard.[45] The final results are given by

$$E_y(\rho,\theta)\approx -k\sqrt{2\pi/k\rho}\,f(\theta)e^{-j(k\rho-\pi/4)}+\sum \text{Res}\,(w_p) \qquad \text{for} \quad \frac{-\pi}{2}<\theta<\frac{\pi}{2}$$

or

$$\approx k\sqrt{\pi/2}\,f''\left(\frac{\pi}{2}\right)\frac{e^{-j(k\rho-\pi/4)}}{(k\rho)^{3/2}}+\sum \text{Res}\,(w_p) \qquad \text{for} \quad \theta=\frac{\pm\pi}{2}$$

$$\left(\text{provided } k\sqrt{\varepsilon-1}\,d\neq\frac{n\pi}{2}\right) \tag{80}$$

where the *far-zone radiation pattern* is given by

$$f(\theta)=\frac{\sqrt{\varepsilon-\sin^2\theta}}{\sqrt{\varepsilon-\sin^2\theta}\cos kd\sqrt{\varepsilon-\sin^2\theta}+j\cos\theta\sin kd\sqrt{\varepsilon-\sin^2\theta}} \times \cos\theta\,\mathscr{E}_y(k\sin\theta) \tag{81}$$

The residues at the surface poles, w_p, are given by

$$\text{Res}\,(w_p)=jk\cos w_p e^{-jk\rho\cos(w_p-\theta)}\mathscr{E}_y(k\sin w_p)$$
$$\times\frac{\sqrt{\varepsilon-\sin^2 w_p}}{\frac{d}{dw}\left[\sqrt{\varepsilon-\sin^2 w}\cos kd\sqrt{\varepsilon-\sin^2 w}+j\cos w\cdot\sin kd\sqrt{\varepsilon-\sin^2 w}\right]\Big|_{w=w}} \tag{82}$$

The poles of the integrand are the solutions to the characteristic equation, (73):

$$\sqrt{\varepsilon-\sin^2 w}\cos kd\sqrt{\varepsilon-\sin^2 w}+j\cos w\cdot\sin kd\sqrt{\varepsilon-\sin^2 w}=0$$

and, as mentioned earlier, denote the excitation of surface waves along the sheath. The $\rho^{-3/2}$ far-zone behavior for $E_y(\rho, 0)$ in (80) is, as explained in Section 2.2.2, characteristic of this polarization of the electric field along or parallel to the ground plane.

In Fig. 8.46, we present results describing the distribution of power among the radiation waves, the surface waves, and the waves reflected back into the excited waveguide illustrated in Fig. 8.43. An incident wave of *unit power*

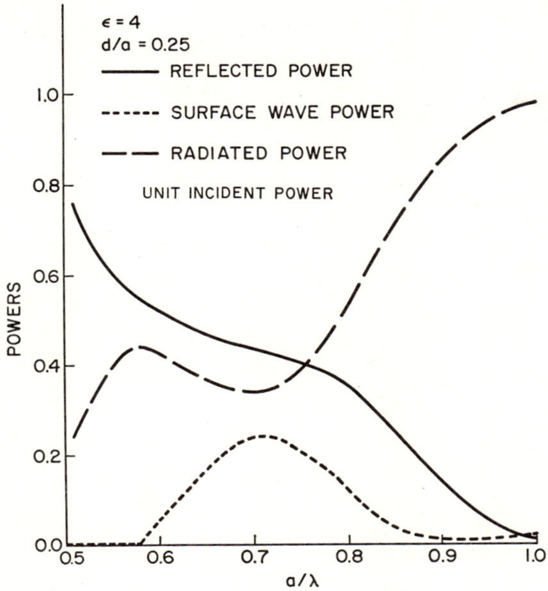

Fig. 8.46. Radiated, reflected, and surface wave powers.

excites the waveguide. The reflected power is obtained by squaring the reflection coefficients from (70). The surface wave power may be obtained by integrating the Poynting vector over the y–z plane (transverse to the direction of surface wave propagation) by computations utilizing the residue contributions in (80)–(82). The radiated power is found by appropriately integrating the radiation portions of (80) and (81).

The results in Fig. 8.46, for $\varepsilon = 4$ and $d/a = 0.25$ as a function of a/λ, indicate that no surface waves are excited until

$$\frac{a}{\lambda} \geq \frac{1}{\sqrt{\varepsilon - 1}} = 0.576$$

for these values of the parameters. This corresponds to the cut-off wavelength of a TE surface wave,[45] as may be determined from the characteristic equation (73):

$$\frac{d}{\lambda} = \frac{1}{4} \frac{1}{\sqrt{\varepsilon - 1}}$$

In Fig. 8.47, the results of computations for "normalized" radiation patterns

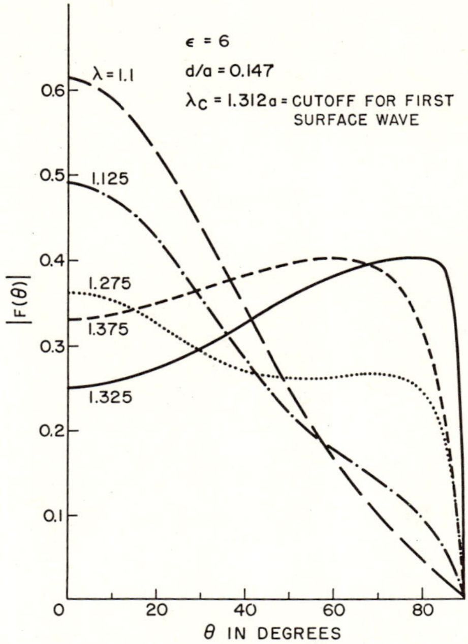

Fig. 8.47. Radiation patterns, $|F(\theta)|$.

are presented for several wavelengths. The normalization is such that

$$\text{Power radiated} = \int_{-\pi/2}^{\pi/2} |F(\theta)|^2 \, d\theta$$

The pattern $F(\theta)$ is of course proportional to $f(\theta)$ in (80) and (81). The cut-off wavelength of the first surface wave mode occurs at $\lambda_c/a = 1.312$ for this set of parameters. We observe that, when $\lambda > \lambda_c$, no surface wave exists, and the pattern shows a maximum radiation occurring at an angle close to the end-fire direction. As the wavelength is increased toward the cut-off value, this critical angle of strong radiation moves closer toward the end-fire direction. When the wavelength is increased beyond cut-off, however, the radiation energy in the end-fire direction is eventually converted into the fields of a surface wave mode. At the same time, higher radiation in the broadside direction is observed.

The actual aperture field distribution, $E_y(x)$, which is computed from a moments solution using the normal waveguide modes as a basis, is illustrated in Fig. 8.48 for several values of λ/a at $\varepsilon = 6$ and $d/a = 0.147$. We note that the frequently used *incident mode* approximation for the aperture field

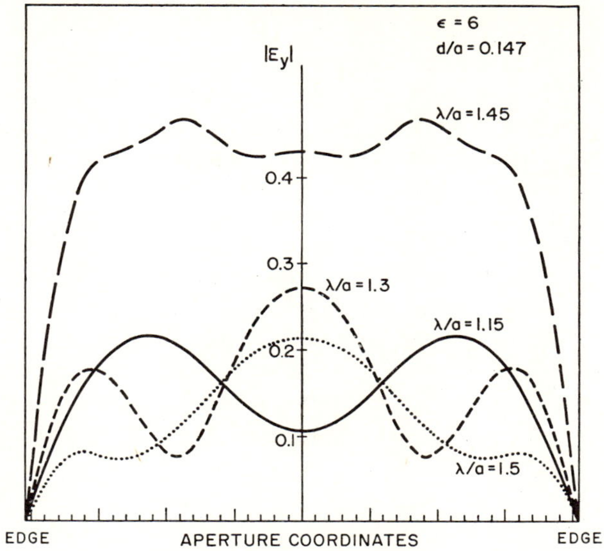

Fig. 8.48. Aperture electric field, E_y.

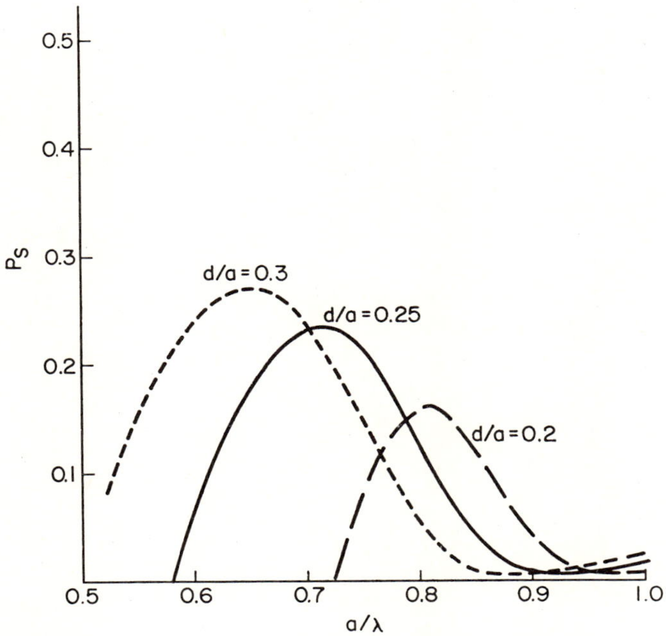

Fig. 8.49. Surface wave powers; P_s vs. a/λ for $\varepsilon = 4$.

neglects higher-order modes that are very strongly excited at some wavelengths. These higher modes are excited in the same approximate wavelength region in which the surface waves are excited.

The significance of the surface wave excitations with regard to array design is of course apparent, insofar as the *edge effects* are concerned. A finite array terminated in a ground plane with a dielectric sheath cover can lose considerable power into the edge, in addition to causing radiation pattern distortion when the surface waves eventually radiate from the finite edge itself. Furthermore, if a number of elements such as are shown in Fig. 8.43 are combined to form an array, the surface wave excitation will surely cause array coupling problems, some of which were discussed extensively in earlier chapters. The distribution of this surface wave power loss or excitation (for unit incident power) is shown in Fig. 8.49 for various sheath thicknesses as a function of a/λ.

APPENDIX I EVANESCENT AND MULTIMODE INTERACTION WITH THE SHORT CIRCUITS

In the discussion of modulated surfaces, we considered, for clarity, that only one mode interacted with the short circuits (or terminal impedances). It is possible to use a method similar to that employed earlier to account for a finite number of modes interacting with the short circuit. These modes may be propagating or evanescent.

To do this we must allow additional higher-order modes to reflect from the short circuit and hence be "propagating" toward the aperture. We must therefore introduce coupling coefficients for these modes. Earlier we introduced, for the periodic phased array, coupling coefficients in the mth mode to the nth waveguide as

$$\mathscr{C}_{nm} = \frac{1}{2\pi} \int_{-\pi}^{\pi} I_m(\psi) e^{-jn\psi}\, d\psi$$

where $I_m(\psi)$ are modal coefficients for the mth mode. These coefficients can be determined either by experiment, by exact analysis, or by numerical solution of the integral equation

$$2Y_0\phi_0(y) = \int_{\text{aperture}} \mathbf{K}(y, y') \cdot \mathbf{E}_{\text{tang}}(y')\, dy'$$

where Y_0 is the modal admittance, $\phi_0(y)$ is the exciting mode, and $\mathbf{K}(y, y')$ is the appropriate array kernel (see Chapter 2).

If we now want to include the $\phi_q(y')$ mode interaction with the short circuits (or terminal impedances), we must solve

$$2Y_q\phi_q(y) = \int_{\text{aperture}} \mathbf{K}(y, y') \cdot \mathbf{E}_{\text{tang}}(y')\, dy'$$

where $\mathbf{K}(y, y')$ *remains the same* and $\phi_q(y)$ is evanescent or propagating. The modal coefficients of this solution may be designated as ${}_qI_m(\psi)$, and new coupling coefficients defined as

$${}_q\mathscr{C}_{nm} = \frac{1}{2\pi}\int_{-\pi}^{\pi} {}_qI_m(\psi)e^{-jn\psi}\, d\psi$$

The coefficients ${}_q\mathscr{C}_{nm}$ designate the coupling from the guide excited by $\phi_q(y)$ into the nth guide in the mth mode. For example,

$${}_q\mathscr{C}_{n0} = \frac{1}{2\pi}\int_{-\pi}^{\pi} {}_qI_0(\psi)e^{-jn\psi}\, d\psi$$

are the coefficients of the zeroth (dominant) mode in the nth removed waveguide due to excitation by the qth mode (possibly evanescent) in the zeroth waveguide.

Let us consider the case in which the zeroth waveguide is excited by a dominant waveguide mode with amplitude $A_0{}^0$. This is the same as the situation considered earlier and depicted in Fig. 8.1. In the present case, however, we will consider *one* higher-order (possibly evanescent) mode to interact with the short circuits as shown in Fig. A.1. (We will refer to this mode as evanescent even though it may be propagating.) This higher mode is seen to have a coefficient of reflection, as indicated and defined in Fig. A.1.

We may proceed to analyze this case as we did that shown in Fig. 8.1. By superposition we may write

$$C_0{}^n = A_0{}^0\mathscr{C}_{n0} + \sum_{m\in M_2} (C_0{}^m \tfrac{1}{2}e^{-\gamma_0 h_m})\mathscr{C}_{n-m,0} + \sum_{m\in M_2} (C_1{}^m \tfrac{1}{2}e^{-\gamma_1 h_m})\,{}_1\mathscr{C}_{n-m,0}, \quad n \in M_1 \quad \text{(A.1)}$$

$$C_0{}^n \tfrac{1}{2}e^{\gamma_0 h_n} = A_0{}^0\mathscr{C}_{n0} + \sum_{m\in M_2} (C_0{}^m \tfrac{1}{2}e^{-\gamma_0 h_m})\mathscr{C}_{n-m,0} + \sum_{m\in M_2} (C_1{}^m \tfrac{1}{2}e^{-\gamma_1 h_m})\,{}_1\mathscr{C}_{n-m,0}, \quad n \in M_2 \quad \text{(A.2)}$$

$$C_1{}^n \tfrac{1}{2}e^{\gamma_1 h_n} = A_0{}^0\mathscr{C}_{n,1} + \sum_{m\in M_2} (C_0{}^m \tfrac{1}{2}e^{-\gamma_0 h_m})\mathscr{C}_{n-m,1} + \sum_{m\in M_2} (C_1{}^m \tfrac{1}{2}e^{-\gamma_1 h_m})\,{}_1\mathscr{C}_{n-m,1}, \quad n \in M_2 \quad \text{(A.3)}$$

The unknown quantities are $C_0{}^m$ and $C_1{}^m$; the known quantities, $\mathscr{C}_{n,m}$ and ${}_q\mathscr{C}_{n,m}$. We note that now there are *two* equations for those $n \in M_2$ (the

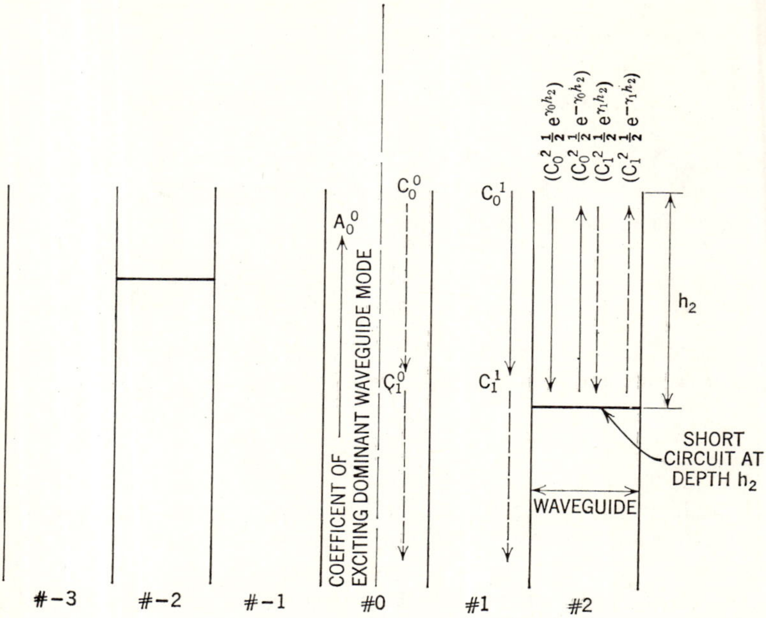

Fig. A.1. Accounting for one evanescent mode; M_1 = number of loaded guides, M_2 = number of shorted guides, $\gamma_0 = -j\beta_0$ = dominant guide propagation constant, γ_1 = first evanescent propagation constant.

shorted waveguides)—one for each mode interacting with the short circuit. Consider, first, the case wherein

$$M_2 = \text{finite} \quad \text{and} \quad M_1 = \infty$$

Then there is a total of $2 \cdot M_2$ unknowns to be found—M_2 unknown $C_0{}^m$ $(m \in M_2)$ and M_2 unknown $C_1{}^m$ $(m \in M_2)$. Equations A.2 and A.3 can be solved, in this case, as a $(2M_2)$-order matrix for these unknowns. The remaining unknown $C_0{}^m$ $(m \in M_1)$ are then *directly* computed from (A.1).

The procedure for solving the case when

$$M_2 = \infty \quad \text{and} \quad M_1 = \text{finite}$$

is similar to that used when $C_1{}^m \frac{1}{2} e^{-\gamma_1 h_m} \approx 0$, that is, no reflected higher or evanescent modes are accounted for. First we note that

$$R(\psi) = \sum_{-\infty}^{\infty} \mathscr{C}_{n0} e^{+jn\psi}, \qquad \mathscr{C}_{n0} = \frac{1}{2\pi} \int_{-\pi}^{\pi} R(\psi) e^{-jn\psi}\, d\psi$$

$$I_m(\psi) = \sum_{-\infty}^{\infty} \mathscr{C}_{nm} e^{+jn\psi}, \qquad \mathscr{C}_{nm} = \frac{1}{2\pi} \int_{-\pi}^{\pi} I_m(\psi) e^{-jn\psi}\, d\psi$$

$$ {}_1I_1(\psi) = \sum_{-\infty 1}^{\infty} \mathscr{C}_{n1} e^{+jn\psi}, \qquad {}_1\mathscr{C}_{n1} = \frac{1}{2\pi} \int_{-\pi}^{\pi} {}_1I_1(\psi) e^{-jn\psi}\, d\psi $$

$$ {}_1I_0(\psi) = \sum_{-\infty}^{\infty} {}_1\mathscr{C}_{n,0} e^{jn\psi}, \qquad {}_1\mathscr{C}_{n,0} = \frac{1}{2\pi} \int_{-\pi}^{\pi} {}_1I_1(\psi) e^{-jn\psi}\, d\psi $$

and also that

$$ \sum_{m \in M_1} + \sum_{m \in M_2} = \sum_{-\infty}^{\infty} $$

An equation is then introduced for the $C_1^{\,n}$ ($n \in M_1$) coefficients [these could, of course, have been introduced earlier, and the $C_1^{\,n}$ ($n \in M_1$) solved for directly once the $C_0^{\,n}$ ($n \in M_2$) and $C_1^{\,n}$ ($n \in M_2$) were solved for]:

$$ C_1^{\,n} = A_0^{\,0} \mathscr{C}_{n,1} + \sum_{m \in M_2} (C_0^{\,m} \tfrac{1}{2} e^{-\gamma_0 h_m}) \mathscr{C}_{n-m,1} + \sum_{m \in M} (C_1^{\,m} \tfrac{1}{2} e^{-\gamma_1 h_m})\, {}_1\mathscr{C}_{n-m,1}, \quad n \in M_1 \qquad \text{(A.4)} $$

If we now multiply (A.1)–(A.4) by $e^{jn\psi}$ and add all the equations in (A.1) and (A.2) (i.e., add all $n \in M_1$ and $n \in M_2$), we obtain

$$ \sum_{m \in M_1} C_0^{\,m} e^{jm\psi} + \sum_{m \in M_2} (C_0^{\,m} \tfrac{1}{2} e^{\gamma_0 h_m}) e^{jm\psi} = A_0^{\,0} R(\psi) + R(\psi) \sum_{m \in M_2} (C_0^{\,m} \tfrac{1}{2} e^{-\gamma_0 h_m}) e^{jm\psi} + {}_1I_0(\psi) \sum_{m \in M_2} (C_1^{\,m} \tfrac{1}{2} e^{-\gamma_1 h_m}) e^{jm\psi} \qquad \text{(A.5)} $$

By similarly combining (A.3) and (A.4) we get

$$ \sum_{m \in M_1} C_1^{\,m} e^{jm\psi} + \sum_{m \in M_2} (C_1^{\,m} \tfrac{1}{2} e^{\gamma_1 h_m}) e^{jm\psi} = A_0^{\,0} I_1(\psi) + I_1(\psi) \sum_{m \in M_2} (C_0^{\,m} \tfrac{1}{2} e^{-\gamma_0 h_m}) e^{jm\psi} + {}_1I_1(\psi) \sum_{m \in M_2} (C_1^{\,m} \tfrac{1}{2} e^{-\gamma_1 h_m}) e^{jm\psi} \qquad \text{(A.6)} $$

Equations A.5 and A.6 form a *pair* of equations equivalent to the *single* equation, (6), for the case with no higher modes taken into account. The same methods can be applied for solving (A.5) and (A.6) as were used for the case in which higher-order interactions with the short circuit are neglected. (Note that $h = h_m$ is a function of m in the above equations, in general.)

For example, let h_m = constant = h. Then (A.5) and (A.6) can be written as

$$ (e^{\gamma_0 h} - Re^{-\gamma_0 h}) \left(\sum_{m \in M_2} C_0^{\,m} \tfrac{1}{2} e^{jm\psi} \right) - ({}_1I_0 e^{-\gamma_1 h}) \left(\sum_{m \in M_2} C_1^{\,m} \tfrac{1}{2} e^{jm\psi} \right) = A_0^{\,0} R - \sum_{m \in M_1} C_0^{\,m} e^{jm\psi} $$

and

$$-(I_1 e^{-\gamma_0 h})\left(\sum_{m\in M_2} C_0{}^m \tfrac{1}{2} e^{+jm\psi}\right) + (e^{\gamma_1 h} - {}_1I_1 e^{-\gamma_1 h})\left(\sum_{m\in M_2} C_1{}^m \tfrac{1}{2} e^{jm\psi}\right) = A_0{}^0 I_1 - \sum_{m\in M_1} C_1{}^m e^{jm\psi}$$

Hence the new dispersion relation which accounts for the first evanescent mode interacting with the short circuits is

$$[e^{\gamma_0 h} - R(\psi)e^{-\gamma_0 h}][e^{\gamma_1 h} - {}_1I_1(\psi)e^{-\gamma_1 h}] - [{}_1I_0(\psi)e^{-\gamma_1 h}][I_1(\psi)e^{-\gamma_0 h}] = 0$$

APPENDIX II GENERALIZATION TO PLANAR ARRAYS

It is easy to show that the discussion of aperiodic arrays and the dispersion relationships derived apply equally well to *two-dimensional* or *planar* arrays. For example, we find that the characteristic determinant, $\mathscr{D}_2$, in (11), for the modulated array with *two* short-circuit depths, at h and C, is also the characteristic determinant (or the dispersion relation when we set $\mathscr{D}_2 = 0$) for the *planar* array of waveguides with two short-circuit depths. In the planar array, the $(m + n)$th element has a short at depth h if $m + n$ is even, and at depth C if $m + n$ is odd. The array lies in the x–y plane. The integer $m = \pm 1, \pm 2, \ldots$ is measured in the x direction, and the integer $n = \pm 1, \pm 2, \ldots$ in the y direction. The reflection coefficient, $R(\psi_x, \psi_y)$, for the planar phased array is a function of the direction cosine angles proportional to ψ_x and ψ_y.[4,5] In this case $R' = R(\psi_x + \pi, \psi_y + \pi)$. The extension to other periodic variations of the short-circuit depth in the plane follows similarly (see Appendix III).

APPENDIX III GENERAL PERIODICALLY MODULATED SURFACES

There exists a systematic and direct approach for determining the appropriate "factored function of R," the dispersion relation for the modulated surface, when the depth of the short, h, is a general periodic function of element position. This method is suggested by the determinantal form of $\mathscr{D}_2$ in (11). An example, demonstrating the "factoring" procedure for a periodic variation of h_m over three elements, will make the general procedure more clear.

Let the shorting position vary periodically so that

$$h_m = h \quad \text{for} \quad m = 1 \pm 3n, \quad n = 0, 1, 2, \ldots$$
$$h_m = c \quad \text{for} \quad m = 2 \pm 3n$$

and

$$h_m = d \qquad \text{for} \quad m = 3 \pm 3n$$

Divide the M_2 elements so that $M_2 = M_2{}^h + M_2{}^c + M_2{}^d$, where if $m = 1 \pm 3n$ the $m \in M_2{}^h$, etc. Since there are three elements per period, we determine three values of ψ:

$$\psi_1 = \psi, \qquad \psi_2 = \psi + \tfrac{2}{3}\pi, \quad \text{and} \quad \psi_3 = \psi + \tfrac{4}{3}\pi$$

from the cubic root of unity. This allows us, with the definitions that

$$R = R(\psi), \qquad R' = R(\psi + \tfrac{2}{3}\pi)$$

and

$$R'' = R(\psi + \tfrac{4}{3}\pi)$$

to write three equations derived from (7) in the following form:

$$\begin{aligned}
&(e^{\gamma_0 h} - R'' e^{-\gamma_0 h}) e^{j(4/3)\pi} \sum_{m \in M_2{}^h} (C_0{}^m \tfrac{1}{2}) e^{jm\psi} \\
&\qquad + (e^{\gamma_0 c} - R'' e^{-\gamma_0 c}) e^{j(2/3)\pi} \sum_{m \in M_2{}^c} (C_0{}^m \tfrac{1}{2}) e^{jm\psi} \\
&\qquad + (e^{\gamma_0 d} - R'' e^{-\gamma_0 d}) \sum_{m \in M_2{}^d} (C_0{}^m \tfrac{1}{2}) e^{jm\psi} \\
&\qquad = A_0{}^0 R'' - \sum_{m \in M_1} C_0{}^m e^{jm\psi} e^{jm(4/3)\pi} \\
&(e^{\gamma_0 h} - R e^{-\gamma_0 h}) \sum_{m \in M_2{}^h} (C_0{}^m \tfrac{1}{2}) e^{jm\psi} + (e^{\gamma_0 c} - R e^{-\gamma_0 c}) \sum_{m \in M_2{}^c} (C_0{}^m \tfrac{1}{2}) e^{jm\psi} \\
&\qquad + (e^{\gamma_0 d} - R e^{-\gamma_0 d}) \sum_{m \in M_2{}^d} (C_0{}^m \tfrac{1}{2}) e^{jm\psi} \\
&\qquad = A_0{}^0 R - \sum_{m \in M_1} C_0{}^m e^{jm\psi} \\
&(e^{\gamma_0 h} - R' e^{-\gamma_0 h}) e^{j(2/3)\pi} \sum_{m \in M_2{}^h} (C_0{}^m \tfrac{1}{2}) e^{jm\psi} \\
&\qquad + (e^{\gamma_0 c} - R' e^{-\gamma_0 c}) e^{j(4/3)\pi} \sum_{m \in M_2{}^c} (C_0{}^m \tfrac{1}{2}) e^{jm\psi} \\
&\qquad + (e^{\gamma_0 d} - R' e^{-\gamma_0 d}) \sum_{m \in M_2{}^d} (C_0{}^m \tfrac{1}{2}) e^{jm\psi} \\
&\qquad = A_0{}^0 R' - \sum_{m \in M_1} C_0{}^m e^{jm\psi} e^{jm(2/3)\pi}
\end{aligned}$$

The appropriate "factored function" of R, R', and R'' is determined by inverting the above matrix equations. This inversion eliminates R, R', and R'' from the sums over M_2, allowing for a general solution of the aperiodic waveguide array that is periodically shorted. The determinant of the above set of equations, set equal to zero, provides us with the dispersion relation for the periodically modulated surface. The generalization to an arbitrary periodicity is clear and leads, by induction, to the generalized determinant given earlier:

$$\mathscr{D}_P = \det \{(e^{\gamma_0 h_m} - R_n e^{-\gamma_0 h_m}) e^{jm(n-1)(2\pi/P)}\}$$

where n = row position, m = column position, P = order of the determinant, and

$$R_n = R\left[\psi + (n-1)\frac{2\pi}{P}\right]$$

REFERENCES

1. V. Galindo, "A Generalized Approach to a Solution of Aperiodic Arrays and Modulated Surfaces," *IEEE Trans. Antennas and Propagation*, Vol. AP-16, No. 4, July 1968, pp. 424–429.
2. Y. T. Lo, "Random Periodic Arrays," *Radio Science*, Vol. 3, No. 5, May 1968.
3. S. W. Lee, "Radiation from an Infinite Aperiodic Array of Parallel-Plate Waveguides," *IEEE Trans. Antennas and Propagation*, Vol. AP-15, pp. 598–606, September 1967.
4. C. P. Wu and V. Galindo, "Properties of a Phased Array of Rectangular Waveguides with Thin Walls," *IEEE Trans. Antennas and Propagation*, Vol. AP-14, pp. 163–173, March 1966.
5. V. Galindo and C. P. Wu, "A Variational Expression for the Dominant Mode Coupling Coefficients Between the Elements in an Infinite Array," *IEEE Trans. Antennas and Propagation*, Vol. AP-14, pp. 637–639, September 1966.
6. V. Galindo and C. P. Wu, "Asymptotic Behavior of the Coupling Coefficients for an Infinite Array of Thin-Walled Rectangular Waveguides," *IEEE Trans. Antennas and Propagation*, Vol. AP-14, pp. 248–249, March 1966.
7. V. Galindo and C. P. Wu, "On the Asymptotic Decay of Coupling for Infinite Phased Arrays," *Proc. IEEE*, Vol. 56, No. 11, pp. 1872–1881, November 1968.
8. C. P. Wu and V. Galindo, "Surface Wave Effects of Dielectric Sheathed Phased Arrays of Rectangular Waveguides," *Bell System Tech. J.*, Vol. 47, pp. 117–142, January 1968.
9. C. P. Wu and V. Galindo, "Surface Wave Effects on Phased Arrays of Rectangular Waveguides Loaded with Dielectric Plugs," *IEEE Trans. Antennas and Propagation*, Vol. AP-16, pp. 358–360, May 1968.
10. R. E. Collin, *Field Theory of Guided Waves*, McGraw-Hill Book Company, New York, 1960, pp. 465–469.
11. D. S. Karjala and R. Mittra, "Radiation from Some Periodic Structures Excited by a Waveguide," *Electronics Letters*, Vol. 1, pp. 111–112, June 1965.
12. A. Hessel and H. Hochstadt, "Plane Wave Scattering from a Modulated-Corrugated Structure," *Proc. International Sci. Radio Union*, Palo Alto, Calif., 1966; see also David Y. Tseng, "Guiding and Scattering of Electromagnetic Fields by Corrugated Structures," Polytechnic Institute of Brooklyn, *Sci. Rept.* PIBMRI-1371-67, July 1967.
13. J. A. Stratton, *Electromagnetic Theory*, McGraw-Hill Book Company, New York, 1941, pp. 361–364.
14. V. Galindo and C. P. Wu, "The Relation Between the Far-Zone Pattern of the Singly Excited Element and the Transmission Coefficient of the Principal Lobe in an Infinite Array," *IEEE Trans. Antennas and Propagation*, Vol. AP-14, pp. 397–398, May 1966.

15. N. Amitay, "Improvement of Planar Array Match by Compensation Through Contiguous Element Coupling," *IEEE Trans. Antennas and Propagation*, Vol. AP-14, pp. 580–586, September 1966.

16. N. Amitay, P. E. Butzien, and R. C. Heidt, "Match Optimization of a Two Port Phased Array Antenna Element," *IEEE Trans. Antennas and Propagation*, Vol. AP-16, pp. 47–57, January 1968.

17a. E. C. Dufort, "Exact Scattering from an Infinite Array of Thick, Non-uniformly Corrugated Plates," *Proc. IEEE International Antennas and Propagation Symposium*, Palo Alto, Calif., 1966, pp. 380–384.

17b. E. C. Dufort, "Design of Corrugated Plates for Phased Array Matching," *IEEE Trans. Antennas and Propagation*, Vol. AP-16, pp. 37–46, January 1968.

17c. E. C. Dufort, "A Scattering Matrix Method for Solving Waveguide Array Impedance Problems," *Radio Science*, Vol. 3 (New Series), No. 5, May 1968, pp. 475–485.

18. R. Tang and N. S. Wong, "Multimode Phased Array Element for Wide Scan Angle Impedance Matching," *Proc. IEEE*, Vol. 56, No. 11, November 1968, pp. 1951–1959.

19. E. G. Magill and H. A. Wheeler, "Wide Angle Impedance Matching of a Planar Array Antenna by a Dielectric Sheet," *IEEE Trans. Antennas and Propagation*, AP-14, pp. 49–53, January 1966.

20. G. F. Farrell, Jr., and D. H. Kuhn, "Mutual Coupling Effects of Triangular-Grid Arrays by Modal Analysis," *IEEE Trans. Antennas and Propagation*, Vol. AP-14, No. 5, pp. 652–654, September 1969.

21. N. Amitay, and V. Galindo, "Characteristics of Dielectric Covered and Loaded Circular Waveguide Phased Arrays," *IEEE Trans. Antennas and Propagation*, Vol. AP–17, No. 6, November, 1969.

22. Y. T. Lo, "Non-uniform Linear Array System for the Radio-Telescope at the University of Illinois," ISRU and IRE Joint Meeting, Washington, D.C., May 1960, subsequently published in *IRE Trans. Antennas and Propagation*, Vol. AP-9, pp. 9–16, January 1961; also Y. T. Lo, "A Spacing Weighted Antenna Array," *IRE International Convention Record*, Part 1, p. 191, March 1962.

23. Y. T. Lo, (1963), "A Mathematical Theory of Antenna Arrays with Randomly Spaced Elements," *IEEE Trans. Antennas and Propagation*, Vol. AP-12, No. 3, pp. 257–268.

24. C. P. Wu, "Numerical Solutions for the Coupling Between Waveguides in Finite Arrays," *Radio Science*, Vol. 4, pp. 245–253, March 1969.

25. E. C. Titchmarsh, *The Theory of Functions*, Oxford University Press, London, 1939, pp. 426–427.

26. N. Amitay and V. Galindo, "Characteristics of Dielectric Covered and Loaded Circular Waveguide Phased Arrays," *Proc. International Science Radio Union*, Boston, Mass., September 1968.

27. N. Amitay and V. Galindo, "The Analysis of Circular Waveguide Phased Arrays," *Bell System Tech. J.*, Vol. 47, No. 9, pp. 1903–1931, November 1968.

28. N. Marcuvitz (ed.), *Waveguide Handbook*, M.I.T. Radiation Laboratory Series, Vol. 10, McGraw-Hill Book Company, New York, 1951, Section 4.6-4.9.

29. C. P. Wu, "Analysis of Finite Parallel-Plate Waveguide Arrays," *IEEE Trans. Antennas and Propagation*, Vol. 18, pp. 328–334, May 1970.

30. R. B. Dybal, R. C. Rudduck, and L. L. Tsai, "Mutual Coupling Between TEM and TE_{10} Parallel-Plate Waveguide Apertures," *IEEE Trans. Antennas and Propagation*, Vol. 14, pp. 574–580, 1966.

31. S. Mikuteit, "Mutual Coupling in a Three-Element Parallel-Plate Waveguide Array by Wedge Diffraction and Surface Integration Technique," *Tech. Rept.* 2485-1, Electro-Science Laboratory, Ohio State University, Columbus, Ohio, 1967.

32. P. M. Morse and H. Feshbach, *Methods of Theoretical Physics*, Part II, McGraw-Hill Book Company, New York, 1953, p. 1323.

33. E. C. Titchmarsh, *The Theory of Functions*, 2nd ed., Oxford University Press, London, 1939, Paragraphs 1.7 and 1.8.

34. N. Amitay and V. Galindo, "Application of a New Method for Approximate Solutions and Error Estimates to Waveguide Discontinuity and Phased Array Problems," *Radio Science*, 3, No. 8, pp. 830–844, August 1968.

35. V. Galindo and C. P. Wu, "Dielectric Loaded and Covered Rectangular Waveguide Phased Arrays," *Bell System Tech. J.*, Vol. 47, No. 1, pp. 93–116, January 1968.

36. W. H. von Aulock, "Properties of Phased Arrays," *Proc. IRE*, Vol. 48, No. 10, pp. 1715–1727, October 1960.

37. C. P. Wu, "Determination of Resonance Conditions in Dielectric-Sheathed or Plug-Loaded Phased Arrays," *IEEE Trans. Antennas and Propagation*, Vol. 16, No. 6, pp. 753–755, November 1968.

38. C. P. Wu, "Resonances in the Radiation Impedances of a Parallel-Plate Waveguide with Dielectric Plug," *Bell System Tech. J.*, Vol. 48, No. 9, September 1969.

39. C. P. Wu, "Characteristics of the Coupling Between Parallel-Plate Waveguides with and without Dielectric Plugs," *IEEE Trans. Antennas and Propagation*, Vol. 18, pp. 180–194, March 1970.

40. R. E. Collin, *Field Theory of Guided Waves*, McGraw-Hill Book Company, New York, 1960, Appendix A.6.

41. R. F. Harrington, *Time Harmonics Electromagnetic Fields*, McGraw-Hill Book Company, New York, 1968.

42. C. P. Wu, "Integral Equation Solutions for the Radiation from a Waveguide Through a Dielectric Slab," *IEEE Trans. Antennas and Propagation*, Vol. AP-17, No. 6.

43. C. T. Tai, "The Effect of a Grounded Slab on the Radiation from a Line Source," *J. Appl. Phys.*, Vol. 22, pp. 405–414, April 1951.

44. S. Barone, "Leaky Wave Contributions to the Field of a Line Source above a Dielectric Slab," Microwave Research Institute, Polytechnic Institute of Brooklyn, *Sci. Rept.* PIBMRI-462, November 1956.

45. R. E. Collin, *Field Theory of Guided Waves*, McGraw-Hill Book Company, New York, 1960, pp. 485–506.

9. Methods of Improving Array Match

1. INTRODUCTION

Mismatch between the active impedance of the phased array elements and their sources of excitation not only represents a decrease in the antenna gain and the system efficiency but also may impose special cooling requirements in high power systems. In certain cases, this mismatch may even affect the frequency stability of the exciting sources if they are not isolated. It is therefore a necessary and significant objective to design the arrays so that they are well matched over a prescribed wide scanning region. This objective is particularly difficult to achieve when mechanical, electrical, and environmental constraints restrict the choice and the design of radiating elements.

An ideal array should possess a perfect impedance match between the array elements and the exciting sources over an assigned scan region and frequency band of the system. Moreover, a specific polarization characteristic of the radiation pattern is usually required by the particular system application. Although a well-matched practical element does not necessarily yield the desired polarization, the polarization characteristic can usually be modified by a suitable choice of the feeding network of the element. Therefore, in this chapter emphasis will be placed on match optimization* of the array elements.

A number of design techniques have been proposed and used for matching array elements. We will briefly review the salient features of these methods and then discuss more general match optimization methods which have been utilized in the design of several types of practical array elements.

Waveguide simulators[1–4] have been used to evaluate the match of array elements at a few discrete scan angles. Arrays which possess certain symmetries can be simulated (see Chapters 2 and 7), and the active match at a few specific scan angles measured. The effects of matching devices placed

* Optimum in the sense that we minimize the mismatch of the active impedance of the element over an assignable scan region and frequency band.

at the aperture of the element or the feed line may be determined in the same manner. However, the limited number of scan angles and polarizations that can be simulated with a reasonably sized simulator restricts the usefulness of this method. Each combination of scan angle and polarization requires a different waveguide simulator.

The use of a thin sheath with a high dielectric constant, placed away from the aperture, has been proposed[5] for wide angle impedance match. Calculations based on transmission line type approximation and simulator measurements have shown that considerable improvement in the active impedance match is possible. However, the presence of forced surface wave resonances in this configuration (see Chapter 6) may not be detected with the limited number of simulator measurements available. In addition, the applicability of this method to various types of array elements has not been demonstrated.

The use of thin metallic fences or baffles[4] and *H*-plane parallel (corrugated) plates[6] has been suggested for improving the match of phased arrays of thin slots, thin dipoles, and rectangular waveguides. The resultant array match is obtained from a transmission line type analysis that accounts for the scattering properties of the parallel plates, as well as the array elements which now radiate into the parallel-plate region. These methods have been applied to the specific elements mentioned above with a particular polarization of excitation.

Multimode excitation of rectangular waveguide or parallel-plate array elements has also been proposed[7] in order to achieve a good match over a wide scan region. The relative complex coefficients of the pertinent modes are controlled, to a certain extent, in order to achieve the match. However, the resultant array match may be frequency sensitive.

In earlier chapters, analytical and numerical methods were employed in the investigation of the characteristics of waveguide planar arrays. These analytical methods are necessary and powerful tools for the design of an optimally matched array element. Of equal importance, however, is a systematic *synthesis* procedure for a matching network which will render an optimum match of the array over a wide scan region and frequency band. Such a synthesis procedure should utilize, whenever possible, the results obtained from the analysis.

Thin dielectric sheaths and/or plugs, as well as thin metallic irises, can be placed at the aperture for matching purposes. In this case, however, one can only rely on the analysis in the design of the optimally matched element. This limitation is due to the fact that the array boundary value problem has to be solved repeatedly for various values of the pertinent parameters of these matching devices.

Ordinarily, it is desirable to insert the matching network very close to the aperture of the element in order to reduce the frequency sensitivity of the match. In many practical cases, however, it turns out to be neither possible nor desirable to match the array element at its aperture. Rather, we can introduce matching networks in the feed lines or at another location where these networks do not perturb the aperture field distribution. In this case we can utilize the numerically obtained reflection characteristics of the element in a transmission line type of synthesis procedure which can determine the parameters of an optimum matching network.

Practical considerations, such as cost, simplicity, and intuition, may sometimes result in the choice of an element with a shape that is not readily amenable to analysis or numerical solution. The active reflection characteristics of such an element can be obtained through a relatively simple measuring procedure.

In this chapter we shall introduce systematic methods to determine the parameters of optimum matching networks which are placed behind the aperture. These synthesis procedures can utilize either experimentally or numerically obtained results. Mutual coupling or active reflection coefficient data may serve as the starting point for the synthesis. Although these methods are based on the infinite array model, we shall demonstrate the excellent agreement between their results and those obtained from finite array experiments. A single-port type of array element (i.e., with a single exciting mode), as well as a two-port element (i.e., with two orthogonal exciting modes), will be considered. In addition, experimental methods for determining the active array match and optimum element design will be discussed.

2. MATCH OPTIMIZATION OF A SINGLE-PORT ARRAY ELEMENT

Consider the infinite planar array of Fig. 9.1. This array consists of a rectangular grid* of single-port type elements. Each element in the array is labeled by two indices (m, n) and is excited by a voltage source, V_{mn}, through a transmission line with a characteristic admittance, Y_0. The array is excited uniformly in amplitude with a linear phase taper so that

$$V_{mn} = Ve^{j(m\psi_x + n\psi_y)} \tag{1}$$

where ψ_x and ψ_y are the steering phases.

The active reflection coefficient of the array elements is

$$R(\psi_x, \psi_y) = \sum_{m=-\infty}^{+\infty} \sum_{n=-\infty}^{+\infty} \mathscr{C}_{mn} e^{j(m\psi_x + n\psi_y)} \tag{2}$$

* A rectangular grid is assumed here for convenience only. The method to be discussed applies to any doubly periodic planar grid.

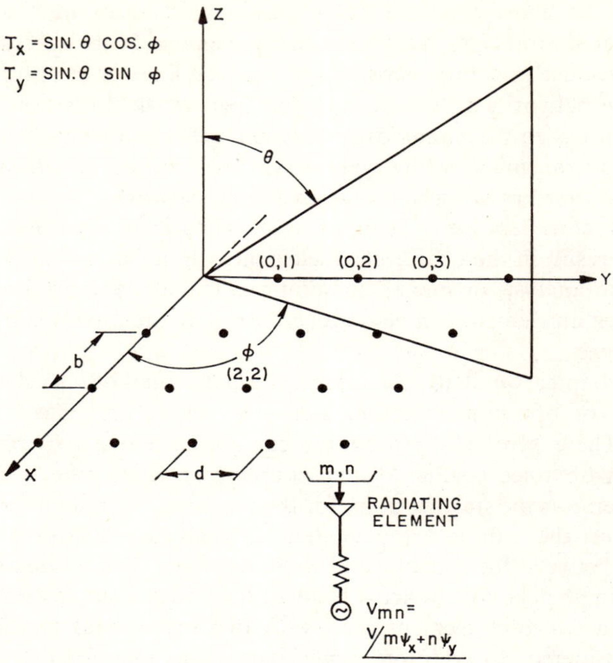

Fig. 9.1. Rectangular grid array model.

where $\{\mathscr{C}_{mn}\}$ are the coupling (scattering) coefficients between the (0, 0)th and the (m, n)th element. These coupling coefficients can be either measured or calculated and will be assumed known for the time being. In principle, (2) is a doubly infinite series. As was shown in previous chapters and will be discussed later in this chapter, only a moderate number of coupling coefficients (say, $M \times N$) is needed to represent R with reasonable accuracy. Thus, to a good approximation (2) can be rewritten as

$$R(\psi_x, \psi_y) = \sum^{M} \sum^{N} \mathscr{C}_{mn} e^{j(m\psi_x + n\psi_y)} \tag{3}$$

Employing the measured (or calculated) coupling coefficients in the finite series of (3) gives a convenient functional representation of the active reflection coefficient at *any* scan angle.

To improve the match, coupling or matching networks can be introduced behind the array face,[8–11] as shown schematically by the straight lines in Fig. 9.2. In this figure, each element (or element feed line) is coupled to its *adjacent neighbors only*. Hannan[11] has shown that an array can be matched over a denumerably infinite number of real scan angles if appropriate

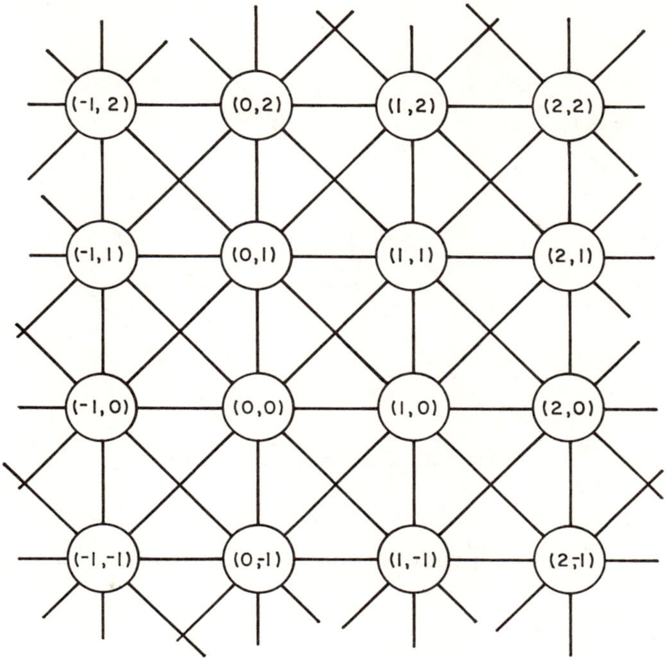

Fig. 9.2. Schematic representation of coupling networks between adjacent elements.

reactances are interconnected between *all possible pairs* of elements. Although this procedure is highly impractical, it shows that a good match over a wide scan region is at least theoretically possible.

The electrical model of the coupling networks is shown in Fig. 9.3. Here, each element or its feed line is interconnected to its adjacent neighbor channel via a lossless (reactive) π network, as shown by the dashed lines. In addition, the characteristic impedance of the transmission lines feeding the elements is allowed to change from $Z_0'' = 1/Y_0''$ to $Z_0' = 1/Y_0'$ at the coupling junction. The electrical location of this junction, $\phi/2 = 2\pi l/\lambda''$, with respect to the arbitrary reference plane A–A, is a variable as well. The effective wavelengths of the transmission line characterized by Z_0' and Z_0'' are λ' and λ'', respectively. In the subsequent discussion the matching networks are interconnected between *adjacent elements only*. The element location indices, as well as the symbolic representation of the coupling networks model, are shown in Fig. 9.4.

When the voltage sources of this infinite array are of constant amplitude with a linear phase taper [i.e., $V_{mn} = Ve^{j(m\psi_x+n\psi_y)}$], the voltages V'_{mn} at the

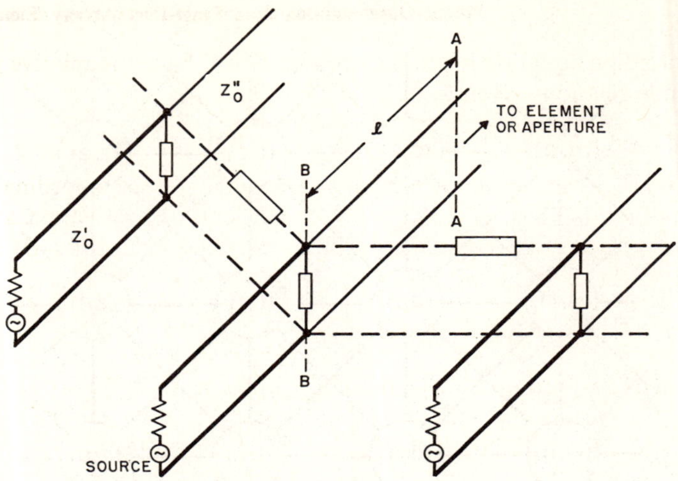

Fig. 9.3. Coupling networks model.

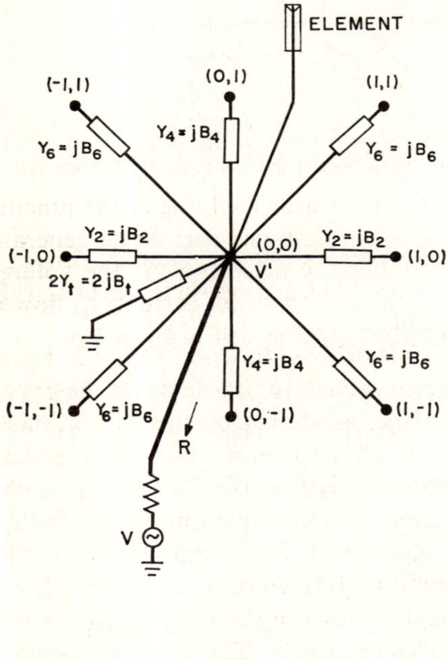

Fig. 9.4. Coupling networks model of the square grid array.

corresponding coupling junctions in Fig. 9.3 will have the relative phase and amplitude relations given by

$$V'_{mn} = V' e^{j(m\psi_x + n\psi_y)} \tag{4}$$

where V/V' is a complex quantity independent of the coordinates of the element (m, n). Thus, if the coupling coefficients are measured at the A–A plane of Fig. 9.3, the reflection coefficient at the junction (B–B) in the line leading to the element will be

$$R''(\psi_x, \psi_y) = e^{-j\phi} \sum_{(m)} \sum_{(n)} \mathscr{C}_{mn} e^{j(m\psi_x + n\psi_y)} \tag{5}$$

The active admittance at the coupling junction is

$$Y''(\psi_x, \psi_y) = Y''_0 \frac{1 - R''(\psi_x, \psi_y)}{1 + R''(\psi_x, \psi_y)} \tag{6}$$

The reflection coefficient in the line leading to the generator, R, at the coupling junction is

$$R(\psi_x, \psi_y) = \frac{1 - X}{1 + X} \tag{7}$$

where X is given by

$$X = j\eta_c + \eta_0 \frac{1 - R''(\psi_x, \psi_y)}{1 + R''(\psi_x, \psi_y)}, \qquad \eta_0 = \frac{Y''_0}{Y'_0} \quad \text{and} \quad j\eta_c = \frac{Y_c}{Y'_0} \tag{8}$$

Here X is the admittance presented to the generator feed line at the junction (normalized to Y'_0), and Y_c is the effective admittance across the generator line at the coupling junction due to the coupling networks. As the voltages at the coupling junctions, V'_{mn}, are given by (4), the total current, i_c, flowing from the coupling junction into the networks of Fig. 9.4 is given by

$$\begin{aligned} i_c = {} & 2V'Y_t + V'Y_2(1 - e^{j\psi_x}) + V'Y_2(1 - e^{-j\psi_x}) \\ & + V'Y_4(1 - e^{j\psi_y}) + V'Y_4(1 - e^{-j\psi_y}) \\ & + V'Y_6[1 - e^{j(\psi_x + \psi_y)}] + V'Y_6[1 - e^{-j(\psi_x + \psi_y)}] \\ & + V'Y_6[1 - e^{j(\psi_x - \psi_y)}] + V'Y_6[1 - e^{-j(\psi_x - \psi_y)}] \end{aligned} \tag{9}$$

By the above definition, $Y_c = i_c/V'$, and

$$\begin{aligned} \eta_c = {} & 2(\eta_t + \eta_2 + \eta_4 + 2\eta_6) - \eta_2(e^{j\psi_x} + e^{-j\psi_x}) - \eta_4(e^{j\psi_y} + e^{-j\psi_y}) \\ & - \eta_6[e^{j(\psi_x + \psi_y)} + e^{-j(\psi_x + \psi_y)} + e^{j(\psi_x - \psi_y)} + e^{-j(\psi_x - \psi_y)}] \end{aligned} \tag{9a}$$

with

$$\eta_t = \frac{B_t}{Y'_0}, \qquad \eta_2 = \frac{B_2}{Y'_0}, \qquad \eta_4 = \frac{B_4}{Y'_0}, \quad \text{and} \quad \eta_6 = \frac{B_6}{Y'_0}$$

Equations (6)–(9) indicate that $R(\psi_x, \psi_y)$ is a doubly periodic function with a basic unit cell (or period) that is a square, $2\pi \times 2\pi$ in the ψ_x–ψ_y plane, as shown in Fig. 9.5. The scanning region for most antenna systems is usually smaller than the unit cell in order to avoid the appearance of grating lobes in real space. The scanning region can assume any desired form within the unit cell. For simplicity, one might try to approximate the scanning region by a rectangle $2\psi_0 \times 2\psi_1$ or by an ellipse with major axis $2\psi_0$ and minor axis $2\psi_1$ within the unit cell (Fig. 9.5).

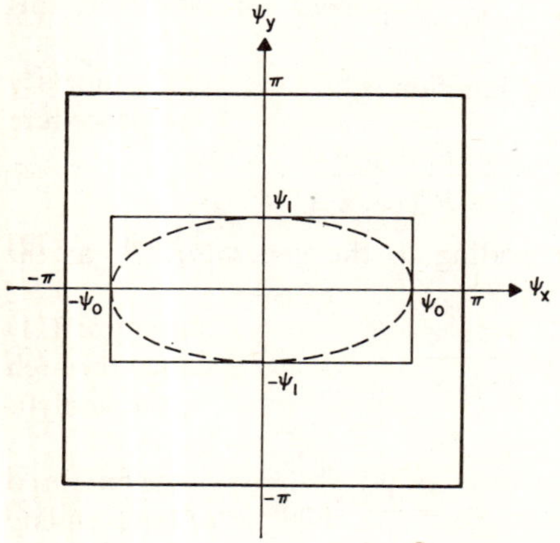

Fig. 9.5. Basic periodic unit cell with elliptical or rectangular scanning regions.

The objective is to minimize $|R|^2$ over the entire scanning region by a proper choice of the parameters available,[12] namely, ϕ, η_0, η_t, η_2, η_4, and η_6. The minimizing procedure will generate a set of equations for the determination of η_0, η_t, η_2, η_4, and η_6.

We will proceed by choosing to minimize*

$$I = \iint_{\text{scan region}} |R|^2 \, d\psi_x \, d\psi_y$$

$$= \iint_{\text{scan region}} \left| \frac{1 - X}{1 + X} \right|^2 d\psi_x \, d\psi_y \tag{10}$$

If $X = 1$ over the scan region, then $I = 0$, the ideal minimum. If an ideal minimum cannot be achieved, however, then for values of X close to unity

* In special cases, one may include under the integral sign in (10) a multiplicative function (of ψ_x and ψ_y) or add additional constraints to the minimizing procedure if necessary or desirable.

the minimum of I will not differ appreciably from the minimum of I_1, where

$$I \approx I_1 = \iint_{\text{scan region}} \left| \frac{1 - X}{2} \right|^2 d\psi_x \, d\psi_y \qquad \text{for} \quad X \approx 1 \tag{11}$$

Thus, if a good array match is feasible, minimization of I_1 in (11) will be equivalent to minimizing I. If an array cannot be matched with the proposed minimization scheme, the minimum of I_1 will not come close to the true minimum of I. In this case I and $|R|^2$ will be high, and a different array element or a different approach may be necessary. So far, however, this approach has not failed in many applications.

From inspection of (5), (8), and (9), it is clear that $(1 - X)/2$ is linearly related to η_0, η_t, η_2, η_4, and η_6. This significantly simplifies the procedure to be followed. In order to obtain the equations for $\{\eta\}$, we set

$$\frac{\partial I_1}{\partial \eta_0} = 0, \qquad \frac{\partial I_1}{\partial \eta_t} = 0, \qquad \frac{\partial I_1}{\partial \eta_2} = 0, \qquad \frac{\partial I_1}{\partial \eta_4} = 0, \qquad \frac{\partial I_1}{\partial \eta_6} = 0 \tag{12}$$

The solution of (12) will yield the values of $\{\eta\}$ which will minimize (11) for a given value of ϕ. The optimum values of $\{\eta\}$ and ϕ are obtained through repeated solutions of (12) for various values of ϕ. In Appendix I, the analytical expressions for (12) are developed; see (A.7)–(A.15).

This procedure can be easily programmed on a digital computer and used as a design tool. Starting from the measurement or calculation of the coupling coefficients at several frequencies across the desired frequency band, one can use this method to obtain the values of $\{\eta\}$ as a function of frequency. It then remains to synthesize these values of $\{\eta\}$ in terms of real or physical networks and to implement them either in the element feed lines or within the elements themselves whenever possible. We may anticipate that some compromise between the optimum (calculated) and realizable values of the coupling networks at the various frequencies will be required.

Figures 9.6–9.8 illustrate the results of the application of the optimization procedure to a square waveguide array at a single frequency. Figure 9.6 shows the (normalized) returned power, $|R|^2$, as a function of the scan angle directional cosines of a typical element in a finite square waveguide array. These plots were redrawn from the computer maps; returned power is quantized to various levels according to the key in the figure. The square waveguide elements, with a wall thickness of $t_p = 0.036\lambda$, are spaced at $b/\lambda = d/\lambda = 0.5714$. They are excited by a linear TE_{10} mode with the electric field parallel to the x axis. The assigned scan region is circular and is defined by $\theta \leq 45°$, where θ is measured from broadside (Fig. 9.1). This particular choice of scan region eliminates the appearance of grating lobes in real space. From Fig. 9.6, it is clear that relatively high values of returned power

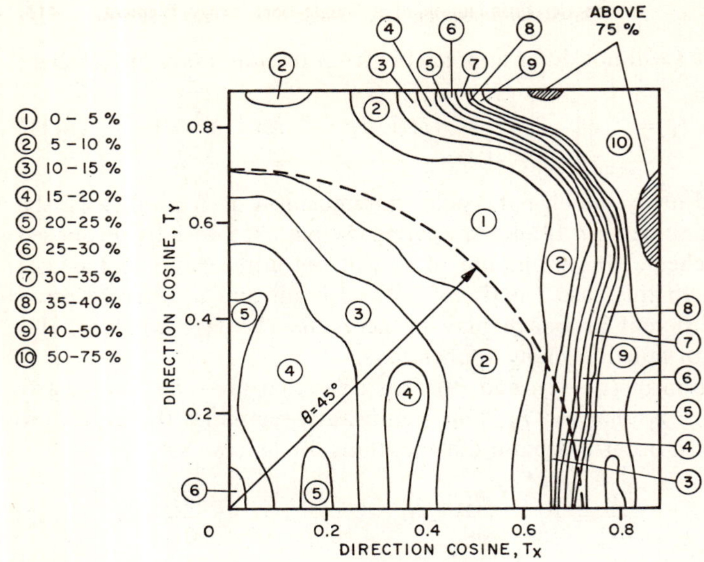

Fig. 9.6. Returned power vs. steering direction in a square waveguide square grid array; $b/\lambda = d/\lambda = 0.5714$, waveguide wall thickness $t_p = 0.036\lambda$.

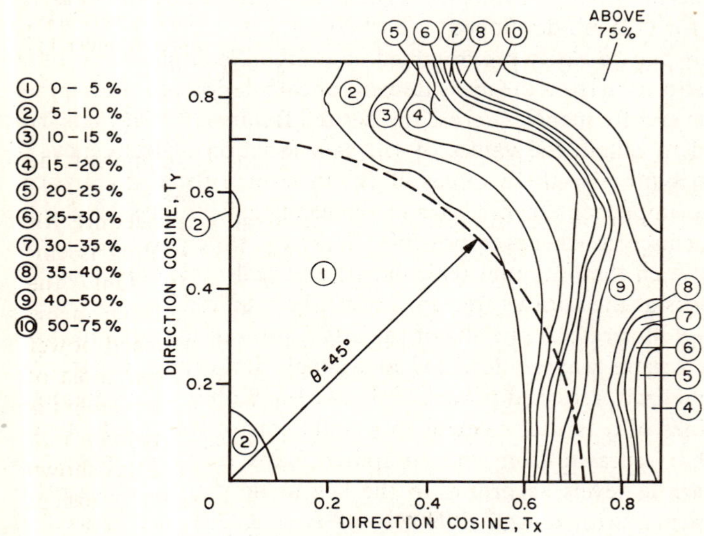

Fig. 9.7. Returned power in square waveguide square grid array with matching networks; $\eta_0 = 0.76$, $\eta_t = 0.265$, $\eta_2 = 0$, $\eta_4 = 0$, $\eta_6 = 0$. Returned power is around 5% over most of the scan region.

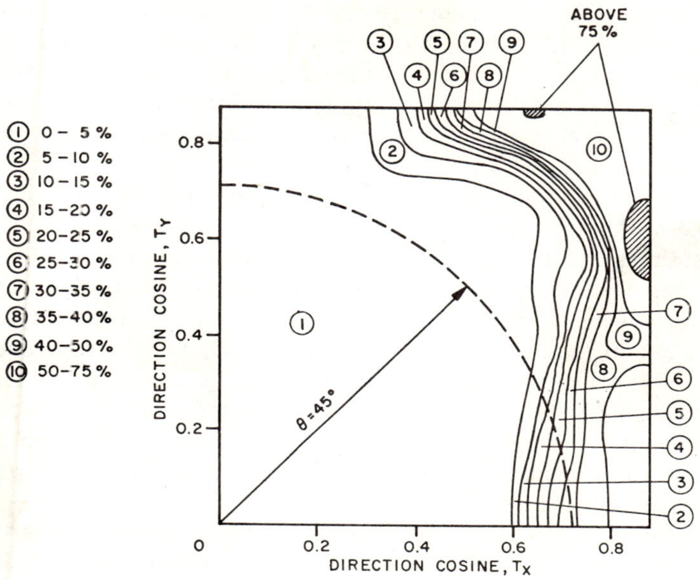

Fig. 9.8. Returned power in square waveguide square grid array with coupling networks; $\eta_0 = 1$, $\eta_t = 0.512$, $\eta_2 = -0.0976$, $\eta_4 = -0.189$, $\eta_6 = 0$. Returned power is under 1% over most of the scan region.

exist within the scan region, especially along the T_x axis (E plane) at the extreme edge of scan.

Figures 9.7 and 9.8 show the results of the optimization procedure for different coupling network configurations. In Fig. 9.7, no coupling is employed between adjacent elements (i.e., $\eta_2 = \eta_4 = \eta_6 = 0$). Only the matching within each element channel or transmission line has been incorporated. As can be seen, there is a marked reduction of the returned power within the scan region, as compared to that seen in Fig. 9.6. Over most of the scan region, the returned power is around 5%. A further improvement is observed in Fig. 9.8, where *coupling between elements having a common wall* has been added (i.e., $\eta_6 = 0$). A close check of the numerical data has shown that over most of the scan region the returned power is less than 1%, except at the extreme of scan ($\theta \approx 45°$) in the E plane (T_x axis).

As only one coupling junction is used in this method, we are able to linearize the reflection coefficient and to obtain a minimum of the average reflected power. This method does *not* yield a theoretical lower bound for the mismatch. As one can visualize, simultaneous use of additional coupling junctions spaced along the feed lines can also be employed. However, minimization of the mismatch under such circumstances becomes more complicated and more frequency sensitive, and the implementation of multiple coupling junctions becomes impractical. The numerical data suggest that the limited number of constraints introduced by this method is sufficient under most circumstances to give a satisfactory match that can be implemented in practice.

So far, the array element has been represented as a one-port device, while the coupling networks were introduced in the transmission lines feeding the elements. This representation is valid for linearly polarized elements such as dipoles or rectangular waveguides with one propagating mode. In fact, the method developed here is still valid for dual-polarization, two-port type elements when one of the ports is terminated by a load, and the matching and coupling networks are introduced in the transmission lines feeding the driven port. Waveguides with orthogonally polarized modes of propagation fall into this category. In this case, an increase in the radiated power through improvement of the match will also increase proportionally the power dissipated in the terminated port.

As was pointed out previously, the voltages at the coupling junctions retain, except for a multiplicative constant, the same relative linear phase and uniform illumination as the generators (for a quasi-infinite array). Therefore, the relative field distribution at the aperture of an excited array, with or without coupling networks, will be preserved except for a multiplicative factor. This means that the polarization characteristic of the array radiation pattern will not be perturbed by the presence of the matching networks. Using the principle of pattern multiplication (i.e., the array radiation pattern equals the product of the array factor and the single-excited-element radiation pattern), we see that, by employing this matching technique, the radiation power pattern of a single excited element in the array is altered, whereas the polarization characteristic is undisturbed.

3. MATCH OPTIMIZATION OF A TWO-PORT ARRAY ELEMENT

In Section 2, a single-port array element was analyzed and the returned power was systematically minimized over a prescribed scan region. It was observed that most of the improvement in the array match was obtained by using matching networks within the elements (or their feed lines) solely. The additional improvement due to interelement coupling networks was not as dramatic. In most applications, and especially in high power systems, interelement coupling networks are costly and even undesirable. We shall therefore omit interelement coupling in the following discussion of the match optimization of a two-port element.

Unfortunately, the minimization procedure discussed earlier cannot be readily extended to the two-port element; specifically, the linearization of R that was used cannot be carried out. Therefore a different approach is adopted in the case of the two-port element.[13] Computer simulation of the active reflection coefficients* is used to yield the design parameters of the matching network which will minimize the match over an assignable scan region. This procedure is an iterative man-machine interaction and is achieved through a judicious variation of the independent parameters of the matching network.

3.1 Terminal Relationships

Figure 9.9 shows a rectangular grid† array of two-port type elements. Ports A and B in each element correspond to two orthogonal polarizations of excitation. The array is excited uniformly in amplitude with a linear phase taper, specified by the steering phases ψ_x and ψ_y. Typically, the array is excited at one set of ports (say, A) while the other set of ports (B) is terminated.

The incident and reflected wave amplitudes in port A of the (m, n)th element (Fig. 9.10) are designated by $I_{mn}{}^A$ and $R_{mn}{}^A$, respectively. The terminal relationships of the array are expressed by the scattering matrix, which may be written in a partitioned form:

$$\begin{bmatrix} [R_{mn}{}^A] \\ [R_{mn}{}^B] \end{bmatrix} = \begin{Vmatrix} \|\mathscr{C}_{mn}{}^{AA}\| & \|\mathscr{C}_{mn}{}^{BA}\| \\ \|\mathscr{C}_{mn}{}^{AB}\| & \|\mathscr{C}_{mn}{}^{BB}\| \end{Vmatrix} \begin{bmatrix} [I_{mn}{}^A] \\ [I_{mn}{}^B] \end{bmatrix} \tag{13}$$

where [ ] denotes a column matrix, and $\| \;\|$ a square matrix. In the partial scattering matrices

$$\|\mathscr{C}_{mn}{}^{AA}\|, \qquad \|\mathscr{C}_{mn}{}^{BA}\|, \qquad \|\mathscr{C}_{mn}{}^{AB}\|, \qquad \text{and} \qquad \|\mathscr{C}_{mn}{}^{BB}\|$$

the first superscript denotes the excited port; the second, the terminated port. The elements of the square scattering matrix are the coupling coefficients.

Let identical matching networks be inserted in the element channels (i.e., either within the elements or in their feed lines) as shown in Fig. 9.11. These networks are assumed to consist of lossless, reciprocal obstacles inserted within each array element. Thus, they can be characterized by a unitary scattering matrix, $\|s\|$ (of the propagating modes), satisfying the following relations:

$$[R^i] = \|s_{il}\| \; [I^l], \quad i, l = 1, 2, 3, 4 \tag{14}$$

* This simulation is based on either the solution of the boundary value problem or the measurement of the coupling coefficients.

† A rectangular grid is assumed again for convenience. The method to be discussed applies, however, to any regular planar grid.

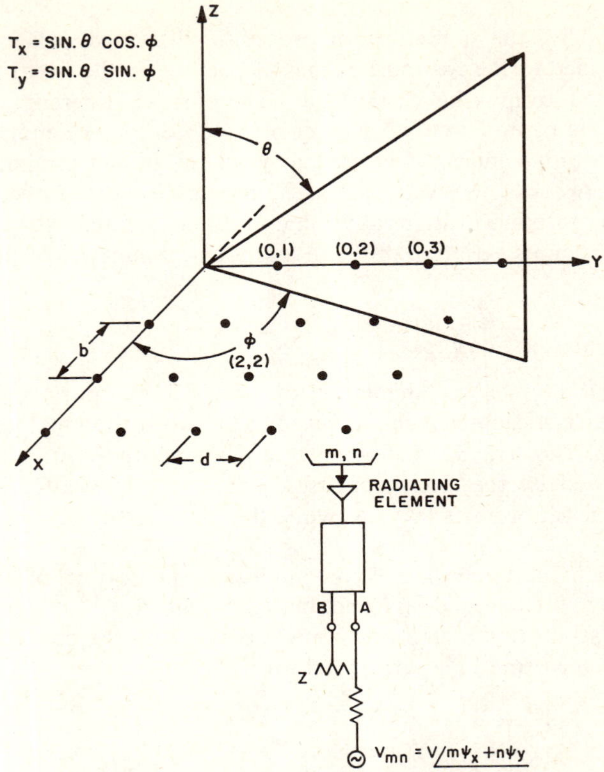

Fig. 9.9. Rectangular grid array model.

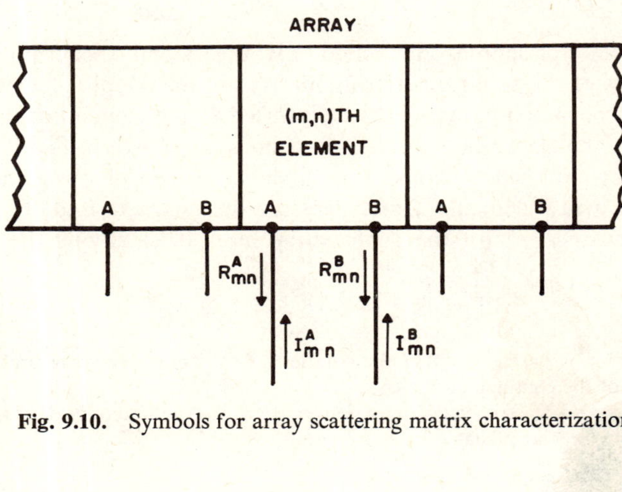

Fig. 9.10. Symbols for array scattering matrix characterization.

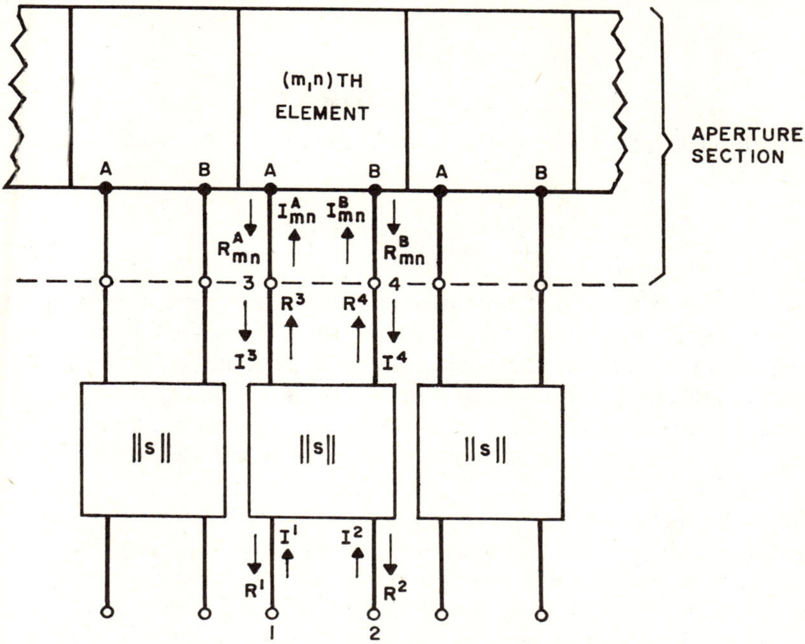

Fig. 9.11. Array elements with matching obstacles.

In Appendix II we derive the relationships between the various sets of ports in terms of the array coupling coefficients and the scattering matrix, $\|s\|$, of the matching obstacles (networks). We prove that, for an infinite array excited uniformly in amplitude and with a linear phase taper, an operation with the $\|s\|$ matrix on the active reflection coefficients of a single element (i.e., inserting a matching obstacle) implies that identical matching obstacles have been inserted in all the elements of the array. Thus the analysis and synthesis of matching obstacles in an infinite array are reduced to a four-port network formalism. This, of course, is the same assumption as was used in single-port matching. The proof is useful, however, in that it substantiates rigorously what appears to be intuitive.

The equations relating the active reflection coefficients at ports 1 and 2 (Fig. 9.11) to the matching obstacle parameters and array coupling coefficients are

$$I^1 D_1 + I^2 D_3 = R^1 D_2 + R^2 D_4 \tag{15}$$

$$I^1 E_1 + I^2 E_3 = R^1 E_2 + R^2 E_4 \tag{16}$$

where

$$\left.\begin{aligned} D_i &= (-1)^i[-G_{2i} + G_{1i}X^{AA} + G_{3i}X^{BA}] \\ E_i &= (-1)^i[-G_{4i} + G_{1i}X^{AB} + G_{3i}X^{BB}] \end{aligned}\right\}, \quad i = 1, 2, 3, 4$$

G_{il} are the elements of the transfer scattering matrix of the matching obstacle [see (A.20)],

$$X^{AA} = \sum_{m=-\infty}^{+\infty} \sum_{n=-\infty}^{+\infty} \mathscr{C}_{mn}{}^{AA} e^{j(m\psi_x + n\psi_y)}, \quad \text{etc.}$$

The quantities X^{AA} and X^{AB} are the active reflection coefficients (versus scan) in ports A and B, respectively, when the array is excited by a matched generator at port A and port B is ideally terminated. Also, X^{BA} and X^{BB} are the active reflection coefficients in ports A and B, respectively, when port B is excited and port A is terminated. Thus X^{AA}, X^{BA}, X^{AB}, and X^{BB} fully characterize the array reflection properties, and a knowledge of $\{\mathscr{C}_{mn}{}^{AA}\}$, $\{\mathscr{C}_{mn}{}^{AB}\}$, $\{\mathscr{C}_{mn}{}^{BA}\}$, and $\{\mathscr{C}_{mn}{}^{BB}\}$ is sufficient to characterize the array match as a function of the scan angle.

So far, the discussion has been limited to infinite arrays with their match characteristics formulated in terms of infinite sums involving their scattering coefficients. As we shall discuss and justify later, the infinite sums can be replaced by finite partial sums of the dominant terms without appreciable loss of accuracy. Therefore the relations derived in (15) and (16) can be used to analyze the inner core elements of large, finite sized arrays as well.

3.2 The Returned Powers in Phased Arrays

In many practical array elements, each port corresponds to a desired polarization of the radiation pattern. Therefore, when the array is in operation, only one port is excited (say port 1, Fig. 9.11), while the other port is ideally terminated (i.e., $I^2 = 0$).* In this case, (15) and (16) reduce to

$$D_1 = \frac{R^1}{I^1} D_2 + \frac{R^2}{I^1} D_4 \tag{17}$$

$$E_1 = \frac{R^1}{I^1} E_2 + \frac{R^2}{I^1} E_4 \tag{18}$$

which are two linear equations for the active reflection coefficients, R^1/I^1 and R^2/I^1. The normalized powers returned to ports 1 and 2 are $|R^1/I^1|^2$ and $|R^2/I^1|^2$, respectively. Similarly, if port 2 is excited and port 1 is ideally

* The case of nonideal termination (i.e., $I^2 \neq 0$) can be analyzed in a similar fashion.

terminated, one obtains

$$D_3 = \frac{R^1}{I^2} D_2 + \frac{R^2}{I^2} D_4 \tag{19}$$

$$E_3 = \frac{R^1}{I^2} E_2 + \frac{R^2}{I^2} E_4 \tag{20}$$

where R^1/I^2 and R^2/I^2 are the active reflection coefficients in ports 1 and 2, respectively, with $|R^1/I^2|^2$ and $|R^2/I^2|^2$ the returned powers.

3.3 Characterization of the Aperture Match

For waveguide and other types of array elements it may be desirable to place the matching network close to the aperture whenever a broad frequency band of operation is desired. It is convenient, whenever possible, to separate the feed network from the rest of the element along an arbitrary reference interface or plane. The latter will be referred to as the aperture section; see Fig. 9.11. Higher-order evanescent modes are assumed to be of negligible amplitude at this reference plane.

Usually, the array coupling coefficients associated with the aperture section, that is, $\{\mathscr{C}_{mn}{}^{AA}\}$, $\{\mathscr{C}_{mn}{}^{AB}\}$, $\{\mathscr{C}_{mn}{}^{BA}\}$, and $\{\mathscr{C}_{mn}{}^{BB}\}$, cannot be measured directly, because the nonideal properties of the feed network (wave-launching probes and sometimes a polarizer) are included in the measurements. The effects of the feed network can be eliminated by recognizing that the $\|s\|$ matrix of Fig. 9.11 can represent any lossless reciprocal network driving the array elements. By making coupling measurements between all sets of ports 1 and 2, the aperture section reflection coefficients can be derived by associating the parameters of the feed network with the $\|s\|$ matrix and mathematically removing the matrix.

Let the required coupling coefficients between ports 1 and 2 be denoted by $\{S_{mn}{}^{11}\}$ and $\{S_{mn}{}^{12}\}$ when port 1 is excited and port 2 is ideally terminated, and by $\{S_{mn}{}^{21}\}$ and $\{S_{mn}{}^{22}\}$ when port 2 is excited and port 1 is ideally terminated. By definition, the quantities R^1/I^1, R^1/I^2, R^2/I^1, and R^2/I^2 from (17)–(20) are given by

$$\frac{R^k}{I^i} = \Gamma^{ik} = \sum_{(m,n)}\sum S_{mn}{}^{ik} e^{j(m\psi_x + n\psi_y)}, \quad k, i = 1, 2 \tag{21}$$

Thus, if Γ^{11}, Γ^{12}, Γ^{21}, and Γ^{22} are known, (17)–(20) can be rearranged to obtain the following four linear equations for the evaluation of the aperture section reflection coefficients, X^{AA}, X^{AB}, X^{BA}, and X^{BB}:

$$H_2 = H_1 X^{AA} + H_3 X^{BA} \tag{22}$$

$$H_4 = H_1 X^{AB} + H_3 X^{BB} \tag{23}$$

$$J_2 = J_1 X^{AA} + J_3 X^{BA} \tag{24}$$

$$J_4 = J_1 X^{AB} + J_3 X^{BB} \tag{25}$$

where

$$\left.\begin{aligned} H_i &= G_{i1} + G_{i2}\Gamma^{11} + G_{i4}\Gamma^{12} \\ J_i &= G_{i3} + G_{i2}\Gamma^{21} + G_{i4}\Gamma^{22} \end{aligned}\right\}, \quad i = 1, 2, 3, 4$$

and G_{ik} are the measured coefficients of the transfer scattering matrix for the wave launcher.

The returned powers, which characterize the match of a typical active element aperture and thereby the array of these apertures, are $|X^{AA}|^2$ and $|X^{AB}|^2$ (with port A excited) and $|X^{BA}|^2$ and $|X^{BB}|^2$ (with port B excited). It should be noted that the returned powers are functions of the two particular orthogonal polarization modes associated with ports A and B. Thus, for example, the match versus scan characteristics of the array are entirely different for linear and for circular polarizations.

3.4 Match Optimization Procedure

The match optimization procedure utilizes a high speed digital computer extensively. The procedure begins with the construction of a small array of trial elements. The choice of the aperture section of the trial element is based on the specific mechanical, electrical, and environmental requirements of the intended system. The terminal coupling coefficients, $\{S_{mn}{}^{ik}\}$, at a single frequency are used to compute the active reflection coefficients, Γ^{ik}, of the trial (initial) array element as a function of the scan angle [see (21)]. These active reflection coefficients, in conjunction with the measured transfer scattering matrix, $\|G\|$, of the feed network, are employed to obtain the active reflection coefficients of the aperture section, X^{AA}, X^{AB}, X^{BA}, and X^{BB} [see (22)–(25)]. The returned powers are calculated as a function of the scan angle from the complex reflection coefficients and are plotted by the computer in the form of a map, similar to the one shown in Fig. 9.6. These maps permit a quick evaluation of the match by visual inspection.

By employing X^{AA}, X^{AB}, X^{BB}, and X^{BA} in further calculations, it is possible to evaluate the match versus scan either for different polarizations (e.g., circular, elliptical) or for a matching obstacle. This is done by associating the $\|G\|$ matrix with either a polarizer or a matching obstacle [see (17)–(20)].

The scattering matrix of an optimum matching obstacle is obtained by varying the independent parameters* about the starting values. Good starting values for some of the independent parameters can be estimated from a knowledge of X^{AA}, X^{AB}, X^{BB}, and X^{BA}. The returned powers for

* In general, four-port networks have many independent parameters; however, by a judicious choice of an obstacle with certain symmetries, the number may be considerably reduced so that it becomes tractable.

each variation are calculated, and computer-plotted maps are obtained. With the visual aid of these maps, a reasonably quick convergence to an optimum obstacle is very often accomplished in an iterative manner. This man-machine interaction procedure is especially suited to on-line computation.

At any frequency within the band of interest, there is a range of parameter values which gives an acceptable performance in terms of the average returned power over the scan region of interest. Optimum performance across the frequency band is achieved when the parameters of an actual obstacle fall within this acceptable range for each frequency.

A prediction of the match versus scan performance of an array of optimized elements is obtained when the scattering matrix of a synthesized actual obstacle is being associated with the $\|G\|$ matrix [see (17)–(20)]. As a final step, the predictions may be confirmed by physically incorporating the matching network into the elements of the array, measuring a new set of terminal coupling coefficients, and computing a final set of returned power maps.

We shall now show how the optimization procedure was applied to a particular case in which a good match characteristic was obtained over a 15% frequency bandwidth. The trial elements were open-ended, coaxial waveguides excited by orthogonal TE_{11} modes. Each aperture contained a thick flat window of high dielectric constant, and the elements were mounted so that the windows were flush with the ground plane. A single trial element was resonably well matched when it was isolated in a ground plane because of an aperture matching network that produced a VSWR of less than 1.4 over the 15% band.

Figure 9.12 shows the total returned power of the aperture section of an array of 91 trial elements before optimization at the high and low edges of the frequency band. A substantial part of the 0–50° scan region has a returned power ranging from 5 to 15%. The portion of imaginary space between the 90° scan circles is of interest, since in an infinite array the returned power in these regions will be 100%. The low frequency plot shows this region most clearly; the returned power approaches 100%. This indicates that a good approximation to the infinite array was obtained.

In order to reduce the independent scattering matrix parameters characterizing the obstacle, a circularly symmetric matching obstacle was chosen. With this choice, only two parameters are independent when the source of excitation and the element feed network are perfectly matched. The range of the parameters which reduced the returned power over the scan region of interest was found to be quite broad at each of the three frequencies. Therefore, the obstacle synthesis for this element was relatively straightforward.

The predicted match versus scan performance of the optimized element is shown in the returned power plots of Fig. 9.13. A substantial improvement

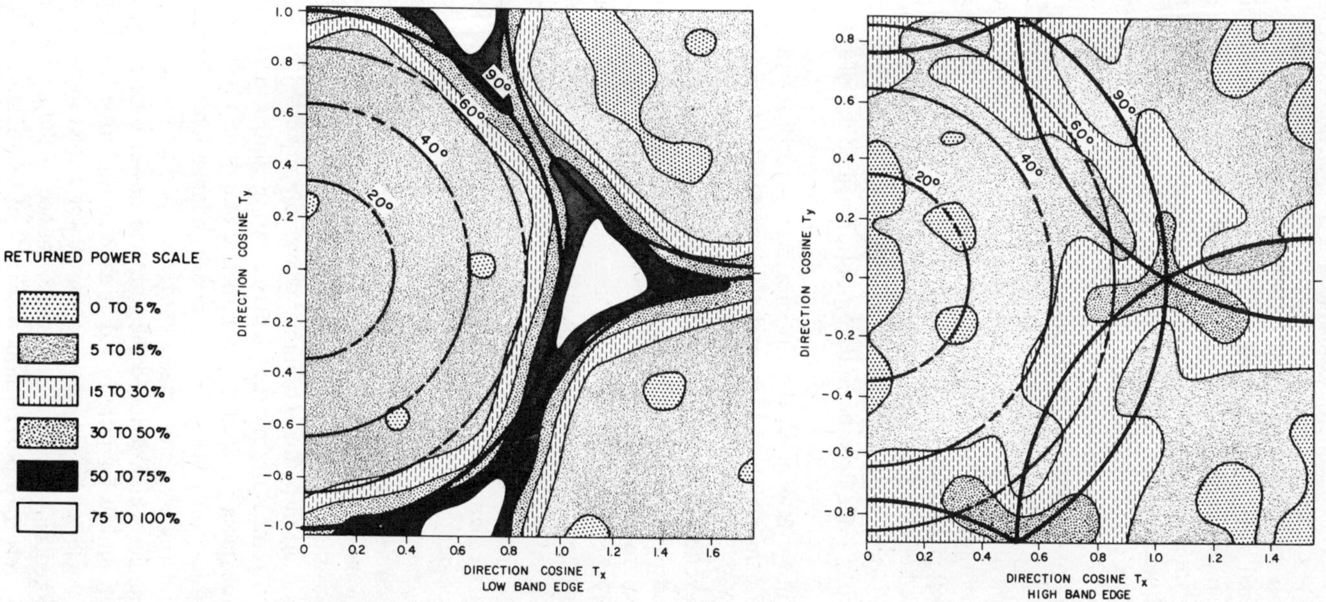

Fig. 9.12. Total returned power of the active trial element aperture section—circular polarization excitation.

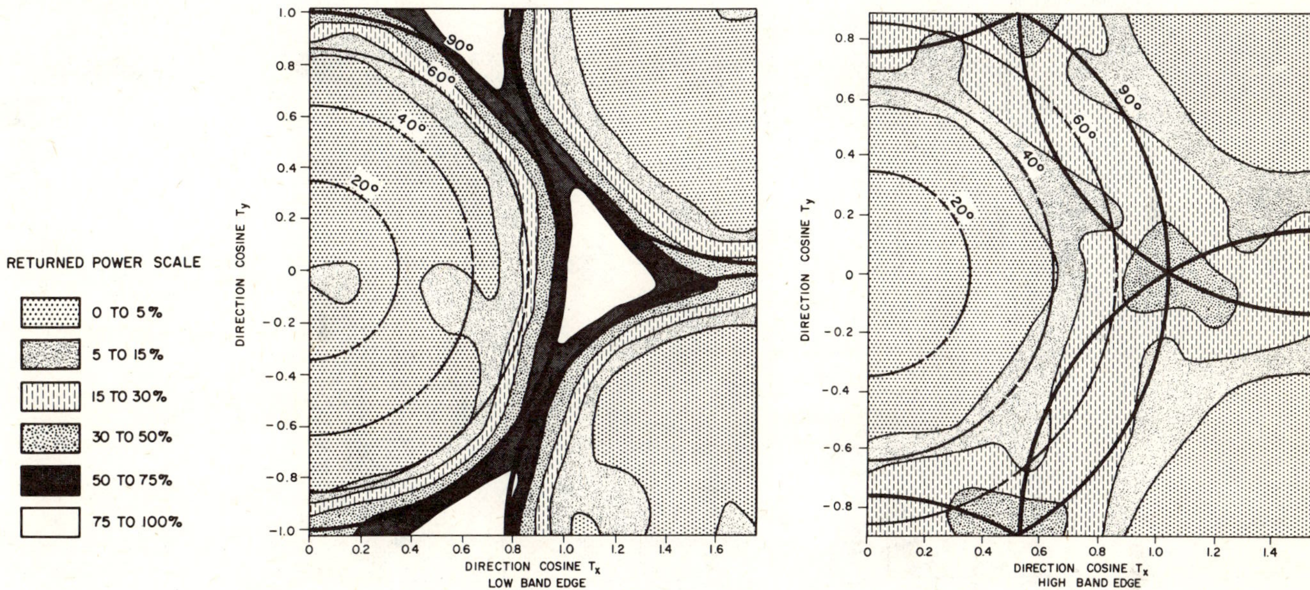

Fig. 9.13. Predicted total returned power of the active trial element aperture section after match optimization—circular polarization excitation.

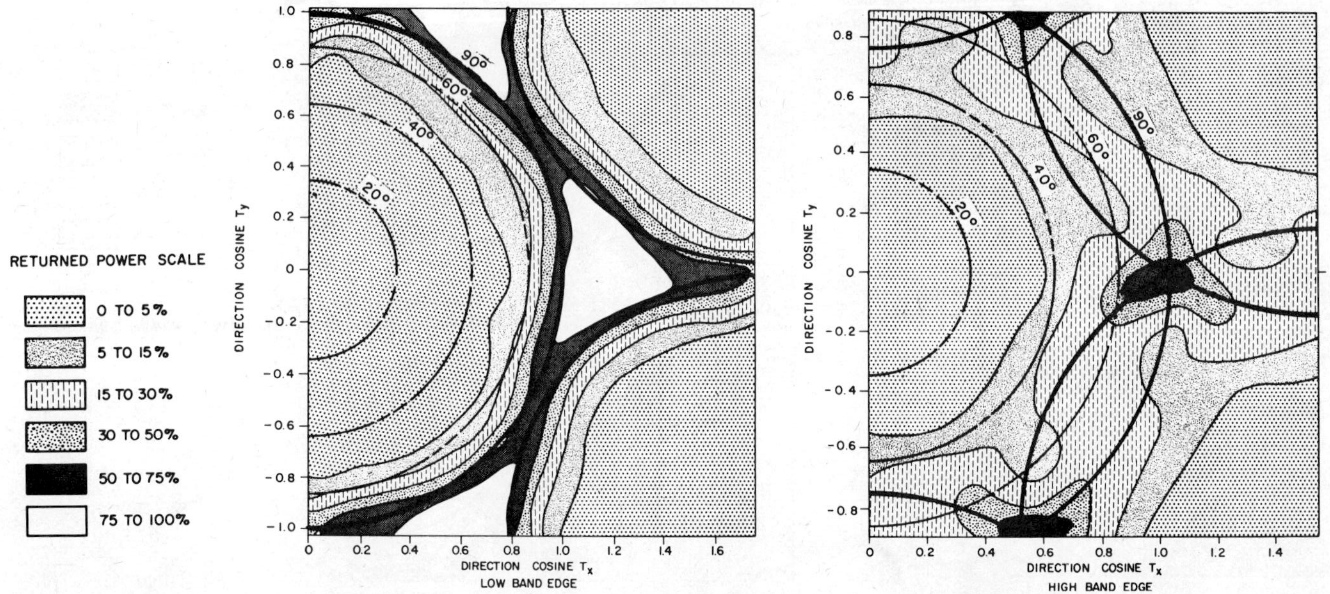

Fig. 9.14. Measured total returned power of the active trial element aperture section after match optimization—circular polarization excitation.

may be noted by comparison with Fig. 9.12. The returned power in the 0° to nearly 40° region has been reduced without increasing the power level in the region between 40° and 60°. The small areas of 5–15% returned power within the 20° circle actually represent just slightly over 5%.

Physically, the synthesized obstacle may be cascaded directly with the aperture section. In the case described, however, the obstacle was combined with the initial impedance matching structure in the coaxial waveguide, and a

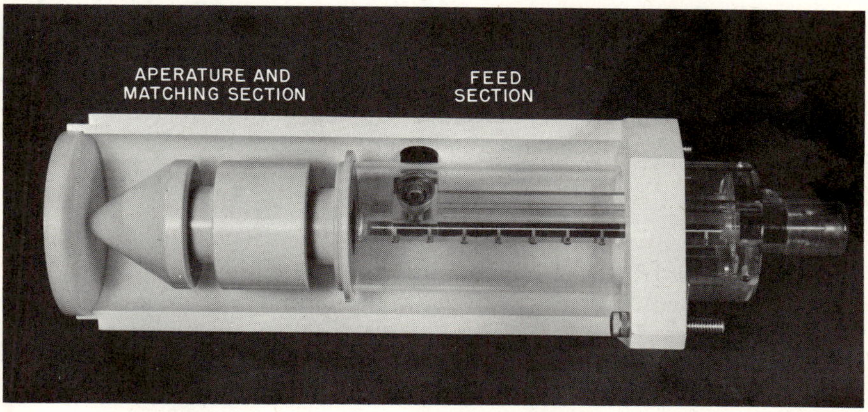

Fig. 9.15. Mock-up of optimized element.

shorter structure, which was nearly electrically equivalent to the cascaded combination, was thus obtained.

Figure 9.14 shows the actual measured results of physically incorporating the new matching network in the array element. A comparison of these results with those of Fig. 9.13 reveals a striking similarity. The contour shapes of the high and low edges of the frequency band are almost identical (with the exception of small areas at the low band edge, which actually differ by only a few per cent and are the result of the quantization of the power levels). The quantization may cause the contours to differ slightly, but a comparison of the numerical data shows that within the 0–40° region the difference in returned power between the two figures is less than 2%.

The element performance was also measured at three other frequencies within the band, and the results were as good as those shown in Fig. 9.14. The relative ease with which this element was matched can be attributed to the initial choice of a reasonably matched isolated trial element. Figure 9.15 shows a mock-up of the final design of the optimized coaxial element.

The match optimization procedure was also applied to an initially "poorly matched" trial element in order to explore the convergence of this procedure in such cases, and excellent results were obtained.[13] Because of the poor initial match, however, the range of values over which the matching obstacle parameters could vary while maintaining a good match was very narrow. Consequently tighter tolerances had to be imposed on the actual matching obstacle.

4. EXPERIMENTAL CHARACTERIZATION OF ARRAY MATCH

The characterization of the terminal relationships of linear multiport networks can be cast in various forms. Two of the most useful and prevalent forms are given by the scattering and admittance matrices.[4] The scattering matrix representation of an infinite planar array can be written as

$$[R_{mn}] = \|\mathscr{C}_{mn}\| \, [I_{mn}] \tag{26}$$

where $[I_{mn}]$ and $[R_{mn}]$ are column vectors representing the incident and reflected wave amplitudes, respectively, at the various ports of the elements. The scattering matrix elements with a single set of subscripts are defined in (A.18) of Appendix II, $\mathscr{C}_{mn}$ being the scattering (coupling) coefficient between the (m, n)th and the $(0, 0)$th elements.* Similarly, the admittance matrix representation of the array is

$$[i_{mn}] = \| Y_{mn}\| \, [v_{mn}] \tag{27}$$

where the column vectors $[v_{mn}]$ and $[i_{mn}]$ are the voltages and currents, respectively, at the various ports of the elements, and Y_{mn} is the mutual admittance between the (m, n)th and $(0, 0)$th elements. Equations (26) and (27) are equivalent representations, and simple relationships among the various vectors and matrices can be easily derived.[4,14]

For a uniform illumination of the array with a linear phase taper, (1), the active reflection or impedance characteristics of all the elements are identical except for a phase factor. Therefore we shall choose the $(0, 0)$th ("center") element to characterize the array with this illumination. From (26) and (27) the active reflection coefficient and admittance of the center element, R_a and Y_a, respectively, are

$$R_a(\psi_x, \psi_y) = \frac{R_{00}}{I_{00}} = \sum_{m=-\infty}^{+\infty} \sum_{n=-\infty}^{+\infty} \mathscr{C}_{mn} e^{j(m\psi_x + n\psi_y)} \tag{28}$$

* For the purpose of clarity, we shall discuss single-port array elements. The extension to two-port elements can be obtained from Appendix II.

and

$$Y_a(\psi_x, \psi_y) = \frac{i_{00}}{v_{00}} = \sum_{m=-\infty}^{+\infty} \sum_{n=-\infty}^{+\infty} Y_{mn} e^{j(m\psi_x + n\psi_y)} \tag{29}$$

Usually, (28) and (29) cannot be evaluated in a closed form and consequently are truncated for numerical or experimental approximate evaluations of R_a and Y_a:

$$R_a \approx \sum_{(M_1)} \sum_{(N_1)} \mathscr{C}_{mn} e^{j(m\psi_x + n\psi_y)} \tag{30}$$

$$Y_a \approx \sum_{(M_2)} \sum_{(N_2)} Y_{mn} e^{j(m\psi_x + n\psi_y)} \tag{31}$$

Once the values of $\{\mathscr{C}_{mn}\}$ or $\{Y_{mn}\}$ are determined experimentally or analytically, the active match of the array is approximately specified by (30) or (31). The degree of approximation will depend on M_1 and N_1 or M_2 and N_2.

The admittance representation, (31), has been used in only a limited number of cases for which approximate analytical solutions for Y_{mn} could be obtained[4] (e.g., arrays of narrow slots or thin dipoles). From an experimental point of view the scattering representation, (30), is superior because, in the ultrahigh and microwave frequency regions, measurements and instruments are oriented toward direct determinations of the scattering coefficients. In addition, in order to measure the mutual admittances or coupling coefficients, we have to construct a finite sized array of the elements involved.

In performing the mutual admittance measurements, we usually excite the center element by a voltage source, short-circuit the terminals of the rest of the elements, and measure the currents (or equivalents) at these terminals. From a knowledge of the currents we can then obtain the various mutual admittances. Under these measurement conditions, however, the array aperture is a purely reactive surface and surface waves may be excited and propagate along it.[14] Because of the finite size of the array, multiple reflections and diffractions of the surface wave may occur at the edge of the array, and consequently the measured values of the mutual admittances may be highly inaccurate. In order to obtain meaningful results, large arrays with special absorbers at the edges may be required. On the other hand, in the measurement of the coupling coefficients the (0, 0)th element is excited by a matched source while the rest of the elements are ideally terminated. The various coupling coefficients are obtained from the measured wave amplitudes at the various elements. Therefore, the array aperture can be viewed as a lossy surface along which no pure surface waves can propagate (see Chapters 7 and 8 for further discussion). Consequently the effects of multiple reflections due to the edges are diminished, and meaningful measurements of the coupling coefficients can be obtained from a comparatively smaller array of elements.

We may expect, therefore, that the coupling coefficients between an identically spaced pair of elements will be rather insensitive to their location in the array, provided that neither element is located right at the array edge. This relative insensitivity of the coupling coefficients was demonstrated in

Fig. 9.16. Experimental array of trial elements.

Chapter 8 for finite arrays of parallel plates embedded in straight or corrugated ground planes. Additional confirmation was obtained experimentally for square and coaxial waveguide planar arrays.

Figure 9.16 shows a small test array of 91 flush-mounted coaxial waveguide elements arranged in an equilateral triangular grid. The coupling coefficients of this array were measured with a precisely calibrated phase and amplitude bridge employing a stabilized frequency source. To minimize effects due to the edge of the ground plane, an absorber was placed around its circular rim. As previously discussed, the coupling coefficients are rather insensitive to their location in the array. Therefore, in measuring the coupling coefficients, an array of n concentric rings of elements (hexagons, rectangles, etc.) surrounding the center element may be "extrapolated" to an array of $2n - 1$ concentric rings. This is depicted in Fig. 9.17. An actual array of 37 elements arranged in an equilateral triangular grid is superimposed in an offset position, in the direction of arrow 1, on an imaginary array of 91 elements.

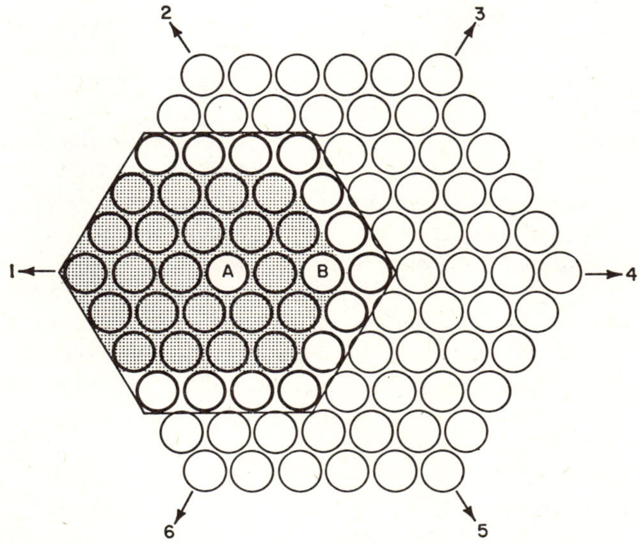

Fig. 9.17. Superposition of an actual 37-element hexagonal array on a 91-element array for extrapolating coupling coefficients.

(As will be noted, the 37-element array has three concentric rings of hexagons, and the 91-element array has five concentric rings.) When the reference element is offset from the center of the 37-element array (A) to the position indicated by B, the couplings between the elements in the shaded area and the new reference element (B) simulate the couplings that would exist in a 91-element array. By doing this in the direction of each of the numbered arrows, a full set of coupling coefficients may be obtained for a much larger array.

This extrapolation technique was fully tested by masking off the two outermost rings of an actual 91-element array and measuring the extrapolated 91 element coupling coefficients in the smaller array. Measurements were then made on the actual (unmasked) 91-element array. It was found that maximum variations in the coupling coefficients between the two methods of measurement were of the order of $\frac{1}{2}$ dB in amplitude and a few degrees in phase. The returned power averages, computed by using the coupling coefficients of the two measurements, were almost identical.

An important aspect of the experimental characterization of the infinite (or very large) array match is the number of coupling coefficients required to yield an acceptable accuracy over a prescribed scan region. This number is directly related to the required size of the experimental test array. Over a scan region free of grating lobes (i.e., only the main beam radiates) the coupling coefficients between the center element and five rings of elements surrounding it adequately represent the basic match properties of a typical element in a large array.

The five-concentric-ring figure described above is not a rigorously proved conclusion for all arrays and elements. Rather, it is an empirical result, based on various items of numerical and experimental evidence, and is meant to serve as a guideline only. In preceding chapters, numerical and experimental results for parallel-plate and square waveguide arrays substantiate that the above result is satisfactory. Similar results were observed for dipole arrays[4] (in admittance matrix representation). In the match optimization procedure for two-port type elements (Section 3.4), we used coupling coefficients of 91 elements which were measured on an actual array of 37 elements. The terminal relationships, (15)–(25), which account for the matching obstacle were derived for infinite arrays. The close agreement between the returned powers of the predicted and measured optimized elements (over the grating lobe free scan region and over a 15% frequency band) further substantiates the adequacy of the statement made above about the number of rings required to represent the match.

APPENDIX I EVALUATION OF THE MATCHING NETWORK ADMITTANCES

From (11), the quantity to be minimized is given by

$$I_2 = 4I_1 = \iint\limits_{\text{scan region}} |1 - X|^2 \, d\psi_x \, d\psi_y \tag{A.1}$$

where

$$\begin{aligned} X = j\{&2(\eta_t + \eta_2 + \eta_4 + 2\eta_6) - \eta_2(e^{j\psi_x} + e^{-j\psi_x}) - \eta_4(e^{j\psi_y} + e^{-j\psi_y}) \\ &- \eta_6[e^{j(\psi_x+\psi_y)} + e^{-j(\psi_x+\psi_y)} + e^{j(\psi_x-\psi_y)} + e^{-j(\psi_x-\psi_y)}]\} \\ &+ \eta_0 \frac{1 - e^{-j\phi} \sum\sum_{(m,n)} \mathscr{C}_{mn} e^{j(m\psi_x + n\psi_y)}}{1 + e^{-j\phi} \sum\sum_{(m,n)} \mathscr{C}_{mn} e^{j(m\psi_x + n\psi_y)}} \end{aligned} \tag{A.2}$$

The quantity

$$\frac{1 - e^{-j\phi} \sum\sum_{(m,n)} \mathscr{C}_{mn} e^{j(m\psi_x + n\psi_y)}}{1 + e^{-j\phi} \sum\sum_{(m,n)} \mathscr{C}_{mn} e^{j(m\psi_x + n\psi_y)}}$$

is a doubly periodic function in ψ_x and ψ_y and can be rewritten as

$$\eta_0 \frac{1 - e^{-j\phi} \sum\sum_{(m,n)} \mathscr{C}_{mn} e^{j(m\psi_x + n\psi_y)}}{1 + e^{-j\phi} \sum\sum_{(m,n)} \mathscr{C}_{mn} e^{j(m\psi_x + n\psi_y)}} = \eta_0 \left[1 - 2 \sum\sum_{(m,n)} D_{mn} e^{j(m\psi_x + n\psi_y)}\right] \tag{A.3}$$

where $-2D_{mn}$ would correspond to the Fourier coefficients of the quantity on the left-hand side of (A.3); $\{D_{mn}\}$ can be obtained by a direct Fourier analysis or, indirectly, by employing the series expansion of (A.3). Note that D_{mn} is a function of $e^{-j\phi}$.

Substituting (A.3) in (A.2) yields

$$\begin{aligned} 1 - X = 1 - \eta_0 + 2\eta_0 \sum\sum_{(m,n)} D_{mn} e^{j(m\psi_x + n\psi_y)} \\ - j\{2(\eta_t + \eta_2 + \eta_4 + 2\eta_6) - \eta_2(e^{j\psi_x} + e^{-j\psi_x}) - \eta_4(e^{j\psi_y} + e^{-j\psi_y}) \\ - \eta_6[e^{j(\psi_x+\psi_y)} + e^{-j(\psi_x+\psi_y)} + e^{j(\psi_x-\psi_y)} + e^{-j(\psi_x-\psi_y)}]\} \end{aligned} \tag{A.4}$$

The quantity $|1 - X|^2$ is

$$|1 - X|^2 = (1 - X)(1 - X^*) \tag{A.5}$$

To obtain a minimum of (A.1) we require

$$\begin{aligned} \frac{\partial I_2}{\partial \eta_i} &= \frac{\partial}{\partial \eta_i} \iint_{\text{scan region}} |1 - X|^2 \, d\psi_x \, d\psi_y \\ &= \iint_{\text{scan region}} \left(\frac{\partial}{\partial \eta_i} |1 - X|^2\right) d\psi_x \, d\psi_y = 0, \quad i = 0, t, 2, 4, 6 \end{aligned} \tag{A.6}$$

Employing (A.4) and (A.5) in (A.6), we obtain the following final equations:

From $\partial I_2/\partial \eta_0 = 0$,

$$\begin{aligned} a_0{}^0 - 2 \sum\sum_{(m,n)} D'_{mn} a_m{}^n \\ = \eta_0 \left[a_0{}^0 - 4 \sum\sum_{(m,n)} D'_{mn} a_m{}^n + 4 \sum\sum_{(m,n)} \sum\sum_{(p,q)} D_{mn} D^*_{pq} a^{(n-q)}_{(m-p)}\right] \\ - 4\eta_t \sum\sum_{(m,n)} D''_{mn} a_m{}^n + 2\eta_2 \sum\sum_{(m,n)} D''_{mn} (a^n_{m+1} + a^n_{m-1} - 2a_m{}^n) \\ + 2\eta_4 \sum\sum_{(m,n)} D''_{mn} (a^{n+1}_m + a^{n-1}_m - 2a_m{}^n) \\ + 2\eta_6 \sum\sum_{(m,n)} D''_{mn} (a^{n+1}_{m+1} + a^{n+1}_{m-1} + a^{n-1}_{m+1} + a^{n-1}_{m-1} - 4a_m{}^n) \end{aligned} \tag{A.7}$$

where

$$D'_{mn} = \mathrm{Re}\,(D_{mn}), \qquad D''_{mn} = \mathrm{Im}\,(D_{mn}) \tag{A.8}$$

and

$$a_m{}^n = \iint\limits_{\text{scan region}} e^{j(m\psi_x + n\psi_y)}\, d\psi_x\, d\psi_y \tag{A.9}$$

Similarly, from $\partial I_2/\partial\eta_t = 0$,

$$\begin{aligned} 0 = {} & -\eta_0 \sum_{(m,n)}\sum D''_{mn} a_m{}^n + \eta_t a_0{}^0 + \eta_2(a_0{}^0 - a_1{}^0) \\ & + \eta_4(a_0{}^0 - a_0{}^1) + 2\eta_6(a_0{}^0 - a_1{}^1) \end{aligned} \tag{A.10}$$

From $\partial I_2/\partial\eta_2 = 0$,

$$\begin{aligned} 0 = {} & \eta_0 \sum_{(m,n)}\sum D''_{mn}(a_{m+1}^n + a_{m-1}^n - 2a_m{}^n) + 2\eta_t(a_0{}^0 - a_1{}^0) \\ & + \eta_2(3a_0{}^0 + a_2{}^0 - 4a_1{}^0) + 2\eta_4(a_0{}^0 + a_1{}^1 - a_1{}^0 - a_0{}^1) \\ & + 2\eta_6[2(a_0{}^0 - a_1{}^0) + a_2{}^1 + a_0{}^1 - 2a_1{}^1] \end{aligned} \tag{A.11}$$

And from $\partial I_2/\partial\eta_4 = 0$,

$$\begin{aligned} 0 = {} & \eta_0 \sum_{(m,n)}\sum D''_{mn}(a_m^{n+1} + a_m^{n-1} - 2a_m{}^n) + 2\eta_t(a_0{}^0 - a_0{}^1) \\ & + 2\eta_2(a_0{}^0 + a_1{}^1 - a_0{}^1 - a_1{}^0) + \eta_4(3a_0{}^0 + a_0{}^2 - 4a_0{}^1) \\ & + 2\eta_6[2(a_0{}^0 - a_0{}^1) + a_1{}^0 + a_1{}^2 - 2a_1{}^1] \end{aligned} \tag{A.12}$$

Finally, from $\partial I_2/\partial\eta_6 = 0$,

$$\begin{aligned} 0 = {} & \eta_0 \sum_{(m,n)}\sum D''_{mn}(a_{m+1}^{n+1} + a_{m-1}^{n+1} + a_{m+1}^{n-1} + a_{m-1}^{n-1} - 4a_m{}^n) \\ & + 4\eta_t(a_0{}^0 - a_1{}^1) + 2\eta_2[2(a_0{}^0 - a_1{}^1) + a_0{}^1 + a_2{}^1 - 2a_1{}^0] \\ & + 2\eta_4[2(a_0{}^0 - a_1{}^1) + a_1{}^0 + a_1{}^2 - 2a_0{}^1] \\ & + 2\eta_6(5a_0{}^0 + a_0{}^2 + a_2{}^0 + a_2{}^2 - 8a_1{}^1) \end{aligned} \tag{A.13}$$

Equations (A.7) and (A.10)–(A.13) constitute a set of five linear equations with five unknowns, η_0, η_t, η_2, η_4, and η_6. These equations, which can easily be solved once their coefficients are determined, are valid if $a_m{}^n$ is real. This is true for the rectangular and elliptical scan regions of Fig. 9.5. Otherwise, a slight modification is necessary.

For the rectangular scan region,

$$a_m{}^n = 4\,\frac{\sin m\psi_0}{m}\,\frac{\sin n\psi_1}{n} \tag{A.14}$$

while for the elliptical scan region,

$$a_m{}^n = 2\pi\psi_0\psi_1\,\frac{J_1(\sqrt{(m\psi_0)^2 + (n\psi_1)^2})}{\sqrt{(m\psi_0)^2 + (n\psi_1)^2}} \tag{A.15}$$

where $J_1(\psi)$ is the Bessel function of the first order and argument ψ.

APPENDIX II TERMINAL RELATIONSHIPS OF INFINITE ARRAYS OF TWO-PORT ELEMENTS

In this appendix we shall derive the array terminal relationships, including the presence of a matching obstacle in terms of the coupling (scattering) coefficients. Any one of the column matrices in (13) (e.g., $[R_{mn}]$ with the appropriate superscript) is of the following form:

$$[R_{mn}]^T = [\ldots R_{2-1}, R_{20}, R_{21} \ldots R_{1-1}, R_{10}, R_{11} \ldots R_{0-1}, R_{00}, R_{01} \ldots R_{-1-1}, R_{-10}, R_{-11} \ldots R_{-2-1}, R_{-20}, R_{-21} \ldots] \tag{A.16}$$

The method of arrangement of an infinite planar array into a single-column matrix is as follows. The values of R_{mn} in each infinite row of elements lying parallel to the y axis of Fig. 9.9 are written into a column. These rows are then written consecutively into one column, which forms the $[R_{mn}]$ matrix. In a similar and consistent fashion, any of the $\mathscr{C}_{mn}$ matrices of (13) is given by

$$\|\mathscr{C}_{mn}\| = \left\| \begin{array}{ccccccccccccc}
 & \vdots & \vdots & \vdots & & \vdots & \vdots & \vdots & & \vdots & \vdots & \vdots & \\
\cdots & \mathscr{C}_{1-1}^{1-1} & \mathscr{C}_{1-1}^{10} & \mathscr{C}_{1-1}^{11} & \cdots & \mathscr{C}_{1-1}^{0-1} & \mathscr{C}_{1-1}^{00} & \mathscr{C}_{1-1}^{01} & \cdots & \mathscr{C}_{1-1}^{-1-1} & \mathscr{C}_{1-1}^{-10} & \mathscr{C}_{1-1}^{-11} & \cdots \\
\cdots & \mathscr{C}_{10}^{1-1} & \mathscr{C}_{10}^{10} & \mathscr{C}_{10}^{11} & \cdots & \mathscr{C}_{10}^{0-1} & \mathscr{C}_{10}^{00} & \mathscr{C}_{10}^{01} & \cdots & \mathscr{C}_{10}^{-1-1} & \mathscr{C}_{10}^{-10} & \mathscr{C}_{10}^{-11} & \cdots \\
\cdots & \mathscr{C}_{11}^{1-1} & \mathscr{C}_{11}^{10} & \mathscr{C}_{11}^{11} & \cdots & \mathscr{C}_{11}^{0-1} & \mathscr{C}_{11}^{00} & \mathscr{C}_{11}^{01} & \cdots & \mathscr{C}_{11}^{-1-1} & \mathscr{C}_{11}^{-10} & \mathscr{C}_{11}^{-11} & \cdots \\
 & \vdots & \vdots & \vdots & & \vdots & \vdots & \vdots & & \vdots & \vdots & \vdots & \\
\cdots & \mathscr{C}_{0-1}^{1-1} & \mathscr{C}_{0-1}^{10} & \mathscr{C}_{0-1}^{11} & \cdots & \mathscr{C}_{0-1}^{0-1} & \mathscr{C}_{0-1}^{00} & \mathscr{C}_{0-1}^{01} & \cdots & \mathscr{C}_{0-1}^{-1-1} & \mathscr{C}_{0-1}^{-10} & \mathscr{C}_{0-1}^{-11} & \cdots \\
\cdots & \mathscr{C}_{00}^{1-1} & \mathscr{C}_{00}^{10} & \mathscr{C}_{00}^{11} & \cdots & \mathscr{C}_{00}^{0-1} & \mathscr{C}_{00}^{00} & \mathscr{C}_{00}^{01} & \cdots & \mathscr{C}_{00}^{-1-1} & \mathscr{C}_{00}^{-10} & \mathscr{C}_{00}^{-11} & \cdots \\
\cdots & \mathscr{C}_{01}^{1-1} & \mathscr{C}_{01}^{10} & \mathscr{C}_{01}^{11} & \cdots & \mathscr{C}_{01}^{0-1} & \mathscr{C}_{01}^{00} & \mathscr{C}_{01}^{01} & \cdots & \mathscr{C}_{01}^{-1-1} & \mathscr{C}_{01}^{-10} & \mathscr{C}_{01}^{-11} & \cdots \\
 & \vdots & \vdots & \vdots & & \vdots & \vdots & \vdots & & \vdots & \vdots & \vdots & \\
\cdots & \mathscr{C}_{-1-1}^{1-1} & \mathscr{C}_{-1-1}^{10} & \mathscr{C}_{-1-1}^{11} & \cdots & \mathscr{C}_{-1-1}^{0-1} & \mathscr{C}_{-1-1}^{00} & \mathscr{C}_{-1-1}^{01} & \cdots & \mathscr{C}_{-1-1}^{-1-1} & \mathscr{C}_{-1-1}^{-10} & \mathscr{C}_{-1-1}^{-11} & \cdots \\
\cdots & \mathscr{C}_{-10}^{1-1} & \mathscr{C}_{-10}^{10} & \mathscr{C}_{-10}^{11} & \cdots & \mathscr{C}_{-10}^{0-1} & \mathscr{C}_{-10}^{00} & \mathscr{C}_{-10}^{01} & \cdots & \mathscr{C}_{-10}^{-1-1} & \mathscr{C}_{-10}^{-10} & \mathscr{C}_{-10}^{-11} & \cdots \\
\cdots & \mathscr{C}_{-11}^{1-1} & \mathscr{C}_{-11}^{10} & \mathscr{C}_{-11}^{11} & \cdots & \mathscr{C}_{-11}^{0-1} & \mathscr{C}_{-11}^{00} & \mathscr{C}_{-11}^{01} & \cdots & \mathscr{C}_{-11}^{-1-1} & \mathscr{C}_{-11}^{-10} & \mathscr{C}_{-11}^{-11} & \cdots \\
 & \vdots & \vdots & \vdots & & \vdots & \vdots & \vdots & & \vdots & \vdots & \vdots &
\end{array} \right\| \tag{A.17}$$

The superscripts of each matrix element $\mathscr{C}_{mn}{}^{pq}$ identify the excited array element; the subscripts, the terminated element.

Because of the periodicity and infinite extent of the array grid, pairs of elements having the same relative spacing with respect to one another will have the same scattering coefficients. In other words, all $\mathscr{C}_{mn}{}^{pq}$ for which $p - m = K_1$ and $q - n = K_2$ (K_1 and K_2 being constants) are identical. Thus, the $\|\mathscr{C}_{mn}\|$ matrix in (A.17) can be rewritten employing subscripts only—$\mathscr{C}_{(p-m)(q-n)}$—as follows:

$$
\|\mathscr{C}_{mn}\mathscr{C}\| = \left\|\begin{array}{ccccccccccccc}
 & \vdots & \vdots & \vdots & & \vdots & \vdots & \vdots & & \vdots & \vdots & \vdots & \\
\cdots & \mathscr{C}_{00} & \mathscr{C}_{01} & \mathscr{C}_{02} & \cdots & \mathscr{C}_{-10} & \mathscr{C}_{-11} & \mathscr{C}_{-12} & \cdots & \mathscr{C}_{-20} & \mathscr{C}_{-21} & \mathscr{C}_{-22} & \cdots \\
\cdots & \mathscr{C}_{0-1} & \mathscr{C}_{00} & \mathscr{C}_{01} & \cdots & \mathscr{C}_{-1-1} & \mathscr{C}_{-10} & \mathscr{C}_{-11} & \cdots & \mathscr{C}_{-2-1} & \mathscr{C}_{-20} & \mathscr{C}_{-21} & \cdots \\
\cdots & \mathscr{C}_{0-2} & \mathscr{C}_{0-1} & \mathscr{C}_{00} & \cdots & \mathscr{C}_{-1-2} & \mathscr{C}_{-1-1} & \mathscr{C}_{-10} & \cdots & \mathscr{C}_{-2-2} & \mathscr{C}_{-2-1} & \mathscr{C}_{-20} & \cdots \\
 & \vdots & \vdots & \vdots & & \vdots & \vdots & \vdots & & \vdots & \vdots & \vdots & \\
\cdots & \mathscr{C}_{10} & \mathscr{C}_{11} & \mathscr{C}_{12} & \cdots & \mathscr{C}_{00} & \mathscr{C}_{01} & \mathscr{C}_{02} & \cdots & \mathscr{C}_{-10} & \mathscr{C}_{-11} & \mathscr{C}_{-12} & \cdots \\
\cdots & \mathscr{C}_{1-1} & \mathscr{C}_{10} & \mathscr{C}_{11} & \cdots & \mathscr{C}_{0-1} & \mathscr{C}_{00} & \mathscr{C}_{01} & \cdots & \mathscr{C}_{-1-1} & \mathscr{C}_{-10} & \mathscr{C}_{-11} & \cdots \\
\cdots & \mathscr{C}_{1-2} & \mathscr{C}_{1-1} & \mathscr{C}_{10} & \cdots & \mathscr{C}_{0-2} & \mathscr{C}_{0-1} & \mathscr{C}_{00} & \cdots & \mathscr{C}_{-1-2} & \mathscr{C}_{-1-1} & \mathscr{C}_{-10} & \cdots \\
 & \vdots & \vdots & \vdots & & \vdots & \vdots & \vdots & & \vdots & \vdots & \vdots & \\
\cdots & \mathscr{C}_{20} & \mathscr{C}_{21} & \mathscr{C}_{22} & \cdots & \mathscr{C}_{10} & \mathscr{C}_{11} & \mathscr{C}_{12} & \cdots & \mathscr{C}_{00} & \mathscr{C}_{01} & \mathscr{C}_{02} & \cdots \\
\cdots & \mathscr{C}_{2-1} & \mathscr{C}_{20} & \mathscr{C}_{21} & \cdots & \mathscr{C}_{1-1} & \mathscr{C}_{10} & \mathscr{C}_{11} & \cdots & \mathscr{C}_{0-1} & \mathscr{C}_{00} & \mathscr{C}_{01} & \cdots \\
\cdots & \mathscr{C}_{2-2} & \mathscr{C}_{2-1} & \mathscr{C}_{20} & \cdots & \mathscr{C}_{1-2} & \mathscr{C}_{1-1} & \mathscr{C}_{10} & \cdots & \mathscr{C}_{0-2} & \mathscr{C}_{0-1} & \mathscr{C}_{00} & \cdots \\
 & \vdots & \vdots & \vdots & & \vdots & \vdots & \vdots & & \vdots & \vdots & \vdots &
\end{array}\right. \tag{A.18}
$$

On a given diagonal parallel to the main diagonal, all matrix elements are equal. Therefore, all the rows of the $\mathscr{C}_{mn}$ matrix contain the same elements ordered in an identical fashion, with adjacent rows being shifted by one position with respect to each other. Upon cascading identical matching networks in the element channels (Fig. 9.11), one obtains the following relations for the (m, n)th channel:

$$\begin{bmatrix} R_{mn}^{3} \\ I_{mn}^{3} \\ R_{mn}^{4} \\ I_{mn}^{4} \end{bmatrix} = \begin{bmatrix} I_{mn}^{A} \\ R_{mn}^{A} \\ I_{mn}^{B} \\ R_{mn}^{B} \end{bmatrix} = \|G\| \begin{bmatrix} I_{mn}^{1} \\ R_{mn}^{1} \\ I_{mn}^{2} \\ R_{mn}^{2} \end{bmatrix} \tag{A.19}$$

where the transfer scattering matrix, G, of the matching obstacle is given by

$$\|G\| = \begin{Vmatrix} 0 & s_{13} & 0 & s_{14} \\ 0 & s_{23} & 0 & s_{24} \\ -1 & s_{33} & 0 & s_{34} \\ 0 & s_{34} & -1 & s_{44} \end{Vmatrix}^{-1} \begin{Vmatrix} -s_{11} & 1 & -s_{12} & 0 \\ -s_{12} & 0 & -s_{22} & 1 \\ -s_{13} & 0 & -s_{23} & 0 \\ -s_{14} & 0 & -s_{24} & 0 \end{Vmatrix} \tag{A.20}$$

The infinite array terminal relationships are given by

$$\begin{bmatrix} [I_{mn}^{A}] \\ [R_{mn}^{A}] \\ [I_{mn}^{B}] \\ [R_{mn}^{B}] \end{bmatrix} = \begin{Vmatrix} G_{11}\|I\| & G_{12}\|I\| & G_{13}\|I\| & G_{14}\|I\| \\ G_{21}\|I\| & G_{22}\|I\| & G_{23}\|I\| & G_{24}\|I\| \\ G_{31}\|I\| & G_{32}\|I\| & G_{33}\|I\| & G_{34}\|I\| \\ G_{41}\|I\| & G_{42}\|I\| & G_{43}\|I\| & G_{44}\|I\| \end{Vmatrix} \begin{bmatrix} [I_{mn}^{1}] \\ [R_{mn}^{1}] \\ [I_{mn}^{2}] \\ [R_{mn}^{2}] \end{bmatrix} \tag{A.21}$$

where $\{G_{pq}\}$ are the elements of the G matrix in (A.20), and $\|I\|$ is an identity matrix whose dimensions are identical to those of $\|\mathscr{C}_{mn}\|$. Upon substitution of (A.21) into (13) and after simple algebraic manipulations, the following two matrix equations are obtained:

$$\begin{aligned} &\{G_{21} - G_{11}\|\mathscr{C}_{mn}^{AA}\| - G_{31}\|\mathscr{C}_{mn}^{BA}\|\}[I_{mn}^{1}] \\ &\quad + \{G_{23} - G_{13}\|\mathscr{C}_{mn}^{AA}\| - G_{33}\|\mathscr{C}_{mn}^{BA}\|\}[I_{mn}^{2}] \\ &= \{-G_{22} + G_{12}\|\mathscr{C}_{mn}^{AA}\| + G_{32}\|\mathscr{C}_{mn}^{BA}\|\}[R_{mn}^{1}] \\ &\quad + \{-G_{24} + G_{14}\|\mathscr{C}_{mn}^{AA}\| + G_{34}\|\mathscr{C}_{mn}^{BA}\|\}[R_{mn}^{2}] \end{aligned} \tag{A.22}$$

and

$$\begin{aligned} &\{G_{41} - G_{11}\|\mathscr{C}_{mn}^{AB}\| - G_{31}\|\mathscr{C}_{mn}^{BB}\|\}[I_{mn}^{1}] \\ &\quad + \{G_{43} - G_{13}\|\mathscr{C}_{mn}^{AB}\| - G_{33}\|\mathscr{C}_{mn}^{BB}\|\}[I_{mn}^{2}] \\ &= \{-G_{42} + G_{12}\|\mathscr{C}_{mn}^{AB}\| + G_{32}\|\mathscr{C}_{mn}^{BB}\|\}[R_{mn}^{1}] \\ &\quad + \{-G_{44} + G_{14}\|\mathscr{C}_{mn}^{AB}\| + G_{34}\|\mathscr{C}_{mn}^{BB}\|\}[R_{mn}^{2}] \end{aligned} \tag{A.23}$$

Equations (A.22) and (A.23) give the terminal properties of the array with the matching obstacle. If the excitation (i.e., $[I_{mn}{}^1]$ and $[I_{mn}{}^2]$) is known, the reflections, $[R_{mn}{}^1]$ and $[R_{mn}{}^2]$, are specified by (A.22) and (A.23).

For the particular excitation of phased arrays in which

$$I_{mn}{}^1 = I^1 e^{j(m\psi_x + n\psi_y)}, \qquad I_{mn}{}^2 = I^2 e^{j(m\psi_x + n\psi_y)} \tag{A.24}$$

the (m, n)th equation of the infinite set of equations described by (A.22) is of the following form:

$$\begin{aligned} e^{j(m\psi_x+n\psi_y)}\Bigg\{ & I^1\Bigg[G_{21} - \sum_{(p,q)}\sum (G_{11}\mathscr{C}_{pq}{}^{AA} + G_{31}\mathscr{C}_{pq}{}^{BA})e^{j(p\psi_x+q\psi_y)}\Bigg] \\ & + I^2\Bigg[G_{23} - \sum_{(p,q)}\sum (G_{13}\mathscr{C}_{pq}{}^{AA} + G_{33}\mathscr{C}_{pq}{}^{BA})e^{j(p\psi_x+q\psi_y)}\Bigg]\Bigg\} \\ = & \Bigg[-G_{22}R_{mn}{}^1 + \sum_{(p,q)}\sum (G_{12}\mathscr{C}_{pq}{}^{AA} + G_{32}\mathscr{C}_{pq}{}^{BA})R^1_{(p+m)(q+n)}\Bigg] \\ & + \Bigg[-G_{24}R_{mn}{}^2 + \sum_{(p,q)}\sum (G_{14}\mathscr{C}_{pq}{}^{AA} + G_{34}\mathscr{C}_{pq}{}^{BA})R^2_{(p+m)(q+n)}\Bigg] \end{aligned} \tag{A.25}$$

A similar equation can be obtained from (A.23). Inspection of the left-hand side of (A.25) shows that the factor $e^{j(m\psi_x+n\psi_y)}$ contains the only dependence on m and n. Therefore, in order for the right-hand side to be consistent with the left, the only acceptable solutions for $R_{mn}{}^1$ and $R_{mn}{}^2$ are given by

$$R_{mn}{}^1 = R^1 e^{j(m\psi_x + n\psi_y)}, \qquad R_{mn}{}^2 = R^2 e^{j(m\psi_x + n\psi_y)} \tag{A.26}$$

and the infinite set of equations in (A.22) and (A.23) degenerates into two linear equations with two unknowns, given by

$$I^1 D_1 + I^2 D_3 = R^1 D_2 + R^2 D_4 \tag{A.27}$$

$$I^1 E_1 + I^2 E_3 = R^1 E_2 + R^2 E_4 \tag{A.28}$$

where

$$\left.\begin{aligned} D_i &= (-1)^i[-G_{2i} + G_{1i}X^{AA} + G_{3i}X^{BA}] \\ E_i &= (-1)^i[-G_{4i} + G_{1i}X^{AB} + G_{3i}X^{BB}] \end{aligned}\right\}, \quad i = 1, 2, 3, 4$$

and

$$X^{AA} = \sum_{m=-\infty}^{+\infty} \sum_{n=-\infty}^{+\infty} \mathscr{C}_{mn}{}^{AA} e^{j(m\psi_x+n\psi_y)}, \quad \text{etc.}$$

The quantities X^{AA} and X^{AB} are the active reflection coefficients (versus scan) in ports A and B, respectively, when the array is excited by a matched generator at port A while port B is ideally terminated. Also, X^{BA} and X^{BB} are the active reflection coefficients in ports A and B, respectively, when port B is excited and port A is terminated. Thus X^{AA}, X^{BA}, X^{AB}, and X^{BB} fully characterize the array reflection properties, and a knowledge of $\{\mathscr{C}_{mn}{}^{AA}\}$, $\{\mathscr{C}_{mn}{}^{AB}\}$, $\{\mathscr{C}_{mn}{}^{BA}\}$, and $\{\mathscr{C}_{mn}{}^{BB}\}$ is sufficient to characterize the array match as a function of the scan angle. The variation of the reflection characteristics of the various elements by a phase factor, (A.26), is consistent with the periodicity of the electromagnetic fields, which was discussed in earlier chapters. The reduction of the infinite set of equations to two linear equations simplifies the analysis of the reflection characteristics and reduces it to a four-port network formalism.

REFERENCES

1. P. W. Hannan, P. J. Meier, and M. A. Balfour, "Simulation of Phased Array Antenna Impedance in Waveguide," *IEEE Trans. Antennas and Propagation*, Vol. AP-11, pp. 715–716, November 1963.
2. P. W. Hannan and M. A. Balfour, "Simulation of a Phased Array Antenna in Waveguide," *IEEE Trans. Antennas and Propagation*, Vol. AP-13, pp. 342–353, May 1965.
3. M. A. Balfour, "Phased Array Simulators in Waveguide for a Triangular Arrangement of Elements," *IEEE Trans. Antennas and Propagation*, Vol. AP-13, pp. 475–476, May 1965.
4. A. A. Oliner and R. G. Malech, in *Microwave Scanning Antennas* (R. C. Hansen, ed.), Vol. II, Academic Press, New York, 1966, Chapters 2–4.
5. E. G. Magill and H. A. Wheeler, "Wide-Angle Impedance Matching of Planar Array Antenna by a Dielectric Sheet," *IEEE Trans. Antennas and Propagation*, Vol. AP-14, pp. 49–53, January 1966.
6. E. C. Dufort, "A Design Procedure for Matching Volumetrically Scanned Waveguide Arrays," *Proc. IEEE*, Vol. 56, pp. 1851–1860, November 1968.
7. R. Tang and N. S. Wong, "Multimode Phased Array Element for Wide Scan Angle Impedance Matching," *Proc. IEEE*, Vol. 56, pp. 1951–1959, November 1968.
8. J. S. Cook and R. G. Pecina, "Compensation Coupling between Elements in Array Antennas," *IEEE G-AP International Symposium Digest*, pp. 234–236, July 1963.
9. P. W. Hannan, D. S. Lerner, and G. H. Knittel, "Impedance Matching a Phased-Array Antenna over Wide Scan Angles by Connecting Circuits," *IEEE Trans. Antenna and Propagation*, Vol. AP-13, pp. 28–34, January 1965.
10. N. Amitay, R. G. Pecina, and C. P. Wu, "Radiation Properties of Large Planar Arrays," *Bell Tel. System Monograph* 5047, February 1965.
11. P. W. Hannan, "Proof that a Phased Array Can Be Impedance Matched for All Scan Angles," *Radio Science*, No. 3, pp. 361–369, March 1967.
12. N. Amitay, "Improvement of Planar Array Match by Compensation through Contiguous Element Coupling," *IEEE Trans. Antennas and Propagation*, Vol. AP-14, pp. 580–586, September 1966.
13. N. Amitay, P. E. Butzien, and R. C. Heidt, "Match Optimization of a Two Port Phased Array Antenna Element," *IEEE Trans. Antennas and Propagation*, Vol. AP-16, pp. 47–57, January 1968.
14. E. C. DuFort, "Finite Scattering Matrix for an Infinite Antenna Array," *Radio Science*, Vol. 2, pp. 19–27, January 1967.

Index